Modulated Structure Materials

NATO ASI Series

Advanced Science Institutes Series

A Series presenting the results of activities sponsored by the NATO Science Committee, which aims at the dissemination of advanced scientific and technological knowledge, with a view to strengthening links between scientific communities.

The Series is published by an international board of publishers in conjunction with the NATO Scientific Affairs Division

A	Life Sciences	Plenum Publishing Corporation
B	Physics	London and New York
C	Mathematical and Physical Sciences	D. Reidel Publishing Company Dordrecht and Boston
D	Behavioural and Social Sciences	Martinus Nijhoff Publishers Dordrecht/Boston/Lancaster
E	Applied Sciences	
F	Computer and Systems Sciences	Springer-Verlag Berlin/Heidelberg/New York
G	Ecological Sciences	

Series E: Applied Sciences – No. 83

Modulated Structure Materials

edited by

T. Tsakalakos

Department of Mechanics and Materials Science
Rutgers University
Piscataway, NJ 08854
USA

1984 **Martinus Nijhoff Publishers**
Dordrecht / Boston / Lancaster
Published in cooperation with NATO Scientific Affairs Division

Proceedings of the NATO Advanced Study Institute on Modulated Structure Materials, Maleme-Chania, Greece, June 15-25, 1983

Library of Congress Cataloging in Publication Data

NATO Advanced Study Institute on Modulated Structure
    Materials (1983 : Maleme, Crete and Chania, Crete)
    Modulated structure materials.

    NATO advanced science institutes series. Series E,
Applied sciences ; 83)
    "Proceedings of the NATO Advanced Study Institute on
Modulated Structure Materials, Maleme-Chania, Greece,
June 15-25, 1983"
    "Published in cooperation with NATO Scientific
Affairs Division."
    Includes bibliographical references.
    1. Layer structure (Solids)--Congresses.
2. Superlattices as materials--Congresses.
I. Tsakalakos, Thomas. II. North Atlantic Treaty
Organization. Scientific Affairs Division. III. Title.
IV. NATO advanced science institutes series. Series E,
Applied sciences ; no. 83.
QD921.N384 1983   530.4'1   84-16603
ISBN-13: 978-94-009-6197-5   e-ISBN-13: 978-94-009-6195-1
DOI: 10.1007/978-94-009-6195-1

ISBN 90-247-3066-X (this volume)
ISBN 90-247-2689-1 (series)

Distributors for the United States and Canada: Kluwer Academic Publishers, 190 Old Derby Street, Hingham, MA 02043, USA

Distributors for the UK and Ireland: Kluwer Academic Publishers, MTP Press Ltd, Falcon House, Queen Square, Lancaster LA1 1RN, UK

Distributors for all other countries: Kluwer Academic Publishers Group, Distribution Center, P.O. Box 322, 3300 AH Dordrecht, The Netherlands

PREFACE

Modulated Structure Materials arise in two basic ways. One is through the natural tendency that certain materials have to develop stable modulations. Typical examples of this category are the long period superlattices, the spinodal alloys and other ordered structures. Another way to introduce modulation into a basic structure is through our own intervention, that is by artificial techniques. Such examples as the composition modulated films and the semiconductor superlattices have recently received appreciable attention not only for their noble and unusual properties but also for their practical applications in high technology areas.

The NATO Advanced Study Institute on Modulated Structure Materials which was held June 15-25, 1983 in Maleme-Chania, Greece, aimed at bringing together international authorities and active researchers to discuss in-depth current knowledge and new developments in both natural and artificial modulated structure materials. Up to this time, the Editor has received indications that the Institute served well its purpose. The fifteen carefully selected invited speakers gave outstanding lectures on all aspects of modulated structures. The lectures were followed by extensive and lively discussions among all participants. It should be noted that on two occasions discussion panels were formed to address some of the fundamental aspects of modulated structures in view of the impressive results of advanced experimental techniques (lattice and structure imaging techniques in high resolution electron microscopy; X-ray and neutron diffraction methods, etc.) and the new theories of statistical mechanics such as the anisotropic next nearest neighbor Ising model and others. Finally a number of successful workshops were held in which specific talks and discussions complemented the invited lectures and provided a framework of exchanging ideas for future work.

Most of these lectures are included in this publication. A particular effort was made by the Editor and authors to avoid specialized terminology and to address fundamental aspects. Because of the interdisciplinary nature of the Institute, the target of this publication is dual. First to transfer knowledge from areas involving natural modulated structure materials (ordering, clustering, etc.) to the relatively new field of synthetic layered structures (metallic or semiconductors) and second to utilize the well defined synthetic modulated structures to answer some difficult questions of the natural state. This dynamic interplay plays a significant role also in the development of new materials of unusual properties about which the field of materials science is centered.

The eight chapters of the book reflect this philosophy. In addition to the fundamental aspects of the modulated structure materials, certain applications in high technology areas are also discussed by some authors.

The Institute was sponsored by the Advanced Study Institute of the NATO Scientific Affairs Division to which the Editor is greatly indebted. In particular the Editor wishes to express his sincere appreciation to Dr. C. Sinclair, Director of the ASI program, not only for his continuous assistance during the difficult period of organizing the Institute but also for his kind visit and talk in Crete which fostered some of the basic targets set by the Director and the ASI.

The director wishes to express his gratitude to the Department of Mechanics and Materials Science for their valuable assistance toward the organization of the Institute and in particular to Professor Yu Chen, Chairman of the Department and Dean Ellis Dill of the College of Engineering of Rutgers University.

The Editor is greatly indebted to Ms. Donna Foster for her excellent and dedicated work in the preparation of this manuscript and Mrs. Renata Joyner for her valuable help during the organization of the conference. Special thanks also to Mrs. Barbara Karl and Mrs. Claudia Kuchinow.

Words are insufficient to express my sincere appreciation to my wife Mary Tsakalakos for her moral support and understanding and the endless assistance she provided during the organization of the conference and the preparation of the proceedings.

Finally the Director expresses his gratitude to the organizing committee, Helen Badekas, Jerry Cohen and Gust Bambakidis for their valuable assistance in bringing this Institute to a success.

Thomas Tsakalakos
Director, NATO Advanced
Study Institute on
Modulated Structure Materials

TABLE OF CONTENTS

Preface. 1

Introduction

    Modulated Structure Materials: A Selected Review . . 11
    S.C. MOSS

Chapter 1: Theoretical Aspects of Modulated Structures

    Modulated Structures in a Simple Ising Model. 23
    W. SELKE

    Composition Modulations in Solid Solutions. 43
    D. DeFONTAINE

    Metal Insulator Transition in Modulated Crystals. . . 81
    C.M. Soukoulis

    A New Theory of Polytypism. 95
    J. SMITH, J. YEOMANS AND V. HEINE

    Densities of States of Compositionally Modulated
    Alloys. 107
    A. GONIS AND N.K. FLEVARIS

    Time Evolution of Phase Separation in Binary
    Mixtures 125
    J. MARRO AND M.H. KALOS

Chapter 2: Crystallography of Modulated Structures

    The Incommensurate Crystalline Phase and its
    Symmetry. 133
    P.M. de WOLFF

    Superspace Transformation Properties of Incommensurate
    Irreducible Distortions 151
    J.M. PÉREZ-MATO, G. MADARIAGA AND M.J. TELLO

    Crystal Structures of Complex Sulfides: From
    Modulated to Modular Structures 159
    E. MAKOVICKY

4

Chapter 3:   Diffraction Methods

    High Resolution Electron Microscopy Study of
    Manganese Silicides 173
    H.Q. YE AND S. AMELINCKX

    The Use of High Resolution Electron Microscopy
    in the Study of Modulated Structures in Alloy
    Systems 183
    S. AMELINCKX, J. VAN LANDUYT AND G. VAN TENDELOO

    Electron Microscopic Study of Modulated Structures
    in $(Au,Ag)Te_2$ 223
    G. VAN TENDELOO, P. GREGORIADES AND S. AMELINCKX

    Incommensurate Structures in the Long-Period
    Ordered Alloys Studied by High-Resolution
    Electron Microscopy 247
    D. WATANABE AND O. TERASAKI

    Analysis of Diffuse Scattering from Composition
    Modulations in Concentrated Alloys. 265
    P. GEORGOPOULOS AND J.B. COHEN

    Neutron Investigation of Modulated Structures 285
    R. CURRAT

    Atom Probe Field-Ion Microscopy Studies of
    Modulated Structures. 309
    S.S. BRENNER, M.K. MILLER AND W.A. SOFFA

Chapter 4:   Mechanics of Modulated Structures

    The Elastic Theory of the Defect Solid Solution . . . 327
    J.W. MORRIS, JR., A.G. KHATCHATURYAN AND
    S.H. WEN

    On the Mechanics of Modulated Structures. 357
    E.C. AIFANTIS

    The Effect of Strain on the Elastic Constants
    of Copper 387
    T. TSAKALAKOS AND A.F. JANKOWSKI

Chapter 5:   Spinodal Structures

    Mechanical Behavior of Spinodal Alloys 411
    L.H. SCHWARTZ

    A Numerical Study of Kinetics in Spinodal
    Decomposition 425
    T. TSAKALAKOS AND M.P. DUGAN

Chapter 6: Composition Modulated Films

Manufacture of Ternary Thin Film Layered Foils by
  Controlled Evaporation 439
  C.N. MANIKOPOULOS AND T. TSAKALAKOS

Theoretical Approaches to Understanding the
  Properties of Modulated Structures - A Review. . . . 455
  P.C. CLAPP

Mechanical and Thermoelectric Behavior of
  Composition Modulated Foils. 465
  D. BARAL, J.B. KETTERSON AND J.E. HILLIARD

Effects of Short Wavelength Composition
  Modulations on Interdiffusion in Silver-Palladium
  Thin Foils 475
  G.E. HÉNEIN AND J.E. HILLIARD

Compositionally Modulated Metallic Glasses 491
  A.L. GREER

Electrical Properties of Multilayered Cr/SiO_2
  Thin Films 501
  D. NIARCHOS, B.J. PAPATHEOFANIS, G. MONFROY,
  M. TANIELIAN AND J. WILLHITE

Chapter 7: Semiconductor Superlattices

Tailored Semiconductors: Compositional and
  Doping Superlattices. 509
  G.H. DÖHLER

Chapter 8: Modulations in Solids

Premartensitic Behavior and Charge Density Waves
  in TiNi Alloys 539
  C.M. HWANG, M. MEICHLE, M.B. SALAMON AND
  C.M. WAYMAN

Electronic Contributions to Mixing- and Gradient-
  Energy of Composition-Modulated Alloy System 557
  H. YAMAUCHI

Mechanical Behavior of Solid Film Adhesives with
  Scrim Carrier Cloths 567
  E. SANCAKTAR

Orientational Phase Transitions in a Quasi-One-
  Dimensional Conductor. 583
  C. MAVROYANNIS

6

Electron-Libron Pairing in Quasi-One-Dimensional
    Conductors. 599
    C. MAVROYANNIS

INTRODUCTION

MODULATED STRUCTURE MATERIALS: A SELECTED REVIEW

S. C. Moss

Department of Physics
University of Houston - University Park
Houston, Texas 77004

ABSTRACT

Modulated structures arise in materials in two basic ways.
One way is through the natural inclination of the material to
suffer a periodic modulation of its basic structure. An example
of this would be the spectrum of thermal fluctuations or phonons
in crystals. Other examples include long-period superlattices
and spinodal decomposition in crystalline or amorphous solids,
where the latter, to be observed, must be present in a metastable
(however durable) condition. The second way in which modulated
structure materials are obtained is through our intervention--
i.e., artificially. Examples of these are the semiconductor
superlattices and the synthetic multilayer metallic structures.

Issues which commonly appear in understanding these materials
include: a) coherence vs. incoherence of the (lattice) structure
being modulated; b) comensurate vs. incommensurate modulation.
This often represents a competition between the driving force for
the modulation and the response or relaxation of the structure
being modulated; c) the influence of defects; d) the origin of
the modulation in the basic thermodynamics or electronic structure of the materials, and e) conversely, the influence of the
induced, or naturally occurring, modulation on the properties.
Selected examples of these issues will be discussed.

1 INTRODUCTION

Since the International Conference on Modulated Strucures [1] was held in 1979 there has been a great deal of activity (and some progress) in this extremely diverse field. The proceedings of that conference [1] bear witness to the diversity and it is not the intention here to convey more than a sense of some of the underlying issues or to provide more than a selected backdrop for dialogues that develop in the following papers.

It seems reasonable to divide the topic of modulated structures into two basic groups, depending upon the origin of the observed modulation. The first of these involves naturally occurring, or intrinsic, phenomena in which, with one or another of the thermodynamic variables, a modulation of a basic lattice - or, for more macroscopic systems, of a homogeneous phase - occurs as an equilibrium structure. This modulation may be of composition or density or it may involve solely (a set of) lattice displacements.

The second category consists, of course, of the artificially modulated materials whose optical, electrical, magnetic, mechanical and structural properties are so interesting and useful. These latter materials are also used to study solid state phenomena on a size (time) scale hitherto inaccessible to us. In the intermediate category, between the two extremes, there are modulated materials of great scientific interest which represent intrinsically metastable states achieved often through selected heat treatment, which, like the martensitic phase of steel, have found their modest place in technology. Spinodally decomposing materials belong to this category. This introductory review will touch upon some aspects of the first two categories.

2 NATURALLY OCCURRING STRUCTURES

The spectrum of thermal fluctuations, or phonons, in crystals serves as the simplest example of a naturally occurring set of modulations of a static structure. They are, by definition, in thermal equilibrium at any temperature and they give rise to a well-defined set of modulation sidebands about the Bragg peaks in the x-ray, electron or neutron scattering pattern, whose envelope is usually referred to as thermal diffuse scattering [2]. By themselves, these excitations are treated simply as harmonic thermal fluctuations. Occasionally, however, one phonon, or a group of phonons centered on a particular wave vector, will undergo a softening with decreasing temperature. In this case, the phonon frequency may approach zero and a phase transition can take place in which the wavelength and eigenvector of the phonon in question serve to describe the atomic displacements in the new low-temperature phase.

The origin of such a structural instability with its periodic-lattice-distortion (PLD) must lie in the electronic structure of the material and its coupling to the phonons. In metals this is often through a Fermi-surface-induced effect, via the electron-phonon interaction, in which the lattice instability occurs at a wavevector $\vec{q} = 2\vec{k}_F$, where $\hbar k_F$ is the Fermi momentum. This charge-density-wave (CDW) instability tends to be most prominent in lower dimensional systems and is responsible, for example, for the incommensurate PLD in the transition metal dichalcogenides (3). The latter thereby represent excellent examples of an electronically induced modulated structure in a quasi-two dimensional (2-D) system.

For metals in which the phonon softening and eventual static distortion wave come at $2\vec{k}_F$, the modulation will be initially incommensurate with the host lattice because $2\vec{k}_F$ need not, in principle, be a rational fraction of a reciprocal lattice vector. But modulated structures may also occur in insulating crystals undergoing lattice softening at some particular incommensurate wave vector. While the physical origin of this structural transformation is less transparent, there are some well-documented examples. K_2SeO_4 studied by Iizumi et. al (4) provides an excellent prototype. This crystal undergoes, on cooling, successive transformations from a paraelectric phase to an incommensurate structure to a commensurate improper ferroelectric. The transition to the incommensurate phase is preceded by a softening of the phonon at $\vec{q} = 1/3 (1 - \delta)\vec{a}^*$ where $\vec{a}^*$ is a reciprocal lattice vector and δ at the transition is ~ 0.07. δ decreases with decreasing temperature dropping discontininously to zero at the lower (commensurate) first-order transition. The phonon softening in the Σ_2 mode above the upper transition was analysed in a straightforward way by Iizumi et al to yield the effective (fictitious) interplanar force constants F_1, F_2 and F_3 for first, second and third neighbor planes. A decrease with temperature of both F_1 and F_2 in the presence of a strong increasing force F_3 accounts well for the measured dispersion curves as a function of temperature and, particularly, for the softening at $\delta \neq 0$. This interesting result demonstrates that planar interactions out to only a few neighbors are required (albeit with particular temperature dependences) to stabilize an incommensurate structure!

While not directly related, this result is none-the-less reminiscent of the recent formal result on the ANNNI model of

Fisher and Selke (5). They show that interacting planes of ordered Ising spins can, for only first and second planar neighbor interactions, undergo transitions from paramagnetic to incommensurate modulated, to ferro - or antiferromagnetic states. The ANNNI model thus provides insight into the competition in a crystal with limited range interactions, between a truly incommensurate modulation and a string of commensurate structures.

Iizumi et al (4) also successfully developed a Landau theory for the phase transition in K_2SeO_4. This treatment required a rather careful specification of symmetry and presents a physical alternative to the use of superspace groups as outlined by de Wolff(6). In fact, the equivalence of the lattice dynamical representation of an incommensurate transition, including the Landau treatment, and the use of superspace symmetry remains an important issue and is discussed in the next paper by de Wolff (7).

As a second example of modulated structures in metals in which the instability at $2\vec{k}_F$ plays a dominant role, we discuss briefly the well-known long period antiphase superlattices in alloys. The definitve work of Sato and Toth (8) showed clearly the trends of varying antiphase domain size with electron concentration, or Fermi surface dimension, in an ordering alloy. The question raised then, which is appropriate still, is how in detail does the actual ordered structure accommodate to the demand for an incommensurate super-period. Already, at that time, Fujiwara (9) had discussed the conditions for the formation of sharp modulation superlattice peaks in (nominally) incommensurate antiphase domain structures. He noted that a regular uniform mixing of integral periods of the correct size would yield sharp diffraction peaks at the average position. Thus, through a mixing of commensurate periods an arbitarily close adjustment could be made by the ordered system to the requirement of a set of diffraction satellites at $2\vec{k}_F$. Elegant confirmation of this idea has come recently from Terasaki (10) who has direct high resolution electron microscope results on the incommensurate 2-D antiphase structure in $Au_{3+}Zn$. His high resolution images are compared with calculated images and the relevant commensurate domain configurations are clearly elucidated, along with both the diffraction patterns and the optical transforms of the microscope images. The relevance to Sato and Toth (8) is noted by Terasaki (10) as is the analogous theoretical treatment for the transition metal dichalcogenides by McMillan (11) in which commensurate domains and phase slips (discommensurations) are shown to account for the (weakly) incommensurate state.

As an alternative to this idea of regular domain mixing with sharp domain walls, we should, however, note the work of Guymont et al (12) who show that the modulated phase of CuAuII cannot be well-described by the Fujiwara model (9). Rather a model of fluctuating ("sinuous") domain walls is required by both the lattice imaging and diffraction data.

Another example of modulated structures in metals which deserves attention is the tendency to form the omega(ω) - phase (13, 14) in the transition metals. This phase may be viewed as an inherent instability of the bcc lattice with respect to the deformation induced by a longitudinal phonon of wave vector 2/3(1,1,1). While never observed in the stable state, the ω-phase represents an intriguing structural instability in a wide class of metals. It is therefore of great interest that a first-principles total energy caculation has recently been made by Ho et al (15) for bcc Mo, Nb and Zr. These authors display the energy vs. the 2/3 (1,1,1) longitudinal displacement for the three metals (at 0°K) and show an increasing softness as one proceeds from Mo to Nb to Zr in detailed agreement with the measured phonon dispersion curves. The potential in Zr, in fact, is quite flat at bcc and has a second lower energy minimum at the ω-phase position. Their calculation has therefore permitted "a detailed analysis of the microscopic mechanisms causing phonon anomalies and soft-mode phase transitions" (15).

3. INFLUENCE OF DEFECTS

A final aspect of naturally occurring modulated structures is the influence of defects. As was perhaps first pointed out by Axe and Shirane (16) in connection with the low temperature lattice instability in Nb_3Sn, lattice defects can serve as nucleation centers for incipient instabilities or modulated phases. The idea was originally (16) invoked in discussing the "central peak" problem in the neutron scattering from soft-mode systems. We mention it here because in any crystal in which there is a particular lattice softness, that softness may manifest itself as a condensation around defects as long as it can couple to the defect displacement field. The process is simple to visualize: a crystal is soft or weak at a particular wave vector to a particular set of forces; a defect, or defect cluster, in the crystal gives rise to a set of forces which have components along the appropriate directions for the soft-mode; the crystal relaxes around the defect with a pronounced modulation at the soft-mode wave vector; the defect has thereby acted as a condensation center for a static modulation at that wave vector. In this fashion, static fluctuations may invariably accompany the dynamical fluctuations that are associated with the (ultimate) formation of modulated structures. They may, in fact, replicate the modulated structure in the absence of a phase transition, especially in crystals with a substantial defect concentration.

4. ARTIFICIAL STRUCTURES

Artificially modulated structures are usually produced through the controlled (shuttered) evaporation of two or more species, where a species can be an element or a chemical compound such as GaAs or InSb. There are several reviews of the structure and properties of these chemically modulated films among which we note the recent articles by Gossard (17) and McWhan (18).

One of the most consistently discussed aspects of these films is the question of the structural coherence that accompanies the concentration profile across the interface between the two species. This is usually discussed within the context of a diffraction experiment performed to evaluate the modulated film. In general we will have films that are grown on a crystal substrate, such as mica, and are textured polycrystals with a prefered crystallographic growth direction. Ideally, however, films may be grown as single crystals of a selected orientation with a composition modulation along the common growth axis. If the crystal lattice is coherent through the modulated (multilayer) film, there can be no variation in lattice spacing in the plane of the film (19). This is, of course, most readily achieved in cases where the two constituents have the same crystal structure and only a small lattice mismatch. Ag-Au Layers, for example, have a mismatch of < 0.2% and form good coherent crystal films. Cu-Ni films have a lattice parameter mismatch of ~2.4% and form coherent films only at shorter modulation wavelengths. The coherency strain energy in these films is a function only of lattice mismatch and modulus and not of thickness while the interfacial energy depends on the number of interfaces per length along the modulation direction (19). There will always be, therefore, some modulation wavelength, λ_m, at which the modulated structure becomes incoherent - i.e. forms a set of incoherent boundaries between the alternating layers through which the perpendicular lattice planes are not continuous. The smaller the lattice mismatch, the larger the cutoff value of λ_m.

As noted above, the quality of the films and the question of their coherency is usually determined from a diffraction experiment - normally performed in symmetrical reflection off the flat film surface. In such cases only the strain and concentration profile along the growth direction (normal to the layers) can be evaluated. The information can be separated into categories (18).

1. The position of the (first-order) satellite reflections about the Bragg peaks of the average lattice yields the modulation wave length, λ_m. [If the film is actually composed of separate layers of macroscopically distinguishable phases, their separate peaks will be observed].

2. The intensity and asymmetry of the set of satellite reflections yields the concentration and lattice parameter variation along the growth direction.

3. The decrease of the average lattice peak with order (as with a Debye-Waller effect) yields the regularity or quality of the film. [If only one Bragg peak is observed, for example, the film is not very good].

For these and other details of the diffraction problem the papers of McWhan [18] and Segmüller and Blakeslee [20] are useful.

A related aspect of the question of coherence has recently been discussed following the observation of Schuller [21] of apparently coherent multilayer films of NbCu in which the bcc Nb (110) planes are parallel to the fcc Cu (111). While the (110) of bcc can be viewed as a distorted hexagonal layer, the planar spacings of the two metals differ by ~ 9%. Lowe et al [21] subsequently showed, for a particular thickness range in these films, that there is indeed an effective average planar spacing perpendicular to the film giving rise to an average Bragg peak plus satellite sidebands due to the concentration modulation. But diffraction patterns taken at increasingly oblique angles to the surface (the diffraction vector becoming increasingly more parallel to the surface) revealed a totally incoherent structure. For those lattice planes normal to the layers there coexisted separate Cu and Nb spacings. The apparent coherence in the growth direction is therefore attributable to a rather uniform incoherent interface of regular thickness separating the Cu and Nb layers. The regularity of this interface can, in fact, be estimated from the dependence on angle of the higher order peaks, as noted above.

In the artificially layered semiconductors, modulated crystals of excellent quality have been produced, especially where the lattice spacing mismatch is in small [17,18]. Where a large mismatch exists between compositions whose average band gap would be desirable, a recent development has proven most ingenious and promising. This method of "strained superlattices", developed by Osbourn and co-workers [23], produces the desired average concentration (band gap) by enriching one of the components gradually with thickness during deposition. At the final concentration the crystalline film is of excellent quality and can then be modulated at will about that pre-selected concentration. The optical and electrical properties of the semiconductor structures are of great interest and are generally discussed by Dohler in these proceedings.

Two additional aspects of modulated metallic films deserve special note here. The first is the remarkable "supermodulus" effect in which the biaxial elastic modulus of compositionally modulated films shows a peak at a wavelength of around 20 (19, 24). This peak represents an enhancement of from 2-5 times the bulk value of the modulus. Explanations of this phenomenon have invoked Fermi surface - Brillouin zone boundary contact (24) on the one hand and the approach of incoherence on the other (25). A full quantitative explanation has yet to be supplied (26). The other aspect of modulated metallic films that shows great promise is their application to the determination of interatomic interaction potentials in alloys. Tsakalakos (27) has developed a method of measuring these pair potentials up to sixth neighbor ultilizing modulated films of (111) and [100] textures. The potentials are extracted from the wavelength dependence of the effective diffusion coefficients on crystallographic orientation, as determined from the decay rate of the x-ray satellite intensities. The Tsakalakas method is a particularly elegant way of obtaining alloy interatomic potentials in a much wider variety of materials than are available as equilibrium single crystals.

REFERENCES

1. Cowley, J.M., J.B. Cohen, M.B. Salomon and B.J. Wuensch, eds., Modulated Structures - 1979, AIP Conference Proceedings, No. 53, American Institute of Physics, N.Y. (1979).
2. James, R.W., The Optical Principles of the Diffraction of X-rays, London, Bell (1948).
3. Wilson, J.A., F.J. DiSalvo and S. Mahajan, Adv. in Physics 24, 117(1965).
4. Iizumi, M., J.D. Axe, G. Shirane and K. Shimaoka, Phys. Rev. B15, 4392 (1977).
5. Fisher, M.E., and W. Selke, Phys. Rev. Lett. 44, 1502 (1980); W. Selke, these proceedings.
6. de Wolff, P.M., Acta Cryst. A33, 493 (1977). A. Janner, T. Janssen and P.M. de Wolff, in Modulated Structures - 1979, AIP Conference Proceedings No. 53, J.M. Cowley, J.B. Cohen, M.B. Salomon and B.J. Wuensch, eds., American Institute of Physics, N.Y. (1979), p. 81.

7. de Wolff, P.M., these proceedings.
8. Sato, H. and R.S. Toth, Phys. Rev. 127, 469 (1962).
9. Fujiwara K., J. Phys. Soc. Japan 12, 7 (1957).
10. Terasaki O., J. Phys. Soc. Japan 51, 2159 (1982).
11. McMillian, W.L., Phys. Rev. B14, 1496 (1976).
12. Guymont, M., D. Gratias and R. Portier, in Modulated Structures - 1979, AIP Conference Proceedings, No. 53, J.M. Cowley, J.B. Cohen, M.B. Salomon and B.J. Wuensch, eds., American Institute of Physics, N.Y. (1979), p. 44.
13. Hickman, B.S., J. Mater. Sci. 4, 554 (1969).
14. Axe, J.D., D.T. Keating and S.C. Moss, Phys. Rev. Lett. 35, 530 (1975).
15. Ho, K.-M., C.L. Fu, B.N. Harmon, W. Weber and D.R. Hamann, Phys. Rev. Lett. 49 673 (1982).
16. Axe, J.D. and G. Shirane, Phys. Rev. B8, 1965 (1973).
17. Gossard, A.C., "Molecular Beam Epitaxy of Superlattices and Thin Films," in Thin Films: Preparation and Properties, K.N. Tu and R. Rosenberg, eds. Academic Press, New York (1982).
18. McWhan, D.B., "Structure of Chemically Modulated Films," in Synthetic Modulated Structure Materials, L.L. Chang and B.C. Giessen eds., Academic Press, New York (in press).
19. Hilliard, J.E. in Modulated Structures - 1979, AIP Conference Proceedings, No. 53, J.M. Cowley, J.B. Cohen, M.B. Salomon and B.J. Wuensch, eds., American Institute of Physics, N.Y. (1979), p. 407.
20. Segmuller, A., and A.E. Blakeslee, J. Appl. Cryst. 6, 19 (1973); 6, 413 (1973).
21. Schuller, I.K., Phys. Rev. Lett. 44, 1597 (1980).
22. Lowe, W., T. Barbee, T.H. Geballe and D.B. McWhan, Phys. Rev., B24, 6193 (1981).
23. Osbourn, G.C., R.M. Biefeld and P.L. Gourley, Appl. Phys. Lett. 41, 172 (1982); G.C. Osbourn, J. Vac. Sci. Technol. B1, 379 (1983).
24. Tsakalakos, T., and J.E. Hilliard, J. Appl. Phys. 54, 734 (1983).
25. Willens, R.H., these proceedings.
26. Clapp, P.C., these proceedings.
27. Tsakalakos, T., Thin Solid Films 86, 79 (1981).

CHAPTER 1: THEORETICAL ASPECT OF MODULATED STRUCTURES

MODULATED STRUCTURES IN A SIMPLE ISING MODEL

WALTER SELKE

Institut für Festkörperforschung der KFA,
5170 Jülich 1, Federal Republic of Germany

ABSTRACT

Complex spatially modulated behavior, as observed experimentally, is discussed in the framework of a simple spin-1/2 Ising model with nearest neighbor interactions augmented by competing next-nearest neighbor couplings acting parallel to a single axis, the "axial next-nearest neighbor Ising" (or ANNNI) model. The three-dimensional version of the model displays infinitely many distinct commensurate low-temperature phases separated by first-order transitions, incommensurate "sinusoidally" ordered structures, and a special multicritical point, the Lifshitz point. In the two-dimensional ANNNI model topological defects such as walls and dislocations play an important role in the physics of the incommensurate phase.

1   INTRODUCTION

Many materials are known to exhibit phases in which the spatial modulation of some local property such as magnetization, electric polarization, charge density, mass density or chemical composition, is related in a complex way, commensurate or incommensurate, to the underlying structure. In this article we address what is one of the simplest statistical mechanical models to exhibit such behavior, namely, in d spatial dimensions, a spin-1/2 Ising model with ferromagnetic couplings within (d-1)-dimensional layers but competing ferromagnetic and antiferromagnetic interactions between nearest ($J_1>0$) and next-nearest layers ($J_2>0$) along one, unique spatial axis. The abbreviation 'ANNNI model' has been proposed (1), and widely accepted, for this anisotropic, or axial next-nearest neighbor Ising model.

Its three-dimensional version on the simple cubic lattice was introduced some years ago by Elliott (2), who was interested in understanding experimental data on erbium, thulium and some of their alloys which display 'sinusoidally' modulated magnetic structures. The interest in that model was renewed in connection with the existence of a uniaxial Lifshitz point, the multicritical point, where ferromagnetic, sinusoidal and paramagnetic phases meet (3,4,5), see also earlier work on the spherical version of the model (6). Results on critical properties close to Lifshitz points can be used to interpret experiments on materials like $RbCaF_3$ (7) and MnP (8). Subsequent theoretical treatments attempted to analyse the spatially modulated phase itself in detail (9,10,11,12); earlier discussions turned out to be either misleading (2) or only preliminary (5). On the basis of Monte Carlo (MC) calculations (9,10) and sophisticated mean-field theories (10,11,12) the apparent features of the phase diagram of the model were found, including two commensurate phases, the ferromagnetic and the (2,2) antiphase (or <2>) structure, as well as the incommensurate phase characterised by a wavevector, q, varying in a complex way with temperature, T, and the ratio $\kappa = -J_2/J_1$. The phase diagram becomes even more interesting close to the multiphase point (T=0, κ=1/2), from which an infinite number of distinct commensurate phases spring (1,13) as has been shown using a systematic low temperature expansion. The relevance of this peculiar behavior to observations on real systems such as cerium monopnictide compounds, in particular CeSb (14) and CeBi (15), has been analysed in the framework of ANNNI models recently (16,17). Applying various methods, the ANNNI model in a field has been studied recently, too (18,19,20).

Its two-dimensional version has been analysed in great detail using a variety of techniques, see, e.g., (21,22,23,24,25,26). In particular, the incommensurate phase (without long range order) can be characterised by topological defects like walls and dislocations (21,22). Its transition to the paramagnetic phase is believed to be driven by the unbinding of pairs of dislocations (21,22). It has been demonstrated convincingly (23) that there is no Lifshitz point in contrast to the three-dimensional case; interesting finite-size effects (22) have been observed. Results are possibly relevant to experiments on monolayers adsorbed on uniaxial substrates like H on Fe(110), Xe on Cu(110), or O_2 on Pd(110) (27,28,29) and also lipid bilayers [30].

In the following part of the article we shall review results on the three-dimensional model. The two-dimensional version is considered in the next section. A brief comparison of the theoretical results with experimental findings is given in the last part.

2   THE THREE-DIMENSIONAL ANNNI MODEL

In ANNNI models, Ising spins $S_i = \pm 1 (+,-)$ are situated on a regular d-dimensional lattice formed of (d-1) dimensional layers of coordination number $q_\perp$ normal to the z-axis. Within the layers each spin is coupled only by nearest neighbor (nn) ferromagnetic interactions, $J_o > 0$. However, along the z-axis, spins are coupled both by nn ferromagnetic interactions, $J_1 > 0$, and by competing, next nearest neighbor antiferromagnetic interactions, $J_2 = -\kappa J_1 < 0$. The parameter κ thus controls the degree of competition. - The three-dimensional version on a tetragonal or simple cubic lattice, $q_\perp = 4$, is depicted in Fig.1.

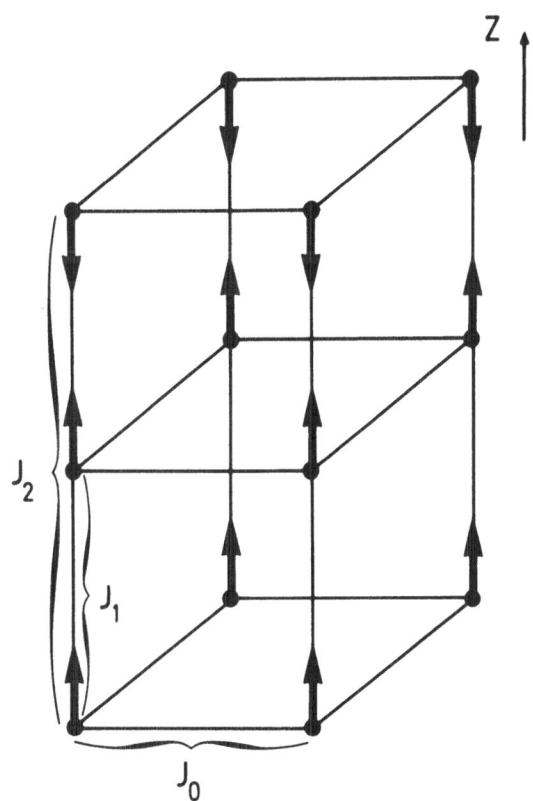

Fig. 1: The three-dimensional ANNNI model.

In the ground state of the model each individual layer is ferromagnetically ordered. For $\kappa < 1/2$ all the layers are aligned parallel forming a ferromagnetic structure: conversely, for $\kappa > 1/2$ the

ground state is a (2,2) or <2> "antiphase" state corresponding to
a layer pattern ..++--++--... At $\kappa=1/2$ the ground state becomes
infinitely degenerate [21,31] with an entropy $S=k_B(\ln[(\sqrt{5}+1)/2])/N$,
where N is the number of spins in a layer, i.e. the ground state
entropy vanishes in the thermodynamic limit for $d \geq 2$.

Now the gross features of the (T,κ) phase diagram for $d >2$ are
shown in Fig.2. The ferromagnetic and $<2>$ antiphase states remain
as stable phases for nonzero T. At high temperatures there is, of
course, a disordered paramagnetic phase for all κ in which, however,
the wavevector-dependent susceptibility, $S(\kappa,T,q)$, exhibits maxima
at <u>nonzero</u> q, if $\kappa \geq 1/4$ [4]. (Only one component of the full wavevector $\underset{\sim}{q}=(0,0,q)$ does not vanish, because there is competition only
along the z-axis). The positions of these maxima, at $\pm q_{max}(\kappa,T)$,
vary continuously with κ and T and, as the temperature falls along
any locus $q_{max}=q_o$, the maximum susceptibility, $S_{max}=S(q_{max})$ eventually diverges on the critical line, $T_c(\kappa)$. The corresponding value
of κ is determined by $q_c(\kappa)=q_o$, where the critical wavevector, $q_c(\kappa)$,
increases continuously (and monotonically) from its zero at the
Lifshitz point, L; $\kappa_L \approx 0.27$ for d=3, $q_\perp=4$, and $J_1=-J_o$ [4,5,9]. In
mean field theory, assuming a <u>purely</u> sinusoidal magnetization pat-
tern below T_c, one simply finds $q_c(\kappa)=\cos^{-1}(1/4\kappa)$ [2] for $\kappa \geq 1/4$, in
good agreement with high temperature series expansions [4] and Monte
Carlo studies [9] for d=3, $J_1=-J_o$.

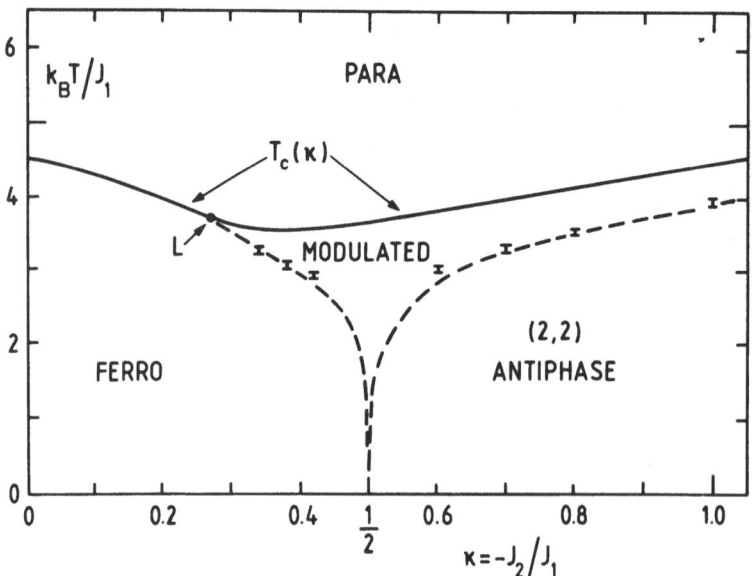

Fig. 2: Main features of the ANNNI model phase diagram as found for
the simple cubic lattice with $J_o=J_1$ by high temperature
series expansions [4], Monte Carlo [5,9] and low temperature
series expansions [13]. The Lifshitz point is denoted L.
(After [9]).

In general, a Lifshitz point is a multicritical point, at which a modulated ordered (described by a non-constant wavevector q), a uniformly ordered ($q=q_u$ is fixed) and the disordered phase meet. L is defined by a special form of the Ginzburg-Landau free energy [3], which lead to additional specific properties, e.g. one classifies Lifshitz points according to the number, n, of components of the order parameter, and the number, m, of axes, along which modulation occurs. In each case there are characteristic critical exponents for the quantities considered usually at phase transitions [32] as well as for the limiting behavior of q as it approaches q_u continuously along the critical line [3]. The ANNNI model exhibits a uniaxial, m=1, L of Ising-like, n=1, character. Supposedly the most reliable estimates for the critical exponents have been obtained using the standard MC method. For example, the exponent of the order parameter, the magnetization, M, which is proportional to the thermal average of the spin s_i at an arbitrary site i, has been found to be $\beta=0.21\pm0.03$ [5].

For $T < T_c(\kappa)$ and $\kappa > \kappa_L$ it is natural to suppose that the system enters some sort of modulated phase, or sequence of phases, with spatially varying magnetic order, $M(z) \approx M_0 \cos qz + ...$, where M(z) is the magnetization per layer, with a characteristic wavevector $q(\kappa,T) \leq q_{<2>} = \pi/2$ (we set the lattice constant in z-direction equal to one). Monte Carlo simulations confirm this general picture. In Fig.3 we show some typical equilibrium configurations, averaged over about 90 Monte Carlo steps per spin (MCS/S), for the magnetization per layer M(z) as a function of temperature at $J_2=-0.6J_1==-0.6J_0$ for a simple cubic lattice of size 6x6xM; M(=40) is the linear dimension along the z-axis. (A profound introduction to the MC method is given in [33]). Because full periodic boundary conditions were imposed the values of the wavevector q are quantized and changes in q occur by jumps of at least $\Delta q=2\pi/M$. Merely by counting the number of wave crests or maxima in Fig.3 one can easily determine the wavevector q(fluctuations in the value of q at given T and κ might occur during very long MC runs; then one should, of course, average over these fluctuations to obtain the mean wavevector; compare to [10] for the three-dimensional ANNNI model and to [22] for the two-dimensional case). We see that the < 2> phase remains stable up to about 77 percent of the temperature T_{max}, which indicates the transition to the paramagnetic phase (strictly speaking T_{max} locates the maximum in the specific heat for a system of a finite size, here 6x6x40. To estimate the true transition temperature, T_c, in the thermodynamic limit, a finite site analysis has to be done [9, 33]). As a consequence of a slight increase in temperature, $T=0.78 T_{max}$, the symmetry of the < 2> phase is broken and a "sinusoidally" modulated pattern of M(z) evolves. M(z) can be fitted quite well by

$$M(z) \simeq M_o \cos qz + M_1 \cos 3qz \qquad (1)$$

with M_0 close to its saturation value, 1, at zero temperature,

$M_1 \ll M_0$, and $q=q_{<2>} \sim -2\pi/40$. Further small changes in temperature lead to further jumps in q by the minimum increment Δq; M_0 decreases very slowly. Only near the transition temperature, T_{max}, q settles down at a value close to the one predicted be mean-field theory [2,4,9], M_0 decreases quite rapidly and the amplitude of the third harmonic, M_1, is no longer detectable within the accuracy of our simulation (mean-field theory, assuming M(z) to be of the form (1), describes the vanishing of M_0 and M_1 in approaching T_c by power-laws: $M_0 \sim t^\beta$ and $M_1 \sim t^{3\beta}$ with the reduced temperature $t=(T_c-T)/T_c$ and the critical exponent $\beta=1/2$ [34]. According to more refined theories, in the context of uniform ordering, β should be about 1/3. At any rate, M_1 appears to vanish much more quickly than M_0).

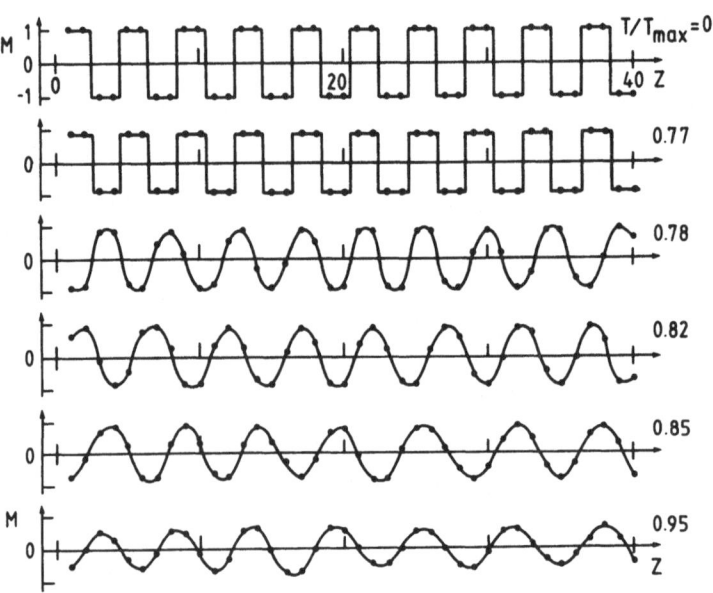

Fig. 3: Equilibrium configurations for the magnetization per layer M(z) (averaged over about 90 MCS/S) for the 6x6x40 system with $J_0=J_1$ and $J_2=-0.6J_1$. (After [9]).

More details of the MC study are given elsewhere [9] including a discussion of dynamical aspects, e.g. the "squeezing effect" describing the emergence of a magnetization pattern with a new wave-vector, which appears to occur also in models describing crystal growth [35]. However, there are some subtle questions on the nature of the modulated region, which cannot be answered using MC techniques (although other methods may have their problems, too; see below). In particular, the precise equilibrium variation of q, not just in jumps of at least $\Delta q=2\pi/M$, is of considerable theoretical interest, not least because of the increasingly large variety of real physical systems which exhibit comparable spatially modulated order (36) and the fascinating ground state properties ("devil's staircase") proven for other models with competing interactions (37,38).

At low temperatures the nature of the modulated phases has been settled on the basis of a systematic series expansion in powers of $w=\exp(-2J_0/k_BT)$ and $(\kappa-1/2)$ valid for $q_\perp$ $J_0>J_1$ and $d>2$ (1,13). The analysis demonstrates that the wedge in the (κ,T) plane, Fig.2, between the ferromagnetic (or $<\infty>$) and $<2>$ phases is filled by a countable infinite sequence of discrete commensurate phases, as depicted in Figs. 4 and 5, with wavevectors

$$q_j = j\pi/(2j+1) \qquad j=1,2,3,\ldots \qquad (2)$$

The jth is characterized by a $<2^{j-1}3>$ pattern of layers magnetized (predominantly) "+" or "-" where, for example, $<2^23>$ denotes the repeating sequence

$$..++--+++--++---..$$

and so on. A plot of the wavevector versus κ, at fixed temperature, see Fig.5, thus varies in a staircase fashion. It is not a true complete devil's staircase in its original meaning (37,39), because the steps for finite value of j occur merely at a few (not all) rational numbers $q/q_{<2>}$, and the phases are separated by first order transitions. However, one might say one has a "devil's top step" (40) at the point where q finally locks into the value $q_{<2>}=q_\infty=\pi/2$, i.e. as $j\to\infty$. There the corresponding phase boundaries, $\kappa_j(T)$, which mark the jump points, pile up closer and closer obeying, roughly, $\kappa_{j+1}-\kappa_j\sim w^{q_\perp j}$. Thus, when $j\to\infty$, the variation of q becomes quasi-continuous, according to

$$q_\infty-q(\kappa,T) \sim 1/\ln\,[\kappa_\infty(T)-\kappa]^{-1} \qquad (3)$$

as κ approaches $\kappa_\infty(T)$, the boundary of the simple commensurate $<2>$ phase. When $T\to0$ all the phases $<2^{j-1}3>$ meet at $\kappa=1/2$, $T=0$, which has thus been styled a multiphase point (1).

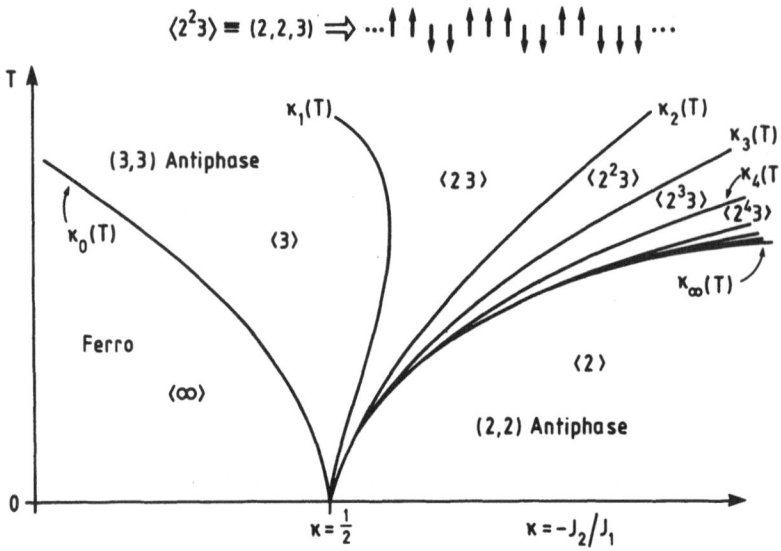

Fig. 4: Schematic phase diagram for the ANNNI model (with d>2) in
the vicinity of the multiphase point, illustrating the infinite sequence of distinct commensurate phases. The reen-
trant form of the < 3 >< 23 > boundary, $\kappa_1(T)$, is confirmed
by the explicit low-temperature expansions. (After [1]).

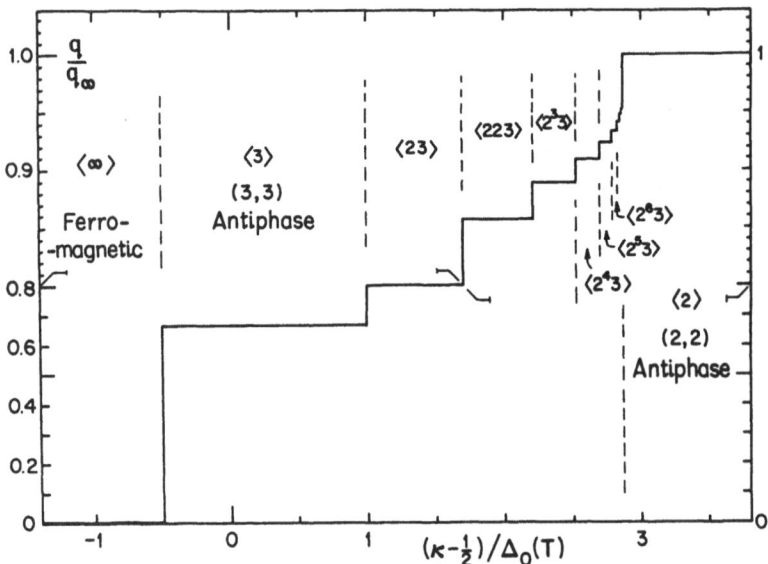

Fig. 5: Variation of the mean wavenumber, $q(T,\kappa)$, at low temperatures (after [1]). A "devil's top step" occurs at the point
where $\bar{q}$ reaches $q_\infty = \pi/2$, at the < 2 > phase.

The principal points of the low temperature analysis are, first, that the free energy expansions about the multiphase point must be calculated for all possible ground states of which, as mentioned, there are an infinite number. Second, at the crucial points the expansion must be carried to indefinitely high order. The analysis (1,13) has been reviewed before (40) and we refrain from presenting further details. It has been applied successfully to chiral Potts or clock models (41), where a finite number of distinct commensurate phases survive at nonzero temperatures, ANNNI models in a field (19,42), an antiferromagnetic Potts model (43), and, in a somewhat disguised form, to ANNNI-type models with competing interactions of longer range (16,17).

In the three-dimensional ANNNI model, the wavevector q appears to change continuously (incommensurate structure) close to the critical line, $T_c(\kappa)$, but at low temperatures it is either constant, q=0 in the ferromagnetic phase and q=π/2 in the <2> phase, or it changes in steps (commensurate phases) close to the multiphase point. Both Monte Carlo simulations and low temperature expansions cannot be applied easily to study the intermediate region precisely: MC, in addidtion to being restricted to jumps $\Delta q = 2\pi/M$, suffers from very slow relaxation of the system to equilibrium as well as fluctuations on large time scales in equilibrium (9,10) due to changes in the magnetization pattern to be done via one-spin-flip kinetics. In particular, one encounters difficulties close to the multiphase point (9). On the other hand, the low temperature analysis becomes quite cumbersome, because any candidate for an equilibrium state must be checked by computing its complete free energy to comparatively high orders. - However, there seems hope to make progress, at least on the understanding of some qualitative aspects in this region, again d>2, using mean-field theories.

Because the in-layer couplings, J_0, in the ANNNI model are ferromagnetic it looks reasonable to construct a mean-field theory by neglecting the fluctuations in the layer magnetization, M_ℓ, where ℓ=..-1,0,1,2,.. labels the layers. This leads to a one-dimensional local mean field theory: the layer magnetization M_ℓ will be coupled only to $M_{\ell+1}$ and $M_{\ell\pm2}$, because the interactions are restricted to spins in nearest and next-nearest layers. Such a theory will be exact in the limit that the interactions in each layer and between adjacent layers become infinitely long-ranged and infinitely weak (keeping the integrated coupling fixed). It might be expected to be correct for the original model at sufficiently low temperatures (a "mean field" result on the low temperature properties close to the multiphase point, in conflict with the analysis given above, is based on additional approximations (44)). Explicitly one obtains (12,45,46,47) for d=3, $q_\perp$=4

$$M_\ell = \tanh (H_\ell/k_B T) \quad \ell=..-1,0,1,..$$

$$H_\ell = 4J_0 M_\ell + J_1 (M_{\ell-1} + M_{\ell+1}) + J_2 (M_{\ell-2} + M_{\ell+2}) \tag{4}$$

The physically stable solution of these equations is the one, which corresponds to the absolute global minimum of the (mean field) free energy, see, e.g. (45,46). Equations (4) can be written as a discrete four-dimensional nonlinear map in the vectors $\underline{X}_\ell = (M_{\ell-2}, M_{\ell-1}, M_\ell, M_{\ell+1})$. Then $\underline{X}_{\ell+1} = f(\underline{X}_\ell)$, where f is a nonlinear operator readily obtained from (4). To find the stable solution, one has, in principle, to iterate all possible initial vectors, $\underline{X}_0$, and to determine the one, which minimizes the free energy. Obviously, this is an extremely trouble-some task, even if one reduces the possibilities using symmetries (46), and usually further approximations are invoked.

In the simplest case (2) one just compares magnetization pattern of the form $M_\ell \sim \cos(q\ell)$. However this approach just gives results close to $T_c(\kappa)$, see also (48), and does not work in the modulated phase itself, which we are interested in.

In more sophisticated treatments equations (4) were solved for <u>finite</u> lattices with periodic boundary conditions (10,11,18). (Similar calculations were done for the three-dimensional chiral Potts model (49), which also displays modulated phases). Obviously, the possible wavevectors are restricted severely, especially for small lattices, as for the MC simulations. Nevertheless, one may confirm the main features of the phase diagram, Fig.2, determine magnetization pattern, and demonstrate a complex behaviour of the wavevector in the modulated regime. In addition to the $< 2j^{-1}3 >$ phases more complicated commensurate structures are observed at higher temperatures (of course, only commensurate configurations are allowed because of the periodic boundary conditions); but no firm conclusions on their emergence could be drawn yet.

An analytic treatment of the mean field equations becomes feasible, if one expands the free energy about the $< 2>$ phase and considers magnetization pattern of the form (12)

$$M(z) = A \cos \left(\frac{\pi}{2} z + \phi(z)\right) \tag{5}$$

with a fixed amplitude, A, and a phase, $\phi(z)$, which may vary continuously (continuum limit). This approximation may give a reasonable description just above the $< 2 >$ phase. $\phi(z)$ is then determined by the static sine-Gordon equation (i.e. the equation of motion of a pendulum), which can be solved in terms of elliptic integrals, see, e.g. (50). A solution immediately above the $<2>$ phase is depicted in Fig.6. We observe that the magnetization per layer, $M(z)$, is almost perfectly of $< 2>$-type for many layers interrupted by a sequence of <u>three</u> layers with magnetization of the same sign, a "soliton" or "wall" (12,36). A similar, soliton-like or wall picture holds for the < 2J⁻¹3> phases found in the low temperature analysis (1,13).

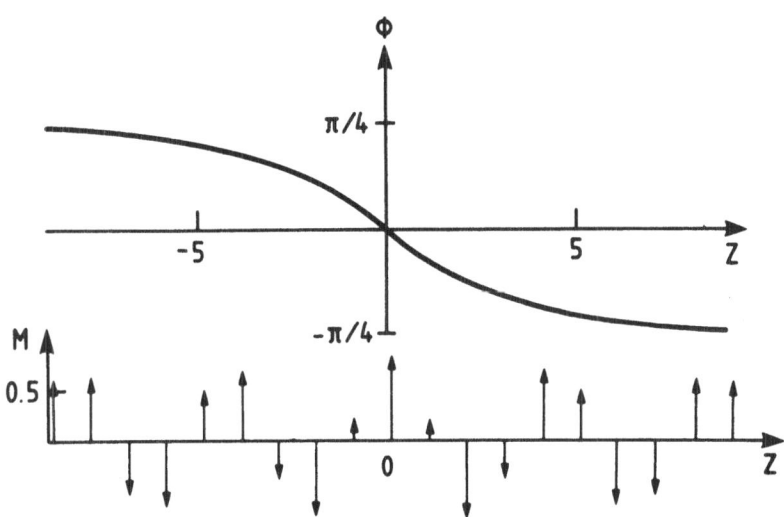

Fig. 6: Solution of the static sine-Gordon equation directly above the <2> phase of the three-dimensional ANNNI model with $J_0=J_1$ and $J_2=-0.7 J_1$. The phase, ϕ, and the magnetization per layer, M, are shown. (After (12)).

While the continuum approximation for $\phi(z)$ yields a wavevector q, which changes continuously with κ and T, the situation becomes more complicated, if one takes into account a discrete phase (ϕ is only defined for discrete layers, ℓ). Then the difference equation for ϕ is identical to the one for the position of a particle, at T=0, in a harmonic chain under the influence of a sinusoidal potential (model of Frenkel and Kontorowa (51) or Frank and van der Merwe [50]). The equation can be interpreted as a two-dimensional map. It has been studied in detail by Aubry [37].The transcription to the ANNNI model initiated a controversy about the stability of chaotic structures (52,37,53). But it is now agreed that such structures are at most metastable (nevertheless, they may have a non-negligible impact on some physical quantities (46)). The wavevector may exhibit, depending on the parameter κ and T, a complete or an incomplete devil's staircase. In the former case, the wavevector locks in only and at all rational values $q/q_{<2>}$ in some finite interval; in the incomplete devil's staircase the wavevector may change continuously, at least in some ranges (i.e. the irrational values cover an interval of non-zero measure).

One should bear in mind that these results are based on an expansion of the free energy about the $<2>$ phase. Presumably they
are only valid close to the $<2>$ phase at large competition ratio
κ. Near the multiphase point the sequence of distinct commensurate
phases, $<2^{j-1}3>$, j=1,2,3,..., may be reproduced only if one considers the higher order "umklapp terms" (12,36) in the expansion. Do-
ing this, one can also calculate the widths, Δ, of the simple commensurate phases, as one approaches $T_c(\kappa)$. Of course, Δ should vanish, because at T_c the wavevector is believed to vary continuously
without any lock-in. Indeed, one obtains $\Delta \sim t^j$ (36), with $t=(T_c(\kappa)-$
$-T)/T_c$. Using renormalization group techniques the vanishing of Δ
has been confirmed, but the exponents have to be modified (54).

A different mean field theory for the three-dimensional ANNNI
model has been published, in which the interactions between spins
in the z-chains are treated exactly, but an average field is assumed for the interactions between z-chains. However, only the critical line to the paramagnetic phase has been calculated (55).

Combining the various results on the phase diagram of the three-
dimensional ANNNI model, we still do not have a complete description.
Therefore our summary is at best heuristic, see also (45): at low
temperatures near the multiphase point the simple $<2^{j-1}3>$ phases,
j=1,2,3,...∞, fill up the wedge between the ferromagnetic and the
$<2>$ phases. The phases extend over non-zero ranges up to $T_c(\kappa)$,
where they vanish with a characteristic power law behavior. At some
non-zero temperatures more complicated commensurate phases may emerge, enlargening the set of possible, rational, wavevectors. Eventually incommensurate structure, with irrational wavevectors, might
show up, possibly at a definite temeprature (46,37) for a given value
of κ (in each interval above that temperature the measure of the incommensurate structures should be non-zero). They will dominate as
one approaches the critical line $T_c(\kappa)$ (along which the commensurate phases form a set of measure zero, as one may see by adding
over all widths and lettig $t\rightarrow0$, see (45)). Near the $<2>$ phase one
may describe the magnetization pattern in terms of solitons or
walls, while close to $T_c(\kappa)$ a description in terms of sine-waves,
with a few higher harmonics of small amplitudes, seems to be appropriate.

3   THE TWO-DIMENSIONAL ANNNI MODEL

In the two-dimensional version the (d-1) dimensional layers
are just chains; i.e. Ising chains of spins, $s_i=\pm1$, with competing
interactions between nearest and next-nearest neighbours, $J_1>0, J_2<0$,
are coupled ferromagnetically, $J_0>0$. Of course, the discussion of
the ground state, given in Sect.2, applies. However, the systematic

low temperature expansion (which converges only for d>2) is no longer valid, and mean field theory gives a qualitatively wrong phase diagram (due to large fluctuations in two dimensions, which destroy much of the incommensurate phase), as can be seen from Fig.7. The phase diagram combines results from MC studies [21,22] and the free fermion approximation (23). Note that the extension of the incommensurate phase for large competition ratio κ (or, in Fig.7, α: we used $J_1=(1-\alpha)J_0$ and $J_2=\alpha J_0$ with $0 \leq \alpha \leq 1$; hence $\kappa=\alpha/(1-\alpha)$) is open to question. It is quite conceivable that a very narrow strip of that phase persists up to $\kappa \to \infty$ ($\alpha=1$) (56) ; in that case the Lifshitz point 'L', would coincide with the

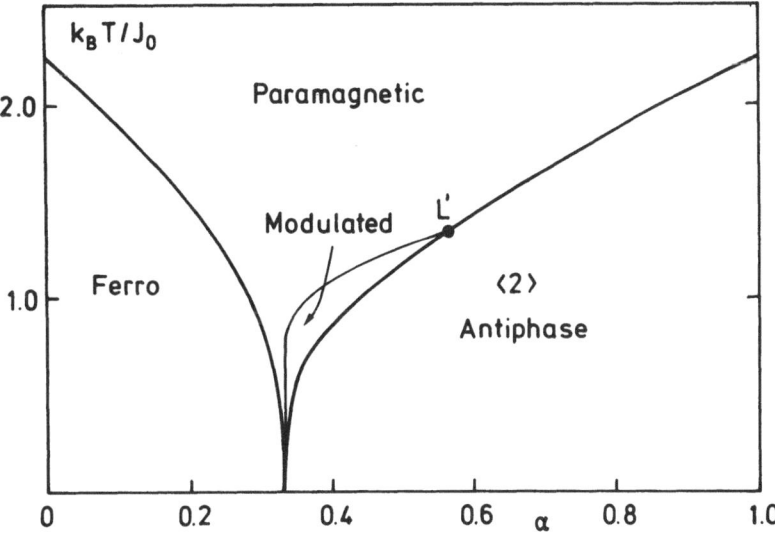

Fig. 7: Phase diagram of the two-dimensional ANNNI model on a rectangular lattice with $J_1=(1-\alpha)J_0$ and $J_2=-\alpha J_0$. Whether the modulated phase extends up to $\alpha=1$ (via a tiny strip) is open to question. (After (22)).

"decoupling point" (57) at $\alpha=1$. Clarification might be expected from MC calculation on a special purpose computer (58) .

There is plenty of evidence that there is no Lifshitz point on the ferromagnetic side of the phase diagram. Using the free fermion approximation, Villain and Bak (23) found that the paramagnetic phase extends down to zero temperature. This claim could be proven in the Hamiltonian limit of the model ($J_0 \to \infty$, J_1, $J_2 \to 0$, κ fixed) (59) . Based on the theory of Kosterlitz and Thouless (60) and results of Schulz (61) it was shown later, using arguments similar to the ones by Villain and Bak, that in general uniaxial commensurate phases of periodicity p melt always into the disordered phase, if $p^2 < 8$ (62). Conflicting evidence stemming from MC computations was revoked by a very careful MC study (22) as due to interesting (but easily misleading) finite size effects. It is worth while to note that even very large lattices, about 200x30, may lead to "phase diagrams" similar to the one shown in Fig.2. This can be understood from the fact that the disorder line (separating regions in the pa-ramagnetic phase with maxima in the wavevector-dependent suscepti-bility $S(\kappa,T,q)$ at zero and non-zero values of q) starts at ($\kappa=1/2$, T=0) and runs very closely to the melting line of the ferromagnetic phase up to about $\kappa_{PL} \approx 0.3$ (22,59).

The most interesting aspect of the two-dimensional ANNNI model is the incommensurate (or floating (23)) phase (variations of the model, such as the mock ANNNI model (63) or ANNNI models on brick-work or even hierarchical lattices (64), do not exhibit this phase). By examining typical MC equilibrium configurations, as shown in Fig.8a, it is seen that the floating phase can be described, at least just above the <2> phase, in terms of walls activated simply by kinks. The corresponding analytic treatment is the free fermion approximation. The walls do not just oscillate a little bit about equilibrium positions, like in the three-dimensional model, but fluctuate appreciably (this behaviour justifies the expression "flo-ating"), as observed during long Monte Carlo runs (22). Analytically the incommensurate phase is characterised by algebraically decaying correlations (23). Along the z-direction spins, separated by a dis-tance r, are correlated by

$$<s_o s_r> \sim r^{-\eta} \cos qz \qquad (6)$$

i.e. there is no long-range order, in agreement with the analogy to the two-dimensional XY-model (21,22,60). A crude argument for this analogy is based on the observation that in the incommensurate phase both amplitude and phase might fluctuate leading to a two-component "order parameter", n=2, for that phase in an Ising model. More precise arguments involve renormalization group techni-ques (65). Because of this analogy one might speculate (21,22) that the transition from the floating to the paramagnetic phase is driven by the unbinding of pairs of some topological defects. In fact, dislocation-like configurations can be identified to play this role (22,27,66,67), see Fig.8b. Again, additional information might be gathered in MC simulations on a special purpose computer (58). Especially the temperature dependence of the critical expo-nent η, equation (6), might be determined using, possibly, pheno-menological renormalization of MC data for $S(\kappa,T,q)$ as has been proposed (and done for small lattices) recently (22).

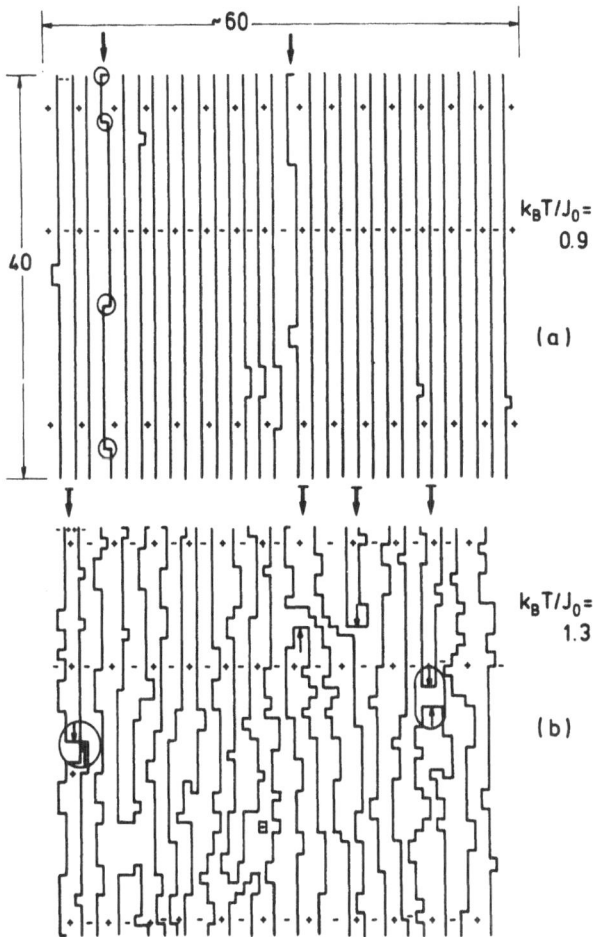

Fig. 8: Typical equilibrium configurations for the ANNNI model of
Fig.7 at α=0.375. Slightly above the <2> phase (a) the
walls (marked by arrows) occur, which are deformed by kinks
(a few examples are marked by circles). Close to the transition to the disordered phase (b) dislocations (some are
marked by arrows) forming pairs (marked by circles) are
seen. About in the centre of the picture a clearly separated pair of dislocations is created. The size of the MC
system is 88x40, but only parts (≈60x40) are shown.
(After (27)).

4 COMPARISON WITH EXPERIMENTS

Of course, ANNNI models seem to be much too simple to describe
specific experiments quantitatively, but they may reproduce crucial
qualitative features observed in real systems exhibiting spatially

modulated structures. We shall discuss the following aspects: (i) Lifshitz points and their critical exponents; (ii) sequences of commensurate phases; (iii) commensurate-incommensurate transitions as well as description of the incommensurate phase in terms of walls or solitons; and (iv) two-dimensional systems with uniaxially modulated structures.

(i) The critical exponents at the Lifshitz point, L, of the three-dimensional ANNNI model are representative for a whole universality class [32], here d=3, n=1, and m=1, independent of the microscopic interactions, which may give rise to such a multicritical point. In particular, the small value of $\beta(=0.21\pm0.03$ [5]) should be experimentally distinguishable from an exponent with a value of about 1/3 at usual second-order transitions. Indeed, such a small exponent has been found in experiments on $RbCaF_3$ [7], where the critical region of the structural phase transition is dominated by a Lifshitz-like behavior, as seen from the flat phonon dispersion relation. Only very closely to T_c this behavior may cross over into a usual critical behavior. Very accurate measurements on MnP show that this magnetic system displays a uniaxial Lifshitz point of Ising-like character in the temperature-field plane [8]. The phase diagram has been discussed in terms of the competition parameters of the ANNNI model [8]. Further experiments on the critical exponents at the L in MnP are done presently. To our knowledge, only on MnP measurements have been performed which demonstrate unambiguously the existence of a L. In particular, the wavevector was determined as one approaches the Lifshitz point [8]. There are quite a few other candidates, i.e. triple points, where a commensurately ordered, an incommensurately ordered and the disordered phases meet, for Lifshitz points, but their critical properties have still to be explored.

(ii) Very complicated phase diagrams in the temperature-field plane were experimentally investigated with fourteen distinct commensurate phases for CeSb [14] and eight distinct commensurate phases for CeBi [15]. Many of the experimental findings can be theoretically interpreted strikingly well [16,17] in the framework of ANNNI models with an intralayer coupling, J_0, much larger than the competing interactions between layers. The analyes are related to the systematic low temperature expansion [1,13] for the ANNNI model. Competing interactions between more distant layers, J_3, $J_4 \neq 0$, and a magnetic field have to be included. However, a fully microscopic theory of these cerium monopnictide compounds might be even more sophisticated, in particular to explain the experimentally observed paramagnetic layers [68]. It is worth mentioning that the low temperature analysis can be easily modified to deal with more complex lattice structures, like hcp [69], yielding similar sequences of phases. Attention should be drawn to different sequences [70], also due to competing interactions.

(iii) The soliton or wall picture of the incommensurate phase close to the second-order commensurate-incommensurate transition, as discussed on the < 2 > side of the ANNNI model, can be applied successfully to describe several experiments on structural phase transitions, at which uniaxially modulated order sets in; for a review, see (71). However, the experimental verification of the theoretically predicted logarithmic behavior of the soliton density at such transitions, see equation (3) or (1), in accordance with a more general, phenomenological theory of commensurate-incommensurate transitions (36), is still a puzzle. In many cases the commensurate-incommensurate transition is of first order, as is the case for the ANNNI model on the ferromagnetic side (12). Actually, the ANNNI model has been studied in connection with the ferroelectric $NaNO_2$ and its sinusoidally modulated phase some years ago (72).

(iv) Two-dimensional systems are, for instance, realized in adsorbed overlayers. Especially chemisorbed gases can be described by lattice gas models, which can be easily transcribed into Ising models. There are various adsorption systems with uniaxially modulated, commensurate and incommensurate, orderings, e.g. H/Fe(110) and O_2/Pd(110) (27,28,29). Indeed, the phase diagrams in the coverage-temperature plane of these two examples can be reproduced very well by models with competing interactions of ANNNI-type (27) or by just the two-dimensional ANNNI model in a field (73), see also (74).
Finally, experimental data on modulated structures in lipid bilayers have been analysed in the framework of a modified version of the two-dimensional ANNNI model (30).

ACKNOWLEDGEMENTS

I wish to thank Prof. Michael Fisher for a very productive and pleasant cooperation. Several discussions, sometimes controversial, but always stimulating, with Prof. Per Bak are gratefully acknowledged. Thanks are due to Dr. Julia Yeomans for a correspondence on the two-dimensional ANNNI model, and to all my coworkers. I also wish to thank Prof. Kurt Binder, who introduced me to the Monte Carlo method, without which relevant parts of the work reported above would not have been done.

REFERENCES

1. Fisher, M.E. and W. Selke, Phys. Rev. Lett. 44 (1980) 1502.
2. Elliott, R.J., Phys. Rev. 124 (1961) 346.
3. Hornreich, R.M., Luban, M., and S. Shtrikman, Phys. Rev. Lett. 35 (1975) 1678; Hornreich, R.M., J. Magn. Magn. Mater. 15-18 (1980) 387.
4. Redner, S. and H.E. Stanley, J. Phys. C10 (1977) 4765; Phys. Rev. B16 (1977) 4901.

5. Selke, W., Z. Physik B29 (1978) 133.
6. Hornreich, R.M., Luban, M., and S. Shtrikman, Physica A86 (1977) 465; Selke, W., Z. Physik B27 (1977) 81.
7. Buzare, J.Y., Fayet, J.C. Berlinger, W., and K.A. Müller, Phys. Rev. Lett. 42 (1979) 465; Müller, K.A., Berlinger, W., Buzare, J.Y., and J.C. Fayet, Phys. Rev. B21 (1980) 1763; Aharony, A. and A.D. Bruce, Phys. Rev. Lett. 42 (1979) 462.
8. Becerra, C.C., Shapira, Y., Oliveira, N.F. and T.S. Chang, Phys. Rev. Lett. 44 (1980) 1692; Shapira, Y., Becerra, C.C., Oliveira, N.F., and T.S. Chang, Phys. Rev. B24 (1981) 2780; Shapira, Y., J. Appl. Phys. 53 (1982) 1914.
9. Selke, W. and M.E. Fisher, Phys. Rev. B20 (1979) 257; J. Magn. Magn, Mater. 15-18 (1980) 403.
10. Rasmussen, E.B. and S.J. Knak-Jensen, Phys. Rev. B24 (1981) 2744; Kawasaki, T., J. Phys. Soc. Jpn 52 (1983) Suppl. 239.
11. von Boehm, J. and P. Bak, Phys. Rev. Lett. 42 (1979) 122.
12. Bak, P. and J. von Boehm, Phys. Rev. B21 (1980) 5297.
13. Fisher, M.E. and W. Selke, Phil. Trans. R. Soc. 302 (1981) 1.
14. Rossat-Mignod, J., Burlet, P. Bartholin, H., Vogt, O., and R. Lagnier, J. Phys. C13 (1980) 6381.
15. Bartholin, H., Burlet, P., Quezel, S., Rossat-Mignod, J., and O. Vogt, J. Physique Colloq. 40 (1979) C5-130.
16. Pokrovski, V.L. and G.V. Uimin, J. Phys. C15 (1982) L-353; ZhETF 82 (1982) 1640.
17. Uimin, G.V., J. Physique 18 (1982) L-665.
18. Yokoi, C.S.P., Coutinho-Filho, M.D., and S.R. Salinas, Phys. Rev. B24 (1981) 4047.
19. Smith, J. and J.M. Yeomans, J. Phys. C15 (1982) L-1053.
20. Öttinger, H.C., J. Phys. A16 (1983) 1483.
21. Hornreich, R.M., Liebmann, R., Schuster, H.G., and W. Selke, Z. Physik B35 (1979) 91.
22. Selke, W., and M.E. Fisher, Z. Physik B40 (1980) 71; Selke, W, Z. Physik B43 (1981) 335; Barber, M.N. and W. Selke, J. Phys. A15 (1982) L-617.
23. Villain, J. and P. Bak, J. Physique 42 (1981) 657.
24. Kroemer, J. and W. Pesch, J. Phys. A15 (1982) L-25.
25. Rujan, P., Phys. Rev. B24 (1981) 6620; Williams, G.O., Rujan, P., and H.L. Frisch, Phys. Rev. B24 (1981) 6632.
26. Barber, M.N. and P.M. Duxbury, J. Phys. A14 (1981) L-251; J. Stat. Phys. 29 (1982) 427; Duxbury, P.M. and M.N. Barber, J. Phys. A15 (1982) 3219.
27. Kinzel, W., Selke, W., and K. Binder, Surface Sci. 121 (1982) 13; Selke, W., Binder, K., and W. Kinzel, Surface Sci. 125 (1983) 74.
28. Imbihl, R., R.J. Christmann, K., Ertl, G., and T. Matsushima, Surface Sci. 117 (1982) 257.
29. Jaubert, M., Glachant, A., Bienfait, M., and G. Boato, Phys. Rev. Lett. 46 (1981) 1679.
30. Pearce, P.A. and H.L. Scott, J. Chem. Phys. 77 (1982) 951.
31. Redner, S., J. Stat. Phys. 25 (1981) 15; Hajudukovic, D. and

S. Milosevic, J. Phys. A15 (1982) L-723.

32. Fisher, M.E., Rep. Prg. Phys. 30 (1967) 615.
33. Binder, K., Monte Carlo Methods in Statistical Physics (Berlin, Springer, 1979).
34. Michelson, A., Phys. Rev. B16 (1977) 577.
35. Kerszberg, M., Phys. Rev. B27 (1983) 3909.
36. Bak , P., Rep. Prog. Phys. 45 (1982) 587.
37. Aubry, S., Soliton and Condensed Matter Physics (Berlin, Springer 1978) p.264; Seminar on the Riemann problem (Berlin, Springer 1981) p.221. See also: Allroth, E. and H. Müller-Krumbhaar, Phys. Rev. A27 (1983) 1575.
38. Bak , P. and R. Bruinsma, Phys. Rev. Lett. 49 (1982) 249.
39. Mandelbrot, B., Form, Chance and Dimension (San Francisco, W.H. Freeman and Company 1977).
40. Fisher, M.E., J. Appl. Phys. 52 (1981) 2014.
41. Yeomans, J.M. and M.E. Fisher, J. Phys. C14 (1981) L-835; Yeomans, J.M., J. Phys. C15 (1982) 7305.
42. Uimin, G.V., JETP Lett. 36 (1982) 250.
43. Samukhin, A.N., Sov. Phys. Solid State 24 (1982) 1297.
44. Villain, J. and M. Gordon, J. Phys. C13 (1980) 3117; Gordon, M. and Villain, J., Physica B108 (1981) 1075.
45. Fisher, M.E. and D.A. Huse, Melting, Localization, and Chaos (Elsevier Science Publishing Company 1982) p.259.
46. Jensen, M.H. and P. Bak, Phys. Rev. B27 (1983).
47. Pandit, P. and M. Wortis, Phys. Rev. B25 (1982) 3226.
48. Smart, J.S., Effective Field Theories of Magnetism (Philadelphia, W.B. Saunders Company 1966).
49. Öttinger, H.C., J. Phys. C16 (1983) L-257.
50. Frank, F.C. and J.H. van der Merwe, Proc. R. Soc. London A198 (1949) 206.
51. Frenkel, J. and T. Kontorowa, Phys. Z. Sowjet. 13 (1938) 1.
52. Bak, P., Phys. Rev. Lett. 46 (1981) 791.
53. Aubry, S., J. Physique 44 (1983) 147; Fradkin, E., Hernandez, O., Hubermann, B.A., and R. Pandit, Nucl. Phys. B215 (1983) 137.
54. Aharony, A. and P. Bak, Phys. Rev. B23 (1981) 4770.
55. Pires, A.S.T., Silva, N.P. and B.J.O. Franco, Phys. Stat. Sol. B114 (1982) K-63.
56. Schulz, H.J., Phys. Rev. B (1983) to be published; Haldane, F.D.M., Bak, P., and T. Bohr, Phys. Rev. B (1983) to be published.
57. Huse, D.A. and M.E. Fisher, J. Phys. C15 (1982) L-585.
58. Hoogland, A., Spaa, J., Selman, B., and A. Compagner, Preprint (1983).
59. Peschel, I. and V.J. Emery, Z. Physik B43 (1981) 241.
60. Kosterlitz, J.M. and D.J. Thouless, J. Phys. C6 (1973) 1181.
61. Schulz, H.J., Phys. Rev. B22 (1980) 5274.
62. Coppersmith, S.N. Fisher, D.S., Halperin, B.I., Lee, P.A., and W.F. Brinkman, Phys. Rev. B25 (1982) 349.
63. Huse, D.A., Fisher, M.E., and J.M. Yeomans, Phys. Rev. B23 (1981) 180.

64. Bideaux, R. and L. de Seze, J. Physique 42 (1981) 371; Morgenstern, I., Phys. Rev. B26 (1982) 5296; Svrakic, N.M., Kertesz, J., and W. Selke, J. Phys. A15 (1982) L-427.
65. Droz, M. and M.D. Coutinho-Filho, AIP Conf. Proc. 29 (1976) 465; Garel, T. and P. Pfeuty, J. Phys. C9 (1976) L-245.
66. Fisher, M.E., Physica A106 (1981) 28.
67. Selke, W. and J.M. Yeomans, Z. Physik B46 (1982) 311.
68. Takegahara, K., Takahashi, H., Yanasa, A., and T. Kasuya, Sol. State Commun. 39 (1981) 857; Kovalenko, A.A. and E.L. Nagaev, Sol. State Commun. 45 (1983) 243; Currat, R., Preprint (1983).
69. Selke, W., J. Phys. C14 (1981) L-17; Nakanishi, K., Preprint (1983).
70. Axel, F. and S. Aubry, J. Phys. C14 (1981) 5433; Mashiyama, H., J. Phys. C16 (1983) 187.
71. Blinc, R., Phys. Reports 79 (1981) 331.
72. Yamada, Y., Shibuya, I., and S. Hoshino, J. Phys. Soc. Japan 18 (1963) 1594.
73. Ertl, G. and J. Küppers, Surface Sci. 21 (1970) 61; Selke, W., Mat. Res. Soc. Symp. Proc. (1983) to be published; Rujan, P., Selke, W., and G.V. Uimin, Preprint (1983).
74. Schaub, B. and E. Domany, Preprint (1983).

COMPOSITION MODULATIONS IN SOLID SOLUTIONS

D. de Fontaine

Department of Materials Science and Mineral Engineering
University of California, Berkeley, California 94720

1 INTRODUCTION

Most crystalline solids contain various atomic or molecular species which may substitute for one another on lattice sites. Hence, in any solid solution, site occupation can be regarded as *composition-modulated*. To describe the state of order of a crystalline phase, one must therefore find a suitable mathematical formulation for its composition modulation. It is further desirable not only to describe the state of order, but also to predict it, or at least to determine theoretically the change of order, long-range (LRO) or short-range (SRO), with respect to temperature, pressure, or average composition (or chemical potential). In this paper, a technique will be presented which accomplishes both tasks rather well, i.e. those of description and of prediction: the Cluster Variation Method (CVM). The CVM was initially proposed by Kikuchi[1] as a hierarchy of statistical mechanical approximations to solve the classical Ising model. The mathematical apparatus was subsequently improved upon by Kikuchi and co-workers[2,3], by Barker[4], Hijmans and De Boer[5], and more recently by Sanchez and co-workers[6-10].

The general theoretical framework will first be given (in a *real space* formulation), followed by a concentration wave approach in *reciprocal space*. Finally, applications will be given to prototype *ordering phase diagram* calculations, then to partial ordering in such systems as Ni_3Mo-based ternary solutions.

The interested reader will find detailed background information in the author's earlier review article[11]. Unfortunately, that article was written before the bulk of recent CVM work had been accomplished; clearly, the earlier review should now be rewritten in the light of recent results. The present paper is but a brief summary of forthcoming comprehensive reviews. Other short reviews on special aspects of this subject are listed in the bibliography[12-15].

2 CVM FORMULATION

Real solid solutions, such as alloys, are very complicated systems generally containing phases separated by incoherent boundaries, misfitting inclusions, dislocations, point defects, and other imperfections. No all-encompassing theory exists today for handling such complicated features, so, for simplicity, only bulk perfect crystals will be treated here. In particular, the following model system will be considered: a perfect rigid lattice, the sites of which may be occupied by either A or B atoms (or + or - spins). Misfit elastic strains are disallowed. The atoms (or spins) are assumed to interact through constant ordering (or exchange) interactions. The theory is currently being generalized to include lattice parameter changes, concentration, and SRO-dependent interaction energies, and multicomponent systems (for the latter, see also earlier work[16,17]). The present formulation of the theory is based upon the work of Morita[18]; more elaborate treatments can be found in the references mentioned above and in unpublished work by Sanchez[19], Ducastelle[20], and the author.

2.1 Averages

The model envisaged is, of course, the well-known Ising model. Despite its simplicity, an exact solution for the three-dimensional case is not available, perhaps never will be. The CVM provides approximate solutions, the nature of which depends upon the method of averaging and point of truncation of various expansions. Consider a fixed lattice of N points. Define the "spin" variable $\sigma(p)$ at site p to be equal to +1 for an A atom, -1 for a B atom. The set of all σ variables, on all lattice sites, completely determines a "configuration" denoted briefly by $\underset{\sim}{\sigma}$. The probability of finding the system of N points in configuration $\underset{\sim}{\sigma}$ will be denoted by $x_{\underset{\sim}{\sigma}}$. The average value of any function g of σ is then given by

$$<g_{\underset{\sim}{\sigma}}> = \sum_{\underset{\sim}{\sigma}} X_{\underset{\sim}{\sigma}} \, g_{\underset{\sim}{\sigma}} \qquad < g_{\underset{\sim}{\sigma}} > = \sum_{\underset{\sim}{\sigma}} X_{\underset{\sim}{\sigma}} \, g_{\underset{\sim}{\sigma}} \qquad (1)$$

where the sum is over all 2^N configurations. At equilibrium, the probability $X_{\underset{\sim}{\sigma}}$ is given by the Boltzmann factor for that particular energy $E_{\underset{\sim}{\sigma}}$, divided by the partition function. The difficulty lies in finding a closed-form approximation of the partition function, or equivalently, of the Helmholtz free energy F = E - TS. Since no volume change and vibrational energy change are assumed on ordering, only the internal energy E and configurational entropy S need be considered.

Let us also define the "point occupation" variable

$$\Gamma_i(p) = \frac{1}{2}[1 + i\,\sigma(p)] \tag{2}$$

equal to 1 if the value of the spin variable at p be equal to the value $+i$ ($i = \pm 1$), and equal to 0 if $\sigma(p)$ be equal to $-i$. A set of n points $\{p_1, p_2, \ldots, p_n\}$ constitutes a "cluster." The probability that such a cluster be found in any one of the 2^n (not necessarily distinct) configurations $\underset{\sim}{\sigma}_n$ is given by

$$X_{\underset{\sim}{\sigma}_n}(p_1, \ldots, p_n) = <\prod_{k=1}^{n} \Gamma_{i_k}(p_k)>. \tag{3}$$

By Eqs. (1) and (2), one then obtains the important formula

$$X_{\underset{\sim}{\sigma}_n}(p_1 \ldots p_n) = \frac{1}{2^n}[1 + \sum_{p_1 \ldots p_n} \nu_{\underset{\sim}{\sigma}_n}(p_1 \ldots p_n; p_i \ldots p_n)\xi(p_i \ldots p_k)], \tag{4}$$

where, by Eq. (3), the summation extends over all possible subsets $\{p_i \ldots p_k\}$ of the cluster set $\{p_1 \ldots p_n\}$, and where the ξ variables are so-called "multiplet correlation functions" given by

$$\xi(p_i \ldots p_k) = <\sigma(p_i) \ldots \sigma(p_k)>, \tag{5}$$

the ensemble average being defined by Eq. (1).

A simpler notation for the ξ variables is sometimes useful: once a cluster of n points, of type s, has been defined, only the lattice point p on which it is "centered" need be defined. One may then write

$$\xi_1(p), \; \xi_{2,s}(p), \; \xi_{3,s}(p), \; \xi_{4,s}(p), \ldots$$

for point, pair, triplet, quadruplet, . . . correlation functions, respectively, the index s referring to the coordination shell for the case of pairs, the type of triangle for the case of triplets, squares or tetrahedra for the case of quadruplets, etc. If all points p are equivalent, as in a lattice with no LRO, then the index p need not be specified at all. As an example, consider a cluster formed by a nearest neighbor triangle in an fcc lattice. The corresponding cluster concentration (or probability) is then given by:

$$x_{ijk} = \frac{1}{8}[1 + (i+j+k)\xi_1 + (ij+ji+ki)\xi_2 + (ijk)\xi_3], \tag{6}$$

where the set i,j,k ($= \pm 1$) denotes the cluster configuration. The coefficients of the ξ variables in Eq. (6) are the elements of the ν-matrix which appears in Eq. (4).

It is clear from the foregoing that the multiplet correlation functions ξ are the fundamental configuration variables which fully characterize the state of order (LRO and SRO) to the level of approximation determined by the largest cluster used. The fundamental nature of these variables is a consequence of the following basic properties:

- the ξ's form a set of linearly independent variables,
- any cluster concentration can be determined uniquely from the ξ's,
- for a given largest cluster chosen in the approximation, the required ξ variables are those which correspond to the cluster itself and all of its distinct subclusters, down to the individual lattice point.

If the largest cluster chosen is the total crystal of N points itself, then any configuration $\underset{\sim}{\sigma}$ is given exactly by the correlation function expansion

$$X_{\underset{\sim}{\sigma}} = \frac{1}{2^n}[1 + \sum_{n,s} \nu_{\underset{\sim}{\sigma}}(n,s) \sum_{p} \xi_{n,s}(p)], \tag{7}$$

the summation extending over all possible clusters in the crystal. This formula can be used for calculating average energy and entropy as shown next[19].

2.2 Configurational Energy

According to Eqs. (1) and (7), the ensemble-averaged configurational energy $E_{\underset{\sim}{\sigma}}$ is given by

$$<E> = \sum_{\underset{\sim}{\sigma}} X_{\underset{\sim}{\sigma}} E_{\underset{\sim}{\sigma}} = E_o + \sum_{n,s} E_{n,s} \sum_{p} \xi_{n,s}(p) \tag{8}$$

in which

$$E_o = \frac{1}{2^n} \sum_{\underset{\sim}{\sigma}} E_{\underset{\sim}{\sigma}} \tag{9}$$

is the energy of the completely disordered state, and where the effective interaction parameters are defined by

$$E_{n,s} = \frac{1}{2^n} \sum_{\underset{\sim}{\sigma}} \nu_{\underset{\sim}{\sigma}}(n,s) E_{\underset{\sim}{\sigma}}. \tag{10}$$

That the latter parameters, for the case of pairs, are equivalent to the well-known ordering (or exchange) pair interactions V(s), of given spacing or coordination shell s, can be shown as follows:

$$E_{2,s} = \frac{1}{2N} \sum_{\sigma_1 = \pm 1} \sum_{\sigma_2 = \pm 1} \sigma_1 \sigma_2 \sum_{\underset{\sim}{\sigma}'} E_{\sigma_1 \sigma_2 \underset{\sim}{\sigma}'} = V_{++} - V_{+-} - V_{+-} + V_{--},$$

where

$$V_{ij} = \frac{1}{2^n} \sum_{\underset{\sim}{\sigma}'} E_{ij\underset{\sim}{\sigma}'}$$

is the average energy of all configurations having atom i at point p and atom j at point (p+s), the notation $\underset{\sim}{\sigma}'$ indicating all configurations excluding those two points. Thus we have

$$E_{2,s} = -2V(s)$$

with

$$V = V_{AB} - \frac{V_{AA} + V_{BB}}{2},$$

in the familiar notation[11]. This procedure can of course be generalized to multiplet, or many-body interaction. In this way, total cohesive energies, obtained for example by band structure calculations, can be parameterized uniquely as a sum of effective pair, triplet, quadruplet, . . . interactions. It is expected that the magnitude of the interactions $\xi_{n,s}$ will decrease rather rapidly for clusters of large n[21] or for pairs of large s. Hence, the ordering energy (second term on the right in Eq. (8), henceforth written as E, for short), can be written approximately as a linear form in the correlation variables:

$$E = \sum_{n,s}' E_{n,s} \sum_{p} \xi_{n,s}(p), \qquad (11)$$

the accent on the summation sign indicating that the expansion in the energy parameters has been truncated at some suitably chosen largest cluster. In Eq. (8), (10), and (11), it is seen that the interaction parameters do not depend on the cluster location p. This is because the elements of the ν-matrix in Eq. (4) actually depend only on the relative positions of cluster and sub-cluster, and thus have the translational symmetry of the lattice of the disordered phase.

2.3 Configurational Entropy

The configurational entropy $\langle S \rangle$ (henceforth designated simply as S) is given exactly by the well-known formula

$$S = -k_B \sum_{\underset{\sim}{\sigma}} X_{\underset{\sim}{\sigma}} \ln X_{\underset{\sim}{\sigma}}, \qquad (12)$$

where k_B is Boltzmann's constant. This expression is of little use, however, because the summation in Eq. (12) is over the unmanageable 2^N configurations. Using the truncated expansion (8) to obtain an approximate entropy also will not do since the logarithm of a sum does not simplify. Hence, another form of series truncation is required, for example that suggested by Morita[1,8]. In this approach, one writes successively higher cluster concentrations as superpositions of lower cluster probabilities multiplied by correction factors, thus:

$$x_1 = y_1$$

$$x_{12} = y_1 y_2 y_{12}$$

$$x_{123} = y_1 y_2 y_3 y_{12} y_{23} y_{31} y_{123} \tag{13}$$

in which the simplified notation $x(p_1) \rightarrow x_1$, $x(p_1 p_2) \rightarrow x_{12}$, . . . has been used. Expressions for the y correction factors can be obtained uniquely by recursion, as follows:

$$y_1 = x_1$$

$$y_{12} = x_{12}/x_1 x_2$$

$$y_{123} = x_{124}/y_{12} y_{23} y_{31} y_1 y_2 y_3$$

$$= x_{123} x_1 x_2 x_3 / x_{12} x_{23} x_{31}$$

etc. (14)

Similarly, the expectation value X_σ of the total configuration σ can be written, in principle, as a product of y-factors pertaining to all possible sub-clusters of the lattice. The important point to be made here is that, as an approximation, the product of y-factors can be truncated at some given maximal cluster x_L, say. Heuristically, one may indeed assume that, for clusters of dimensions greater than the typical correlation length, the corresponding cluster probability x will be well approximated by a direct superposition of lower cluster probabilities; in other words, the y correction factors for large clusters will tend to unity. Such is the basic philosophy of the Cluster Variation Method. Unfortunately, no rigórous criteria are available for the selection of clusters to be retained in the expansion; as a general rule, the approximation improves as the clusters used are larger, provided that they are sufficiently compact and symmetric, in some sense. Thus, whereas an expansion in small clusters and sub-clusters may suffice at high temperatures, such an expansion would be invalid in the vicinity of a critical point where the correlation lengths become very large. Indeed, the CVM tends to do rather poorly in the immediate vicinity of a critical point, and yields classical critical exponents.

We have still to couch the configurational entropy in tractable form. To that end, we write, symbolically,

$$\ln\ X = \ln \Pi\ y \overset{\sim}{=} \sum{}' \ln\ y = -\sum{}'\ \gamma\ \ln\ x, \qquad (15)$$

the accent on the summation indicating truncation after some suitable maximal cluster X_L. In Eq. (15), the γ coefficients are those obtained by recursion from equations such as those of Eq. (14). These so-called Kikuchi-Barker coefficients are in fact given by[4, 5].

$$\gamma_L = -m_L$$

$$\gamma_\ell = -m_\ell - \sum_{j=\ell+1}^{L} m_{\ell,j} \gamma_j \qquad (16)$$

where m_k is the number of clusters of type ℓ (global index replacing n,s) per lattice point, and $m_{\ell,j}$ is the number of subclusters of type ℓ contained in a cluster of type j. By Eqs. (12) and (15), one finally obtains the approximate CVM configurational entropy as

$$S = k_B \sum_{\ell > L} \gamma_\ell \sum_\sigma \sum_p X_\sigma \ln x_\ell = k_B \sum_\ell \gamma_\ell \sum_\sigma \sum_p x_\ell \ln x_\ell. \qquad (17)$$

A more explicit form of the entropy will be given in the next section.

Equation (17) has a general form, so that to make use of it, one must merely specify the Kikuchi-Barker coefficients, which are positive or negative integers. Values for these coefficients have been given for various crystal structures, such as fcc, bcc, hcp, and for various cluster approximations such as the *point* (equivalent to Bethe)[1], *square* on a square lattice (equivalent to Kramers-Wannier)[1], various clusters up to the bcc cube in the bcc lattice[6], various approximations in the fcc lattice such as the *tetrahedron, tetrahedron-octahedron, double-tetrahedron-octahedron*, and *fcc-cube*, and various cluster approximations in the hcp structure[10].

2.4 Configurational Free Energy

From the foregoing, it was seen that both multiplet correlations ξ and cluster concentrations x depended on the set of cluster points $\{p_1...p_n\}$, although the x depended on the configuration σ, whereas the ξ did not. In general, then, both ξ and x must depend on cluster type (n,s) and location p. However, any two points p and p' which are crystallographically equivalent in a given structure must have the same values of x and ξ associated with them. Thus, considerable economy results in limiting the sums over p encountered in Eqs. (11) and (17), for example, to sums over the crystallographically non-equivalent positions of the unit cell, i.e. over the sublattices (α = 1,..., M) of the ordered superstructure considered.

The CVM free energy is then obtained by combining the energy and entropy expressions (11) and (17) obtained above, yielding the explicit expression for the free energy per lattice point

$$f = \frac{1}{M} \sum_{\alpha=1}^{M} \sum_{n,s} [\varepsilon_{n,s} \xi_{n,s}(p_\alpha) - k_B T \sum_{\sigma} x_{n,s}(\sigma,p_\alpha) \ln x_{n,s}(\sigma,p_\alpha)] \quad (18)$$

All clusters appearing in the energy must also appear in the entropy, but not conversely; in actual practice, the configurational energy is often expressed as a sum over near neighbor pair interactions. If pairs of given separation s are considered, clusters used in the entropy must be large enough to contain them. For example, if next-nearest neighbor pairs are used in the energy of the fcc lattice, the entropy must contain at least octahedron cluster probabilities. This imposes a rather severe constraint on the range of interactions allowed: In the fcc tetrahedron-octahedron approximation, there are 10 independent ξ variables in the disordered phase; in some of the ordered superstructures that number can be of the order of a hundred.

By means of Eq. (4), the free energy (18) can now be expressed as a function of the independent ξ functions alone:

$$f = f(\xi). \quad (19)$$

Appeal must then be made to the variational principle which states that the best approximation of the (equilbrium) free energy, for given functional (19), is that which is stationary for small variations in the ξ variables. Minimization of the free energy f with respect to the independent ξ correlations results in a set of simultaneous non-linear algebraic equations which must be solved numerically by iteration techniques. Currently, two methods are in use: The Natural Iteration (NI) method proposed by Kikuchi[2], which operates on the redundant set of x variables, and a Newton-Raphson (NR) method proposed by Sanchez[6], which operates on the set of independent ξ variables.

The latter appears to be more convenient when the number of variables is large. Nevertheless, severe numerical problems are often encountered in the minimization process: The NR method requires the calculation of the matrix of second derivatives of $f(\xi)$, and each element of this matrix turns out to be given by a sum of products of two ν-matrix elements divided by a cluster probability x. Particularly at low temperatures and low average concentrations, some cluster concentrations become exceedingly small, and the second-derivative matrix becomes ill-conditioned. Computer codes have been developed to perform the required minimization[22], but numerical convergence remains as the most severe problem which limits the application of the CVM to complex systems.

2.5 Ordered Ground States

A sum over M sublattices appears in the fundamental free
energy formula (18). Which sublattice scheme to select for given
disordered-phase lattice and which set of energy parameters ε is
a *ground state problem*, i.e. one of determining the ordered configuration of A and B atoms on lattice sites which gives the configurational energy E its lowest value, and hence that which minimizes
the free energy at absolute zero of temperature, for given average
concentration x_B.

Clearly, the ground state structure will depend on the nature of
the interactions ε: If these are long-range, complex crystal
structures may result with very large unit cells. Likewise, if
multiplet interactions are important, a richer set of ground state
structures will result than that obtained from pair interactions
alone. The general ground state problem is a straightforward one
to pose but, except in very simple cases, it has proved to be an
extremely difficult one to solve. The method outlined below is
based on the cluster expansions (4) and (8).

Values of cluster concentrations x must lie between 0 and 1
so that, by Eq. (4), the following linear inequalities in correlation variables must be obeyed in the disordered state[23]:

$$1 + \sum_{\ell} \nu_{\sigma}(L,) \, \xi_{\ell} \geq 0 \tag{20}$$

where, in this simplified notation, L labels the largest cluster
used, and ℓ labels any cluster or sub-cluster of the largest
cluster retained in the approximation. Inequalities (20) delimit
in ξ-space a convex polyhedral region which, in principle, encom-
passes all possible states of order realizable. Hence, any state
of (partial) order can be fully characterized, at the given level
of approximation, by a vector ξ, completely contained in the multi-
dimensional convex region defined by inequalities (20), a region
denoted by the term *configurational polyhedron*. Alternately, the
vector ξ may be expressed in terms of barycentric coordinates ρ_k
as follows:

$$\xi = \sum_{k=1}^{K} \rho_k \, \xi^{(k)}, \tag{21}$$

where $\xi^{(k)}$ denotes any one of the K vertices of the configurational
polyhedron, and where ρ_k are non-negative numbers such that

$$\sum_{k=1}^{K} \rho_k = 1. \tag{22}$$

By Eq. (11), the ordering energy can also be expressed in terms of barycentric coordinates as

$$E = \sum_{k=1}^{K} \rho_k E^{(k)} \tag{23}$$

where $E^{(k)}$ is the energy $\sum \epsilon_\ell \xi_\ell^{(k)}$ of the kth vertex of the configurational polyhedron. It then follows directly that the lowest internal energy, obtained by minimizing E in Eq. (23), subject to condition (22), is that of the vertex which has lowest value of $E^{(k)}$ for the interaction parameters E_ℓ selected. Thus, in principle, the vertices of the configurational polyhedron are the ground states of order sought, i.e. the ordered superstructures.

The search for ground states thus reduces to a problem in linear programming: The very difficult one of enumerating all vertices of a convex polyhedron determined by a set of linear inequalities. An account of this procedure, for the case of ground states in the fcc lattice under the tetrahedron-octahedron cluster approximation was given elsewhere[23]. Briefly, a vertex enumeration code based on the Simplex algorithm, and developed by M. J. Carrillo of the RAND Corporation, was used to list the $\xi_\ell^{(k)}$ coordinates of all vertices. A total of 43 vertices were found, of which 26 were truly distinct (under the operation of A-B atom interchange). Of these 26 solutions, one corresponded to pure A (or B), and 8 were those of the ground states determined previously by Kanamori[24] and by Cahn and collaborators[25,26] for the case of first and second-neighbor *pair* interactions only. Of the remaining 17 vertices, 7 could be constructed as bonafide crystal structures[23]. Since the analysis of Kanamori and Cahn exhausts all possible superstructures of fcc stable under 1st and 2nd neighbor pair interactions, the 7 new superstructures must necessarily be stabilized by multi-site interactions, such as tetrahedral or octahedral forces.

The 10 remaining vertices were shown to correspond to non-constructible structures, thereby indicating that equalities (20) are not sufficiently "tight" to insure that only geometrically consistent sets of $\xi_\ell^{(k)}$ coordinates are obtained. At present, it is now known how to remedy this deficiency of the method. In simple cases, though, it is possible to determine all superstructures of a given lattice which must be stable for various ratios of interaction parameters and at various stoichiometries.

2.6 Phase Diagram Calculations

The ground state structures described in the previous section are always superstructures of the parent disordered state structure, since only interchanges of A and B atoms are allowed on fixed

crystallographic sites. It follows that resulting (fully or
partially) ordered phases in equilibrium must be completely coherent
with one another, in the sense that the underlying lattice must be
continuous across interphase boundaries. The term *coherent* will be
used to denote this type of equilibrium along with the resulting
phase diagrams.

In elementary treatments of binary phase equilibrium, the
common tangent construction is used to derive   (or at least to
rationalize) phase diagrams. In the present case of free energies
which are function of a great many independent variables, the
elementary tangent construction is not appropriate. For this
reason, Kikuchi[2,34] proposed a method of phase equilibrium determination based on minimizing what he called the *grand potential* defined
by

$$\omega = f - \mu\xi, \tag{24}$$

where ξ is actually a sum of point correlation variables over the
various sublattices and where μ is an appropriate chemical potential
shortly to be specified. To make the meaining of Eq. (24) more
explicit, consider a new function f_A defined as the free energy f
in which $(2x_A - 1)$ has been substituted for ξ_1, all of the other
ξ variables having their equilibrium values. Another function f_B
is defined similarly. The classical "intercept rule" may be
written in two alternate ways:

$$\mu_A = f - x_A \frac{df_A}{dx_A}, \tag{25a}$$

$$\mu_B = f - x_B \frac{df_B}{dx_B}, \tag{25b}$$

where also

$$\frac{df_A}{dx_A} = \mu_A - \mu_B = - \frac{df_B}{dx_B} \tag{26}$$

in which μ_A and μ_B are the classical chemical potentials of A and
B respectively. Making use of Eq. (26) and summing Eqs. (25a) and
(25b), we obtain immediately Eq. (24), where ω now appears as a
sum and μ as a difference of chemical potentials:

$$\omega = \frac{\mu_A + \mu_B}{2}, \quad \mu = \frac{\mu_A - \mu_B}{2}.$$

Calculation of phase equilibrium then proceeds as follows:
For fixed values of temperature T and chemical potential μ, the
grand potential ω is minimized with respect to the independent ξ
variables, the free energy f to be used being the CVM free energy

appropriate for the ordered (or disordered) phase under consideration, i.e. the one for which the ν-matrix reflects the proper sublattice structure. Next, the minimum values of ω are plotted as a
function of μ for that phase and for other phases of interest.
Points at which two ω vs. μ curves intersect determine phase equilibrium. The lowest intersections found are then used to construct
the phase diagram: An intersection between curves for phases α
and β, say, yields two values of ξ, hence two values of equilibrium
concentrations x_α and x_β, for example, at the chosen temperature.
Two sets of all other correlation variables are also obtained, one
set for each phase, thereby completely determining the state of
order of each of the two phases in equilibrium. This procedure is
repeated for all temperatures and for all ordered phases predicted
by the ground state analysis. The calculated loci of phase boundary
points, in $T-x_B$ or $T-\mu$ space constitutes the required *coherent* (or
ordering) equilibrium phase diagram.

The grand potential method is obviously equivalent to the
traditional one since, at a point of intersection of ω curves,
both the sum ω and the difference μ of chemical potentials are
the same, hence the chemical potentials themselves are equal in
the two phases, as required. It is also seen that the grand
potential can be obtained from the free energy f by a Legendre
transformation, μ then appearing as the intensive (field) variable
conjugate to the extensive (density) variable ξ. Examples of
calculated "coherent" phase diagrams will be given in Section 4.1.

3   CONCENTRATION MODULATIONS

It is often of interest to test the stability of a given state
by varying slightly the configurational variables which define it.
The varied structure may then be described in terms of "concentration waves,"[1,27] the natural representation of which is that of
Fourier (or reciprocal) space. Concentration modulations may be
virtual, as used for stability analysis, may be of dynamical origin,
as in equilibrium fluctuations, or may be of kinetic origin, as in
the early stages of decomposition of a solution of uniform composition.

3.1 Stability Analysis

Consider a disordered solid solution with equilibrium values
of the correlation variables denoted by $\xi_\ell{}^0$. To test the stability
or instability of this state, let us perturb the $\xi_\ell{}^0$ to new values
$\xi_\ell(p)$, different, in general, for each lattice point. To second
order, the resulting change in free energy is then given by the
quadratic form

$$F_2 = \frac{1}{2} \sum_{\ell,\ell'} \sum_{p,p'} f_{\ell\ell'}(p,p') \, \delta\xi_\ell(p) \, \delta\xi_{\ell'}(p') \tag{27}$$

with

$$\delta\xi_\ell(p) = \xi_\ell(p) - \xi_\ell^0. \tag{28}$$

Since the internal energy is linear in ξ_ℓ, only the entropy contributes directly to the coefficients of Eq. (27). Differentiation is performed by the chain rule with the help of Eq. (4). The matrix elements $f_{\ell\ell'}(p,p')$ are those of the second derivative matrix referred to above.

It is convenient to impose periodic boundary conditions on the sample crystal of N lattice points. Then, a first diagonalization of Eq. (27) may be achieved by performing Fourier transforms:

$$F_2 = \frac{N}{2} \sum_k \underset{\sim}{X}(-k) \underline{F}(k) \underset{\sim}{X}(k) \tag{29}$$

where $\underline{F}$ is the matrix of Fourier transforms $F_{\ell\ell'}(k)$ of $f_{\ell\ell'}(p - p')$ and $\underset{\sim}{X}$ is the vector of Fourier transforms $X_\ell(k)$ of the correlation deviations $\delta\xi_\ell(p)$, given by Eq. (28). The matrix $\underline{F}$, dependent on wave vector $\underset{\sim}{k}$, is Hermitian and may be diagonalized by an orthogonal transformation to yield the sum of squares

$$F_2 = \frac{N}{2} \sum_k \sum_{\ell=1}^{L} \Lambda_\ell(k) \left| Z_\ell(k) \right|^2 \tag{30}$$

where Λ_ℓ are the (real) eigenvalues of $\underline{F}$, the Z_ℓ being, in a sense, normal modes of the correlation deviations.

At high temperatures, the disordered state is expected to be stable so that all eigenvalues Λ_ℓ are positive for arbitrary wave vector $\underset{\sim}{k}$. As the temperature is lowered, the disordered phase, if necessary prevented by kinetic constraints from nucleating a coherent ordered structure, could become unstable to a particular harmonic mode Z_ℓ of wave vector $\underset{\sim}{k}^0$, say, as soon as the associated eigenvalue $\Lambda_\ell(\underset{\sim}{k}^0)$ goes through zero. For given average composition x_B, the highest temperature at which an instability is found is called the instability, or spinodal temperature T_0, the vanishing eigenvalue defining the unstable mode (or eigenvector) and instability (or ordering) wave vector $\underset{\sim}{k}^0$. In (x_B, T) space, i.e. phase diagram space, the stability locus, or limit of incipient instability, or spinodal for ordering wave $\underset{\sim}{k}^0$ is thus determined very simply by the vanishing of the determinant of the matrix $\underline{F}$ (the Hessian of f):

$$\Delta(\underset{\sim}{k}^0; x_B, T) = 0. \tag{31}$$

3.2 Special Points

The matrix elements $f_{\ell\ell'}(p - p')$ possess the point group symmetry of the lattice, hence their Fourier transforms $F_{\ell\ell'}(k)$ have the same point group symmetry and the translational symmetry of the reciprocal lattice[28]. The gradient of any function possessing this symmetry, say $F(k)$, may be regarded as a vector whose magnitude must necessarily vanish at any point in k-space at which two or more symmetry elements intersect at a point, otherwise the invariance of $\nabla F(k)$ would be violated. Hence the general conclusion: At such "special points" any function $F(k)$, having the symmetry of the reciprocal lattice, must present a maximum, minimum, or saddle point. Of course, the function may, indeed must, present additional extrema elsewhere.

Use was first made of special points (SP), as defined above, by Lifshitz[29] in his study of the conditions for second-order phase transitions. To find SP in cases where the crystal structure is simply a Bravais lattice, one may look up the space group of the reciprocal lattice in the International Tables of X-ray Crystallography[30], and note those Wyckoff positions which contain only fixed indices. Those points, with indices translated into reciprocal space notation, are the required special points. Table I lists these for fcc and bcc lattices. The symmetry of the wave vector group is also indicated.

TABLE I

Special Points

fcc		bcc	
k-space point	point group symmetry	k-space point	point group symmetry
<000>	m3m	<000>	m3m
<100>	4/mmm	<100>	m3m
$<\frac{1}{2}\frac{1}{2}\frac{1}{2}>$	$\overline{3}$m	$<\frac{1}{2}\frac{1}{2}\frac{1}{2}>$	$\overline{4}$3m
$<1\frac{1}{2}0>$	$\overline{4}$2m	$<\frac{1}{2}\frac{1}{2}0>$	mmm

Recently[31], complete sets of SP have been determined for all possible crystallographic groups. The procedure is as follows: For an arbitrary crystal structure of given space group G_x, construct a new space group made up of the direct product of the point group of the structure, say G_p, a center of inversion $\overline{1}$ (if G_p does not already contain it), and the translation group T_g of the lattice which is the reciprocal of the direct lattice of the crystal structure in question. The fixed-index Wyckoff positions of the new space group

$$G_\omega = G_p \; 0 \; \overline{1} \; 0 \; T_g$$

are the required special points, after transformation to reciprocal space notation. The space groups given by this general formula, by construction, have no glide planes of screw axes: They are the centered symmorphic groups, of which there are only 24. A complete table of these groups and their SP has been given in [31].

It follows from the foregoing, and from the more general derivation[31], that the eigenvalues $\Lambda_\ell(k)$ of the stability matrix $F(k)$ must present extrema at all special points, by symmetry reasons alone, independently of the physical parameters which enter into the free energy. These SP are thus expected to play a funda-mental role: Indeed, the unstable mode occurs for k^0 which gives the smallest eigenvalue Λ_m its minimum value. Although accidental minima may be found at other than special points in k-space, their position will depend on the actual ratio of interaction parameters, on temperature, and average concentration. Special point minima, on the contrary, are invariant, and the search for minima of eigen-values, i.e. for unstable modes, should always begin by the con-sideration of SP's. Which points of the set of SP yield the lowest minimum (which is of interest) and which yield other extrema (which are not of interest in the present context) depend on the values of the interaction energy parameters ϵ_ℓ.

3.3 Fluctuations

It is shown, for example by Landau and Lifshitz[32], that the probability of occurrence of a fluctuation $\{\delta\xi\}$ is given by

$$P = \exp[-\Delta F/k_B T] \tag{32}$$

where ΔF is the difference in free energy between the uniform (disordered) state F_0 and the varied (modulated) state. Let us expand f in powers of $\delta\xi_\ell$:

$$f = F_0 + F_1 + F_2 + \ldots \tag{33}$$

The first order term vanishes so that, for small fluctuations, $\Delta F = f - F_0$ is simply given by the second order term F_2, itself given by Eq. (27).

The probability P thus represents a Gaussian distribution of $\delta\xi_\ell$ values about the average ξ_ℓ^0, or by Eq. (29), a Gaussian distri-bution of amplitudes $X(k)$ of concentration waves in k-space about zero amplitude. By standard mathematical manipulations[32], one readily obtains the expectation value of a product of amplitudes, at k,

$$<X_\ell X_{\ell'}> = \frac{k_B T}{N} (F^{-1})_{\ell\ell'} \qquad (34)$$

given, as it is seen, by the corresponding element of the matrix inverse to $\underline{F}(\underline{k})$:

$$(F^{-1})_{\ell\ell'} = M_{\ell\ell'}/\Delta \qquad (35)$$

where $M_{\ell\ell'}$ is the minor associated with the element $(\ell\ell')$ and Δ is the determinant of the matrix, as in Eq. (31). Although Eq. (34) was derived by making the standard Gaussian approximation, Sanchez[33] has recently shown it to be a rigorous result, provided that the free energy matrix $\underline{F}$ can be formulated rigorously, which in turn requires an exact free energy.

Fluctuation amplitude-squared, or intensity, can also be expressed in terms of normal modes as

$$<|Z_\ell(\underline{k})|^2 = \frac{k_B T}{N\Lambda_\ell(\underline{k})} \qquad (36)$$

in which the symbols have the same meaning as the corresponding ones of Eq. (30). The connection between fluctuations and instability is now quite apparent: Fluctuation intensity diverges at an instability, and maximum intensity will be found at those $\underline{k}^0$ positions in k-space which will yield unstable modes at lower temperatures. The implications of these theoretical concepts will now be discussed.

4   RESULTS OF CVM CALCULATIONS

The CVM may not be appropriate for the study of critical phenomena *per se*, but the use of the method for calculating phase diagrams has been highly successful. Thus far, no claim can be made for the actual calculation of realistic phase diagrams from first principles; for that, much more elaborate composition- and SRO-dependent energy data than are currently available would have to be input into the CVM codes. What has been done, however, is to calculate a set of typical *Prototype Ordering Phase Diagrams* for the purpose of illustrating the power of the method and also for providing a first broad classification of Ising model-based phase diagrams. Somewhat surprisingly, when our first T-O (Tetrahedron-Octahedron) ordering phase diagram appeared in print[7], it turned out to be, to the author's knowledge, the very first realistic (albeit approximate) such calculation for the Ising model lattice with nearest <u>and</u> next-nearest neighbor (nn and nnn) pair interactions on an fcc lattice. Results of some recent phase diagram calculations will now be described.

4.1   Prototype Ordering Phase Diagrams

The term *prototype* means here that the calculated diagrams
are not intended to represent real binary systems, but rather,
typical cases illustrating the variety of "coherent" equilibria
which may be found between ordered superstructures of a given
parent lattice.   The fcc lattice was chosen because of its frequency
of occurrence in binary metallic solid solutions.   In the energy,
nn and nnn pair interactions were chosen for the following fundamental reasons: First, it is only for nn and nnn pairs that the
ordered ground states have been determined completely and rigorously[24,25].   Second, the use of longer-range effective pair interactions in the energy expression (11) would entail the use of rather
large clusters in the configurational entropy, Eq. (17), thereby
increasing both the total number of independent ξ variables and the
complexity of ordered superstructures beyond what is presently
manageable.   Fortunately, many of the most commonly occurring fcc
superstructures are shown to be stable with just nn and nnn (ε_1 and
ε_2) effective pair interactions.   For simplicity, the ε_1 and ε_2
energy parameters will be treated as constant, so that the only
physical parameter which enters the calculation is the constant
ratio

$$\alpha = \varepsilon_2/\varepsilon_1 \ . \tag{36}$$

It has been shown that[11], with constant ε_1 and ε_2, the unstable
wave $\underset{\sim}{k}^0$ is unique for all compositions $0<x_B<1$, and that the ordering
wave vector is always at a special point (SP).   It is therefore
particularly convenient to classify all prototype phase diagrams
in just four basic *special point families*.   The family to which a
system belongs is determined uniquely by the values of α and the
sign of ε_1, as indicated in Table II (see also Fig. 8 of Ref. 11).

TABLE II

fcc Special Points	Typical Systems	Conditions on $\varepsilon_1, \varepsilon_2$
$\langle 000 \rangle$	Al-Zn	$\varepsilon_1 < 0, \ \varepsilon_2 < \lvert \varepsilon_1 \rvert$
$\langle 100 \rangle$	Cu-Au	$\varepsilon_1 > 0, \ \varepsilon_2 < 0$
$\langle 1\frac{1}{2}0 \rangle$	Ni-Mo , Ni-V	$\varepsilon_1 > 0, \ 0<\varepsilon_2/\varepsilon_1 < \frac{1}{2}$
$\langle \frac{1}{2}\frac{1}{2}0 \rangle$	Cu-Pt	$\begin{cases} \varepsilon_1 > 0, \ \ \varepsilon_2/\varepsilon_1 > \frac{1}{2} \\ \varepsilon_1 < 0, \ \ \varepsilon_2/\lvert \varepsilon_1 \rvert > 1 \end{cases}$

As stated above, the fcc ground states for first and second-
neighbor interactions have been completely determined.   The resulting ordered superstructures themselves can be grouped into the
four families just defined, as shown in Table III.   Whereas (with
constant α) a binary system belonging to a given family is charac-

terized by a unique instability, the ordered ground states expected
will also depend on the average concentration, i.e. certain phases
are favored over others at certain stoichiometries, as indicated
in the last column of Table III. At values of α which delimit the
three ordering families, several ordered phases will have same
energy. Thus, at $\alpha = 0$, ordered superstructures designated as
$L1_2$ and A_2B_2 have same energy, and so has the pair $L1_2$ and DO_{22}.
At $\alpha = \frac{1}{2}$, there is degeneracy between $L1_1$ and A_2B_2 at $x_B = \frac{1}{2}$, and
between A_2B (C2/m) and Pt_2Mo at $x_B = 1/3$. The structure A_5B (C2/m)
is common to both $\langle 1\frac{1}{2}0 \rangle$ and $\langle \frac{1}{2}\frac{1}{2}\frac{1}{2} \rangle$ families. Structure A_2B_2 is not
a stable phase in known binary alloys, but has been observed experi-
mentally in magnetic ordering under zero field. The complicated
monoclinic structures C2/m have not been observed experimentally;
presumably, simpler superstructures, known to be stable if higher-
neighbor pair interactions are taken into account, will turn out to
have lower free energies, particularly at temperatures where atomic
mobility is high enough to cause ordering.

An earlier investigation[35] had shown the inadequacy of the
Bragg-Williams model for reproducing even such a relatively simple
coherent phase diagram as the Cu-Au. Correct topology was obtained
by the CVM in the first-neighbor tetrahedron approximation, and
a very good fit to the experimentally determined diagram was even
obtained by introducing multiplet (tetrahedron) interactions[35].
The same approximation was used to construct a ternary coherent
Cu-Ag-Au phase diagram based on the Cu-Au binary previously determined, a coherent miscibility gap on the Cu-Ag side and an almost
ideal Ag-Au binary[17].

TABLE III

Fcc-based ordered ground state superstructures
stable for first and second-neighbor pair interactions

Family, $(\alpha=\epsilon/\epsilon_1)$ $(\epsilon_1>0)$	Structurbericht Symbol	Symmetry Class	Int. Table	Prototype
$\langle 100 \rangle$	$L1_0$	s. tetragonal	$P4/mmm$	CuAu
$\alpha<0$	$L1_2$	s. cubic	$Pm3m$	Cu_3Au
$\langle \frac{1}{2}\frac{1}{2}\frac{1}{2} \rangle$	$L1_1$	rhombohedral	$R\bar{3}m$	CuPt
$\alpha>0.5$	—	s.c. monoclinic	$C2/m$	A_2B
	—	s.c. monoclinic	$C2/m$	A_5B
$\langle 1\frac{1}{2}0 \rangle$	—	b.c. tetragonal	$I4_1/amd$	A_2B_2
$0<\alpha<0.5$	—	b.c. orthorhombic	$Immm$	Ni_2V, Pt_2Mo
	DO_{22}	b.c. tetragonal	$I4/mmm$	Ni_3V, Al_3Ti
	—	s.c. monoclinic	$C2/m$	A_5B

Unfortunately, the tetrahedron (T) approximation of the CVM cannot contain second-neighbor pair interactions, essential if other ordered superstructures than Ll$_0$ and Ll$_2$ are to be included. A cluster containing second neighbor pairs is required, such as the regular octahedron (0). Preliminary tests showed that the tetrahedron-octahedron combination (TO) of the CVM gave very good values for the fcc ferromagnetic transition temperature, as compared to

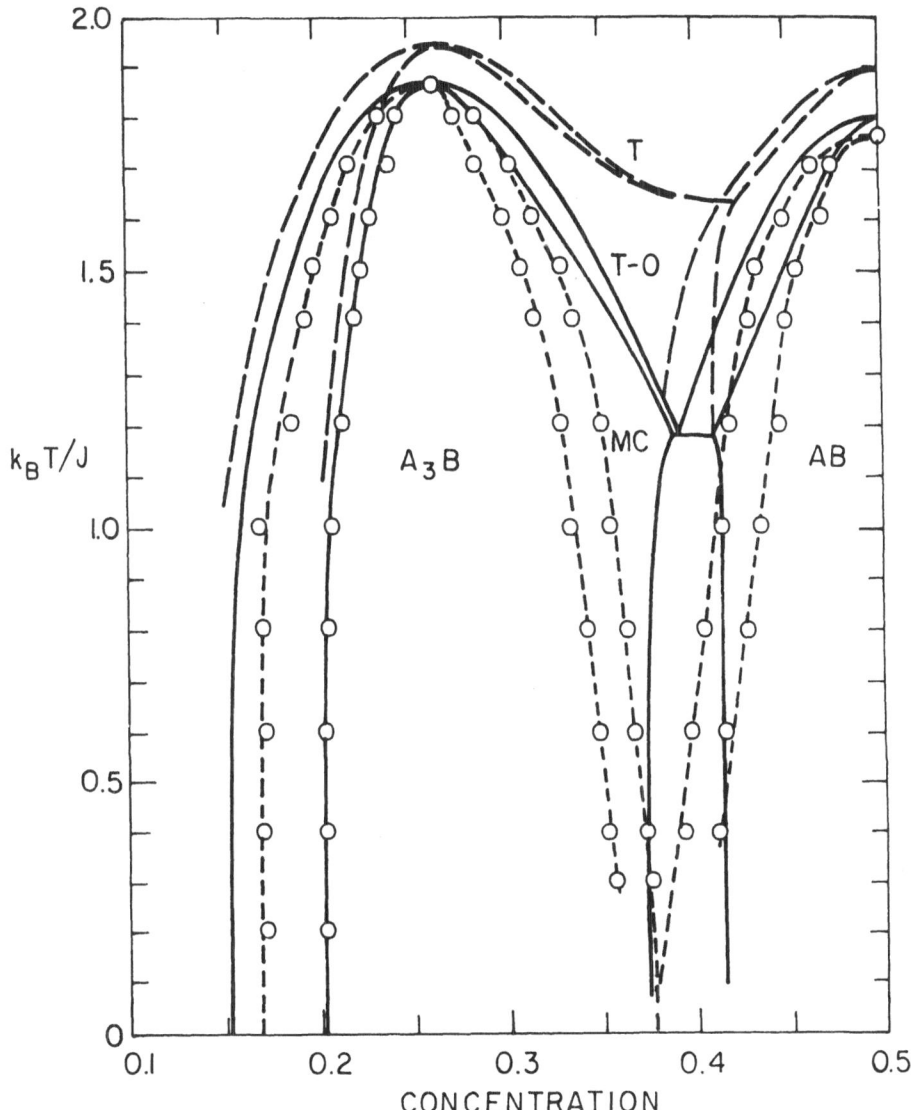

Fig. 1. Comparison of phase diagram for α=0 (common limit of <100> and <1½0> families) calculated in the TO-CVM (full line) and T-CVM (long dashes) approximations and Monte Carlo method (short dashes).

the best known high-temperature expansion of the Ising model[6]. The TO-CVM approximation was, therefore, used to compute prototype coherent phase diagrams based on the rigorously determined ground states of Table III. Particular emphasis was placed on the $\langle 1\frac{1}{2}0 \rangle$ family. The resulting diagram for $\alpha = 0$ is shown in Figure 1[9], with an earlier CVM tetrahedron approximation[36,37] and a Monte Carlo simulation[38] of the phase diagram shown for comparison. Assuming that the latter gives the "correct" Ising model phase diagram, it is seen that the TO-CVM does extremely well, particularly near the first-order transitions at stoichiometries A_3B (or AB_3) and AB. All diagrams calculated with concentration-independent pair interactions are symmetric about $x_B = \frac{1}{2}$, so that only the A-rich portion is shown here and in subsequent figures.

As some second-neighbor interactions is added, the ordered phase around $x_B = \frac{1}{2}$ becomes the A_2B_2 and that around $x_B = \frac{1}{4}$ becomes the DO_{22}. As shown in Figure 2[7], calculated with $\alpha = 0.25$, the

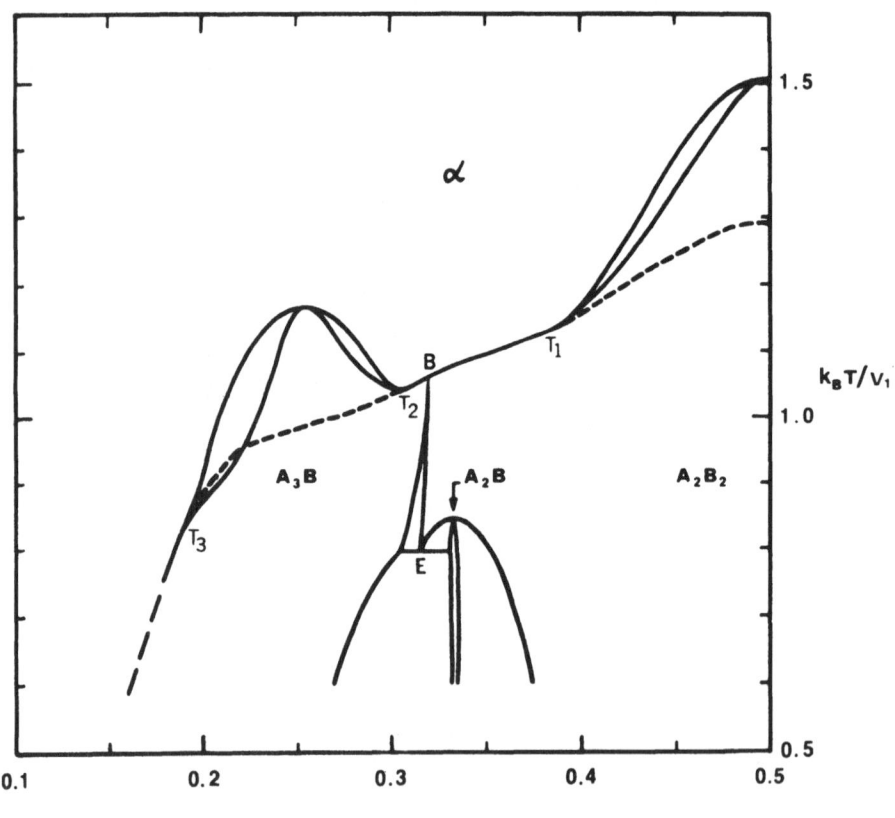

Fig. 2. Phase diagram for $\alpha = 0.25$ calculated in the TO-CVM approximation, A_5B phase field not shown. Dashed line is $\langle 1\frac{1}{2}0 \rangle$ spinodal.

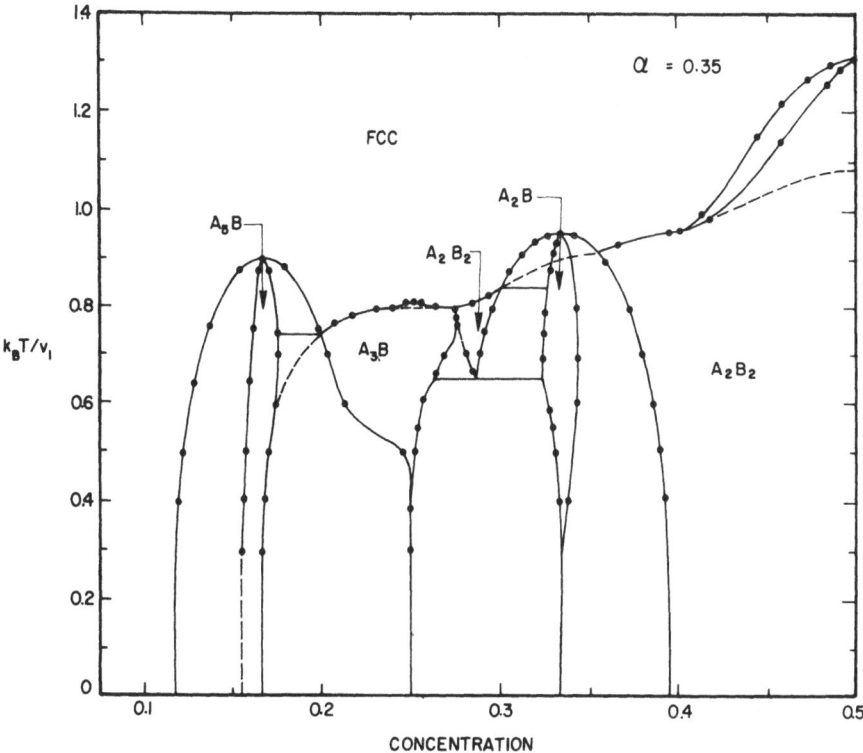

Fig. 3. Phase diagram for $\alpha = 0.35$ calculated in the TO-CVM approximation. Dashed line is $\langle 1\frac{1}{2}0 \rangle$ spinodal.

A_2B "Pt$_2$Mo" orthorhombic structure begins to appear, its top-most transition temperature lying well below the $\langle 1\frac{1}{2}0 \rangle$ instability (dashed line). The latter locus, or "ordering spinodal," was calculated by use of Eq. (28). Whenever phase boundaries and spinodal coincide, a second-order transition is predicted between the disordered solid solution (labeled α) and the lower-lying ordered phases. The diagram of Figure 3 is incomplete towards the low concentration end where the A_5B phase is expected because of numerical convergence problems. Recent Monte Carlo simulations performed with the same value of $\alpha = 0.25$[39] show remarkably close agreement with the TO-CVM calculation, particularly in the vicinity of the stoichiometric compounds.

When the value of α is increased to 0.35 (Figure 3), then to 0.45 (Figure 4), the A_2B and A_5B phase fields take up increasingly large portions of the phase diagram, while the A_3B phase field tends to get squeezed out of existence between A_2B and A_5B[8]. The transitions producing the latter phases are strongly first-order, the overall first-order character of the transitions from the

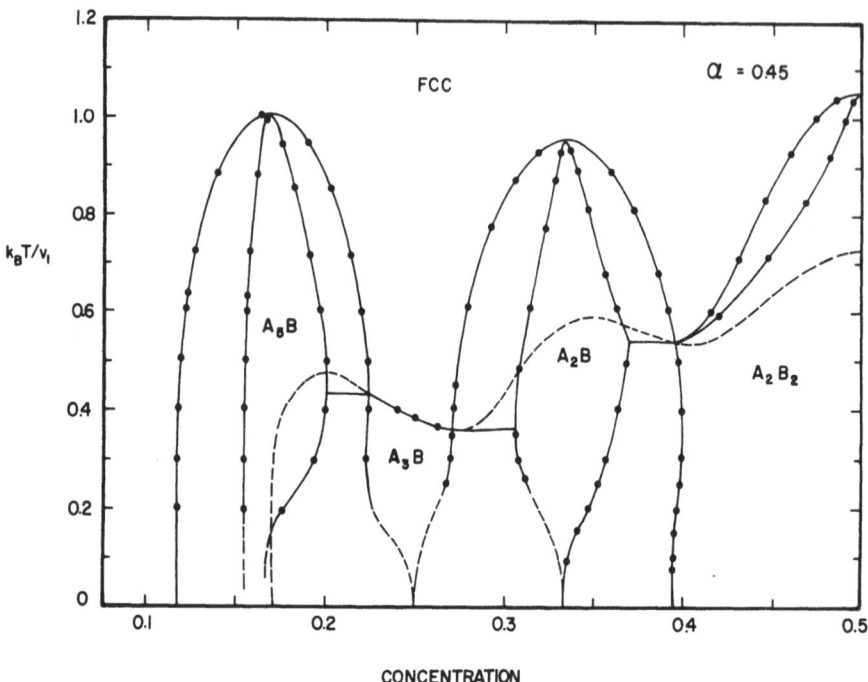

Fig. 4. Phase diagram for $\alpha = 0.45$ calculated in the TO-CVM approximation. Dashed line is $\langle 1\frac{1}{2}0 \rangle$ spinodal.

disordered to the ordered phases apparently increasing as the ratio of pair interactions α increases from 0.25 to 0.5. Note, however, that at $\alpha = 0.35$, the A_2B_2 phase field is split by A_2B, with the A_2B_2 transition around $x_B = 0.3$ being second-order.

Recent calculations[40] indicate that, at $\alpha = 0.55$ (which places the diagram into the $\langle \frac{1}{2}\frac{1}{2}\frac{1}{2} \rangle$ family), all second-order transitions have disappeared, leaving only first-order transitions from the disordered solution to A_5B, A_2B (C2/m), and AB (L1$_1$). On the other side of the $\langle 1\frac{1}{2}0 \rangle$ family, at $\alpha = -0.2$, the diagram (which now belongs to the $\langle 100 \rangle$ family) resembles that of Figure 1, but the eutectoid between A_3B (L1$_2$) and AB (L1$_0$) is not as deep, and the first-order character of the transitions becomes somewhat less pronounced.

As the value of ε_1 is made to decrease, the ratio α will tend to infinity, and it is anticipated that the two simple cubic sublattices which make up the fcc will tend to decouple, leading to parallel phase transitions on each sublattice with almost second-order character. The resulting phase diagrams will then probably start to resemble those derived by the Bragg-Williams approximation.

4.2 Order Parameters

The various ordered superstructures differ from one another and from the parent disordered phase by their different sublattice structure; it follows that, in the language of crystallographic group theory, the symmetry (space) group of any ordered superstructure must be a sub-group of that of the parent disordered structure[32]. Thus, in the disordered phase, any multiplet correlation variable $\xi_{n,s}(p)$ will have same value at all lattice sites p, while their values in general will differ at different sites within the unit cell of the ordered phases. Hence, in particular, different point correlation variables $\xi_1(p)$ must be defined for the different superlattices which describe the ordered structure considered. It is then customary to characterize the degree of *long-range order* by one or more *order parameters* η which are taken as linear combinations of ξ_1 variables so as to represent differences of composition on the various sublattices. To these order parameters must be added a parameter ξ denoting the average concentration of each phase in equilibrium; its value is obtained by averaging all sublattice concentrations. It follows that the number of independent order parameters for a given ordered phase is equal to the number of independent sublattices, which is itself equal to the number of non-equivalent crystallographic sites in the unit cell. This new description of the state of order is thus seen to result from a linear transformation of the set of independent ξ_1 point correlation variables to a set of ξ and η order parameters.

As an example, consider structures $I4_1/amd$ and $I4/mmm$ (DO_{22}), referred to as A_2B_2 and A_3B, respectively, in Figs. 2 to 4 (see also Table III). It was shown elsewhere that these structures could be described conveniently in terms of four planar sublattices which are successive <420> lattice planes of the disordered phase[7,28,41]. Label these planes p_1, p_2, p_3, and p_4. The independent order parameters may then be defined as follows:

$$A_2B_2 \text{ structure} \quad \eta = \frac{1}{4}[\xi_1(p_1) + \xi_1(p_2) - \xi_1(p_3) - \xi_1(p_4)], \quad (37)$$

$$A_3B \text{ structure} \quad \eta_1 = \frac{1}{4}[\xi_1(p_1) + \xi_1(p_2) - \xi_1(p_3) - \xi_1(p_4)], \quad (38a)$$

$$\eta_2 = \frac{1}{4}[\xi_1(p_1) - \xi_1(p_2) + \xi_1(p_3) - \xi_1(p_4)], \quad (38b)$$

$$\text{for both} \quad \xi = \frac{1}{4}[\xi_1(p_1) + \xi_1(p_2) + \xi_1(p_3) + \xi_1(p_4)], \quad (39)$$

there being two non-equivalent crystallographic positions in A_2B_2, and three in A_3B. Likewise, there are zero order parameters in the disordered structure, one in $L1_0$, $L1_2$, $L1_1$, and Pt_2Mo ($Immm$), and

four in the C2/m structures. Note that the order parameters
defined by Eqs. (37) and (38a) also represent the amplitude of a
$<1\frac{1}{2}0>$ ordering wave, and that defined by Eq. (38b) represents the
amplitude of an ordering wave of half the wavelength, i.e. of the
$<210>$ wave, equivalent to the $<100>$. A CVM calculation of η_1 and
η_2 in the A_3B phase clearly shows the different temperature dependences of these two LRO parameters[7].

The CVM free energy f, Eq. (18), can be used as a basis for a
Landau expansion in the order parameters: ξ, η_1, η_2, . . . For
that, it suffices to continue the Taylor's expansion in $\delta\xi$ variations initiated in Eq. (30)

$$\Delta F = F_2 + F_3 + F_4 + F_5 + F_6 \cdot \cdot \cdot, \tag{40}$$

F_n representing the nth order term. Usually the expansion is
performed about the disordered state so that the nth order derivatives appearing in Eq. (40) will have the translational symmetry of
the disordered lattice. The variables of the expansion must be the
set of independent correlation variations $\delta\xi$ appropriate for the
ordered phase which will evolve from the parent at the transition.
Since, in this scheme, the number of variables is typically very
large, it may be preferable to proceed as follows: In the free
energy functional, first perform a linear change of point variables
from the $\xi_{n,s}$ set to the $\{\xi,\eta\}$ set, then eliminate all other (multiplet) correlation variables by minimizing the free energy with
respect to the latter set, holding fixed the chosen values of the
$\{\xi,\eta\}$ set. In principle, there will then result a new free energy

$$\overline{f} = \overline{f}(\xi,\eta_1,\eta_2, . . .), \tag{41}$$

the overscore indicating that the multiplet correlation variables
have been eliminated by the process of constrained minimization
described above. One may then proceed with a Landau-type expansion
in temperature, composition, and order parameter variations. The
order parameter variables in the expansion will appear in the form
of nth-order invariants reflecting the symmetry of the "daughter"
ordered phase; the coefficients of these invariants will, because
of the "hidden" multiplet correlation dependence, be function of
the state of short range order associated with the $\{\xi,\eta\}$ values.
Clearly, such an expansion may differ considerably from one derived
simply from, say, a Bragg-Williams free energy model.

Since, in the present context, an explicit free energy is
available through the CVM approximation, there is not much point
in carrying out an elaborate Landau expansion. Qualitatively,
however, order parameter expansions are useful for the purpose of
classifying the nature of the transitions and of the critical and
multicritical points which may occur. Allen and Cahn[42] recently
undertook such a classification for binary systems. Interestingly,

most of the multicritical points described by these authors are
found in the prototype phase diagrams shown here in Figs. 2 to 4.
These diagrams have been calculated by the tetrahedron-octahedron
approximation of the CVM, according to the procedure described in
Section 2.6. A Newton-Raphson iteration was used to solve the set
of non-linear equations resulting from the free energy minimization
condition. In these three cases, the $<1\frac{1}{2}0>$ ordering instability
(or ordering spinodal) was plotted as dashed lines, which are the
loci defined by the instability condition (31). Since the constant
ratio value of $\alpha = \varepsilon_2/\varepsilon_1$ was chosen to lie between 1 and $\frac{1}{2}$, all
three phase diagrams belong to the same $<1\frac{1}{2}0>$ family (see Table II)
and the ground state superstructures are the appropriate ones listed
in Table III. Numerical convergence problems were often encountered
at low concentrations and low temperatures; it appears that a
prototype system experiences as much trouble attaining numerical
convergence as a real alloy system experiences attaining thermodynamic equilibrium.

Let us recall that two ordered phases differing by the values
of their respective order parameters are said to evolve from one
another by a *second-order transition* if the difference of their
order parameters goes to zero continuously as the transition is
approached. If the order parameter change is discontinuous, the
transformation is said to be *first-order*. In a phase diagram,
equilibrium phases may be separated by continuous lines of second-
order critical points or by first-order transition loci. In a
temperature-composition phase diagram, a first-order transition
locus is actually a pair of lines of conjugate points joined by
tie-lines. A *tricritical point* is one at which a line of critical
points merges into a pair of first-order transition lines. At a
bicritical point, three phases coexist, two pairs of which are
separated by lines of second-order transitions, the third pair
being separated by conjugate first-order transition lines. At
least two LRO parameters η_1 and η_2 must be involved. At a *tetracritical point,* four lines of second-order critical points merge at
a point of four-phase coexistence. Finally, a *critical end point*
occurs where a line of critical points intersects the solvus line
of, say, a miscibility gap. For more detailed descriptions, the
reader is referred to the article by Allen and Cahn[42]. Here, we
shall only point out some of the relevant features on the CVM-
calculated prototype phase diagrams.

It is clearly apparent, for instance in Fig. 2, that lines of
second-order critical points are present, as mentioned earlier;
along these lines, phase boundaries (solid lines) and instabilities
(dashed lines) coincide. This must be so because, according to
the Landau theory[32], a second-order transition takes place when
the second-order term in the free energy expansion vanishes at
equilibrium, i.e. whenever the instability condition (28) is satis-
fied. The α (disordered phase) to A_2B_2 transition at $x_B = \frac{1}{2}$ ($\xi_1 \approx 0$)

is seen to be strongly first-order, whereas a Bragg-Williams model would necessarily predict (incorrectly) a second-order transition. Hence, the CVM free energy expansion (symmetric about $\xi_1=0$) must be taken to at least sixth order in the order parameter η defined by Eq. (37):

$$\Delta F = A\eta^2 + C\eta^4 + D\eta^6 , \tag{42}$$

the coefficient C being negative to the right of point T_1 in Fig. 2, positive to the left. T_1 is therefore a tricritical point. The figure is drawn slightly incorrectly, as pointed out by Cahn[43]: The line of critical points should in fact extend into the two-phase field. T_2 is also a tricritical point. Actually, if A_3B had only a single order parameter, the transition would have been necessarily first-order because the third-order term in the Landau expansion could not possibly vanish for an asymmetric sublattice structure (A_3B) at any asymmetric composition. The fact that a second-order transition is possible is due to the existence of two LRO parameters in this phase, with third-order invariants cancelling out exactly over a range of compositions. A bicritical point B is found between the tricritical points, as expected from the definition. Presumably, the line of critical points extends smoothly through B because order parameter η_1 of A_3B is exactly the same as η of A_2B_2, as seen in Eqs. (37) and (38a).

Phase diagrams of Figs. 3 and 4 exhibit similar features but, in addition, boast of several *critical end points* where lines of second-order transitions intersect the outer conjugate line of a two-phase field. These features may be regarded, in a sense, as degenerate eutectoids or peritectoids. No tetracritical points appear to be present. The Monte Carlo simulation[38] appears to predict a multicritical point at absolute zero for the $\varepsilon_2/\varepsilon_1 = 0$ prototype (Fig. 1), whereas the CVM yield a eutectoid, which is presumably incorrect.

At a given temperature, Eq. (41) represents a surface in f vs. order parameter space. Equilibrium states can be obtained by minimizing the (modified) grand potential

$$\bar{\omega} = \bar{f} - \mu\xi ,$$

similar to Eq. (24). Only the average point correlation variable ξ appears explicitly, because the η order parameters do not have as conjugate variables actually applicable external fields such as the chemical potential μ. Graphically then, the equilibrium condition would be obtained by constructing common tangent planes to the f surface, these planes being maintained normal to the f-ξ plane. Projected on this plane, the construction would look for all the world like the classical two-dimensional common tangent construction, but the points of tangency would in general be located outside the f-ξ (or f-x_B) plane, at points yielding non-zero values of the η order parameters. Of course, this graphical construction need not be performed in practice: The CVM minimization procedure described in Section 2.6 takes care of everything analytically.

4.3 Application to Alloy Systems

The original motivation for undertaking CVM phase diagram calculations was to provide a firm basis for a *spinodal ordering* theory developed some years ago by the author[28] for the purpose of explaining some puzzling effects discovered by Okamoto and Thomas[41] on quenched Ni-Mo alloys in the Ni_4Mo-Ni_3Mo composition range. At the time, it was only possible to calculate a semblance of spinodal from a Bragg-Williams model, but the latter was incapable of providing an adequate phase diagram to go with the calculated instability. It was then that van Baal[36] showed how to calculate an ordering phase diagram on the fcc lattice in the nearest-neighbor tetrahedron approximation of Kikuchi's CVM. Since Ni-rich NiMo solutions belonged to the $<1\frac{1}{2}0>$ family, the stability of which requires at least non-zero nnn pair interaction, further progress had to await the development of a higher-order CVM approximation[6,7], namely the tetrahedron-octahedron approximation, which could take mmm pair interactions into account.

Probably the best illustration of some of the theoretical ideas presented in this paper is provided by the results of a recent investigation performed by P. L. Martin[44] on $Ni_3(MoX)$ alloys, i.e. alloys on A_3B stoichiometry, with A referring to Ni and B referring to Mo with small substitutional additions of a third element X, that symbol itself referring to Al, V, W, Ta, or Nb. For brevity, only the Al (3 at.%) and W (5 at.%) alloys will be discussed here. A full account of experimental findings and their interpretation will be found in the doctoral dissertation of Dr. Martin (henceforth referred to as PLM) and in forthcoming publications.

Electron diffraction patterns from these $Ni_3(MoX)$ alloys quenched from the disordered state and aged for various times at 700° C and 800° C are shown in Figs. 5 to 8, taken from PLM. These patterns may be indexed with the help of Fig. 9 (PLM). As was already apparent from earlier work on the Ni-Mo binary[41], it is clear that only diffuse intensity at the $<1\frac{1}{2}0>$ special points appears immediately upon quenching; even after some aging has taken place, no other features are present in the diffraction patterns (except for the W-alloys aged at 800° C, to be discussed later). This paradoxical behavior may be interpreted in terms of the free energy model presented here. For that purpose, consider the expansion given in Eq. (40). Imagine that all terms in powers of deviations $\delta\xi_{n,s}$ of correlations variables from their values in the uniform disordered solution have been written in their Fourier representation: the second-order term has already been given in matrix notation in Eq. (29), and is seen to consist of a sum of weighted amplitude-squared $|X|^2$, or concentration wave intensities. Higher-order terms do not have such a simple form, however: The third-order term contains products of three complex amplitudes, the fourth order term contains products of four complex amplitudes, etc. These higher-order terms are impossible to deal with numerically, but are useful in qualitative discussions, as will presently be shown.

At any temperature above a phase transition, only SRO intensity will be present, the expectation value of its Fourier components being given (exactly) by the first element $<|X_1|^2>$ of the matrix equation (34)[33]. According to the normal mode equation (36), this SRO intensity will obviously peak at points in reciprocal space whose corresponding k vectors minimize the k-space coefficient $\Lambda_\ell(\underset{\sim}{k})$ in the quadratic term F_2, Eq. (30) of the free energy expansion, this peaking generally occurring at special points (see Table I). In the present case, intensity maxima are therefore expected at or near $<1\frac{1}{2}0>$. During rapid quenching, or during the very early stages of aging, it may be assumed that the correlation variable deviation amplitudes will be small so that, to a first approximation, the quadratic term F_2 will dominate the free expansion. It follows that, if the quenching operation be fast enough to avoid nucleation of the equilibrium superstructure, the unstable wave will grow in amplitude as the ordering spinodal is crossed. The important point to make is that the concentration waves which receive maximum amplification in this *spinodal ordering mechanism* are precisely the ones having maximum SRO intensity, since both fluctuations and instability are governed entirely by the second-order term F_2, more properly by the eigenvalue $\Lambda_\ell(k)$, as are early-stage kinetics. Consequently, the short-time aging electron diffraction patterns of PLM exhibit only diffuse intensity at the "unstable" special points $<1\frac{1}{2}0>$, just as in the calculated diffraction pattern of Fig. 10, to b discussed shortly.

In the case of $<1\frac{1}{2}0>$ family systems, the superstructure wave vectors are generally not located at the special points, so that, as aging proceeds, new intensity maxima must appear elsewhere in reciprocal space. Indeed, away from concentration $x_B = \frac{1}{2}$, the $<1\frac{1}{2}0>$ wave cannot produce a correct ordered superstructure[28]. What then happens is this: As the intensity of SRO waves increases, the quadratic term F_2 no longer dominates the free energy, and higher-order terms come into play through anharmonic coupling of non-special point waves. This produces intensity transfer[45] away from the $<1\frac{1}{2}0>$ diffuse maxima, towards the superlattice reflections of the expected ordered ground states.

Consequently, one would expect intensity transfer to take place after some aging time from $<1\frac{1}{2}0>$ to $2<1\frac{1}{2}0> = <210>$, equiva-

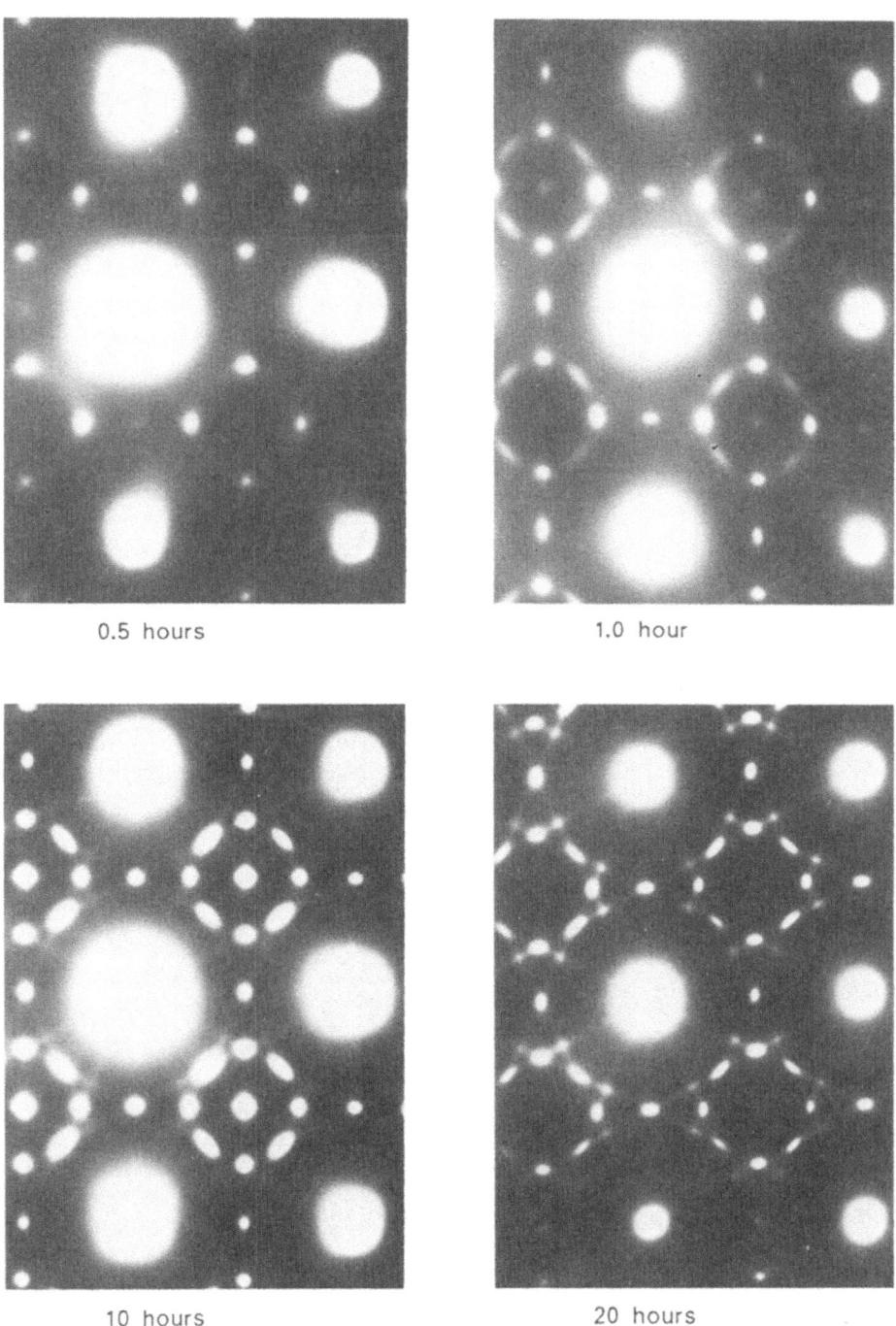

0.5 hours

1.0 hour

10 hours

20 hours

Fig. 5. Al alloy aged at 700°C, [001] zone patterns. (PLM)

72

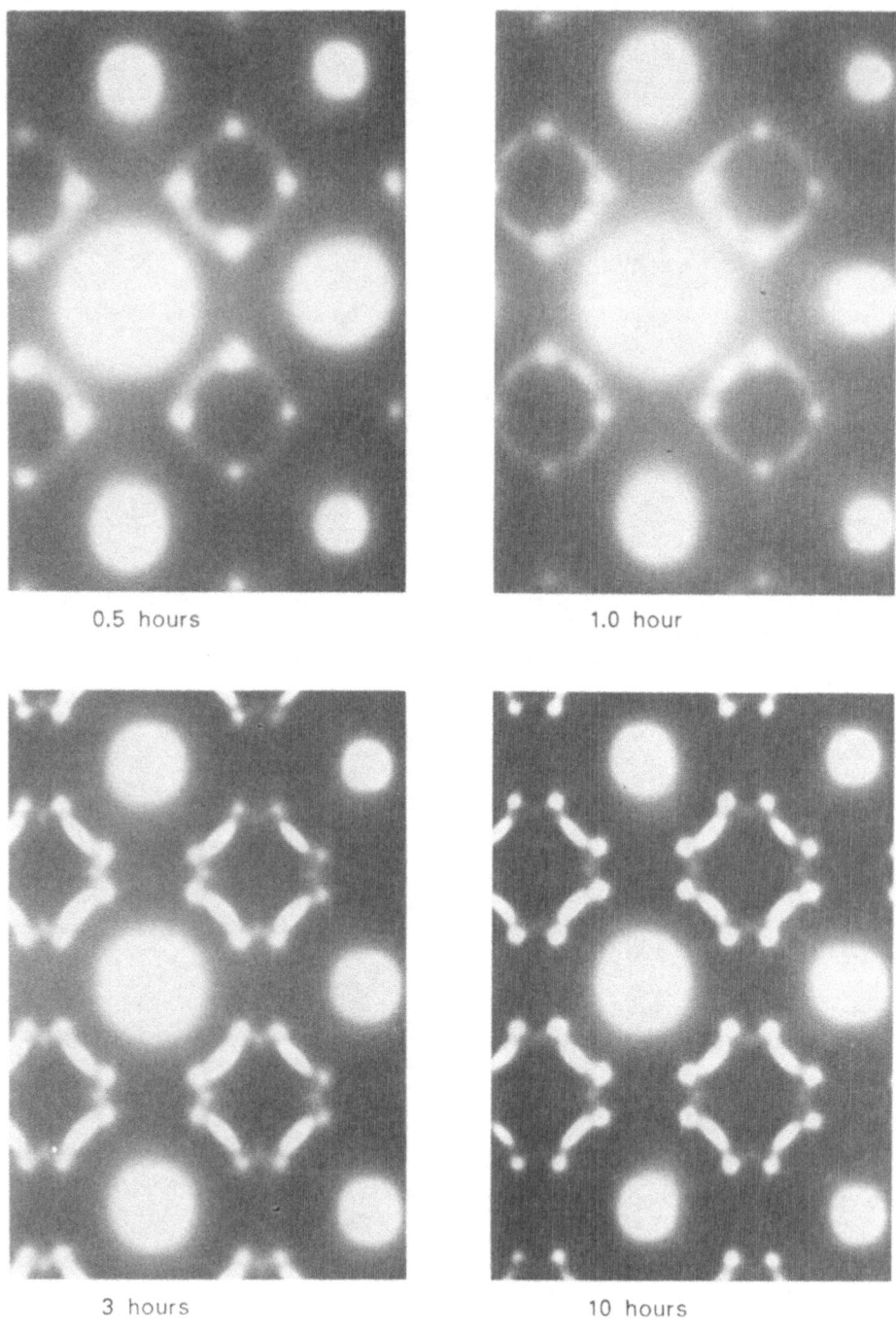

0.5 hours

1.0 hour

3 hours

10 hours

Fig. 6. W alloy aged at 700°C, [001] zone patterns. (PLM)

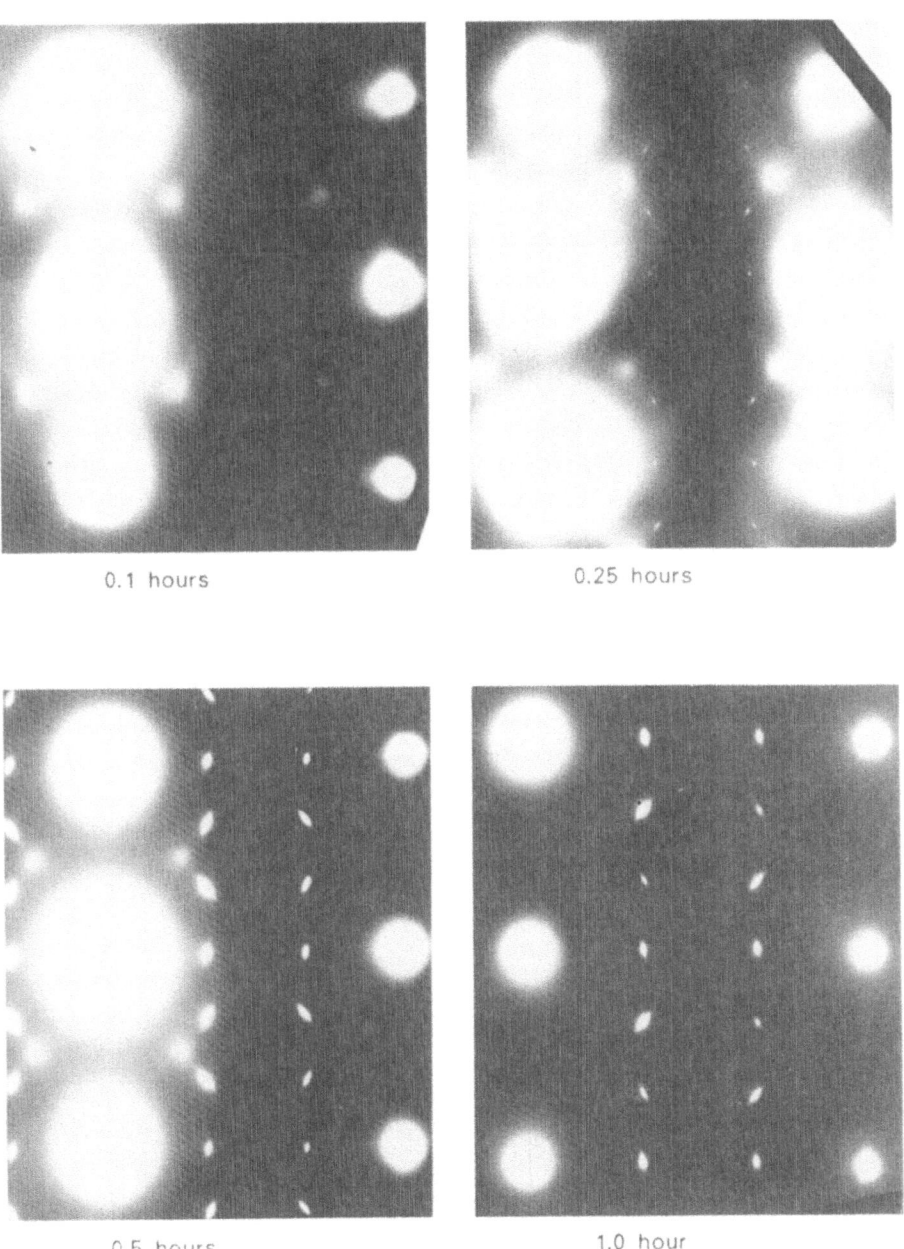

0.1 hours

0.25 hours

0.5 hours

1.0 hour

Fig. 7. Al alloy aged at 800°C, [112] zone patterns. (PLM)

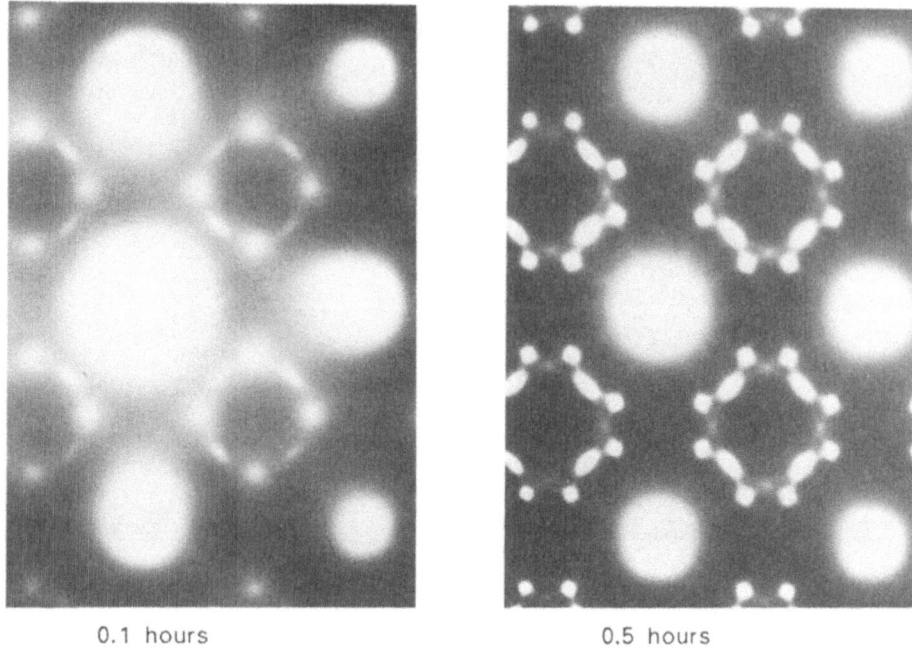

0.1 hours   0.5 hours

Fig. 8. W alloy aged at 800° C, [001] zone patterns. (PLM)

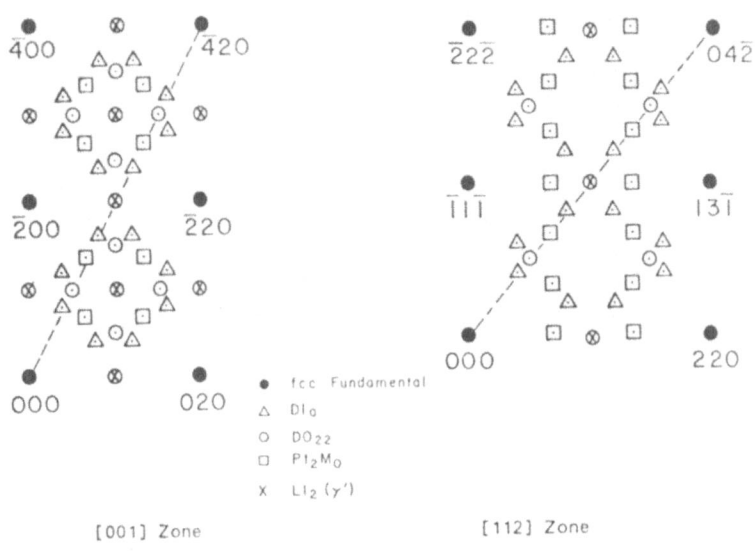

fcc Fundamental
△ DIₐ
○ DO₂₂
□ Pt₂Mₒ
X Li₂ (γ')

[001] Zone   [112] Zone

Fig. 9. Composite zone patterns for the case when all AₓB super-structures are present. Only one quadrant is shown for each zone.
(PLM)

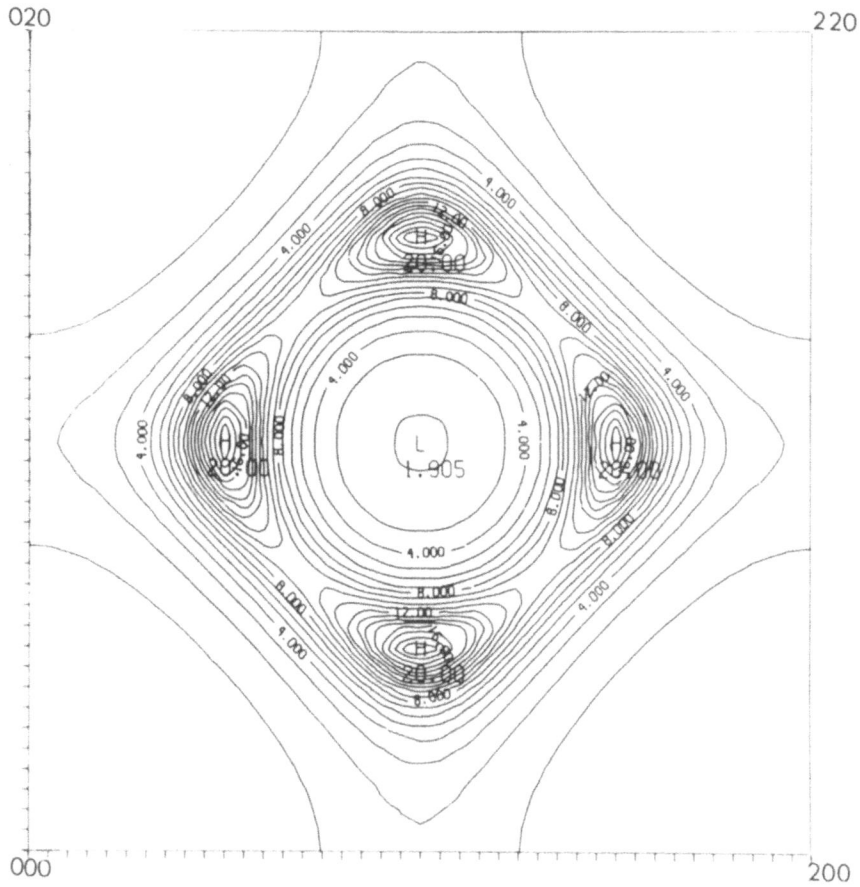

Fig. 10. SRO intensity contours in (001) plane calculated by Krivoglaz-Clapp-Moss formula with $\varepsilon_2/\varepsilon_1 = 0.4$, all other pair interactions set to zero, $T_0/T = 0.95$.

lent to <100>, thereby producing the A_3B superstructure of the <1½0> family, i.e. the DO_{22} structure. Surprisingly this is not what happens: A glance at Figs. 4b-d (700° C aging, Al-alloy) indicates that the "Pt_2Mo" reflections appear just as prominently as the DO_{22}, in fact more prominently after 20 hours at this temperature. At that time, the DO_{22} even seems to be on its way out, whilst the Dl_a structure (A_4B) begins to come into the picture. There are several reasons why no satisfying explanation for this complex kinetic behavior can be given at this time: First, experimentally observed ground states for the <1½0> family are A_4B (Dl_a), A_3B (DO_{22}), and A_2 (Pt_2Mo), whereas the ground state analysis with nn and nnn pair interactions predicts A_5B (C2/m), A_3B, A_2B, and A_2B_2. The discrepancy occurs because, as shown by Kanamori[24], fourth neighbor interactions would be required to favor the Dl_a tetragonal phase over the less symmetric (monoclinic) C2/m phase; taking such long-range interaction into account in the CVM simply would be too laborious at this time. Secondly, no tractable kinetic theory is available for predicting ordering behavior of systems evolving far from equilibrium.

At 700°C, the W-alloy also runs the gamut of "coherent" ground states, although the A_3B phase hardly appears at all, whilst the A_4B phase quickly enters the picture. Based on these results, one might say that the Al-alloy is more nearly represented by a phase diagram such as that of Fig. 2 (ratio $\alpha = 0.25$), whereas the W-alloy is best represented (at this level of approximation) by the diagram of Fig. 4 ($\alpha = 0.45$), since the extent of the A_3B phase field appears to decrease in the $<1\frac{1}{2}0>$ family as the ratio $\alpha = \varepsilon_2/\varepsilon_1$ increases.

Aging at the higher temperature of 800°C introduces an interesting new feature, at least in the case of the Al-alloy. Again, as seen in Fig. 7, the $<1\frac{1}{2}0>$ spinodal ordering reaction cannot be avoided on quenching, but soon after the alloy has been brought to the 800°C aging temperature, extremely sharp A_2B (Pt_2Mo structure) begin to appear (Fig. 7b). This characteristically different behavior surely results from nucleation of discrete, almost fully ordered Ni_2(MoAl) particles, rather than by the mechanism of "intensity transfer" typical of the more "continuous" transformation observed at the lower temperatures. From these results, it was concluded that the $<1\frac{1}{2}0>$ spinodal lay between 700 and 800°C in the Al-alloy, though the actual instability temperature T_0 must be history-dependent and theoretically model-dependent. The W-alloy gives almost no evidence of spinodal ordering (see Fig. 8) at 800°C. Instead, rather diffuse A_2B reflections are quickly seen in the diffraction patterns, with no further evolution towards the A_3B phase.

As stated above, values of fluctuation SRO intensity, or expectation value of point-point correlation amplitude-squared, are given by the first element of the matrix Eq. (34). It would thus be of interest to plot values of $(F^{-1})_{11}$ in some section of reciprocal space to see how well experimental intensity could be reproduced. Unfortunately, we do not have fully operational computer codes for this purpose, at this time, so that, for illustration, we have had to fall back on the simpler Bragg-Williams-based Krivoglaz-Clapp-Moss formula (see for example Ref. 11). Following Das et al.[46], we have selected the value $\alpha = 0.4$ and $T = 0.95\ T_0$ to produce the contour plot of Fig. 10, which represents diffuse SRO intensity in an [001] section in k-space. The resemblance with the electron diffraction patterns of quenched Ni_3(MoX) alloys obtained by PLM is striking. Diffraction patterns for later aging times cannot be obtained in this manner, however, even with the CVM formulation. What is lacking, of course, is a realistic non-linear kinetic theory at the same level of approximation as the T-O CVM used here for equilibrium calculations. Kikuchi's Path Probability Method[47] is fully compatible with the CVM, but is of such complexity that it cannot be reasonably applied to the case of interest here.

Still, the calculated diffraction pattern of Fig. 10 may offer some clue as to the prevalence of the A_2B structure in all but the very early-stage aging diffraction patterns of PLM. Surely, one would expect, as end product of the reaction, the DO_{22} phase (A_3B) rather than the A_2B which has the wrong stoichiometry, and which does not even appear on the (experimentally determined) Ni-Mo equilibrium phase diagram.

A tentative, qualitative explanation is as follows: intensity transfer does not take place readily from $\langle 1\frac{1}{2}0 \rangle$ to $\langle 100 \rangle$ presumably because, as is apparent from the low-value contours in the center of the reciprocal unit cell shown in Fig. 10, the $(F^{-1})_{11}$ coefficient at $\langle 100 \rangle$ presents a *maximum* and is therefore kinetically unfavorable. Instead, the system appears to prefer to transfer intensity to the $1/3\langle 420 \rangle$ positions which are located roughly at the saddle points of the intensity of a "nearby" ground state. At these saddle points, $(F^{-1})_{11}$ has reasonably low values, so that points $1/3\langle 420 \rangle$ may in fact be more readily accessible kinetically, at early times, than the lofty $\langle 100 \rangle$ (equivalent to $1/2\langle 420 \rangle$). Why a system subsequently prefers the A_4B to the A_3B structure presumably depends on the value of the effective ratio $\alpha = \varepsilon_2/\varepsilon_1$, as was mentioned above, but more cannot be said about this without delving deeply into non-linear kinetics, which is simply not feasible at this time.

In closing, let us mention the interesting work on *in situ* electron irradiation of Ni-Mo alloys[48,49]. Briefly, it was found that low-temperature irradiation could not only destroy LRO, but also $\langle 1\frac{1}{2}0 \rangle$ SRO. Upon slight reheating, it was not the expected ground state which reappeared, but the $\langle 1\frac{1}{2}0 \rangle$ "spinodal state," showing that, regardless of how the initial non-equilibrium state of the sample was prepared, either by quenching or by irradiation, as long as concentration wave amplitudes were small, the F_2 quadratic term dominated in the free energy, and spinodal ordering prevailed.

5. CONCLUSION

"If you know the free energy, you know everything," the problem is knowing the free energy. It was suggested here that the CVM provides probably the best method of approximating the free energy of alloy solid solutions, in the sense that the model is both realistic and tractable. In particular, it was shown how to obtain prototype phase diagrams which, even under the crude simplifications used, could provide insight into the behavior of real alloys. A reformulation of the CVM in Fourier space, although not yet completed, will prove to be very helpful in understanding diffuse intensity patterns resulting from fluctuations and early-stage kinetics, especially spinodal ordering. The next planned improvements concern the introduction of more realistic internal energy models in the CVM free energy functional.

ACKNOWLEDGEMENTS

The author has benefitted greatly from many helpful conversations with Drs. J. M. Sanchez and T. Mohri. The former has authorized the use in these pages of his Eq. (10), which has not been published yet. The help of Dr. Mohri in the preparation of this paper, particularly in obtaining Fig. 10, is gratefully acknowledged. The author also wishes to thank Dr. P. L. Martin for permission to use his diffraction patterns, Figs. 5-9, prior to publication. This work was supported by the Army Research Office (Durham).

REFERENCES

1. R. Kikuchi, Phys. Rev. $\underline{81}$, 998 (1951).
2. R. Kikuchi, J. Chem. Phys. $\underline{60}$, 1071 (1974).
3. M. Kurata, R. Kikuchi, and T. Watari, J. Chem. Phys. $\underline{21}$, 434 (1953).
4. J. A. Barker, Proc. Roy. Soc. A216, 45 (1953).
5. J. Hijmans and J. De Boer, Physica 21, 471, 485, 499 (1955); 408 (1956).
6. J. M. Sanchez and D. de Fontaine, Phys. Rev. B $\underline{17}$, 2926 (1978).
7. J. M. Sanchez and D. de Fontaine, Phys. Rev. B $\underline{21}$, 216 (1980).
8. J. M. Sanchez and D. de Fontaine, Phys. Rev. B $\underline{25}$, 1759 (1982).
9. J. M. Sanchez, D. de Fontaine, and W. Teitler, Phys. Rev. B $\underline{26}$, 1465 (1982).
10. J. M. Sanchez, D. Gratias, and D. de Fontaine, Acta Cryst. A $\underline{38}$, 214 (1982).
11. D. de Fontaine, Solid State Physics $\underline{34}$ (1979).
12. D. de Fontaine, Physica (Utrecht) $\underline{103B}$, 57 (1981).
13. D. de Fontaine, Course on Mechanical and Thermal Behavior of Metallic Materials, Enrico Fermi School of Physics, G. Cagliotti and A. Ferro Milone, Eds., Vol. 82 (1981).
14. D. de Fontaine in the Proceeding of the International Conference on Solid-Solid Phase Transformation, Pittsburgh, PA (1981).
15. D. de Fontaine, to be published in the Proceedings of the MRC Annual Meeting (1982).
16. R. Kikuchi, Acta Metall. $\underline{25}$, 195 (1977).
17. R. Kikuchi, J. M. Sanchez, D. de Fontaine and H. Yamauchi, Acta Metall. $\underline{28}$, 651 (1980).
18. T. Morita, J. Phys. Soc. Japan $\underline{12}$, 753 (1957); J. Math. Phys. $\underline{13}$, 115 (1972).
19. J. M. Sanchez, unpublished.
20. F. Ducastelle, private communication.
21. A. Bieber, F. Gautier, G. Treglia, and F. Ducastelle, Solid State Comm., $\underline{39}$, 149 (1981).

22. J. M. Sanchez, unpublished research at UCLA and U.C. Berkeley.
23. J. M. Sanchez and D. de Fontaine in Structure and Bonding in Crystals, M. O'Keeffe and A. Navrotsky, Eds., (Academic, N.Y. 1981) Vol. II, pp. 117-132.
24. J. Kanamori, Progr. Theor. Phys. 35, 66 (1966); J. Kanamori and Y. Kakehashi, J. Phys. (Paris) 38, C7-274 (1977).
25. M. J. Richards and J. W. Cahn, Acta Metall. 19, 1263 (1971).
26. S. M. Allen and J. W. Cahn, Acta Metall. 20, 423 (1972); Scripta Metall. 7, 1261 (1973).
27. A. G. Khachaturyan in Progress in Materials Science, J. W. Christian, P. Haasen, and T. B. Massalski, Eds. (Pergamon, Oxford 1978).
28. D. de Fontaine, Acta Metall. 23, 553 (1975).
29. E. M. Lifshitz and L. P. Pitaevskii, Statistical Physics, 3rd Edition, Part 1 (Oxford 1980).
30. International Tables for X-ray Crystallography, Symmetry Groups, edited by N. F. M. Henry and K. Lonsdale, Vol. 1, Kynoch Press, Birmingham (1952).
31. J. M. Sanchez, D. Gratias, and D. de Fontaine, Acta Cryst. A 38, 214 (1982).
32. L. D. Landau and E. M. Lifshitz, Statistical Physics, Chapter XII, Addison-Wesley, Reading, Massachusetts (1958).
33. J. M. Sanchez, Physica 111A, 200 (1982).
34. R. Kikuchi and D. de Fontaine in NBS Publication SP-496, 967 (1978).
35. D. de Fontaine and R. Kikuchi in NBS Publication SP-496, 999 (1978).
36. R. Kikuchi and C. M. van Baal, Scripta Metall. 8, 425 (1974).
37. C. M. van Baal, Physica (Utrecht) 64, 571 (1973).
38. K. Binder, J. L. Lebowitz, M. K. Phani and M. H. Kalos, Acta Metall. 29 1655 (1981).
39. R. A. Bond and D. K. Ross, J. Phys. F, 12, 597 (1982).
40. J. M. Sanchez, T. Mohri, and D. de Fontaine, to be published.
41. R. P. Okamoto and G. Thomas, Acta Metall. 19, 825 (1971).
42. S. M. Allen and J. W. Cahn, to be published in the Proceedings of the MRC Annual Meeting (1982).
43. J. W. Cahn, private communication to D. de Fontaine.
44. P. L. Martin, Ph.D. Dissertation, Dept. of Metallurgical Eng. and Materials Sc., Carnegie-Mellon University, Pittsburgh, PA (1982); P. L. Martin and J. C. Williams, to be published.
45. W. A. Soffa and D. E. Laughlin, pp. 159-183, Procedings of an International Conference on Solid-Solid Phase Transformations, H. I. Aaronson, D. E. Laughlin, R. F. Sekerka and C. M. Wayman, Eds., The Metallurgical Society of AIME, Warrendale, Pennsylvania (1982).
46. S. K. Das, P. R. Okamoto, P. M. J. Fisher, and G. Thomas, Acta Metall. 21, 913 (1973).
47. R. Kikuchi, Progr. Theor. Phys. 535, 1 (1966).
48. G. Van Tendeloo and S. Amelinckx, pp. 305-309, Proceedings of an International Conference on Solid-Solid Phase Transforma-

tions, H. I. Aaronson, D. E. Laughlin, R. F. Sekerka and C. M. Wayman, Eds., The Metallurgical Society of AIME, Warrendale, Pennsylvania (1982).

49. S. Banerjee, U. Urban, and W. Wilkens, pp. 311-315, Proceedings of an International Conference on Solid-Solid Phase Transformations, H. I. Aaronson, D. E. Laughlin, R. F. Sekerka and C. M. Wayman, Eds., The Metallurgical Society of AIME, Warrendale, Pennsylvania (1982).

METAL INSULATOR TRANSITION IN MODULATED CRYSTALS

Costas M. Soukoulis

Corporate Research-Science Laboratories, Exxon Research and
Engineering Company, P.O. Box 45, Linden, New Jersey 07036

1. INTRODUCTION

In recent years there have been many studies of crystals
containing a modulating periodic potential of a period different
from that of the underlying lattice (1,2). Such modulations occur
naturally in crystals containing charge density waves or spin
density waves (2) as well as in ionic conductors(3) and in certain
alloys (4). It has also become possible to grow crystals with such
modulations by molecular-beam epitaxy(5) (superlattices). Such
periodic modulations can be either commensurate or incommensurate
with the underlying lattice. Both cases are of considerable
interest theoretically and experimentally.

A crystal with a commensurate modulation with a period
different from, and much longer than, the period of the underlying
crystal in an important case for study. The energy spectrum and
nature of energy eigenfunctions (within the one-electron
approximation) of such a system are well known, because the Bloch
theorem can be applied to the new long period of the system. The
eigenfunctions are extended throughout the crystal and the energy
spectrum consists of a series of mini-bands separated by mini-gaps,
as determined by a wave vector varying within a mini-Brillouin zone,
much smaller than the Brillouin zone of the unmodulated crystal.
Such an energy spectrum affords an opportunity to measure a number
of interesting effects that an electric field will have on such a
system (6). For example, α strong electric field can induce negative
differential conductivity as it pushes the electrons toward the
region of the energy dispersion curve that begins to bend as it
approaches the mini-Brilouin zone boundary. Furthermore, for higher
electric fields one should begin to see oscillations of the electric

current due to the Stark ladder, and the interband or Zener, tunneling. All these effects have been considered for a long time in connection with the study of unmodulated crystals (6). However they are very difficult to observe experimentally, as they require extremely strong electric fields and very long relaxation times. The long-period modulated crystals will, however, bring all these effects within the possibility of experimental test, because the width of the mini-bands and the size of the mini-gaps are considerably smaller than those of an unmodulated crystal.

A crystal with an incommensurate modulation presents an interesting case, in that, strictly speaking, it does not possess translational order. In this respect, it is like a disordered solid. However, unlike a disordered solid, it possesses lengths over which it <u>almost</u> repeats. Thus, an incommensurate crystal presents a case intermediate between an ordered and a disordered solid. Since real crystals are of finite size and contain imperfections, it would be difficult in practice to distinguish an incommensurate crystal from a high order commensurate one. Thus, one might expect to see in them effects similar to those we discussed above for commensurate crystals.

However, true incommensurability has attracted considerable attention, being, as it is, a first step towards disorder. It has been argued(7-14) that, within the one dimensional (1D) one-band tight bonding model, with incommensurate modulation there exists a "metal-insulator" transition at a critical strength of the modulation, i.e. all the energy eigenfunctions are localized for strengths of the modulation above the critical value, while they become extended just below it. Some features of the spectrum of an appropriate commensurate modulation is preserved in the metallic regime. Such behavior is unlike that of a disordered solid, for which all states are localized in one dimension or of a commensurate crystal, for which all states are extended. Such a transition in the nature of the eigenfunctions--which may be induced experimentally in the same sample by, for example, external pressure--presents a very interesting theoretical possibility that is worth investigating. For the one-dimensional case, the electronic transport properties in an electric field in the vicinity of the transition could provide an experimental test. In the localized regime, we expect zero conductivity, while in the extended regime, we anticipate a metallic-type conductivity.

It is the purpose of this paper to examine the nature of the eigenstates of this model by studying the transmission coefficient T of the 1D system of size N. This technique is employed in the problem of electrical conductance of a 1D disordered crystal with very interesting results (15-17). In the course of the present study, a detailed density of states (DOS) calculation is made, and the spatial dependence of the eigenstates are examined. Before we present our results, a review of the previous theories will be made.

The model we consider is

$$\varepsilon_n c_n + t(c_{n+1} + c_{n-1}) = E c_n \tag{1}$$

where the energy at site n is

$$\varepsilon_n = V_0 \cos(Qn + \theta) \tag{2}$$

c_n is the amplitude at site n, t is the hopping matrix element, V_0 is the modulation potential strength, θ is a phase factor, Q is the wave vector of the modulation (we will be interested in the case of an irrational multiple of π) and the lattice constant is taken to be unity.

2. CONTINUED FRACTION EXPANSION OF ENERGY SPECTRUM

First we will discuss some of the peculiarities of the band structure of an incommensurate system. For this purpose consider first a commensurate crystal. We are dealing with a commensurate system when Q is commensurate with 2π, i.e.

$$Q = 2\pi \, L/M \tag{3}$$

where L and M are integers. Note that after M sites the modulation function repeats itself and therefore the unit cell has at most M cites. This means that in this case we can solve for the energy levels by making use of lattice translational invariance of the system. In general, for a 1-D lattice with a unit cell of M sites the energy spectrum has M bands that are separated by M-1 gaps. The length of the Brillouin zone (BZ) is 1/M that of a regular chain.

The problem that occurs when we are dealing with an incommensurate system is obvious. In that case $M \to \infty$ and it is impossible to define a BZ. This is another way of saying that the unit cell of an incommensurate system is as large as the system itself. Experimentally, it has been shown, in general that the wave vector of modulation varies smoothly with temperature with respect to the underlying lattice. Therefore, we expect there will be some characteristic properties of the incommensurate phase that will not depend on the commensurability or incommensurability of Q. If this is the case, we may approximate $Q/2\pi$ by a rational number. On the other hand, one could as well expect rather chaotic behavior for the distribution of the gaps over the energy spectrum as $Q/2\pi$ is varied. For example, near 3/50 lies 4/50. In the first case, the maximum number of gaps is 49, in the second, the number of gaps is 24. After 4/50, we get 5/50, having only 9 gaps. If we take a larger denominator M which is not prime, so that it can be decomposed into smaller factors, the chaotic behavior of number of gaps is even more stricking since the numerator L runs from 1 to M-1. Every time L and M have a common factor, the maximum number of gaps is reduced. The maximum number of gaps is infinite when $Q/2\pi$ is irrational. Therefore we see that a smooth variation of the wave vector of modulation could lead to a wildly changing distribution of gaps of the energy spectrum. The variation of the ratio L/M is "smooth" if we consider it as a decimal fraction. But it will be no longer be "smooth" if we represent real numbers as continued fractions (18). In fact, both Azbel(19) and later Hofstadter(20) have shown that the structure of the gap distribution is determined essentially by the successive stages of the expansion $Q/2\pi$ as a continued fraction. The energy spectrum of the tight-binding model in Eq. (1) splits into clusters of bands, which in turn split into subclusters, whereupon they split into sub-subclusters and this splitting continues ad infinitum, when the modulation wave vector is incommensurate. In the commensurate case, i.e. for $Q/2\pi$ = L/M (L and M are relatively prime), the final splitting is only M-fold. For a given value of Q, the splitting is strongly related to the continued fraction expansion of $Q/2\pi$. This continued fraction expansion can be written as

$$Q/2\pi = 1/(a_1 + p_1/q_1)$$

$$= 1/[a_1 + 1/(a_2 + p_2/q_2)]$$

$$= 1/a_1 + 1/[a_2 + 1/(a_3 + p_3/q_3)] \qquad (4)$$

where a_i, p_i, and q_i are integers (note that in general $p_{i-1} = q_i$). This rather simple connection between the cluster pattern for give Q and its continued fraction expansion is due to the fact the modulation function $\varepsilon_n = V_0\cos(Qn)$ that Hofstadter considered has only one Fourier component. When there are more Fourier components involved, the splitting might get more complicated. In fact, it is very interesting to study if the modulated Kronig-Penney model(14) exhibits similar features.

3. NATURE OF ELECTRONIC STATES

In the last section we discussed the relation between the cluster pattern of the energy spectrum and the continued fraction of the wave vector of the modulation. When $Q/2\pi$ is a rational number ($Q/2\pi$ = L/M), the energy spectrum of Equation (1) consists of M bands and the corresponding states have Bloch type-wave functions. The system looses its periodicity for an extended value of $Q/2\pi$ and an interesting question arises: Are the states localized or extended?

Aubry and André(7) have studied the case with $\varepsilon_n = V_0\cos(Qn)$.

This model is called self-dual because the Fourier coefficients of c_n, g_m obey the same type of equation as Equation(1). In particular

$$g_{m+1} + g_{m+1} + \frac{4t}{V_0} \cos(\Omega m + k) = \frac{2E}{V_0} g_m \tag{5}$$

where k is an arbitrary constant. The region $V_0 > 2t$ is transformed to the region $V_0 < 2t$ by this duality and therefore the $V_0 = 2t$ is the self dual point. By using a formula for the exponetial decay of the wave function due to Thouless(21), they show that all solutions of Equation (1) are localized for $V_0 > 2t$ and all solutions of Equation (5) are obviously extended if $4t > 2V_0$ or $V_0 < 2t$. Aubry(7) has also shown than when Q is irrational, the DOS at coupling constant V_0 is up to an overall scaling of energy by a factor of $V_0/2t$ identical to the DOS at coupling constant $4t/V_0$. While it has certainly not been proven rigorously, there is considerable evidence as we will see below that when Q is irrational this model has only localized states if V_0 is larger than 2t and only extended states if V_0 is smaller than 2t. Thus for a value of V_0 very close to 2t one has DOS that agree up to an essentially identity scale factor even though one DOS describes the extended region and the other the localized region.

Sokoloff(9), using Anderson's locator method, found for the same model agreement with Aubry for modulation amplitudes larger than the critical value, 2t. For $V_0 < 2t$ he found there exists only localized states for $|E|>|2t-V_0|$ and only extended states for $|E|<|2t-V_0|$, in disagreement with Aubry(7). Later it was recognized(10) that for $|E|>|2t-V_0|$ the widths energy gaps are very small and a careful calculation of the DOS is needed in order to find the positions and the widths of gaps and bands, and afterwards the nature of the eigenstate was determined. Note that a state in the gap can erroneously be interpreted as a localized state.

For the Kronig-Penney model Azbel(8) found a mobility edge between localized and extended states when the modulation strength is small, using semi-classical methods. This picture was confirmed recently by de Lange and Janssen(14), who by direct calculations of the wave functions show that there exist both localized and extended states in the Kronig-Penney model. Independently Bellissard et al(11) by constructing the Poincaré map of the Kronig-Penney model also obtain mobility edges. In particular, the Kronig-Penney model

$$[-\frac{d^2}{dx^2} + \sum_n b_n \delta(x-n)]\Psi(x) = E\Psi(x) \tag{6}$$

can be mapped, by the Poincaré map representation to the difference equation of the form

$$\Psi_{n+1} + \Psi_{n-1} - b_n \frac{\sin\sqrt{E}}{\sqrt{E}} \Psi_n = (2\cos\sqrt{E})\Psi_n \tag{7}$$

where $\Psi_n = \Psi(x = n^+)$. This map is exact and nothing is lost when we pass from Eq. (6) to Eq. (7). For instance in the periodic use; where all $b_n = b_0$, one recovers the complete band structure of the Kronig-Penney model. Belissand et al.(11) recognized that the form of Eq. (7) is similar to that of Aubry's model(7) if one takes $b_n = V_0\cos(Qn)$ with Q an irrational multiple of π. The only difference is the E-dependent coefficients. The study of this model proceeds by the same methods used to study the Aubry model (7). For that model, as we discussed above, the existence of extended states is established for $V_0 < 2t$ and localized states for $V_0 > 2t$. For the Kronig-Penney model or its Poincare map Eq. (7) we find localized states for $V_0 > 2\sqrt{E}/\sin\sqrt{E}$, which is energy dependent. Therefore for a given V_0 we have regions of energy with localized states and regions with extended states. The occurrence of localized states is more likely to occur in the lowest bonds of the energy spectrum.

Soukoulis and Economou(10) by combining for the first time the Green's function technique for calculating the DOS (so the positions and widths of gaps and bands are well known); direct determination of eigenstates and calculation of the length dependence of the transmission coefficient, avoid the common mistake of erroneously interpreting the gaps as localized states. We found for Eq. (1) with $\varepsilon_n = V_0\cos(Qn)$ results in agreement with Aubry that for $V_0 > 2t$ all states are localized and for $V_0 < 2t$ all states are extended. From experience with random systems, one expects the states at the end of the band to become localized more easily than those at the center. On the other hand for $\varepsilon_n = V_0\cos(Qn)$, the site energy spacing becomes smaller as we move towards the band edges and this facilitates propagation. It seems, that for the simple sinusoidal case, the two opposing tendencies cancel each other and the localization is independent of energy. For more general modulations (even the sum of the cosines) we find mobility edges, i.e. for a given modulation strength some of the states are localized and some extended.

Finally another interesting model has been proposed by Grempel et al.(12) in which the potential is unbounded and has the form $\varepsilon_n = V_0\tan(Qn)$. They argue that all the states are localized for incommensurate Q but their results were challenged by Luban (12).

4. NUMERICAL RESULTS

To determine the nature of the eigenstates for the model given in Eq. (1), we must first accurately calculate the DOS so that the positions and widths of bands and gaps are known. With the DOS three independent tests are used to determine the nature of eigenstates:

(i) The transmission coefficient T of the system is determined as the size of the system N increases for a given energy E, modulation strength V_0, and wave vector Q. For localized eigenstates $T \to 0$ as $N \to \infty$, while for extended one $T \to 0$ as $N \to \infty$. We would like to mention that in this problem the study of the transmission coefficient T is considerably simpler than for the 1D random potential as we are dealing with a definite potential and thus all questions of ensemble averages(16,17) do not arise. However, because some of the bands are extremely narrow, special care is required in order to avoid calculating T at a gap and erroneously interpreting the result as showing the existence of localized eigenstates.

(ii) By directly diagonalizing(16) Eq. (1) for a given N, V_0, and Q, one obtains the eigenvalue E and the corresponding eigenstate. In the present study, because the system is one-dimensional and has rigid boundary conditions, the matrix form of Eq. (1) is tridiagonal. Thus the eigenvalues and corresponding eigenstates(16) can be easily calculated for systems up to size 10,000. The accuracy of our eigenvalues and eigenstates is roughly sixteen and eight significant figures, respectively, using double precision. From the spatial behavior of the eigenstates, one can decide their nature, i.e., whether they are localized or extended. The application of this diagonalization method requires special care when there is almost degeneracy. We shall report elsewhere the related problems and the way we handled them.

(iii) One can approximate Q by $2\pi L/M$ (L,M integers without a common factor) so that after M sites the potential almost repeats. Consider the nth group of M consecutive sites (nth block). Within this group one can define $\varepsilon_n^{(1)}$ as the eigenvalue closer to the energy under consideration and $t^{(1)}$ as the effective hopping-matrix element between wave functions of neighboring blocks. ($t^{(1)} = \langle \psi_n | H | \psi_{n+1} \rangle$, where $|\psi_n\rangle$ is the wave function of the nth block.) Since $|\psi_n\rangle = \Sigma_i c_{ni} |i\rangle$, where the ith summation is over the sites of the nth block, one obtains that $t^{(1)} = t c_{nM}^+ c_{(n+1)M}$, c_{nM} is the amplitude at the last site of the nth block, and $c_{(n+1)M}$ the amplitude at the first site of the (n+1)th block. In all the cases we examined $\varepsilon_n^{(1)}$ is, within numerical uncertainties, of the form

$V_0^{(1)} \cos (Q^{(1)}n + \phi)$. This shows that under this transformation the original Hamiltonian maps to itself with transformed values of the parameters $V_0^{(1)}$ and $Q^{(1)}$. Our numerical results showed that for all values $V_0/2t > 1$, $V_0^{(1)}/2t^{(1)} > V_0/2t$, while for all values

$V_0/2t < 1$, $V_0^{(1)}/2t^{(1)} < V_0/2t$ independent of the values of E and Q. Hence by repeating this transformation it follows that for $V_0 > 2t$ ($V_0 < 2t$) the Hamiltonian maps finally into

$\lim_{n \to \infty} V_0^{(n)}/t^{(n)} \to \infty$ (0). This means physically that in the first case the states are localized while in the second case they are extended. This is a very interesting idea that has to be explored some more.

By combining for the first time the Green's-function technique for calculating the DOS, direct determination of eigenstates, and calculation of the transmission coefficient T, we avoid the common mistake of erroneously interpreting the gaps as localized states. It is indeed the consistency of the three complementary techniques which allow us to decide with confidence about the nature of eigenstates.

For the case where $\varepsilon_n = V_0 \cos(Qn)$ and Q is an irrational multiple of π our results show that $V_{oc}/t = 2$ is the critical modulation strength independently of E and Q in agreement with previous work(7,9). For $V_0 > 2t$ all the states are localized while for $V_0 < 2t$ all the states are extended. In the localized regime one can define a decay localization length l_c either from the spatial decay rate of the eigenstate or from the length dependence of T as $L \to \infty$, which agrees with that proposed by Aubry(7), i.e., $1/l_c = \ln(V_0/2t)$. It is surprising that V_{oc} is independent of E. From experience with random systems, one expects the states at the end of the band to become localized more easily than those at the center. However, for $\varepsilon_n = V_0 \cos(Qn)$, the site energy spacing becomes smaller as we move towards the band edges and this facilitates propagation. It seems that, for the simple sinusoidal case, the two opposing tendencies cancel each other and the localization is independent of the energy. To check this physical

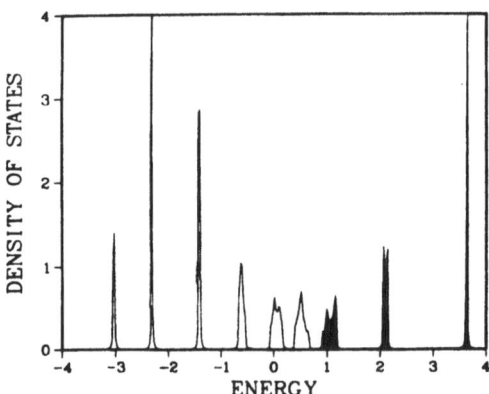

Fig. 1. The density of states (DOS) per sites as a function of energy with $N = 20,000$, $Q = 0.70$, $V_0 = 1.9t$, and $V_1 = 1/3$. The shaded areas denote localized states.

explanation, we also considered more complicated modulations, e.g.,

$$\varepsilon_n = V_0[\cos(\Omega n) + V_1\cos(2\Omega n)] \tag{8}$$

with $V_0/t = 1.9$, $V_1 = 1/3$, and $Q = 0.7$. The above argument suggests that, for ε_n given by Eq. (8), the eigenstates corresponding to high energies are easier to localize than those for low energies. By using the three methods which we described before we found that this is actually the case and that mobility edges exist. By using method (iii) we found that $\varepsilon_n^{(1)}$ is given by $V_0^{(1)} = \cos(Q^{(1)}n + \phi)$, but this time $V_0^{(1)}$ depends on energy E. For a given set of V_0, V_1, and Q (with V_0 less than a critical value) one has some of the eigenstates localized and some of them extended. In particular, as seen from Figure 1, the states above $E = 0.70t$ are localized, while the rest are extended. Thus, our tentative conclusion is that the mobility edge lies in a gap. The form of the DOS cannot be used to differentiate subbands of localized states from those of extended states (see Figures 2(a) and 2(c)). A difference in the DOS exists in this particular case; the subband of localized states has a smooth DOS (Figure 2(c)) while the subband of extended states has a strong structure in the DOS. However, it is definitely not a general feature. As a counterexample, we mention that the subband around $E = -1.4t$ (Figure 1) which corresponds to extended states has an even smoother DOS than the one shown in Figure 2(c) which corresponds to localized states. The spatial behavior of

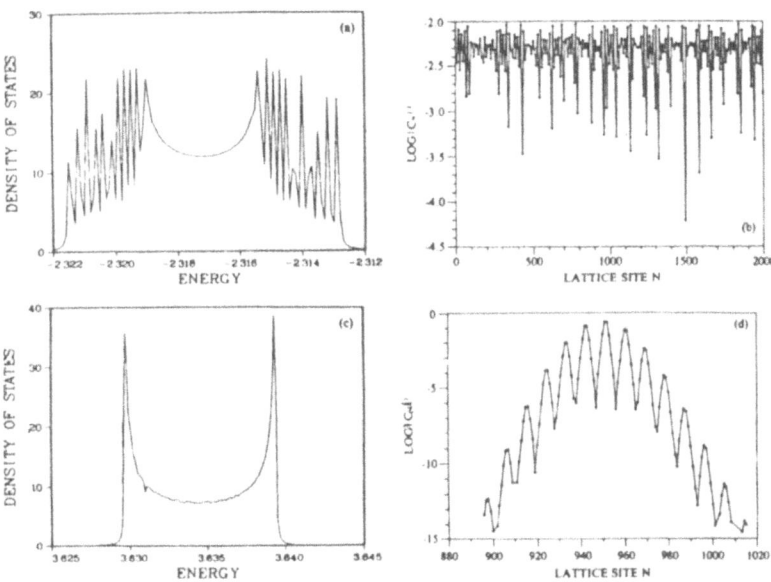

Fig. 2. (a), (c): Blowup of two subbands of Fig. 1. Typical (b) extended and (d) localized eigenstates corresponding to the centers of the subbands shown in (a) and (c), respectively.

90

corresponding eigenstates (see Figures 2(b) and 2(d)) changes
drastically as we cross the mobility edges. Because we analyzed a
<u>large</u> number of subbands, we were able to reach the conclusion that
the DOS per se cannot be used to distinguish whether the states are
localized or not. This does not mean that differences between
individual cases do not exist. It simply means that no general rule
seems to exist for distinguishing the two cases.

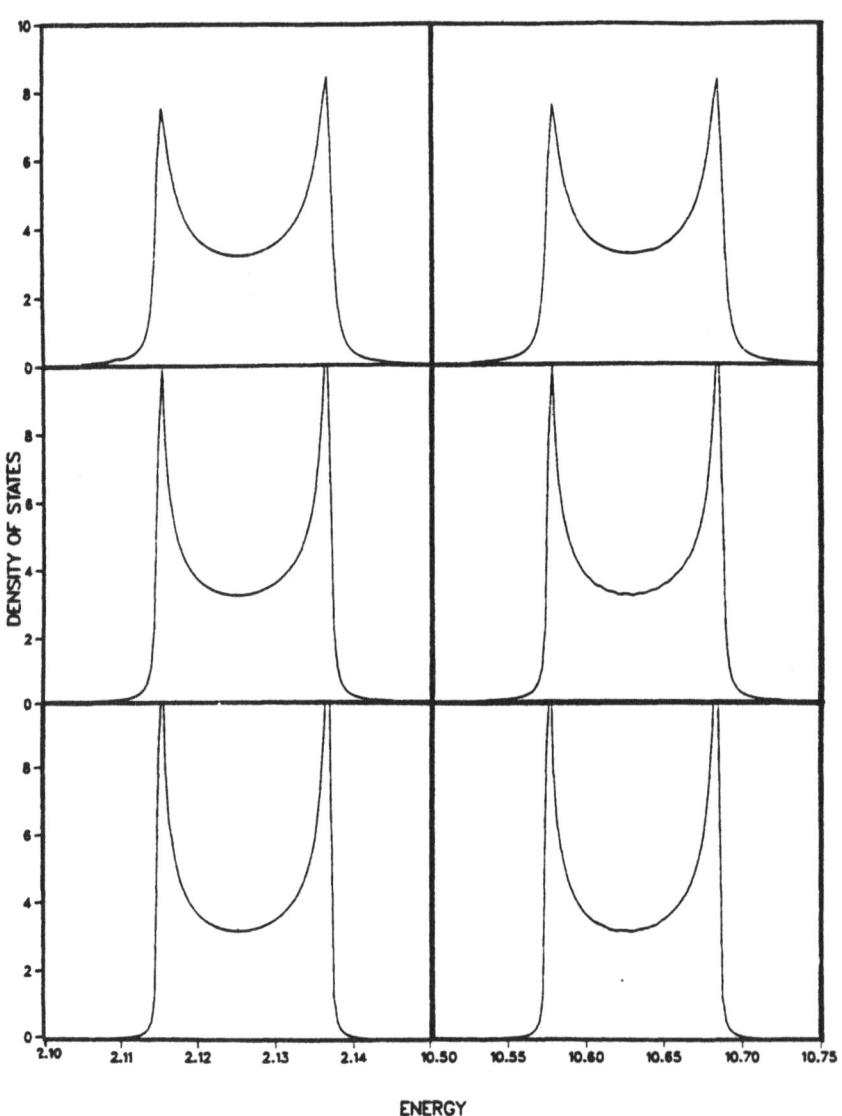

Fig. 3. DOS for an extended subband (left column, V_0 = .4) and for
its dual localized subband (right column, V_0 = 10) vs E for Q = .7
and N = 10^4 (upper row), 2 X 10^4 (middle row) 4 X 10^4 (lower row).

We have also checked the "Aubry duality"[7] which states that the DOS for localized states are essentially identical to the DOS of extended states as $N \to \infty$. It is claimed[7] that when the sites energies $\varepsilon_n = V_0 \cos(Qn)$, the DOS at V_0 is the same (apart from an energy sealing factor $V_0/2t$) as the one at $4/V_0$, when $N \to \infty$. We have examined whether or not this equality is obeyed at finite N by comparing cases of V_0 and $V_0^1 = 4/V_0$ ($V_0 = 0.4$ and 1.9) for $N = 10,000$, 20,000, and 40,000. The $V_0 = 0.4$ (extended states) and $V_0 = 10$ (localized states) results (shown in Figure 3) are essentially identical for every N we studied. This provides strong evidence in support of Aubry duality [7]. There is in our results, a detail worthwhile to mention. In the localized case, in contrast to the extended case, a very weak but real structure appears for $N = 20,000$ or 40,000. On the basis of our results for $V_0 = 1.9$ ($V_0^1 = 2.11$), it seems that this weak structure will appear in the extended case as well, but for much larger length N.

5. EXPERIMENTAL CONSEQUENCES

In conclusion the main result of all this work that has been discussed is that the presence of incommensurate modulations produces, in general, mobility edges in one dimensional systems[22]. Thus experiments in (clean) thin wires with incommensurate modulations in which the Fermi level can be made to be either in extended regions or in localized regions, e.g. by applying pressure, offer the rare opportunity to study the role of localization (versus electron-electron correlations) in effects like the low-temperature excess resistance observed in disordered systems[23].

In the metallic regime, a mini-band of energy levels in the presence of uniform and steady electric field will be reorganized in the so-called Stark ladder[24], describing an oscillatory motion of the electron due to the Bragg reflections at the edges of the mini-band[6]. However the electric field will also induce Zener tunneling among the various mini-bands. Although these effects have been considered[24] for normal crystalline solids, particularly semiconductors, the period of Stark oscillation is too large and the probability of Zener tunneling is too small for these effects to be seen in these materials. However, the long-period modulated crystals provide a better opportunity to experimentally see these effects. To observe the Stark oscillations[24] the frequency $eE/\hbar k_{max}$, where e is the electric charge, E is the electric field and k_{max} the width of a Brillouin zone (which can be very narrow for incommensurate systems) must be much larger that the electron scattering time. Finally, it is also possible to observe the Zener current depending on the mini-gaps and the relaxation time. In a general way we expect that the most favorable conditions are for high electric fields, large periods of modulation, not too small energy gaps and large relaxation times. However, it must be pointed out that not only collisions with impurities but also the process of Zener tunneling would dampen the Stark oscillations, and thus obscure the observation of the Stark Ladder. Although as far as we know, no such oscillations have been seen in modulated crystals; we believe that further experimental investigations in different modulated crystals will be fruitful in observing these effects.

REFERENCES

1. J. P. Pouget, G. Shirane, J. M. Hastings, A. J. Heeger, M. D. Miro, and A. G. MacDiarmid, Phys. Rev. B18, 3645 (1978).
2. J. A. Wilson, F. J. DiSalvo, and S. Mahajan, Adv. Phys. 24, 117 (1975).
3. H. V. Beyeler, Phys. Rev. Lett. 37, 1557 (1976).
4. Modulated Structures, ed. by J. M. Cowley, J. B. Cohen, M. B. Salainon, and B. J. Wrench, AIP Conference Proceedings, No. 53 (1979).
5. A. C. Gossard, Thin Solid Films 57, 3 (1979); L. L. Chang and L. Esaki, Progr. Cryst. Growth 2, 3 (1979).
6. L. Esaki and R. Tsu, IBM Journal of Res. and Rev. 14, 61 (1970).
7. S. Aubry and G. André in Annals of the Israel Physical Society ed. by C. G. Kuper (Adam Hilger, Bristol, 1979) Vol. 3, p. 133.
8. M. Ya Azbel, Phys. Rev. Lett. 43, 1954 (1979).
9. J. B. Sokollof, Phys. Rev. B22, 5823 (1980); 23, 2039 (1981); 23, 6422 (1981).
10. C. M. Soukoulis and E. N. Economou, Phys. Rev. Lett. 48, 1043 (1982).
11. J. Bellissard, A. Formoso, R. Lima, and D. Testard, Phys. Rev. B26, 3024 (1982).
12. D. R. Grempel, S. Fishman, and R. E.Prange, Phys. Rev. Lett. 49, 833 (1982); M. Luban unpublished.
13. A recent review from the mathematical point of view is given by B. Simon, Adv. Appl. Math. to be published.
14. C. deLange and T. Janssen, Electrons in Incommensurate Crystals: Spectrum and Localization, to be published.
15. R. Landauer, Phil. May. 21, 863 (1970).
16. E. N. Economou and C. M. Soukoulis, Phys. Rev. Lett. 46, 618 (1981) and Solid State Comm. 37, 409 (1981).
17. B. S. Andereck and E. Abrahams, J. Phys. C13, L383 (1980); E. Abrahams and M. J. Stephen, J. Phys. C13, L377 (1980); V. I. Mel'nikov, Pisma Zh. Eksp. Teor. Fiz 32, 244 (1980) [JETP Lett. 32, 225 (1980)]; P. W. Anderson, D. J. Thouless, E. Abraham and D. Fisher, Phys. Rev. B22, 3519 (1980); J. Sak and B. Kramer, Phys. Rev. B24, 1761 (1981).

18. Ya A Khintchine, Continued Fractions (Groningen: Noordholf 1963).
19. M. Ya Azbel, Zh. Eksp. Teor. Fiz. 46, 939 (1964) [Sov. Phys. - JETP 19, 634 (1964).
20. D. R. Hofstadter, Phys. Rev. B14, 2239 (1976).
21. D. J. Thouless, J. Phys. C5, 77 (1972).
22. C. M. Soukoulis and A. N. Bloch in "Localization Effects in Doubly Periodic Incommensurate Nearly 1-D Conductor" (to be published) argue that the localization theories are realized in the organic conductor TTF-Br$_x$.
23. N. Giordano, Phys. Rev. B22, 5635 (1980).
24. J. Zak, Solid State Physics (Ed. H. Ehrenrerch, F. Seitz, and D. Turnbull) (Academic Press, N.Y. 1972) Vol. 27, p. 8 and references therein.

Jonathan Smith and Julia Yeomans[+]

Department of Physics, The University,
Southampton, SO9 5NH, U.K.

·and

Volker Heine

Cavendish Laboratory, Madingley Road,
Cambridge, CB 3 OHE, U.K.

1. INTRODUCTION

In any close packed structure each stacking layer of atoms may occupy one of three positions conventionally denoted A, B and C (1). No two successive layers may occupy the same position but all other stacking sequences, AB,BA,AC,CA, are allowed.   The two most common patterns are ...ABABAB... and ...ABCABC... which give rise to the hexagonal and the cubic structures respectively.

However, in some compounds, for example silicon carbide and cadmium iodide, a large number of other stacking sequences are observed, many of which have remarkably long periodicities.   For example, more than a hundred different forms of SiC are known in some of which the repeat distance of the stacking sequence exceeds 10^3 nm.   This is the phenomenon of polytypism (2,3):the different structures of a given compound are called polytypes.

[+]Address from October 1983:Department of Theoretical Physics,
 1, Keble Road, Oxford OX1 3NP, U.K.

A number of theories have been advanced to explain this curious behaviour, but so far these have been, almost entirely,based on non equilibrium mechanisms such as crystal growth from dislocations (4). However, there is considerable evidence to suggest that polytypism is,at least partially, an equilibrium phenomenon (5). In this paper, we describe how many of the essential features follow from the equilibrium properties of a simple model, originally studied in the context of magnetic phase transitions, the ANNNI model (6).

In the next section of the paper we describe the structure of polytypic materials in greater detail and give a brief outline of existing theories. The ANNNI model is introduced in section 3 and its phase diagram described. In section 4 we discuss the relevance of the model to polytypic behaviour and a brief summary in section 5 concludes the paper.

2. POLYTYPISM

The principal examples of polytypism in stoichiometric compounds are the micas and the so called classical polytypes (7), which are related to close packed structures. Here we consider only binary compounds of the forms MX and MX_2 (for example SiC and CdI_2) since these are, crystallographically, the simplest.

In both the MX and MX_2 compounds each M atom is tetrahedrally coordinated with X atoms which are stacked in a close packed array. The MX compounds can be considered as alternating layers of M and X atoms. If the A,B,C stacking sequence of the X layers is fixed, then the entire crystal structure is uniquely specfied.

In MX_2 polytypes, however, the M atoms occupy only alternate planes of intersticies between the layers of X atoms. Hence the compound is made up of X-M-X "sandwiches". Again the structure can be described by giving the A,B,C position of the layers of X atoms but now to gain a unique specification the first layer is conventionally chosen to lie to the left of a layer of M atoms (2,3).

The most common polytypes of a given compound have short period stacking sequences: for example in ZnS, the principal structures are those of zinc-blende (...ABCABC...) and wurtzite (...ABABAB...). When describing longer period structures, however, it soon becomes impractical to specify the full ABC sequence, so more compact notations have been devised. Compactness is achieved by specifying only the relative orientation of successive stacking planes. The most useful of these systems for our purposes is the Zdhanov notation (8).

In this scheme, each pair of consecutive layers in the stacking

sequence is labelled according to whether the progression is cyclic
(A→B, B→C or C→A) or anticyclic(A→C, C→B or B→A), using the symbols
+ and - respectively. For example, the structure

$$\ldots A \; C \; A \; B \; A \; C \; A \; C \; B \; C \ldots \tag{1}$$

is denoted

$$\ldots - + + - - + - - + + \ldots \tag{2}$$

Interchanging the roles of the + and - symbols is immaterial since
this amounts to interchanging say B and C in the stacking sequence,
or equivalently rotating the crystal through $\pm 60^{\circ}$ about the
stacking direction.   Thus, it is only necessary to specify the
lengths of successive "bands" of the + and - symbols, where a band is
defined as a string of neighbouring + (-) symbols terminated by
- (+) symbols.   For example, the structure (1) is represented by
a Zdhanov symbol $(122)_2$ because the + - sequence comprises one
'1-band' followed by two '2-bands' and is repeated twice before the
stacking sequence repeats.   Note that we have mapped a three-state
system, A,B,C, onto a two-state system by associating a variable,
+ or -, with each pair of neighbouring stacking layers.

The Zdhanov symbols of a typical selection of CdI_2 and SiC
polytypes are given in Table 1. It is at once apparent from these
results that only a highly restricted subset of all the possible
polytypes are observed for each compound.   The SiC polytypes
correspond almost entirely to Zdhanov symbols comprising only 2-
bands and 3-bands whereas for CdI_2 almost all the symbols contain
only 1- and 2-bands.   Indeed structures with Zdhanov symbols including longer bands are rare in all classical polytypes. Any theory
of polytypism must account not only for the large number of long
period structures, but also for the predominence of the structural
elements represented by the 1-, 2- and 3- bands in the Zdhanov
symbols.

Theories of polytypism are of two general types: those based
on crystal growth mechanisms and those employing thermodynamic
considerations.   We now briefly review the main ideas and their
relationship to the experimental picture.

2.1 Growth Theories

Frank's theory of polytypism (4) assumes that the crystal
initially grows from a screw dislocation in the stacking planes, so
that the period of the polytype is determined by the step height of
the growth spiral.   Using this mechanism it is possible to generate
almost any polytype from shorter period structures, especially if
the theory is modified to include growth about systems of inter-
weaving spirals with different step heights (3).   However these
mechanisms are certainly not universal since many samples of long
period polytypes show no evidence of growth spirals.   A further

SiC		CdI$_2$		PbI$_2$	
Structure	Period	Structure	Period	Structure	Period
11	2	11	2	11	2
∞	3	22	4	22	4
22	4	2211	6	2211	6
33	6	33	6	∞	6
44	8	22(11)$_2$	8	(11)$_3$22	10
3223	10	(121)$_2$	8	(31)$_3$	12
(22)$_2$33	14	1232	8	(11)$_5$22	14
(32)$_3$	15	(22)$_2$11	10	(11)$_7$2112	20
(33)$_2$22	16	222123	12		
(22)$_3$33	18	(21)$_2$(12)$_2$	12		
22(23)$_3$	19	(22)$_2$(11)$_2$	12		
...		...			

Table 1 : Typical Zdhanov sequences of some simple polytypes (3).

drawback is that these theories make no predictions about the relative stabilities of the various structures.

2.2 Equilibrium Theories

Equilibrium theories assume that a given polytype represents the thermodynamically stable structure, under given external conditions. There are two major difficulties in testing such a hypothesis: firstly, the polytypic structure obtained is found to depend strongly on impurity concentrations thus obscuring any correlation of structure with more easily controlled variables such as temperature and pressure. Secondly, different polytypic structures are usually separated by large energy barriers and therefore transformations between them may be extremely slow. It is well established that the zinc-blende and wurtzite structures are the low and higher temperature phases respectively of both ZnS and SiC. Other temperature structure relations are more controversial (3). However, recently a number of transformations between the shorter period polytypes have been observed (5) many of which are reversible. These results are summarised in Table 2. The existence of reversible transformations between polytypes provides strong evidence that they are, at least partially, equilibrium structures.

The only equilibrium theory of polytypism so far is due to Jagodzinski(9) who argues that the different polytypes arise from the ordering of stacking faults. This follows from assuming a relationship between the vibrational entropy and the extent of the

Transformation	Temperature ($^{\circ}$C)	Other conditions
11 $\rightarrow$ ∞ ∞ $\rightarrow$ 11	1400 - 1600 1700 - 2300	Hot pressed with AℓN at $\sim$ 350 Atm
∞ $\rightarrow$ 33 33 $\rightarrow$ ∞	2100 - 2300 2500	10-30 Atm N_2
(2,3)$_3$ $\rightarrow$ 3	-	Pure powder samples
22 $\rightarrow$ 33	1800 - 2250	10 Atm of N_2 or Ar, (2,3)$_3$ also produced when T < 2000°C.
33 $\rightarrow$ 22	1980	Hot pressed with B

Table 2: Transformations between SiC Polytypes (5).

disorder in the stacking direction. An essential prediction of this theory is that long period polytypes should have highly disordered stacking sequences but this is contrary to many observations. It has also now been shown that the vibrational entropy differences between polytypes are too small to contribute significantly towards their stabilisation(10).

So far, the crystal growth theories have received far greater attention largely because of the difficulty in finding an equilibrium mechanism which can lead to a large number of stable structures and very long ranged periodicities. In the next section we describe a simple model with only short ranged interactions which produces both these features.

3. THE ANNNI MODEL

Our aim is to construct a simple model of polytypism. To this end we recall that the Zdhanov notation introduced in the previous section allowed us to describe polytypes as a sequence of two-state systems by associating a variable + or - with neighbouring pairs of stacking layers according to their relative orientation. This is very reminiscent of an array of the Ising spins, $S_i = \pm 1$, which appear in simple theories of magnetic phase transitions.

Therefore let us model polytypism by considering a three-dimensional array of Ising spins and constructing a Hamiltonian describing the interaction between the spins which has the following properties:

1. The model must have strong ferromagnetic interactions within each two-dimensional layer to reproduce the integrity of the layers in the polytype. However these interactions cannot be infinitely strong as transitions between different polytypes must be allowed.

2. The interactions between layers must be comparatively weak to allow different stacking sequences to have similar energies. If we consider only interactions between first neighbours no long period phases are stable and therefore we also include a second neighbour interaction.

3. There must be no magnetic field, to preserve the + ⟷ - symmetry of the Zdhanov notation.

A model which obeys these criteria is the ANNNI (axial next nearest neighbour Ising) model which comprises Ising spins, $S_i = \pm 1$, on each of the sites i of a three dimensional, layered lattice with interactions specified by the Hamiltonian

$$\mathcal{H} = -J_o \sum_{nn}^{b} S_i S_j \quad -J_1 \sum_{nn}^{z} S_i S_j \quad -J_2 \sum_{nn}^{z} S_i S_j \tag{3}$$

The interactions are illustrated in figure 1. The first term is a ferromagnetic coupling, $J_o > 0$ between nearest neighbour spins within each layer. Along the z axis, nearest neighbour spins are

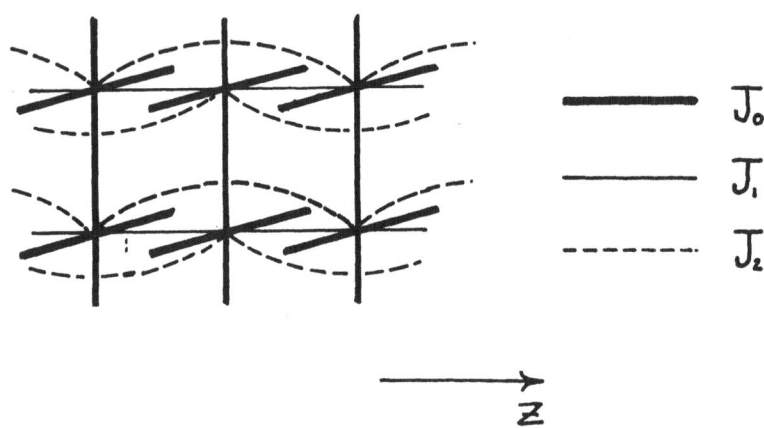

Figure 1: Interaction scheme for the ANNNI model on a simple cubic lattice.

coupled both by a first neighbour interaction J_1 and by a second neighbour interaction J_2.

Here we consider only the low temperature region, $J_o \gg k_B T$, where each layer of spins perpendicular to the z direction is ferromagnetically aligned. Hence we may distinguish the phases of the model by specifying the ordering in each layer using a notation (6) similar to the Zdhanov symbols introduced in section 2. We denote a layer of spins $S_i = + 1$ and $S_i = - 1$ by the symbols $+$ and $-$ respectively, so that, for example, a typical configuration, made up of the repeating sequence,

$$\ldots \ ++--++--- \ \ldots \tag{4}$$

is represented by the symbol <2223> or equivalently <$2^3$3>.

The equilibrium properties of the ANNNI model in the low temperature limit have been rigorously calculated by Fisher and Selke (6) using a combination of series expansions and linear programming techniques. The method is rather complicated, so here we concentrate on describing the results of their analysis and giving a physical interpretation.

The phase diagram of the ANNNI model at a fixed, low temperature is shown in figure 2 drawn schematically for the case $J_o \gg k_B T \gg |J_1|$. The model exhibits two infinite sequences of phases. One sequence includes phases with only one and two-bands; the other, phases with only two and three-bands. Note that states with arbitrarily long range periodicity appear in the sequences and that no phases which include bands of length four or more appear in the phase diagram.

All phase transitions are first order and, for $J_2 < 0$, the phase boundaries are displaced from the lines $J_1 = 2J_2$ and $J_1 = -2J_2$ by distances of the order $w^{q_\perp}$, where

$$w = \exp(-2J_o/k_B T)$$

and $q_\perp$ is the coordination number of the J_o-coupled layers (for example $q_\perp = 4$ if the ANNNI model is defined on a simple cubic lattice). The regions of stability of the phases $< 2^k 3 >$ and $< 2^{k+1} 1 >$, k = 0,1,2 ... , become exponentially narrower with increasing k, controlled by a factor $w^{k q_\perp}$.

The symmetry of the phase boundaries about $J_1 = 0$ arises because the Hamiltonian (3) is left unchanged by a reversal in the sign of J_1, provided alternate spins along the z axis are simultaneously reversed.

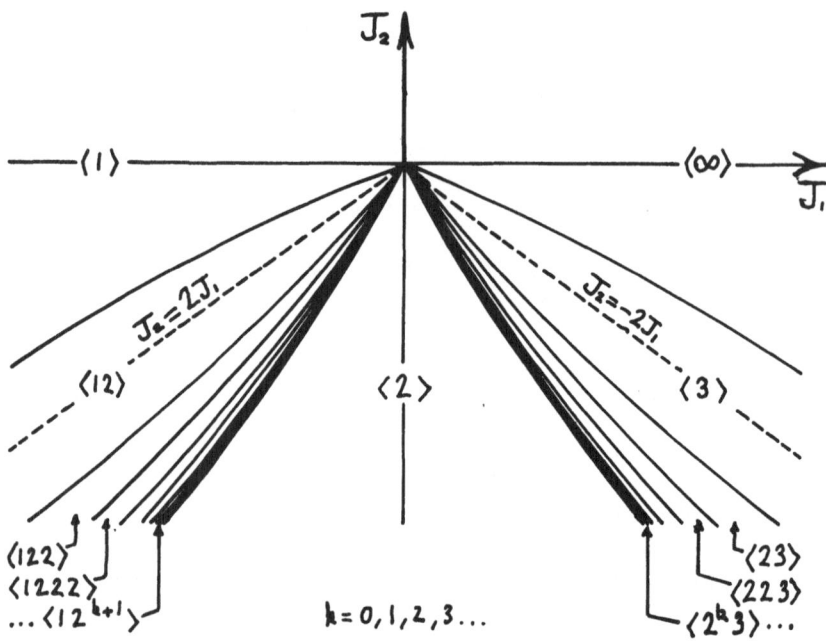

Figure 2: Phase diagram of the ANNNI model at a fixed temperature, $k_B T \ll J_o$.

All phase boundaries meet at the point $J_1 = J_2 = 0$ where the model reduces to an ensemble of non-interacting Ising planes. We note that these results are qualitatively unchanged by the inclusion of further neighbour interactions within each plane provided these are also ferromagnetic.

The existence of complicated phase sequences in a simple model with short ranged interactions is a consequence of a delicate balance between the energy and entropy terms in the free energy. At zero temperature, the ground state of the ANNNI model comprises only three phases of finite extent, $\langle\infty\rangle$, $\langle 1 \rangle$ and $\langle 2 \rangle$. However, on the boundary lines between these states, many different structures can have the same energy. On the boundary $J_1 = 0$, $J_2 > 0$ only the neighbouring states $\langle 1 \rangle$ and $\langle\infty\rangle$ are degenerate. However when $J_2 < 0$, the competing interactions, J_1 and J_2, lead to an infinite degeneracy. On the line $J_1 = 2J_2$ between the $\langle 1 \rangle$ and $\langle 2 \rangle$ states all structures comprising only 1-bands and 2-bands (for example $\langle 12 \rangle$ and $\langle 12122\rangle$) have the same energy. Similarly on the line $J_1 = -2J_2 > 0$, all states comprising 2,3,4 ... ∞ - bands are degenerate. As the temperature is increased from zero the elementary excitations, spin flips $S_i \rightarrow -S_i$, become important and this leads to the preferential stabilisation of a subset of the degenerate structures. This subset of structures form the two infinite sequences of phases and as they are entropy stabilised, their regions of stability expand with increasing temperature.

4. A MODEL OF POLYTYPISM

We now interpret the properties of the classical polytypes in terms of the ANNNI model. As anticipated, we identify each phase of the ANNNI model with a polytype via the corresponding Zdhanov symbol. Thus, for example, the phase <∞> represents the zinc-blende structure for an MX compound.

After taking this step, two striking features are immediately apparent: Firstly, the structures which have by far the greatest regions of stability are the short period polytypes represented by the phases <∞>, <1> and <2> (corresponding, in MX compounds, to the zinc-blende, wurtzite and 4H structures respectively). Longer period structures have much narrower regions of stability and are therefore less likely to be observed, in clear agreement with experiment.

The second important feature is that the structures generated by the ANNNI model have Zdhanov symbols comprising 1- and 2-bands ($J_1 < 0$) or 2- and 3-bands ($J_1 > 0$). This is remarkably similar to the patterns of structures in Table 1.

So far, we have said nothing about the physical basis for the interactions J_1 and J_2 when the ANNNI model is compared to polytypic materials. We make no attempt to calculate these parameters from first principles but it is clear that the effective interactions correspond to small differences in phonon and electronic free energies of different polytypic structures. These free energies will certainly depend on temperature and will be sensitive to impurity concentrations; thus in describing transformations between polytypes we may allow the effective interactions J_1 and J_2 to vary. This is conveniently visualised in terms of figure 2 by considering the state of a crystal of polytypic material to be represented by a point with coordinates J_1, J_2 and $k_B T/J_o$. As the temperature varies, this point describes a trajectory which may pass through many different phases. The form of this trajectory, and hence the sequence of stable phases will depend on impurity concentration and temperature through $J_1(T)$ and $J_2(T)$. Thinking in these terms, the wurtzite-zinc-blende transition which occurs in ZnS and AgI and, more importantly, all the observed reversible transformations in SiC can be accounted for (see Table 2). The occurrence in SiC of the transformations corresponding to the sequence <1> ↔ <∞> ↔ <3> ↔ <23> is particularly striking. Failure to observe any of the longer period <$2^k 3$> phases in the <3> ↔ <2> transformation may be accounted for by the very small region of stability of these phases.

Although the ANNNI model succeeds in describing broad features of polytypism it fails to produce many of the observed structures. However a high level of realistic detail is not to be expected from such an idealised model. There are two major reasons why the ANNNI model is not a faithful representation of any of the classical polytypic materials. Firstly the elementary excitations of these crystals are only very crudely represented by the spin flips of the three dimensional Ising model. Clearly, if, in the ANNNI model, a whole layer of spins is flipped, this corresponds to a shift in relative orientation of two neighbouring stacking layers of X atoms. It is however not obvious that there is a simple analogue of a single spin flip in the real system.

Secondly, the symmetry properties of the ANNNI model and those of the real classical polytypes are different. For example consider the phase <123>. In the ANNNI model, all phases described by band sequences which are cyclic permutations or inversions of <123>, for example <321> , <231> and <312>, have identical energies. The symmetry properties of the Zdhanov notation have been analysed in detail by Trigunayat and Jain (11) who show that for classical polytypes the inversion symmetry does not exist. Thus, $(123)_2$ and $(321)_2$ represent different crystal structures which can, in general, be expected to have different free energies and regions of stability. In MX compounds, for example, this occurs because each layer of M atoms is not symmetrically distributed between the neighbouring layers of X atoms.

Further, we note that the ANNNI model takes no account of the 'sandwich' structure of MX_2 compounds, in which alternate X layers interact only through weak Van der Waals forces.

5. SUMMARY

To conclude, we have shown that many of the important features of polytypism can be explained by assuming that polytypes are equilibrium structures. Representing the system by an Ising model, we find that the large number of polytypic structures can arise from a delicate balance of short-range competing interactions. The structures are predominantly entropy stabilised. The theory accounts for the 1:2 and 2:3 pattern of the Zdhanov symbols, the narrow regions of stability for the long period structures and all known reversible transformations between polytypes. We speculate that a more sophisticated model incorporating a more realistic description of the excited states and the symmetry of these crystals will generate more of the observed polytypic structures.

ACKNOWLEDGEMENT

We should like to thank M.F.Thorpe for helpful discussions.

1. Kittel, C. "Introduction to Solid State Physics", Wiley N.Y. 1976.
2. Verma, A.R. and Krishna, P. "Polymorphism and Polytypism in crystals", Wiley N.Y. 1966.
3. Trigunayat, G.C. and Chada, G.K. (1971) Phys.Stat.Sol.(a) $\underline{4}$ 9-42 and 282-303.
4. Frank, F.C. (1951) Phil.Mag. $\underline{42}$ 1014.
5. Jepps N.W. and Page T.F. (1983) to be published.
6. (a) Fisher, M.E. and Selke,W.(1981) Phil.Trans.R.Soc. $\underline{302}$ 1.
   (b) Fisher, M.E. and Selke,W.(1980) Phys.Rev.Lett. $\underline{44}$ $\overline{1502}$.
7. Thompson,J.B.(Jr.) (1981) Struct.Bond.Cryst.II 167.
8. Zdhanov,G.S. and Minervina Z. (1945) J.Phys.$\underline{9}$ 151.
9. Jagodzinski,H. (1954) Neves Jahst.Mineral. Mh. $\underline{3}$ 49.
10. Weltner, W.(Jr.) (1969) J.Chem.Phys. $\underline{51}$ 2469.
11. Trigunayat, G.C. and Jain, P.C. (1977) Acta.Cryst. $\underline{A33}$ 257.

DENSITIES OF STATES OF COMPOSITIONALLY MODULATED ALLOYS

A. Gonis and N.K. Flevaris

Department of Physics and Astronomy and Materials Research
Center, Northwestern University, Evanston, Illinois 60201

1. INTRODUCTION

In this lecture, we will present methods that allow the
calculation of the one-particle properties, in particular the
density of states (DOS), in a large class of disordered systems.
These methods are applicable to substitutionally disordered
electronic systems, such as intermetallic alloys of two or more
components, either random or exhibiting various kinds of short-
range order (SRO). Even though certain of the methods to be discussed here are directly applicable or can be extended to alloys
with arbitrary degrees of SRO, we will consider explicitly compositionally modulated alloys, (CMA), in which the concentrations
of the constituents vary periodically in a given direction. The
physical properties of CMA in many cases have been found to
deviate substantially from those of homogeneous alloys of similar
average composition.

Experimental studies have shown that the properties of CMA
may be functions of both the amplitude and wavelength of modulation. For example, the transport properties of CMA CuNi have been
found (1) to depend strongly on the amplitude of modulation. The
biaxial elastic modulus of several CMA, such as CuNi, CuPd, AgPd,
and AuNi, has been shown (2-5) to increase by several hundred percent for certain values of the wavelength of modulation. The band
structure of many semiconducting CMA may be significantly different
from that of a random alloy, as is indicated by experimental
studies (6) of the excitation spectrum of two-phonon absorption
in layered crystals. Experimental studies of semiconducting CMA are
reviewed in an article by Gossard (7).

In addition to the mechanical properties, the magnetic properties of many CMA may depend (8-14) on the characteristics of the
modulation. For example, the Curie temperature and the magnetization density of ferromagnetic CMA can differ substantially from
those of corresponding homogeneous alloys. Finally, the properties
of superconducting CMA have been observed (15-19) to differ from
those of homogeneous alloys. A detailed review of the observed
properties of many CMA is given in the article by Schuller and
Falco (19).

An understanding of the band structure of CMA is essential for
an understanding of many of the experimental results mentioned
above. For example, the increase in the elastic modulus in some CMA,
for particular values of the wavelength, as well as the absence of
such increase in other compositionally modulated systems has recently been explained (20) at least qualitatively in terms of the
geometry of the Fermi surface of the corresponding homogeneous
alloys. It is conjectured that elastic hardening occurs if the
Fermi surface of the homogeneous alloy can become tangent to the
mini Brillouin zone boundaries that are induced by the modulation.
The results that one deduces from this picture and the calculated
(21) Fermi surfaces of CuNi and AgPd alloys seem to be in accord
with the observed behavior of the physical properties of the corresponding CMA. Also, an understanding of the magnetic properties of
CMA requires a knowledge of the band structure, and in particular,
of the DOS of these materials.

Several attempts have been made in the literature (22-26) to
study the DOS and CMA theoretically. The DOS and the magnetism of
CuNi coherent modulated structures, i.e., periodically repeating
sets of planes containing pure Cu and pure Ni, have been studied
(22) using accurate band structure techniques. Also, pseudopotential
methods have been used (23-26) sometimes with conflicting results,
to study the DOS, dielectric function, and charge density of semiconductor CMA.

The theoretical methods just mentioned cannot take account of
the effects of chemical disorder that is present in CMA. In a
CMA, the planes perpendicular to the direction of modulation do not
contain atoms of only one species, but are planes of alloys
characterized by different concentrations of the various constituents. Thus, any theory of CMA should treat properly the effects of
substitutional disorder and should become exact as the disorder
approaches zero, i.e., as the CMA approaches a coherent modulated
structure.

In order to study the effects of disorder on the DOS and other
one-particle properties of CMA, we have generalized the coherent
potential approximation (27-29) (CPA), which is the most satisfying
single-site theory for the study of homogeneous, substitutionally

disordered random alloys. The desirable properties of the CPA include (i) analyticity of all calculated quantities such as the
electronic self-energy and Green's function, (ii) preservation of
all symmetry properties of the underlying lattice, and (iii) physically meaningful (nonnegative) DOS and spectral weight functions.
There exist several excellent reviews (30-32) of the CPA both for
systems describable by tight binding (TB) (30) Hamiltonians and for
systems describable by muffin-tin (MT) (31-32) Hamiltonians.

The remainder of the lecture is divided into several sections
as follows. In section II, we present a brief derivation of the CPA
for random, substitutionally disordered alloys. In Section III,
this derivation is extended in a straightforward way to compositionally modulated alloys. Numerical results obtained in this
generalization are presented in Section IV, for one-dimensional as
well as three-dimensional systems. A summary and a discussion of
the presentation are given in Section V.

2. THE COHERENT POTENTIAL APPROXIMATION

As mentioned above, the CPA is applicable to systems describable by TB or MT Hamiltonians. Since TB systems are somewhat easier
to discuss than MT systems, we will confine our attention to TB
alloys. The usual single-band tight-binding model Hamiltonian
assumes the form,

$$H = \sum_i \varepsilon_i \, a_i^\dagger \, a_i + \sum_{i,j} W_{ij} \, a_i^\dagger \, a_j, \tag{2.1}$$

in a site (or Wannier) representation. In this equation, the $a_i^\dagger$
(a_i) are creation (destruction) operators for an electron at site i.
The ε_i are site energies which vary with the chemical occupation
of site i. Thus, for a binary alloys, $A_c B_{1-c}$, with atoms of type
A and B distributed over the N sites of a lattice with concentration
c and 1-c respectively, ε_i can be equal to ε_A or ε_B. Due to charge
transfer effects, the ε_i may depend on the occupation of sites
near site i as well as on the occupation of site i itself. The CPA
has recently been generalized (33) to include charge-transfer
effects. However, for ease of presentation we will consider only
the variation of ε_i with respect to the atom occupying site i,
known as diagonal disorder. The transfer (or hopping) terms W_{ij} describe the tendency of an electron to propagate (hop) from site i to
site j. The W_{ij} in general depend on the occupation and the local
environment of sites i and j. The dependence of the W_{ij} on the
chemical configuration of an alloy is known as off-diagonal disorder (ODD). The CPA has been generalized (34,35) to systems characterized by ODD where the W_{ij} are functions of the chemical occupation of sites i and j. For our purpose, it is sufficient to con-
sider transfer terms W_{ij} that depend on the distance $|\vec{R}_{ij}|$ between

sites i and j but are independent of the chemical occupation of sites i and j. Thus, we will discuss alloys characterized only by diagonal disorder.

The experimentally determined physical properties of an alloy are ensemble averages over all alloy configurations. The single-particle properties can be conveniently expressed in terms of the ensemble average, <G>, of the single-particle Green's function operator for a complex energy z,

$$G = (z-H)^{-1}, \tag{2.2}$$

in units such that $\hbar = 1$. For example, the DOS is given by the expression,

$$n(\varepsilon) = -\frac{1}{\pi N} \text{Im Tr} <G>, \tag{2.3}$$

where Tr denotes the trace (sum of diagonal elements) of an operator. It is clear that <G> must be evaluated in some approximate way. In the CPA one considers that the real disordered material is replaced by a translationally invariant effective medium characterized by a site diagonal self-energy, σ. The self-consistency condition determining σ may be expressed in various ways and involves equalities between ensemble averaged quantities, such as G_{ij}, and the corresponding quantities evaluated in the effective medium. We will now derive one form of the CPA self-consistency condition.

The Hamiltonian operator, Eq. (2.1), can be written in the form,

$$H = \varepsilon + W, \tag{2.4}$$

where ε is site-diagonal. Then the Green's function operator becomes,

$$(z - H) G = I, \tag{2.5a}$$

or

$$(z - \varepsilon - W) G = I, \tag{2.5b}$$

or

$$(z - \varepsilon) G = I + W G, \tag{2.6}$$

where I denotes an N x N unit matrix. In a site representation (2.6) yields the equation of motion of the Green's function,

$$G_{ij} = g_i \delta_{ij} + g_i \sum_k W_{ik} G_{kj}, \tag{2.7}$$

where

$$g_i = (z - \varepsilon_i)^{-1} \qquad (2.8)$$

is the bare locator for site i. We wish to obtain an expression for G_{ii}, the site-diagonal element of the Green's function operator. To do this we can proceed by iterating Eq. (2.7) in a formal way (36), treating the second term on the right-hand side as a perturbation. One obtains the expression,

$$G_{ii} = (z - \varepsilon_i - \Delta_i)^{-1} \qquad (2.9)$$

where Δ_i, the renormalized interactor, is given by the expression,

$$\Delta_i = \sum_{j \neq i} W_{ij} g_j W_{ji} + \sum_{\substack{j \neq i \\ k \neq i,j}} W_{ij} g_j W_{jk} g_k W_{ki} + \ldots . \qquad (2.10)$$

Thus, Δ_i is the correction to the energy of site i which arises from the interaction of site i with the surrounding material. It is clear from expression (2.10) that Δ_i represents the sum of all paths through the material that start and end on site i, but avoid site i in all intermediate steps.

Up to this point no approximation has been introduced, and Eq. (2.9) for G_{ii} is exact. In the spirit of the CPA, we now consider that the material surrounding site i is replaced by an effective medium characterized by a self-energy σ and hence a bare locator,

$$\bar{g} = (z - \sigma)^{-1} . \qquad (2.11)$$

Replacing every g_j in the expression for Δ_i, Eq. (2.10), with $\bar{g}$, we can write

$$G_{ii} = (z - \varepsilon_i - \bar{\Delta})^{-1} , \qquad (2.12)$$

where $\bar{\Delta}$ is the renormalized interactor corresponding to the effective medium, and G_{ii} in Eq. (2.12) is the site diagonal element of the Green's function corresponding to a real atom embedded in that medium at site i. The self-consistency condition that determines σ can now be expressed in the form,

$$\langle G_{ii} \rangle_{SS} = \bar{G}_{ii} , \qquad (2.13)$$

where $\langle \ldots \rangle_{SS}$ denotes the single site (SS) average over the occupation of site i, and $\bar{G}_{ii}$ is the site-diagonal element of effective medium Green's function. Clearly, $\bar{G}_{ii}$ is obtained from Eq. (2.12) upon the replacement of ε_i with σ, i.e., $\bar{G}_{ii}$ is given by the expression,

$$\bar{G}_{ii} = (z - \sigma - \bar{\Delta})^{-1}. \tag{2.14}$$

The CPA self-consistency condition, Eq. (2.13), involves two effective medium parameters, σ and $\bar{\Delta}$. Either of these can be eliminated in terms of the other. Note that from Eq. (2.14), we have,

$$\bar{\Delta} = z - \sigma - \bar{G}_{ii}^{-1}. \tag{2.15}$$

Now, an expression for $\bar{G}_{ii}$ can be obtained from the equation of motion of the effective medium Green's function. With g_i replaced with $\bar{g}$, Eq. (2.7) yields,

$$\bar{G}_{ij} = \bar{g}\,\delta_{ij} + \bar{g}\,\sum_k W_{ik}\,\bar{G}_{kj}, \tag{2.16}$$

an equation for the translationally invariant quantities $\bar{G}_{ij}$ that can be solved by means of Fourier transforms (FT). Introducing the quantities,

$$\bar{G}(\vec{k}) = \frac{1}{N^2}\sum_{i,j} G_{ij}\, e^{-i\,\vec{k}\cdot\vec{R}_{ij}}, \tag{2.17}$$

and

$$W(\vec{k}) = \frac{1}{N^2}\sum_{i,j} W_{ij}\, e^{-i\,\vec{k}\cdot\vec{R}_{ij}}, \tag{2.18}$$

we readily obtain,

$$\bar{G}(k) = (z - \sigma - W(\vec{k}))^{-1}. \tag{2.19}$$

The elements of $\bar{G}$ in a site representation are obtained as the inverse FT of Eq. (2.19),

$$\bar{G}_{ij} = \frac{1}{\Omega_{BZ}} \int_{BZ} \bar{G}(\vec{k})\, e^{i\,\vec{k}\cdot\vec{R}_{ij}}\, d^3k, \tag{2.20}$$

where the integration extends over the first Brillouin zone (BZ) of the lattice and Ω_{BZ} denotes the volume of the zone in reciprocal space. In particular, $\bar{G}_{ii}$ is given by the expression,

$$\bar{G}_{ii} = \frac{1}{\Omega_{BZ}} \int_{BZ} \bar{G}(\vec{k})\, d^3k. \tag{2.21}$$

Using Eqs. (2.15) and (2.21), one can eliminate either of σ or $\bar{\Delta}$ from the CPA self-consistency condition, Eq. (2.13). In terms of σ, Eq. (2.13) becomes,

$$\langle(\sigma - \varepsilon_i + \overline{G}_{ii}^{-1}(\sigma))^{-1}\rangle = \overline{G}_{ii}(\sigma), \tag{2.22}$$

where the dependence of $\overline{G}_{ii}$ on the self-energy σ is explicitly indicated. For a binary alloy, Eq. (2.22) takes the form,

$$c(\sigma - \varepsilon_A + \overline{G}_{ii}^{-1}(\sigma))^{-1} + (1-c)(\sigma - \varepsilon_B + \overline{G}_{ii}^{-1}(\sigma))^{-1} = \overline{G}_{ii}(\sigma). \tag{2.23}$$

The CPA self-consistency condition, Eq. (2.22), is a nonalgebraic equation for σ which in general must be solved by numerical means. Once σ has been determined, the DOS associated with an A or B atom as well as the total DOS are readily obtained from the corresponding Green's functions.

3. THE CPA FOR MODULATED ALLOYS (MODCPA)

In a compositionally modulated alloy, the concentration of a constituent depends on the position inside the sample. Consider a CMA with a wavelength of modulation, λ, which we take to be equal to an integral number of interplanar spacings in the direction of modulation. If c_J denotes the concentration of one of the constituents on a plane J perpendicular to the direction of modulation, then we have,

$$c_J = c_{J+m}, \tag{3.1}$$

where m is the number of planes included in λ. We wish to examine the effects on the DOS of CMA produced by three different factors: (i) increasing scattering strength, (ii) increasing modulation strength, and (iii) direction of modulation.

For a binary alloy describable by a TB Hamiltonian the scattering strength δ is defined as the ratio $|\varepsilon_A - \varepsilon_B|/w$, where $\varepsilon_{A(B)}$ is the site-energy associated with atoms of type A(B), and w is half the band width of either the pure A or pure B material. In random alloys, increased scattering strength may cause increased structure in the DOS including the creation of gaps. The CPA becomes exceedingly inexact in the strong scattering limit, and the behavior of any extension or generalization of the CPA in that limit is of particular interest. It should be noted that although it is inexact in the strong scattering limit, the CPA does provide a reliable interpolation scheme for the band structure of strong scattering alloys.

Modulation strength can be defined somewhat loosely as the measure of concentration variation along the direction of modulation. For a sinusoidal modulation of a binary CMA, for example, modulation strength can be defined as the ratio $(c_{max} - c_{min})/\lambda$, where c_{max} and c_{min} are the maximum and minimum concentrations respectively, of the A (or B) atoms on planes perpendicular to the direction of modulation. Modulation strength can be expected to play a crucial role in the determination of the DOS and the spectral-weight functions of CMA since the structure in these quantities is determined primarily by concentration fluctuations within rather compact clusters of atoms. Although independent effects of the amplitude and the wavelength of modulation on the DOS of CMA can be expected, modulation strength, which could cause strong concentration fluctuations in the local environment of a given site, is a relevant parameter for monitoring such effects.

Concentration fluctuations for CMA also depend on the ratio $r_J = n_J/n$, where n_J is the number of near neighbors of a site that are located in the same plane, J, as the site itself, and n is the coordination number of the lattice. The ratio r_J may be strongly dependent on the direction of modulation for some lattices.

In a tight binding formalism, a CMA can be described by the Hamiltonian of Eq. (2.1). The concentration of atoms of a given kind, say A, instead of being random, is now dependent on position; on plane J perpendicular to the direction of modulation, atoms of type A are distributed randomly with concentration c_{AJ}. A mathematically detailed generalization of the CPA to CMA has been given elsewhere (37). In the following, we present a somewhat more intuitive approach.

The development for homogeneous random systems following Eq. (2.1) can be carried over to modulated alloys in a straightforward way. In particular, one can easily show that the site-diagonal element of the Green's function operator for a site i in a plane J has the general form,

$$G_{ii}^J = (z - \varepsilon_i^J - \Delta_{ii}^J)^{-1}. \tag{3.2}$$

Charge transfer effects may cause the site energies ε_i to depend on the plane J containing site i. Such a dependence is easily incorporated into the theory through the replacement of ε_i by ε_i^J in Eq. (2.25). If the disordered material surrounding site i in plane J is replaced with an effective medium, to be described presently, then the Green's function for that site takes the form,

$$G^J = (z - \varepsilon^J - \overline{\Delta}^J)^{-1}. \tag{3.3}$$

Here, the subscript i is omitted to denote the equivalence of all sites in the same plane and to unburden our notation somewhat. The self-consistency conditions that determine the $\overline{\Delta}^J$, are direct generalizations of the CPA self-consistency condition, and are given by

$$<G^J>_J = \overline{G}^J, \text{ for all J in } \lambda. \tag{3.4}$$

Note the Eq. (3.4) represents a set of m equations where m is the number of planes included in λ. Also, $< \ldots >$ denotes a single-site average over the concentrations appropriate to plane J.

As is indicated in Eq. (3.4), there exists m renormalized interactors $\overline{\Delta}^J$, one for each of the m planes in a modulation wavelength. We can determine the $\overline{\Delta}^J$ in terms of the self-energy of the effective medium as follows. We assume that the self-energy consists of site diagonal quantities σ^J one for each plane in λ. The effective medium can be viewed as a collection of cells or clusters, C, each of which contains at least one of each kind of plane in λ. The equation of motion of the Green's function describing such a medium can be easily solved by FT and we obtain,

$$(\overline{G}_{CC}) = \frac{1}{\Omega_{BZ}} \int_{BZ} \overline{G}_{CC}(\vec{k}) \, d^3k$$

$$= \frac{1}{\Omega_{BZ}} \int_{BZ} [\overline{g}_C - W_C(\vec{k})]^{-1} \, d^3k, \tag{3.5}$$

where $\overline{G}_{CC}$ denotes the cluster diagonal part of the Green's function operator. The bare cluster locator $\overline{g}_C$ can be defined by the expression,

$$\overline{g}_C = (z - \overline{H}_C)^{-1}, \tag{3.6}$$

where

$$(\overline{H}_C)_{ij} = \sigma^J \delta_{ij} + W_{ij}, \qquad \begin{array}{l} i,j \text{ in cluster } C \\ i \text{ in plane } J \end{array} \tag{3.7}$$

is the intracluster effective medium Hamiltonian for cluster C, and σ^J is the self-energy associated with plane J. The FT of the intercluster matrix elements is defined by the expression

$$\tag{3.8}$$

$$W_C(\vec{k}) = \frac{1}{N_C} \sum_{(\vec{R}_0 - \vec{R}_C)} W_{OC} \, e^{-\vec{k} \cdot (\vec{R}_0 - \vec{R}_C)}$$

where N_C denotes the number of clusters in the material, $\vec{R}_C$ denotes the position of cluster C, and W_{OC} is the intercluster transfer matrix with matrix elements,

$$(W_{OC})_{ij} = W_{ij}, \; i \; \varepsilon \; 0, \; j \; \varepsilon \; C'. \tag{3.9}$$

The site-diagonal matrix elements of $\overline{G}_{CC}$ for sites on the various planes J are the quantities $\overline{G}^J$ shown in the self-consistency equations (3.4). In a manner analogous to that used in connection with the ordinary CPA, we may easily obtain the expressions,

$$<[\sigma^J - \epsilon^J + (\bar{G}^J(\{\sigma^K\}))^{-1}]^{-1}> = \bar{G}^J(\{\sigma^K\}) \tag{3.10}$$

where the dependence of each $\bar{G}^J$ on the full set of self-energies σ^K is explicitly indicated. The set of Eqs. (3.10) can be solved numerically to determine the self-energies, σ^J, and hence the DOS for each type of atom on each plane,

$$n_\alpha^J = -\frac{1}{\pi} \text{Im}<G^J{}_\alpha>. \qquad \alpha = A \text{ or } B \tag{3.11}$$

The DOS for each plane J is then,

$$n_\alpha^J = \sum_{\alpha = A,B} C_{\alpha J} n_\alpha^J, \tag{3.12}$$

and the total DOS is given by the expression,

$$n = \frac{1}{m} \sum_J n^J. \tag{3.13}$$

4. NUMERICAL RESULTS

The MODCPA was used to calculate DOS in one-dimensional and three-dimensional binary CMA describable by TB Hamiltonians, for different strengths and directions of modulated. In all cases the average concentration was kept at 0.50.

Figure 1 shows the density of states for a one-dimensional binary CMA with 3-site modulation (m = 3), with concentrations $C_{A1} = 0.3$, $C_{A2} = 0.5$ and $C_{A3} = 0.7$, and for the relatively strong scattering strength $\delta = 2.0$. The solid curve represents the DOS obtained in the MODCPA and the dashed curve is the CPA DOS for a random alloy with C = 0.5. The histogram represents exact DOS obtaine using eigenvalue counting methods (38). Figure 2 shows corresponding results for a much stronger modulation case, with $C_{A1} = 0.1$, $C_{A2} = 0.5$ and $C_{A3} = 0.9$. Comparison of the exact results in these figures shows that the DOS are dependent upon modulation strength. Increasing modulation amplitude causes the weaker peaks to collapse with the spectral weight being transferred to the three dominant peaks, but causes no observable shifts in the positions of these peaks. As expected, the single site CPA is structureless and gives only a poor representation of the exact DOS. In contrast, the MODCPA results are sensitive to the modulation strength and correspond remarkably well to the exact results in the case of strong modulation. This accuracy of the MODCPA for strongly modulated alloys can be understood in terms of the properties of the CPA itself. The CPA becomes exact in the limit as c approaches 0 or 1. Hence, the self-

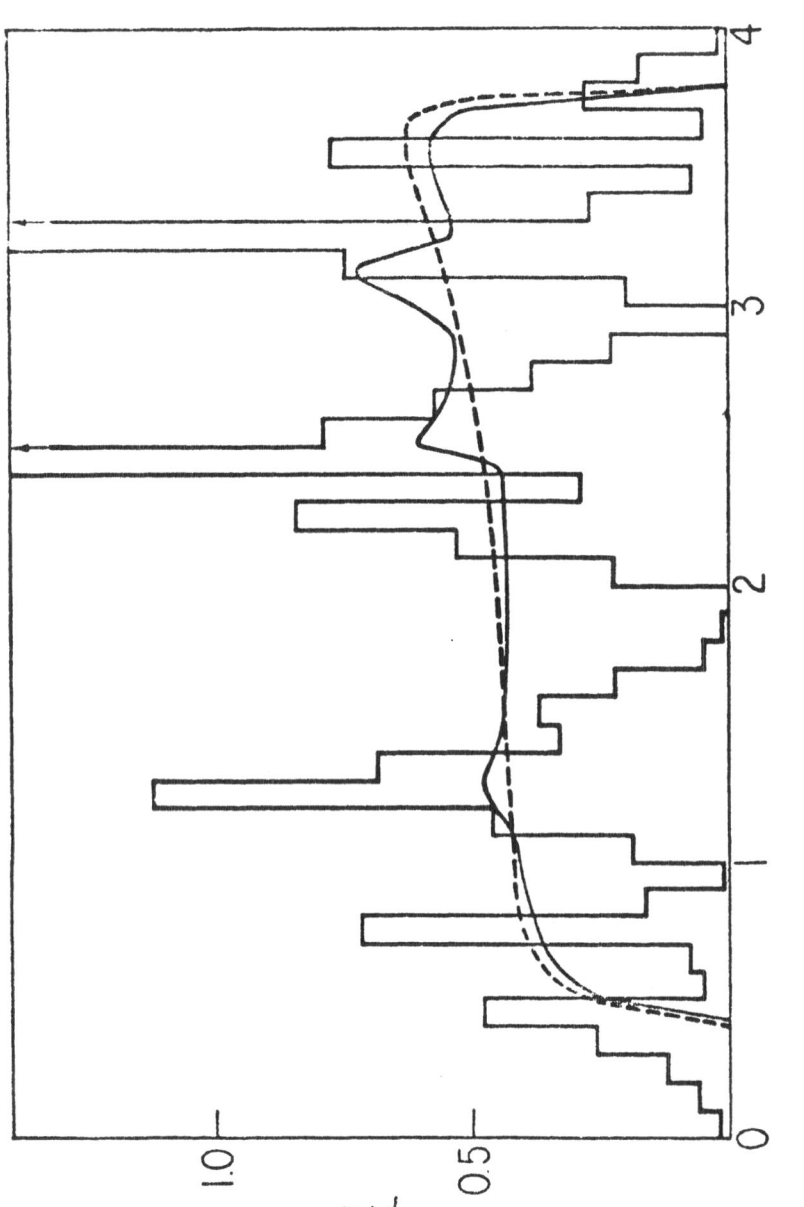

Fig. 1. Density of states for a one-dimensional binary CMA system AB, with $C_{A1} = 0.3$, $C_{A2} = 0.5$, and $C_{A3} = 0.7$ obtained with the generalization of the CPA reported in the text, compared with an exact density of states histogram. The CPA density of states for $C_A = 0.5$ is also shown for comparative purposes.

118

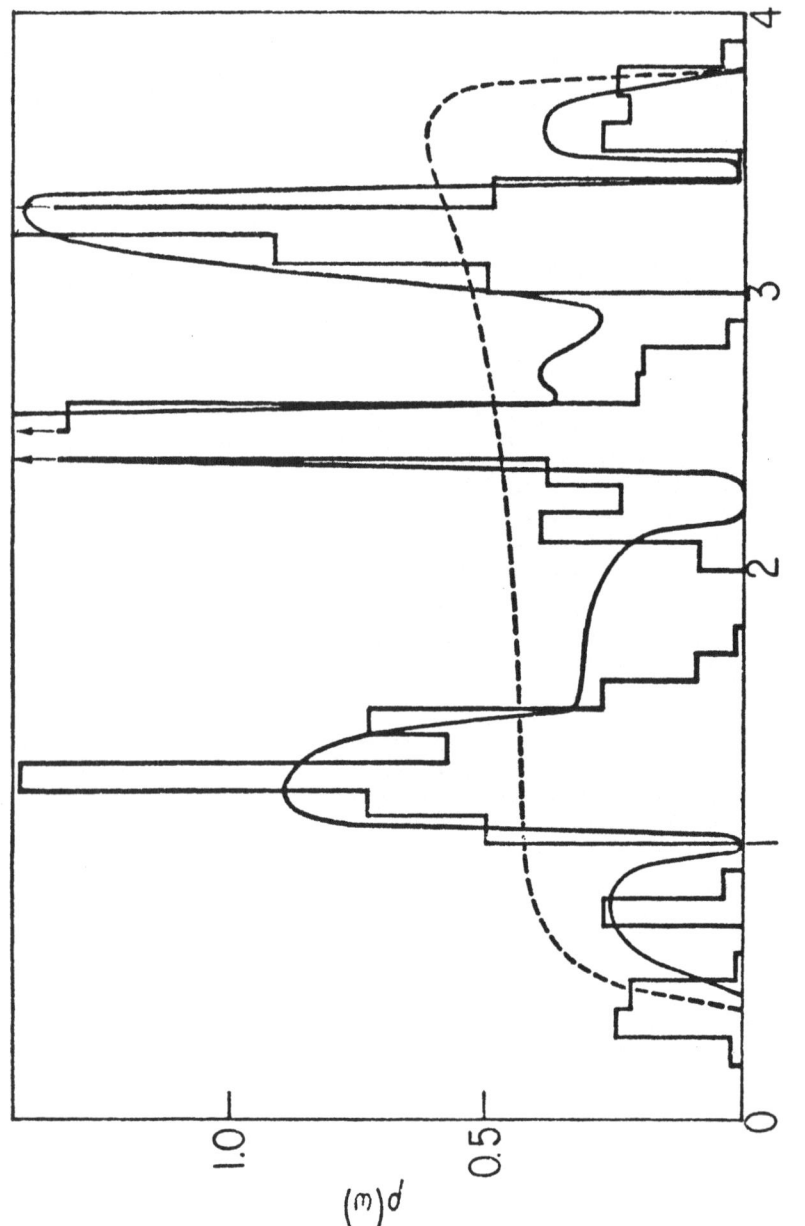

Fig. 2. Results of the type depicted in Fig. 1, but with $C_{A1} = 0.1$, $C_{A2} = 0.5$, and $C_{A3} = 0.9$.

energy on sites characterized by small or large values of C is given very accurately. Thus, the overall effect is a quite faithful representation of the exact DOS.

The DOS of modulated simple cubic alloys with 3-site modulation along the [111] direction are shown in Figs. 3 and 4, for the weak scattering strength $\delta=0.667$, but for both weak and strong modulation. In the weak modulation case, Figure 3, the concentrations were $C_{A1} = 0.3$, $C_{A2} = 0.5$ and $C_{A3} = 0.7$. The figure shows total and plane decomposed DOS: The solid curve represents the total DOS obtained in the MODCPA, whereas the dotted and dash-dotted curves represent the DOS associated with the planes with A-atom concentrations 0.3 and 0.7, respectively. The CPA DOS, dash curve, for a random alloy with C = 0.5 are also shown. Analogous results are shown in Fig. 4 for the strong modulation case, $C_{A1} = 0.1$, $C_{A2} = 0.5$ and $C_{A3} = 0.9$. Note the formation of a gap in the DOS for the strongly modulated alloy in Fig. 4. The ordinary CPA, which for weakly modulated alloys yields results comparable to those of the MODCPA, Fig. 3, fails to reproduce the gap in the DOS of the strongly modulated alloys, and yields an overall inaccurate representation of the DOS. Further numerical calculations (37) have shown that for cubic alloys modulated along the [100] direction, the ordinary CPA gives results comparable to those of the MODCPA for both weak and strong modulation, and a wide range of scattering strengths. Thus, both the strength and the direction of modulation can affect profoundly the DOS of CMA.

5. DISCUSSION AND CONCLUSIONS

We have presented a single-site theory which is a generalization of the coherent potential approximation to compositionally modulated alloys. This generalization preserves many of the desirable properties of the CPA, such as uniqueness and analyticity, and always yields physically meaningful results. The model numerical calculations presented here indicate that in general the MODCPA provides a rather accurate description of the DOS in CMA. These results also show that modulation strength and direction may have much more pronounced effects on the DOS of CMA than scattering strength. For realistic three-dimensional CMA, especially CMA of transition metals, modulation strength is expected to play a stronger role than the direction of modulation in determining the DOS since such alloys have predominantly an fcc structure for which the ratio r_J is not affected drastically by changes in direction.

The generalization of the CPA presented above in conjunction with TB systems can also be formulated (37) for systems describable by more realistic, muffin-tin Hamiltonians. Also, it can be used with cluster techniques (37) in order to incorporate the effects of statistical fluctuations in the local environment of a site.

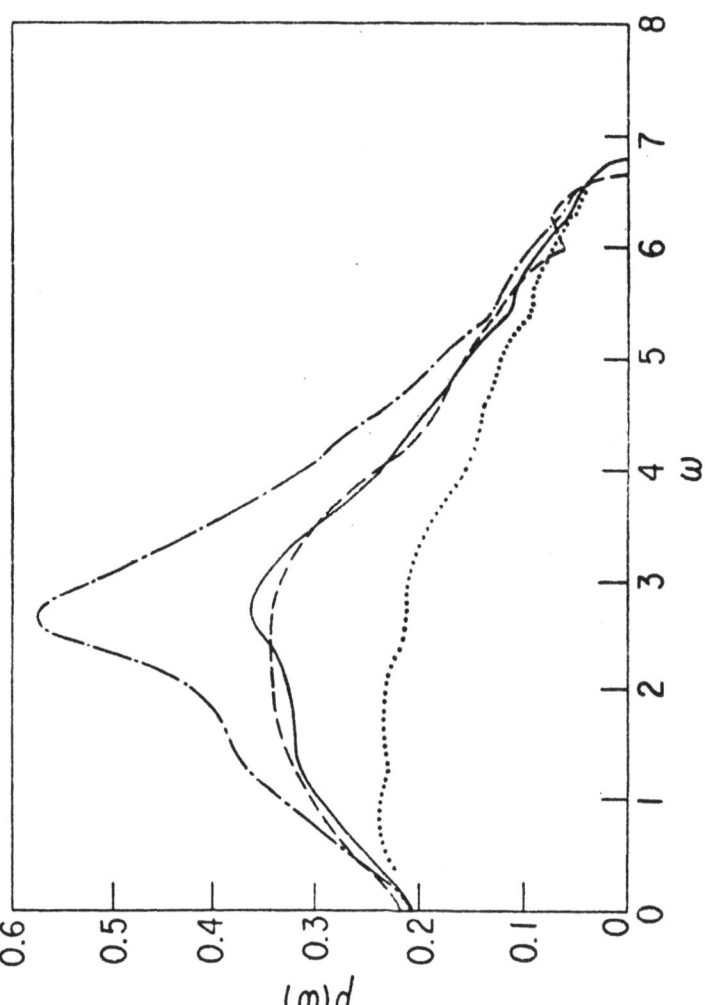

Fig. 3. Total and plane decomposed densities-of-states obtained within the MODCPA for a weak scattering, = 0.667 cubic alloy modulated along the [111] direction with C_{A1} = 0.3, C_{A2} = 0.5, and C_{A3} =
0.7. The solid curve corresponds to the total density of states obtained in the MODCPA, and the
dotted and dash-dotted curves to the densities of states on planes with A-atom concentrations 0.3 and
0.7, respectively. The CPA density of states C_A = 0.5, dash curve, is also shown. The plane density of
states for C_{A2} = 0.5 is very close to the total and CPA densities of states and is not shown.

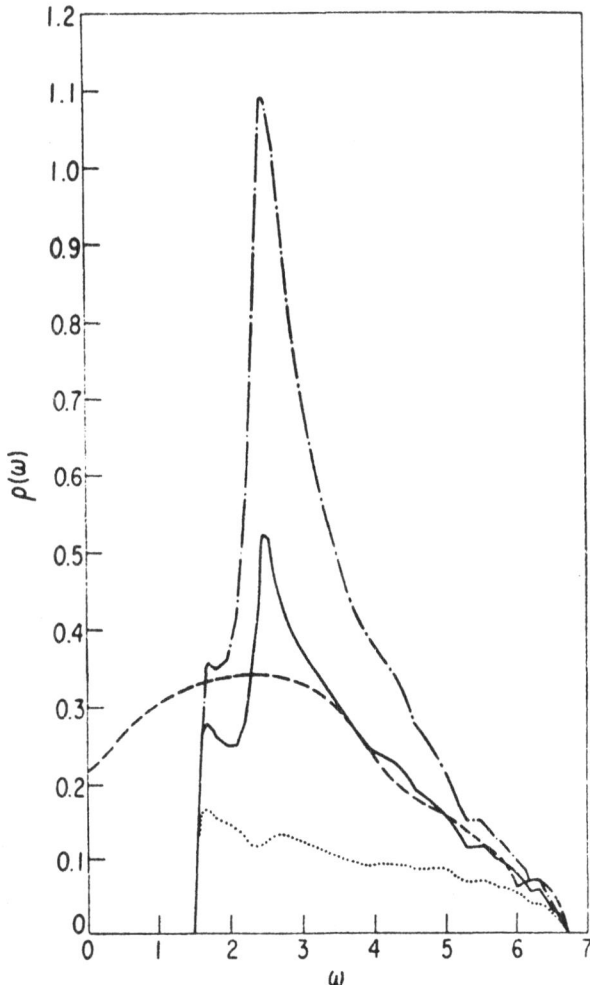

Fig. 4. Results of the type depicted in Fig. 3, but with C_{A1} = 0.1, C_{A2} = 0.5, and C_{A3} = 0.9.

122

REFERENCES

1. Schuller, I., C.M. Falco, J.E. Hilliard, J. Ketterson, B. Thaler, R. Lacoe, and R. Dee in Modulated Structures - 1979 (Kailua Kona, Hawaii), Proceedings of the International Conference on Modulated Structures, edited by J.M. Cowley, J.B. Cohen, M.B. Salamon, and B.J. Wuensch (American Institute of Physics, New York, 1979), p. 417.
2. Yang, W.M.C., T. Tsakalakos, and J.E. Hilliard, J. Appl. Phys. 48, 876 (1977).
3. Henein, G. and J.E. Hilliard, J. Appl. Phys. (in press).
4. Tsakalakos, T. and J.E. Hilliard, J. Appl. Phys. (in press).
5. Testardi, L.R., J.T. Krause, D.B. McWhan, and S. Nakahara, J. Appl. Phys. 52, 510 (1981).
6. Van der Ziel, J.P. and A.C. Gossard, Phys. Rev. B17, 765 (1978).
7. Gossard, A.C., in Thin Films: Preparation and Properties, edited by K.N. Tu and R. Rosenberg (Academic Press, New York).
8. Thaler, B.J., J.B. Ketterson, and J.E. Hilliard, Phys. Rev. Lett. 41, 336 (1978).
9. Gyorgy, E.M., J.F. Dillon, Jr., D.B. McWhan, L.W. Rupp, Jr., L.R. Testardi, and P.J. Flanders, Phys. Rev. Lett. 44, 57 (1980)
10. Zheng, J.Q., C.M. Falco, J.B. Ketterson, and I.K. Schuller, Appl. Phys. Lett. 38, 424 (1981).
11. Zhou, Wen-sheng, H.K. Wong, J.R. Owens-Bradley, and W.P. Halperin, Physica 108B + C, 953 (1981).
12. Flevaris, N.K., D. Baral, and J.E. Hilliard, in Metal Hydrides, edited by G. Bambakidis (Plenum, New York, 1981), pp. 301-312.
13. Flevaris, N.K., J.B. Ketterson, and J.E. Hilliard, Proceedings of the American Institute of Metallurgical Engineers Annual Meeting, 1980.
14. Brodsky, M.B. and A.J. Freeman, Phys. Rev. Lett. 45, 133 (1980).
15. Bulaevskii, L.N., Usp. Fiz. Nauk 116, 449 (1975) [Sov. Phys. Usp. 18, 514 (1976)].
16. Schuller, I.K. and C.M. Falco, Inhomogeneous Superconductors - 1979, (Berkeley Springs, West Virginia), Proceedings of the Conference on Inhomogeneous Superconductors, edited by D.U. Gubser, T.L. Francavilla, J.R. Leibowitz, and S.A. Wolf (American Institute of Physics, New York, 1979), p. 197.
17. Granquist, C.G. and T. Claeson, Solid Commun. 32, 531 (1979).
18. Zheng, J.Q. and J.B. Ketterson, Prog. Low Temp. Phys. (in press).
19. Schuller, I.K. and C.M. Falco, in Microstructure Science and Engineering, edited by N.G. Einspruch (Academic Press, New York, 1981), Vol. 4.
20. W.E. Pickett, J. Phys. F12, 2195 (1982).
21. Pindor, A.J., W.M. Temmerman, B.L. Gyorffy, and G.M. Stocks, J. Phys. F10, 2617 (1980).
22. Jarlborg T. and A.J. Freeman, Phys. Rev. Lett. 45, 653 (1980).
23. Caruthers E. and P.J. Lin-Chung, Phys. Rev. B17, 2705 (1978).

24. Andreoni, W., A. Baldereschi, and R. Car, Solid State Commun. 17, 821 (1978).
25. Schulman N. and T.C. McGill, Phys. Rev. B21, 6341 (1979).
26. Andreoni, W. and R. Car, Phys. Rev. B21, 3334 (1980).
27. Soven, P. Phys. Rev. 156, 809 (1967).
28. Taylor, D.W., Phys. Rev. 156, 1017 (1967).
29. Velicky, B., S. Kirkpatrick, and H. Ehrenreich, Phys. Rev. 171, 725 (1968).
30. Elliott, R.J., J.A. Krumhansl, and P.L. Leath, Rev. Mod. Phys. 46, 465 (1974).
31. Gyorffy, B.L. and G.M. Stocks, in Electrons in Disordered Metals and Metallic Surfaces, edited by P. Phariseau, B.L. Gyorffy and L. Scheire, (Plenum Press, New York, 1978) p. 89 and references therein.
32. Faulker, J.S., in Progress in Material Science, edited by J.W. Christian, P. Hassen and T.B. Massalski (Pergamon Press, New York, 1982), Numbers 1 and 2.
33. Winter, H. and G.M. Stocks, Phys. Rev. B27, 882 (1983).
34. Blackman, J.A., D.M. Esterling, and N.F. Berk, Phys. Rev. B4, 2412 (1971).
35. Gonis, A. and J.W. Garland, Phys. Rev. B16, 1495 (1977).
36. Gonis, A. and J.W. Garland, Phys. Rev. B16, 2424 (1977).
37. Gonis, A. and N.K. Flevaris, Solid State Commun. 37, 595 (1981); Phys. Rev. B25, 7544 (1982).
38. Dean, P., Rev. Mod. Phys. 44, 127 (1972).

TIME EVOLUTION OF PHASE SEPARATION IN BINARY MIXTURES[*]

J. Marro** and M.H. Kalos***

**Departamento de Física Teórica, Universidad de Barcelona, Diagonal 647, Barcelona-28, Sapin, ***Courant Institute of Mathematical Sciences, New York University, New York 10012

ABSTRACT

We describe some general features of the process of phase separation in binary mixtures which seem to characterize the behavior of many real and model systems.

1   INTRODUCTION

Binary mixtures suddenly cooled from a homogeneous state in the one-phase region to a temperature inside the miscibility gap undergo a process of phase separation [1] which may influence various physical properties of the sample such as hardness, flexibility, resistivity, etc. [2].

The process, variously described as nucleation, spinodal decomposition, coarsening, Ostwald ripening, Smoluchosky coagulation, etc., can be observed in alloys such as Al-Zn, Fe-Cr, Au-Pt or Cu-Ni, liquid mixtures such as water with lutidine, water with isobutyric acid or the quantum liquid mixture ^{3}He-^{4}He, (quasi binary) glasses such as B_2O_3-Pb O -(Al_2O_3), protein solutions similar to the ones in the human eye, etc.; see Refs. [3,4] for a bibliography. Those observations use electron or field ion microscopy to determine the properties of the grains or droplets in the precipitate, scattering of X-rays light or neutrons in order to monitor the structure function $S(k,t)$ at time t after quenching, or indirect methods like resistivity or calorimetry; see Refs. [3,5] . The process of phase separation can also be simulated on a computer by using a ferromagnetic Ising or binary alloy model with Kawasaki dynamics [6,8].

[*] Work supported by NSF Grant DMR81-14726-01 and DOE Grant AC02-76ER03077.

Recent analyses have shown that the behavior of segregating binary alloys, fluid mixtures, glasses and other materials and the behavior of a ferromagnetic kinetic Ising model are very close to each other. Some other related models seem to exhibit a behavior similar in many aspects. We briefly describe in this note a number of general features which seem to characterize the time evolution of phase separation in all the above systems.

2 CAHN'S THEORY

Theory, experiments and computer simulations very often refer to the structure function defined alternatively as the Fourier transform of the composition correlation function of the mixture or as the intensity observed in a scattering experiment. Let $S(k,t)$ be the sphericallized structure function normalized to unity in the first Brillouin zone [8].

The behavior of $S(k,t)$ just after quenching can be described by means of a simple equation first proposed by Cahn and analysed by Hilliard and Cook [1,9]. This gives the time derivative of the structure function as

$$S(k,t) = -2Mk^2[\Omega S(k,t) - 1] \tag{1.a}$$

where $M > 0$ is a mobility and

$$\Omega = A + B k^2 \tag{1.b}$$

Here A and $B > 0$ are two parameters related to the properties of the local free-energy density of the system in the homogeneous state and to the inhomogeneity corrections respectively [1,6].

It was shown (9) that Eq. (1) is formally consistent with the computer simulation data corresponding to "deep" quenches (e.g. $T = 0.6T_c$, $\varrho \gtrsim 0.2$ and $\varrho = 0.5$, $T \lesssim 0.8T_c$ where ϱ is the density of the minority species in the mixture and T_c is the critical temperature) at short times, $t < t_C = B(1+A)/MA^2$, and when one sets a constant value for B (=0.2), independent of T and ϱ. The fit of Eq. (1) to the data produces values for A, which are consistent with the expected tail, say $S_{coex} \sim (\varrho^2 + k^2)^{-1}$ (see section 3), and values for M which set the appropriate time scale. The parameter t_C corresponds to an observable time, e.g. of the order of hours at $T = 0.6T_c$ in the case of a real alloy.

There is also some agreement between Eq. (1) using the same value for B and the computer simulation data in the case of "shallow" quenches (which are close to the coexistence line, e.g. $T = 0.6T_c$, $\varrho \lesssim 0.075$ and $\varrho = 0.5$, $T \gtrsim 0.9T_c$); nevertheless, the fit is here definetely worse [9]. In either case, Eq. (1) predicts a growth of

$S(k,t)$ much faster than observed in experiments for times $t > t_C$.

3 DYNAMICAL SCALING

Let $S_{eq}(k)$ be the equilibrium value of $S(k,t)$ when the (macroscopic) system is fully segregated into two pure phases,

$$S_{eq}(k) = (m_o^2 - m^2)\ \overline{\delta}(k) + S_{coex}(k:T). \tag{2}$$

Here $m = 1-2\rho$, m_o equals m at the coexistence line, $\overline{\delta}(k)$ is the sphericallized delta function at $k = 0$ corresponding to the macroscopic droplets rich in the minority species at equilibrium, and $S_{coex}(k;T)$ is the equilibrium structure function on the coexistence line at the temperature of quenching. The latter, say with an Ornstein-Zernike form $S_{coex} \sim (\xi^2 + k^2)^{-1}$, corresponds to the "vapor" phase rich in the minority species; this is expected to be relatively small at low enough temperatures.

Let $S_1(k,t) = S(k,t) - S_{coex}(k;T)$ with goes to $(m_o^2 - m^2)$ $\overline{\delta}(k)$ as $t \to \infty$. Recent analysis have shown that $S_1(k,t)$ scales with time after some transient time, say $t > t_s (> t_C)$ [8] . More precisely, $S_1(k,t) = [R(t)]^3\ F(x;t)$ with $F(x;t) \simeq F(x)$ a smooth function of x independent of time for $t > t_s$:

$$S_1(k,t) = [R_i(t)]^3\ F_i(x_i), \qquad t > t_s . \tag{3}$$

Here $x_i = k\ R_i(t)$ and $R_i(t)$ is some characteristic scaling length in the system to which one may associate a scaling function F_i .

In order to include the above facts into description (1a) one may write

$$\Omega(k,t) = [1 + S_1(k,t)\rho(k,t)]\ S(k,t)^{-1} \tag{4}$$

with $\rho(k,t)$ the undetermined function (which for $t < t_C$ is defined by comparison between Eqs. (1b) and (4)). It then follows that

$$S_1(k,t)\ \exp[-2Mk^2 \int_o^t \rho(k,t)dt] . \tag{5}$$

This is a very convenient equation with which to discuss assumptions on Ω (which can be associated with an effective drift causing the interdiffusion of species) leading to the condition (3) of dynamical scaling, in the spirit of some recent approaches [10].

4   SCALING LENGTHS

The most convenient scaling length when analysing simulations
data with reference to Eq. (3) is the inverse of $k_1(t)$, the first
moment of $S_1(k,t)$. $R_1(t) = 1/k_1(t)$ defines $F_1(x)$ whose shape
has been analysed in Refs. (4,8). It seems that $\bar{F}_1(x)$ corresponding to deep quenches (see section 2) presents a shape slightly different than in the case of shallow quenches. The similarity of
$F_1(x)$ near the critical temperature T_c and near the coexistence
curve for small ς would appear to suggest some sort of "spinodal
line" criticality.

Assuming $R_1(t) \sim t^a$, the exponent a shows a slow tendency
to increase with time. The values for a thus obtained, interpreted
according to familiar grain dynamics approaches (6,7), suggest that
the effective diffusion and coagulation of the grains is the predominant mechanism in the case of deep quenches and also during the
early stages of the evolution at almost any location inside the coexistence line while single atom processes tend to dominate when the
quench is close to the coexistence line late in the evolution. As a
matter of fact, the k_1 values corresponding to all the phase points
seem to follow after time t_s the Lifshitz-Slyozov prediction that
R^3 evolves linearly in time.

The length $R_2(t) = 1/k_{max}(t)$, where $k_{max}(t)$ represents the
location of the maximum scattered intensity, seems to lead, within
experimental errors, to a scaling function $F_2(x)$ quite similar to
$F_1(x)$ (8,4) . This suggests that $k_1(t) \sim k_{max}(t)$ as is also confirmed in some experiments.

The scaling regime can be extended even to earlier times using
the Guinier radius $R_G(t)$ as the basic scaling length. This is defined from a plot $-\ln S_1(k,t)$ versus k^2 for intermediate values
of k (5) and setting $R_G^2/5$ equal to its slope. The scaling function $F_3(x)$ can then be deduced from (3), along with the proporcio-
nality between R_G , R_1 and R_2 which are in turn in good agreement
with the droplets sizes determined directly (7). Furthermore the
Guinier radius, which is identified as 5/3 times the radius of gyra-
tion of the droplet (i.e. the geometrical radius in the case of a
spherical droplet) gives a good description of the clusters even before the scaling regime sets in and even in cases where the approximations usually made to justify in the Guinier model are not valid.

5   INTERPRETATION OF F(x)

The scaling lengths R_G , R_1 and R_2 may not determined accurately in this way because not all the information contained in
the curve $S_1(k,t)$ against k is used. In order to get the best
possible statistics, a graphical method, using the whole curve
$S_1(k,t)$ against k , is presented in Ref. (5) which provides the

mean droplet size and density as well as the scaling function $F(x)$. When F is normalized to F_3 it appears that both the computer simulations and some real systems satisfy

$$F(x) = \Phi(x) \cdot \Psi[x \cdot \delta(\rho, T)] \tag{6}$$

where $x = k \cdot R_G(t)$.

In analogy to classical two phase models and molecular liquids, Ψ may be interpreted as a droplet interference function and Φ as a single droplet function. The parameter δ seems to indicate how deep inside the miscibility gap the experiment is performed. At high values of k , $\Psi \equiv 1$ and F is determined uniquely by the Guinier radius $R_G(t)$. The function Φ is universal, independent of density, temperature, and even the substance investigated. This universal function behaves as $\Phi(x) \sim 1/x^4$ for large x .

REFERENCES

1. For early reviews see J.W. Cahn. Trans. Metall. Soc. AIME, 242,166(1968); J.E. Hilliard, in Phase Transformations (edited by H.I.Aaronson), p.497, Am. Soc. for Metals, Metals Park, Ohio (1970).
2. B. Ditcheck and L.H. Schwartz. Ann. Rev. Mater-Sci. 9,219(1979).
3. W.I.Goldburg, in Scatt. Tech. Applied to Supramolecular and Nonequilibrium Systems (edited by S. Chen, B. Chu and R. Nossal). p.383, Plenum Publ. Co., New York (1981), and references therein.
4. J. Lebowitz, J. Marro and M. Kalos. Comm. on Solid State Physics 10,201(1983), and references therein.
5. P. Fratzl, J. Lebowitz, J. Marro and M. Kalos, Acta Metall., to be published, and references therein.
6. J. Marro, A. Bortz, M. Kalos and J. Lebowitz. Phys. Rev. B12, 2000(1975); K. Binder, M. Kalos, J. Lebowitz and J. Marro. Adv. Colloid. Interface Sci. 10,173(1979).
7. O. Penrose, J. Lebowitz, J. Marro, M. Kalos and A. Sur. J. Stat. Phys. 19,243(1978); O. Penrose et al., to be published.
8. J. Marro, J. Lebowitz and M. Kalos, Phys. Rev. Lett. 43,282 (1979); Acta Metall. 30,297(1982).
9. J. Marro and J.L. Vallés. Phys. Lett. 95A,443(1983).
10. See for instance H. Furukawa. Preprint (January 1983).

CHAPTER 2: CRYSTALLOGRAPHY OF

MODULATED STRUCTURES

THE INCOMMENSURATE CRYSTALLINE PHASE AND ITS SYMMETRY

P.M. de Wolff

Lab. v. Techn. Natuurkunde, Postbus 5046, 2600 GA
Delft, Netherlands

1 MODULATED STRUCTURES

The term "incommensurate structure" has become the general
name of a certain class of modulated structures. Strictly speaking
it is an inappropriate indication, since "incommensurate" (in the
sense of incompatible, or having no common measure) clearly should
refer to at least two things and not to a single object such as a
crystal structure. The two things, in this case, are the two
elements of a modulated structure as defined, for instance, by
Cowley (1): the basic structure on the one hand, and the periodic
perturbation of this basic structure resulting in the modulated
structure on the other hand. Hence the proper, though inconvenient
term is: incommensurably modulated structures. We shall however
stick to the incorrect but short term used in the title of this
paper.

The term "modulated structure" itself cannot be taken for
granted either. As shown by the above reference, no sharp delimi-
tation of this concept was given at the conference in 1979, nor has
any been accepted since then. For our present purpose -and quite
apart from the question of incommensurability- we need a much
narrower definition than the one given by Cowley. The two elements,
however, have to be retained, and we shall treat them separately:

(i) the basic structure. We shall use this term in the sense
of a normal crystal structure, called BS, used as a reference
structure for the description of the modulated crystal. The
structure BS may or may not also exist as the structure of a non-
modulated phase of the same compound. It is infinitely large in all
directions and free of defects. Only thermal motion can be incor-

porated in the form of a spatial smearing, as usual in the treatment of this effect in structure analysis. However for most purposes it suffices to treat BS as a structure of point atoms.

This definition, together with the next (given below) precludes the use of "basic structure" for a structure recognizable in domains, but not coherent throughout the modulated crystal. Examples: many interface modulated structures as discussed by Amelinckx (2) or the complex superstructures of sulfide minerals described by Wuensch (3)

(ii) The modulation, or periodic perturbation. Consider an atom of the basic structure situated at (X, Y, Z) with respect to some fixed origin and coordinate basis. In general the axes of BS will be used, though it must be stressed that XYZ are not fractional coordinates but depend also upon the position of the cell of BS in which the atoms lie. The atom is characterized, moreover, by a scalar P, for instance its weight or its scattering power. We define the modulation as an effect changing position and/or weight of the atom:

$$X \rightarrow X+u \quad Y \rightarrow Y+v \quad Z \rightarrow Z+w \quad P \rightarrow P+p,$$

in a manner depending on the kind of atom. This "kind" is not only the chemical nature, but it also differentiates between symmetry-related atoms within a primitive unit cell of BS. So if $i = 1 \ldots N$ numbers all the atoms within such a cell, there are N kinds of atoms; atoms of the same kind are always translation-equivalent in BS.

The definition of modulation is completed now by stating that the value of u, v etc. for each atom of the i-th kind is given by finite, periodic (not necessarily harmonic) functions of the position (XYZ) of that atom in BS:

$$u = u_i (X,Y,Z) \qquad v = v_i (X,Y,Z) \tag{1}$$

such that all these functions (N for u,v,w and p each) share the same periods in X,Y,Z.

2 TYPES OF MODULATION IN DIRECT SPACE

A very simple example is shown in Fig. 1. It is a case of dis-placive modulation $(p=0)$ of O-atoms in a hypothetical two-dimensional compound XO, in which the X-atoms are not modulated at all. So if $X=1$ and $O=2$ (the primitive unit cell of BS contains just these two atoms) we have here $u_1 = v_1 = v_2 = p_1 = p_2 = 0$, and, if $u_2(Y)$ happens to be harmonic, $u_2(Y) = \ell \cos (2\pi Y/\lambda + \theta)$ where ℓ and θ are amplitude and phase parameters belonging to the coordinate u_2. The function $u_2(Y)$ can be plotted in this case as a locus of displaced positions. In Fig. 2, a component v_2 has been added, and it is seen that the locus wave line now would be ambiguous. One actually needs the complete vector function in order to find the displacement (u_2, v_2) for a given O-atom.

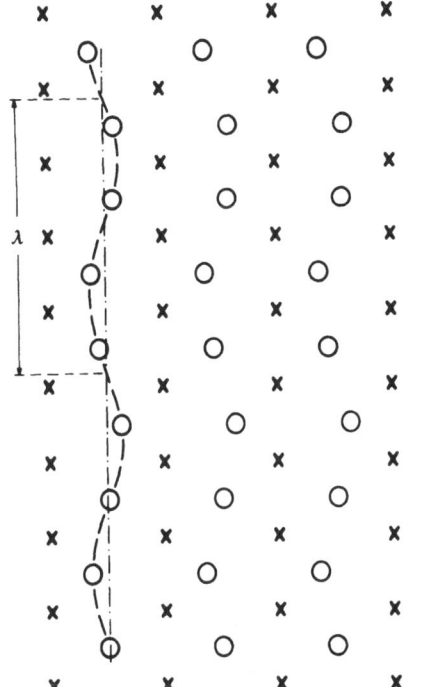

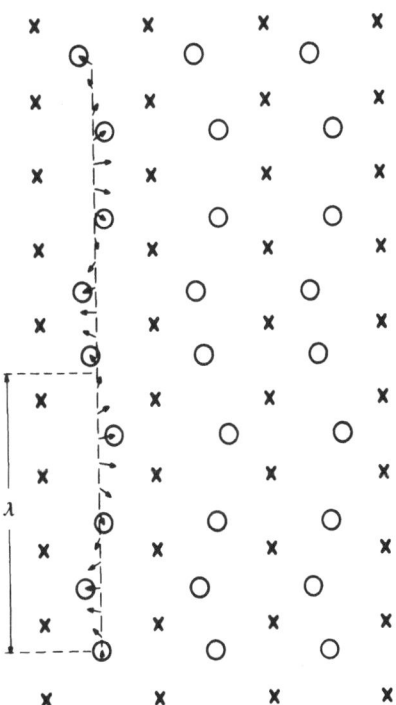

Fig. 1. Example of displacive modulation with only transverse displacements. Basic structure: Two-dimensional compound XO, orthogonal basis a (horizontal), b(vertical). Modulation: only of O-atoms. Wave length λ.

Fig. 2. Example of general displacive modulation. Data as in Fig. 1.

While it is not difficult to visualize these displacive modulations, a modulation of the weight P (scalar modulation) is, at first sight, quite a mystery. Indeed it can happen only in a statistical manner for partially occupied positions in BS. For instance in the basic structure of Figs. 1 and 2 it could be that the chance of an O-atom position to be actually occupied depends periodically upon the Y-coordinate of that position. Then the average occupation density of a statistical ensemble will actually show the kind of wave-pattern represented in Fig. 3. Scalar modulation thus applies only to statistical averages. However artificial this kind of modulation may appear, it is physically relevant and by no means rare (though seldom occurring without additional displacements). A well-known example is the modulated phase of $NaNO_2$ between 162 and $164^{\circ}C$.

136

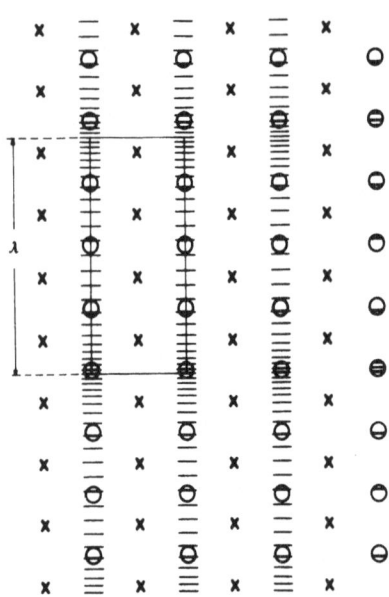

Fig. 3. Example of scalar (occupation density-)modulation.
Data as in Fig. 1.

When considering symmetry, scalar modulation is much simpler
than the displacive type, since u, v and w are vector components
rather than scalars. Therefore it should be mentioned that a scalar
interpretation of displacive modulation is possible. To this end,
the unit cell of BS is subdivided in volume elements, small enough
to consider the density of matter in each element as constant. (The
model of point atoms is replaced here by a continuous electron
distribution; this allows even non-spherical symmetry of atoms).
Each element can now be imagined to take over the role of an atom
in the former picture, only its position is constant. Hence any
type of modulation for the atoms becomes a purely scalar density
modulation for the volume elements, with as many functions p(XYZ)
as there are elements in the BS-unit cell. This is illustrated
schematically in Fig. 4 for a purely displacive modulation of one-
dimensional atoms.

In contrast to the statistical model discussed above, the
last mentioned interpretation is indeed artificial, and it is
useless, for e.g., solving a displacively modulated structure.
However it shows that any general discussion of modulation can be
modelled on a scalar (density-)modulation type without loss of
generality.

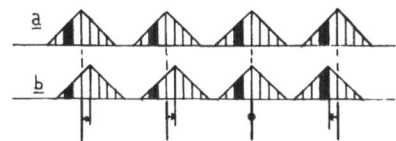

Fig. 4. Representation of a displacive modulation by a scalar one.
a: Basic structure consisting of equidistant atoms on a line.
b: Modulated structure. For each volume element, such as the
hatched one, the modulation is of the scalar type.

3 MODULATION IN RECIPROCAL SPACE

Modulation is easily detected by looking at diffraction
patterns obtained by using either X-rays, neutrons or electrons.
The characteristic features of the pattern are: strong main
reflections corresponding to the basic structure, and "satellites":
weak spots arranged in a regular fashion between the main reflections. These features follow at once from the fact that

- the pattern is essentially a Fourier transform of the structure.
- the structure, according to the scalar model introduced above,
  can be considered as the product of the basic structure BS
  multiplied by a weighing function p(XYZ) which is periodic in
  space, or as the sum of parts of BS each multiplied by such a
  function.
- the Fourier transform of a product is the convolution of those of
  the factors
- hence if the basic structure has diffraction maxima at points $\vec{H}$
  in reciprocal space, and p(XYZ) on its own would produce maxima
  at $\vec{H}_p$, then the modulated structure has maxima at points
  belonging to the set $\vec{H}+\vec{H}_p$.

The last conclusion can be illustrated by the structure of
Fig. 3. Its basic structure has diffraction maxima at the points
$\vec{H}=ha^* + kb^*$ in reciprocal space (two-dimensional; a* = 1/a, b* =
1/b; h and k integer). The modulation function has a transform
consisting of points at $\vec{H}_p =m\beta$ b* (m = integer), if b=βλ. Hence
the maxima for the modulated structure lie at ha*+(k+mβ)b*, with
m=o for the main reflections. As shown in section 2, this conclusion is independent of the type of modulation; it is therefore
equally valid for the pure displacive cases shown in Figs. 1 and 2.

So the displacive and the substitutional (scalar) types of
modulation are indistinguishable in reciprocal space when only the
position of reflections is taken into account. The intensities of
reflections can give a clue, since those of the satellites decrease

when H→o if the type is displacive. In practice, however, this feature is often unobtrusive; only a full structure analysis gives certainty.

On the other hand, diffraction measurements can yield very precise values of the modulation period(s), such as λ (or $\beta=b/\lambda$) in the above example. Thereby they provide a direct answer to the question of commensurability of the structure.

4 COMMENSURATE MODULATION: SUPERSTRUCTURES

As a definition of commensurability we assume that the modulation parameter, β in the above example, is equal to a simple rational fraction. Moreover the rational value of β should be valid in at least a finite interval of temperature and pressure. For instance, let $\beta=1/4$, which means $\lambda=4b$. Several conclusions can be drawn:

(i) The modulated structure is a normal periodic crystal (be it a two-dimensional one in our example), with a unit cell ax4b. (The cell ax4b is called a supercell and the structure is called a superstructure. That term is sometimes used to indicate the presence of modulation generally, but we shall use it only for the commensurate case). The period b of the basic structure can at most represent an approximate period of the actual structure.

(ii) Since the modulation functions (1), represented graphically in Figs. 1-3, are now sampled only in points repeating at a distance of $b=\lambda/4$, each function can be replaced by a set of just 4 numbers. Thus the concept of a modulation function has no real significance for the commensurate case.

(iii) Satellite reflections cannot be indexed uniquely. For instance, the satellite at $ha^* +(k+m\beta)b^*$, abbreviated as (h, k, m), for m=2 becomes (h,k,2), and for $\beta=1/4$ it clearly coïncides with (h,k+1,-2) as well as with (h,k+2,-6) etc.

It follows from (i) that the description of the superstructure as a modulated one is irrelevant for its exact symmetry, however useful that description may be for understanding its properties or its behaviour at phase transitions.

5 INCOMMENSURATE MODULATION

For the incommensurate case, all statements in the last section must be reversed. Still one should be careful when interpreting the reversed meaning. We shall begin by considering the rationality of β - to stay with our two-dimensional example -

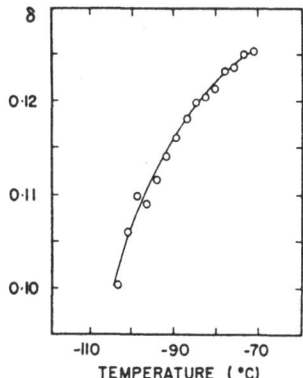

Fig. 5. Early measurements (4) of the modulation parameter δ of thiourea as a function of temperature.

which, when reversed, would mean that we now consider cases in which β has an irrational value. Clearly no such conclusion can ever be drawn from a straightforward measurement of β, resulting in a number with a known and finite statistical error.

A more practical indication is the word "simple" used in section 4: the parameter β should be a simple rational number. In reversing that definition, we arrive at the condition that "complex fractions" for β allow us to classify the crystal as incommensurate? Again there is no criterion to make the distinction. The measured value of β can always be represented as a fraction u/v. Within the error limit, there is an infinite number of such fractions but it is possible to find the simplest, that is, the fraction with the smallest denominator v. Even so, what value of v is acceptable as simple? For some time, many people thought that superstructures would not have a multiplicity (which is what v really means) above 5.

This is where the second part of the definition comes in: a superstructure should have a constant value of β (or any other modulation parameter). Measurements of β as a function of T and/or p should, therefore, show at least a finite plateau at the value of β supposed to correspond to a superstructure. If one finds such a plateau in a certain interval of T or p, then the existence of the superstructure is well established for that interval, and the plateau value of β, expressed as the simplest possible fraction in the above sense, determines the character of the modulation (including the multiplicity v). Outside the plateau interval, the measured values of β will appear to obey a continuous function of T or p, and in these regions the crystal can be said to be incommensurate.

How careful one must be in stating this is shown by the experimental results for thiourea. This structure, like our examples of Figs. 1...3, is modulated in just one direction; so there is just one modulation parameter, say δ. Thiourea is one of the first compounds for which a displacive modulation has been discovered. Fig. 5 shows a plot of δ against T according to the earliest measurements (4). One might very plausibly assume that these represent a smooth function. Recent measurements by Denoyer et al. (5) have shown, however, that there is a small plateau at the level of $\delta = 1/9$ signifying that the continuously changing incommensurate phase is interrupted by a superstructure with a 9-fold supercell. (These and other phase transitions in thiourea are also discussed in R. Currat's contribution to the present volume).

The fact remains that in thiourea as well as in many other compounds, very precise measurements indicate a continuous dependence of the modulation parameter upon temperature within finite, often very large intervals of T. By that evidence, incommensurate modulation must be accepted as a new type of crystalline matter. We shall now discuss the consequences of the inversion of the statements (i)-(iii) from the last section, in order to find the main characteristics of this new type.

6   DETERMINACY OF MODULATION PARAMETERS; Q-EQUIVALENCE

In relation to statement (iii), it might seem that incommensurate structures will yield no problem in the indexing of satellite reflections since there will be no coincidences like those mentioned under (iii). Indeed there is, in practice, no problem - and this is remarkable since in theory there should be a new difficulty far greater than the former. For instance, for a (virtually) irrational β, the positions of satellites given in our example by $k+m\beta$ in the b-direction would yield a crowd of points with unlimited density if k and m could both assume all integer values - instead of the very small range of m observed actually.

Thus the mere fact that β can be determined at all, resulting in unambiguous satellite indices m, is a very fundamental feature of incommensurate structures. In direct space, it means that the modulation functions (1) are not only continuous but actually quite smooth. Then their Fourier components vanish rapidly for increasing order, and this is the necessary condition for satellites also to vanish when their order m increases.

A very trivial ambiguity exists for β, because if β_0 is a value allowing complete indexing of all reflections then clearly $1-\beta_0$ does the same. Generally speaking, a "modulation vector" $\vec{q}$ can be defined as a basis vector of the reciprocal lattice consisting of points $\vec{H}_p$ in reciprocal space (the Fourier transform of the modulation pattern only, disregarding the basic structure, which was defined in section 3). Then the ambiguity meant here is caused by the fact that any reciprocal lattice vector $\vec{H}$ of the basic structure can be added to $\vec{q}$ while conserving integer satellite indices. Also $\vec{q}$ can be replaced by $-\vec{q}$, so the most general form is $\pm\vec{q}+\vec{H}$. Symmetry groups for different choices among this set can differ widely from each other but are considered equivalent. In order to characterize that situation one may use the term "q-equivalence" for such different descriptions of the same symmetry.

7 NEW SYMMETRY OPERATIONS IN INCOMMENSURATE CRYSTALS

Symmetry operations of an object are movements which bring the object into the coïncidence with itself. For a given object they form a group. The set of these symmetry groups depends upon what one allows as a "movement", and several extensions of that concept have in the past led to new kinds of symmetry groups. A classical example is the admission of improper operations such as reflection or inversion, which almost a century ago led from Sohncke's 65 groups to the 230 space groups. Since then the latter have been recognized as the complete set for three-dimensional crystals, because the distance-preserving operations in space were exhausted.

When considered in this light, incommensurate modulation causes a loss of symmetry even worse than that discussed in section 4 for superstructures. In the example with $\beta = \frac{1}{4}$, at least the symmetry translations 4b, 8b.... were conserved; but if β is irrational, none at all are left in the direction of $\vec{b}$. This severe conclusion seems inescapable, in spite of the fact that pictures like Figs. 1-3 suggest a perfect order parallel to the Y-axis.

However a totally different result, viz. conservation of all symmetry translations of BS, can be obtained by a suitable exten-sion of the allowed "movements." It is based upon the modulation functions (1) which in the incommensurate case are of course fully relevant (instead of degenerating into a few numbers, as for superstructures). The new movement is described most clearly for a one-dimensional crystal, or for a vertical row in Fig. 3. It consists of a movement of the modulation function as a whole with respect to the BS-atoms. In fig. 6a this is illustrated in reverse: the function p(Y), shown by a variable hatching, is left where it is, but the atoms (open circles) are shifted to the dotted posi-tions. Now if β is irrational, any shift of arbitrary amount of course changes the local value of p for a given atom - but the same sequence of p-values will turn up elsewhere, within any specified degree of accuracy. Since we assumed BS to be infinitely large, this means that the new movement is in fact a distance-preserving operation entirely comparable to the classical movements like translation, rotation or reflection of the whole crystal. Therefore it can freely be combined with the latter to produce symmetry operations as defined in the start of this section.

The application to symmetry translations of BS is straight-forward. Such a translation nb, n=integer, when applied to the modulated structure, brings the atoms into coïncidence but does not reproduce the value of p for any given atom, since p(Y) has moved by nb as well. However it suffices to shift p(Y) backwards over nb (the new movement!) in order to achieve complete coïn-cidence, for any integer value of n. Therefore nb is a symmetry translation - in the above sense - for the incommensurate structure as well.

Applications of these new symmetry operations leads to an equally new set of symmetry groups. Many of these are found to occur in actual modulated structures. Indeed, if a structure cannot be classified in this way it is because of lack of experimental data rather than by shortcomings of the classification, since the latter is exhaustive. The actual occurrence of the new groups merely corroborates the validity of the starting point - the equation (1) - which is demonstrated independently by the general features of diffraction patterns of incommensurate structures. Different ways to express their symmetry will now be discussed.

8. COMPARISON OF SYMMETRY DESCRIPTIONS

8.1. Superspace Groups

This description is shown schematically in Fig. 6b for the simple one-dimensional crystal depicted in Fig. 6a. The principle consists in adding an extra dimension, and using it in such a way that the actual modulated crystal appears as a section through a periodic structure, the "supercrystal." In Fig. 6b the supercrystal is the two-dimensionally periodic pattern with the dashed parallel-ogram as its unit cell. The actual structure of Fig. 6a appears on a section along the horizontal line drawn in full. The added vertical dimension corresponds to the new movement (section 7): dotted atoms on a dotted line at a lower level can be recognized as having the same value of p (density of hatching) as the dotted atoms in Fig. 6a.

The strength of this approach - not readily apparent from a schematic picture - is the fact that the complete symmetry of the incommensurate structure is contained in that of the supercrystal: the superspace group. In actual application the structure is three-dimensional, and there can be any number $\underline{d}$ of extra dimensions

a

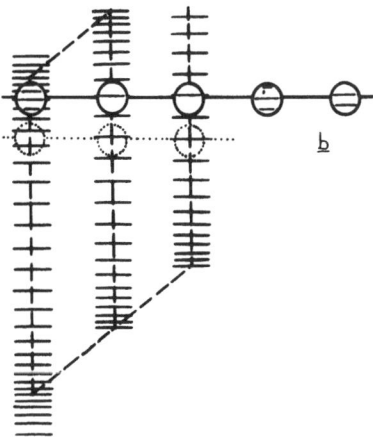

b

Fig.6 a) One-dimensional crystal
with scalar modulation, before
(full circles) and after (dotted
circles) movement of BS with
respect to modulation pattern.

Fig.6.b) Same crystal obtained
as a horizontal section through
the supercrystal.

(this number d is related to reciprocal rather than direct space:
it is the smallest number of modulation vectors $\vec{q}$, required as
basis vectors to index all reflections), so the supercrystal is
(3+d)-dimensional. It covers all imaginable kinds of incommensurate
structures. Therefore the term "superspace group" will also be used
here in a general sense to indicate the symmetry of such a structure, even though other interpretations not using superspace are
also possible. Since the supercrystal is a normally periodic
structure and since its symmetry operations have to obey certain
restrictions, listing of the ensuing groups is feasible and has
indeed been performed for the case d = 1, to which the majority of
incommensurate structures belong. See (6) which also gives more
details and references to earlier papers.

The superspace approach originated from the interpretation of
the diffraction pattern in a space of higher dimensions. It is,
therefore, particularly suited to the analysis of incommensurate
structures in terms of reciprocal space. Interpretation in direct
(crystal-)space is also possible so that complete structure
analysis can be made on this basis, as shown in several applications, for instance, by Yamamoto (7).

8.2. The Dualistic Approach

Symmetry operations involving the new type of movement (section 7) are not readily visualized when using the higher-than-three dimensional superspace. A more direct interpretation of these operations is afforded by the dualistic approach (8). Here the two elements of an incommensurate structure are used explicitly in the description of that structure and of its symmetry, viz.
- the basic structure BS introduced in section 1;
- the "modulation pattern" MP.
The latter is a new concept which has been hinted at in section 3, where the scalarly modulated structure was described as a product of BS with that pattern. For instance, in Fig. 3 the pattern MP consists of the modulation function p(Y) represented by the variable hatching. The modulated structure is indeed obtained here by multiplying the density of O-atoms in BS by this function p(Y). The X-atoms are not modulated and should therefore be dealt with separately; this means that the product BS × MP must be taken for each atom in the unit cell of BS in turn, and these products summed over the atoms.

In dealing with symmetry, however, such a synthesis of the modulated structure is not the main object. It is the symmetry of MP, next to that of BS, which interests us. An important feature of MP is that it can be drawn as a set of linear elements such as the strips in Fig. 3. Drawing MP as separate strips rather than as a single wave does not make any difference for situations like that of Fig. 3. However, in Fig. 7 we observe a case where the pattern MP has a centred orthogonal translation lattice (as indicated by its centred unit cell), while BS still has a primitive one. Such a peculiar translational symmetry is not arbitrary ; it is one of the centred Bravais types in superspace.

Centred Bravais types for three-dimensional incommensurate crystals occur quite frequently. As illustrated for two dimensions in Fig. 7 these apparently exotic types of superspace lattice can be decomposed in the familiar types in three dimensions: some kind of centring for MP and primitive for BS; or vice versa; or different kinds of centring for each.

Moreover, not only lattices but also the full superspace groups can be split in the same way. Fig. 8 shows a two-dimensional incommensurate crystal with a rather intricate symmetry. The separate patterns of BS and MP, however, have normal space groups pgg and pmg, of which the (glide-)reflection lines are shown in the border. Together, these two groups define the complete symmetry; their combination is thus fully equivalent to the relevant superspace group.

The above dualistic treatment of symmetry has been modelled

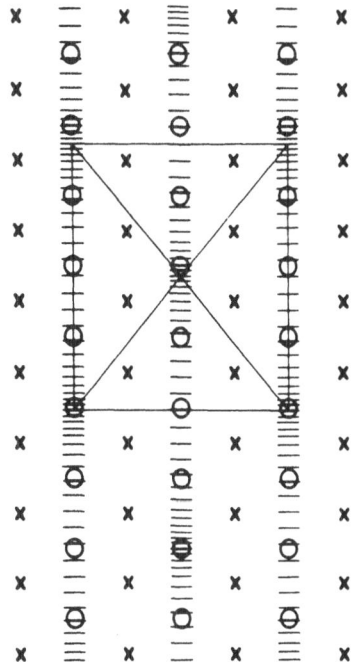

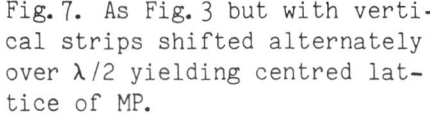

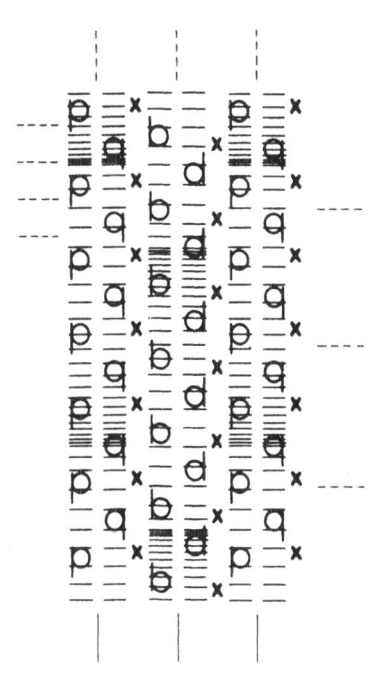

Fig. 7. As Fig. 3 but with vertical strips shifted alternately
over $\lambda/2$ yielding centred lattice of MP.

Fig. 8. BS: compound Xp_2, plane
group pgg, glide reflection lines
as at left and on top. MP: scalar, shown by the vertical strips,
plane group pmg, (glide-)reflec-
tion lines at right and below.

upon cases of scalar modulation. It is, however, equally applicable
to the displacive type of modulation. In that case, the pattern MP
consists of vector functions like those depicted in Fig. 2; symmetry
operations of MP act on the displacements as on vectors. There is a
true restriction to the dualistic approach in the number d of
independent q-vectors (cf. section 8.1): it can deal only with d <
4. Also these vectors must not be collinear or, for d = 3, copla-
nar. The superspace groups for d=1 - that is, for the majority of
known incommensurate structures - are entirely parallelled by the
dualistic classification.

8.3   Irreducible Representation of the BS Space Group

The use of these representations for describing the symmetry
of incommensurate structures is wide-spread among solid state
physicists. Obviously there must be a link with the approaches of
section 8.1 and 2, but very little has been written about this
relation. The paper (9) is a first effort to analyse it from the
viewpoint of Landau theory. Here we will attempt a brief analysis
in the opposite direction, with the aim to show how a superspace

group leads to a representation of G, the spacegroup of the basic structure.

The analysis could be performed generally, but for clarity's sake we shall deal only with the case d = 1; so there are just two modulation vectors, $\vec{q}$ and $-\vec{q}$. The symmetry operations of the space groups of BS either transform each of these two in itself or interchange them (up to vectors $\vec{H}$ of the reciprocal lattice of BS). Those of the first kind, transforming $\vec{q}$ into $\vec{q} + \vec{H}$, form the "group of $\vec{q}$", G_q. This group plays an important rôle: its irreducible representations lead directly to those of G. In the present case d = 1, either G_q = G or G_q has just one coset in G.

In the superspace group H we find exactly the same distinction between two kinds of elements (6). Actually each element h of H is composed of one operation g of G plus an operation in the extra dimension. The latter is characterized by two parameters. One parameter is called ε. It has the value +1 for those elements h of which the g-part belongs to G_q. For the remaining elements h (if any) ε =-1. In the dualistic language, these are the operations which reverse the modulation functions, e.g. the glide reflections transforming the functions for the p-atoms into those for the b-atoms in Fig. 8.

The other parameter is called τ. For the elements having ε = +1, the value of τ is 0 or $\pm 1/\nu$ (modulo integers), where ν = 2,3,4 or 6. Hence to each element g_q of G_q belongs a definite value of τ, viz. the value - modulo integers - pertaining to the elements h which contain g_q. Again switching to dualism: τ is the axial shift, expressed in units of λ, of the symmetry operation of MP which accompanies g_q. In Fig. 7 for instance, BS has vertical mirror lines passing through the X-atoms. The corresponding reflections belong to G_q, and they have $\tau = \frac{1}{2}$ since these lines are glide reflection lines for the vertical strips of MP (the modulation axis is vertical). For each operation g_q the two-line symbol (6) for the superspace groups yields τ at once, since it shows the generating elements of G on the upper line and the corresponding value of τ - in a coded form - on the lower line.

Now it can easily be shown, in either language, that if $g_1 = g_2 g_3$, g_2 and g_3 being elements of G_q, the corresponding τ's satisfy $\tau_1 = \tau_2 + \tau_3$ (modulo integers), or $\exp(2\pi i \tau_1) = \exp(2\pi i \tau_2) \cdot \exp(2\pi i \tau_3)$. Therefore the numbers $\exp 2\pi i \tau$ constitute a one-dimensional - and thus irreducible - representation of the point group of G_q. Standard procedures lead from here to the corresponding irreducible representation of G_q for the wave vector $\vec{q}$, and to that of G as well. If the lattice translations t_B of BS have components in the extra dimension (that is, if they correspond to centring translations of MP) the value of τ for these translations does not vanish. Then the operations g and $g t_B$ have different τ-values so that there is no single τ for the corresponding element of the point group. A consistent representation of the latter can still be constructed by assigning τ = 0 to all elements for which this value occurs at all.

Thus the incommensurate structure is indeed transformed as an irreducible representation simply related to its superspace group. There are, of course, discrepancies as well. The modulation functions can be anharmonic; then different Fourier components need not have the same representations. Also the equivalence of superspace groups is defined very differently from that of representations. On the whole the latter are ill-suited to structure analysis, just as the former do not easily fit into Landau theory. Nevertheless one should be aware of the fact that the mutual relation is a very close one, as shown by (9) and by the above considerations.

8.4. Approximate Superstructures

As mentioned in section 5, the modulation parameter of an incommensurate structure can be approximated by a rational number u/v, u and v being mutually prime integers. This leads to a superstructure with a v-fold unit cell. Since it also means that the modulation function is replaced by v points in the period, obviously v should not be too small if one wishes to determine the function in detail. Expressed differently: v is equal to twice the highest order of Fourier component which will be determined significantly. On the other hand, no great error results if u/v lies somewhat outside the error range, so one may choose for v quite a small number when only one or two harmonics are expected to be important.

The symmetry of the superstructure is of course determined by that of the incommensurate phase. It depends also, however, on v. An unfortunate choice of v may cause a severe degradation of symmetry, resulting in an increase in the number of structural parameters by a factor 2 or more. The reason is the following: Symmetry operations of the superstructure are operations which are common symmetry operations of BS and MP. A trivial example is the translation 4b in the case of Figs. 1-3, when β = 1/4; or generally vb when β = u/v, since the translation vb of BS then equals the translation $\mu\lambda = ub/\beta$ of MP. All true symmetry operations of the incommensurate structure, such as the horizontal translations and the vertical mirror lines in these figures, are also common to BS and MP. In contrast to the former example that holds even in the incommensurate structure, whereas vb becomes a normal translation only for rational β = u/v.

Anyhow all these symmetry operations are valid for the superstructure independent of the value of v. This is not true

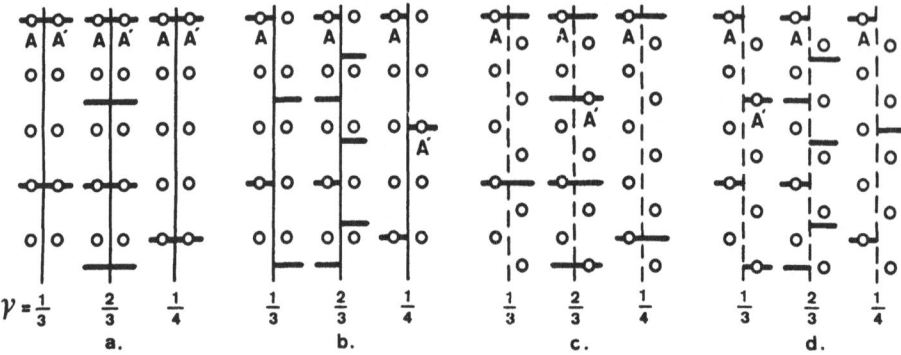

Fig. 9. The vertical line in each figure is a mirror line (m) or glide mirror line (g) for BS and for MP. In that order, the combinations are a)mm, b)mg, c)gm, and d)gg. Open circles are atoms of BS. Horizontal bars mark levels of equal phase in MP, so their interdistance is λ. The fractions are u/v, the value of the modulation parameter γ. Symmetry operations common to BS and MP are those which transform A into A'.

however, for operations involving a non-integral component in the direction of modulation. An example is the vertical glide mirror in Fig. 7, passing through X-atoms, where the glide is ½λ; or the centring translation of MP in the same figure. They can be, but need not be, symmetry operations of BS, and this depends upon v. The glide mirror is also a glide reflection for BS when ½λ is a multiple of b; this yields ½(v/u)b = nb or v = 2nu = even; hence u = odd. The same condition applies to conservation of the centring of MP as a centring of the superstructure.

Recently these conditions have been shown in a graphical manner (10) in a different context, viz. that of lock-in phase transitions. For all three parity combinations of u and v (odd-odd, even-odd, odd-even) Fig. 9 illustrates a symmetry element which is: a) a mirror both for BS and MP; b) a mirror m for BS but a glide reflection line g for MP: "mg"; c) gm, and d) gg. It is seen that in the last three cases only one parity combination yields glide reflections common to BS and MP. If a "wrong" parity is chosen, the superspace symmetry element (which in the dualistic view consists of the simultaneous action of the corresponding operations in BS and MP) is lost in the superstructure.

9. OUTLOOK

The above approximate treatment of symmetry is often used in structure determinations. As such, it is less economical than the superspace approach, for which computer programs are now available (7). Apart from that, superspace groups are known in which the different symmetry elements yield contradictory conditions for u and v, so that no superstructure spacegroups can render the complete superspace symmetry. There exist incommensurate phases for which such symmetries are quite probable.

On the other hand, it has been shown that even real crystal structures which can be described as (commensurate) superstructures may be analyzed using superspace groups (7). This seems to contradict our earlier conclusion stating that the symmetry of a superstructure can be described only by its normal space group. The two standpoints are reconciled, however, in cases where only a few orders of satellites occur so that they can be indexed in a unique way even for rational values of β. It may be that this situation - which occurs quite often- will have interesting physical implications, because it means that the modulation pattern MP is not fixed with respect to BS but can be shifted by any amount without affecting the diffraction pattern - just as in the "new movement" for incommensurate structures.

Finally, it can be expected that superspace groups and the largely equivalent dualistic approach will be helpful in the interpretation of Landau theories of phase transitions in terms of crystallographic symmetry groups, rather than by mere representations of the basic structure's spacegroup.

ACKNOWLEDGEMENT

The author is indebted to Dr. T. Janssen (Nijmegen) for his valuable advice concerning the presentation of section 8.3.

REFERENCES

1. Cowley, J. M. Retrospective introduction: What are modulated structures? A. I. P. Conference Proceedings Nr. 53 (New York, American Institute of Physics, 1979) 1-9.
2. Amelinckx, S. Survey of modulated structure phenomena. Ibid. 102-113.
3. Wuensch, B. J. Superstructures in sulfide minerals. Ibid. 337-354.
4. Futama, H. , Y. Shiozaki, A. Chiba, E. Tanaka, T. Mitsui and J. Furuichi. Satellite X-ray scattering by thiourea. Phys. Letters 25A (1967) 8-9.
5. Denoyer, F. , A. H. Moudden, R. Currat, C. Vettier, A. Bellamy and M. Lambert. The effect of hydrostatic pressure on the modulated structures in thiourea. Phys. Rev. B25 (1982) 1697.

6. De Wolff, P. M. , T. Janssen and A. Janner. The superspace groups for incommensurate crystal structures with a one-dimensional modulation. Acta Cryst. A37 (1981) 625-636.
7. Yamamoto, A. Structure factor of modulated crystals. Acta Cryst. A38 (1982) 87-92.
8. De Wolff, P. M. Dualistic interpretation of the symmetry of incommensurate structures. Acta Cryst. A. In the press.
9. Pérez-Mato, J. M. , G. Madariaga and M. J. Tello. Superspace transformation properties of incommensurate irreducible distortions. This volume.
10. Hogervorst, A. C. R. and P. M. de Wolff. Selection rules for superstructures in thiourea, Rb_2ZnBr_4 and Rb_2ZnCl_4. Sol. St. Comm. 43 (1982) 179-182.

SUPERSPACE TRANSFORMATION PROPERTIES OF INCOMMENSURATE IRREDUCIBLE
DISTORTIONS

J.M. Pérez-Mato, G. Madariaga and M.J. Tello

Departamento de Física, Universidad del Pais Vasco,
Apdo 644, Bilbao, Spain

The so-called superspace groups have recently been introduced
to characterize the symmetry properties of incommensurate structures
(1-5), allowing a unified approach to selection rules and other
regularities observed in this kind of phases. Essentially the
theory defines an imaginary (3+d)-dimensional structure called a
supercrystal which is determined by the real three dimensional
structure. In this way a translational lattice symmetry is
"recovered". The space group associated with this "supercrystal"
is the so-called superspace group of the corresponding incommensurate structure. The supercrystal is therefore defined in a space
which apart from the three dimensional one includes a d-dimensional
"internal" space, d being the number of independent incommensurate
modulation wave vectors in the structural distortion.

The theory has been particularly successful in the interpretation of diffraction patterns in incommensurate phases, which are
viewed under this approach as the diffraction amplitudes corresponding to a supercrystal section. In fact the origin of the theory is
closely related to this interpretation (1,3). Recent X-ray diffraction analysis of several incommensurate compounds (6-8) have demonstrated the adequacy of the superspace approach for this kind of studies,
allowing complete structural determinations in this type of system.

Although the theory has been claimed to be also useful to
interpret spectroscopic selection rules in these modulated structures (9), until now its use has not become common in this field,
nor in the symmetry breaking theory for the transitions sequence
normally observed in this kind of compound. One of the possible
reasons for this delay is probably, from a physical point of view,
the rather unconvincing "picture" that normally accompanies the

introduction of superspace symmetry, in particular, the mentioned definition of a (3+d) dimensional supercrystal.

Here we present a different approach to superspace symmetry, that pretends to stress its physical basis and avoids the supercrystal picture as an essential piece of the reasoning. First we describe the group of transformations, which can be defined in an incommensurate displacive distorted structure while keeping its free energy invariant. The superspace group is then defined as the subgroup of transformations keeping the structure invariant.

Let us consider an incommensurate structure resulting from a distortion over a basic commensurate one with space group G. We suppose the distortion is described by a displacive mode transforming according to a physically irreducible representation (IR) of G, $D_0(\vec{k}_1)$, where $\vec{k}_1$ is an arbitrary representative of the IR star of wave vectors and is incommensurate with the translation lattice in the basic structure. We shall see below that this last assumptio of the distortion irreducibility, although convenient does not restrict the validity of the demonstration for a more general case. Let $\{Q_{\epsilon j}\}$, $\epsilon=1,2\ldots n$, $j=1,2,\ldots p$, be the amplitudes (complex in general) of the distortion mode, where n is the number of wave fectors in the IR star $\{\vec{k}*\}$ and p the dimension of the small representation corresponding to $D_0(\vec{k}_1)$, the dimension of the IR $D_0(\vec{k}_1)$ being s=n x p.

The transformation properties of $\{Q_{\epsilon j}\}$ with respect to space operations $\{R|\vec{t}\}$ belonging to G are described by the corresponding matrices $D_0(\vec{k}_1,\{R|\vec{t}\})$ associated with the mentioned IR:

$$Q'_{\eta i} = \sum_{\epsilon j} D_0(\vec{k}_1,\{R|\vec{t}\})_{\eta\epsilon ij} Q_{\epsilon j} \tag{1}$$

Let us now consider the free energy of the distorted phase as an expansion with respect to the basic structure free energy, in terms of the distortion amplitudes, in a similar manner as it is done in the Landau theory of phase transitions (the actual existence of the transition: basic structure - incommensurate structure is however not essential). This free energy expansion is restricted by the G space symmetry of the basic structure to be invariant with respect to the amplitude transformations (1), and can be written in its most general form as:

$$\Phi = \Phi_0 + \tfrac{1}{2}A \sum_{\epsilon j} Q_{\epsilon j} Q_{\epsilon' j} + \sum_{\substack{\epsilon_1\ldots\epsilon_m \\ j_1\ldots j_m \\ m>2}} V\left(\begin{matrix}\vec{k}_{\epsilon_1} & & \vec{k}_{\epsilon_m} \\ j_1 & \ldots & j_m\end{matrix}\right) Q_{\epsilon_1 j_1} \ldots Q_{\epsilon_m j_m}$$

$$\Delta(\vec{k}_{\epsilon_1}+\ldots+\vec{k}_{\epsilon_m}) \tag{2}$$

where $\Delta(\vec{g})$ is unity when $\vec{g}$ belongs to the reciprocal lattice of the basic structure and zero in any other case. The second term of the right hand side corresponds to the only second order invariant, where $\vec{k}_{\varepsilon'} = -\vec{k}_{\varepsilon}$.

The incommensurability of the IR wave vector star allows to consider certain translations of the amplitude phases as generalized transformations of the distortion which also keep this free energy expansion. They constitute a kind of hidden symmetry of the basic structure. In this respect, if we consider a general translation of the amplitude phases, described by the set $\{\alpha_{\varepsilon j}\}$:

$$Q'_{\varepsilon j} = \exp(i2\pi\alpha_{\varepsilon j})Q_{\varepsilon j} \tag{3}$$

the free energy expansion (2) shall be invariant to it, if the following conditions are satisfied:

$$\alpha_{\varepsilon j} = -\alpha_{\varepsilon' j} \pmod{Z} \; \forall j, \; k_{\varepsilon'} = -k_{\varepsilon} \tag{4}$$

$$\alpha_{\varepsilon_1 j_1} + \alpha_{\varepsilon_2 j_2} + \ldots + \alpha_{\varepsilon_m j_m} = 0 \pmod{Z},$$

$$\forall m, \; \forall\{j_i\}, \; \forall\{\varepsilon_i\}/\sum_i \vec{k}_{\varepsilon_i} = 0 \tag{5}$$

All the phase equations in the following will be mod Z and we shall obviate it. Condition (5) is based on the fact of the incommensurability of the k-star, which assures that no umklapp terms exist in (2). Given a set of wave vectors $\{\vec{k}_{\varepsilon_i}\}$ fulfilling $\sum_i \vec{k}_{\varepsilon_i} = 0$, condition (5) should be satisfied for any set $j_1 \ldots j_m$, and therefore identical equations will result for any of the sets, so that we can put:

$$\alpha_{\varepsilon j} = \alpha_{\varepsilon j'} \equiv \alpha_{\varepsilon} \qquad \forall \varepsilon, \; \forall j,j' \tag{6}$$

Equation (5) reduces then to:

$$\alpha_{\varepsilon_1} + \alpha_{\varepsilon_2} + \ldots + \alpha_{\varepsilon_m} = 0, \qquad \forall m \geq 3, \; \forall\{\varepsilon_i\}/\sum_i \vec{k}_{\varepsilon_i} = 0 \tag{7}$$

Let $\{\vec{k}_1, \ldots, \vec{k}_d\}$ be a set of wave vectors from the representation star $\{\vec{k}*\}$ constituting a minimal base of independent vectors in the sense that any member of the star can be uniquely written as:

$$\vec{k}_{\varepsilon} = \sum_{i=1}^{d} n_{\varepsilon i}\vec{k}_i \qquad\qquad n_{\varepsilon i} \; e \; Z \tag{8}$$

while

$$\sum_{i=1}^{d} n_i \vec{k}_i = 0, \quad n_i \, \epsilon \, Z \text{ only if } n_i = 0 \, V \, i \tag{9}$$

From (9) it is straightforward that (8) does not restrict the possible values of the phase translations for $\{\vec{k}_1,\ldots,\vec{k}_d\}$, which we call $\{\alpha_1,\ldots,\alpha_d\}$, while the rest of them are determined by these:

$$\alpha_\epsilon = \sum_{i=1}^{d} n_{\epsilon i} \alpha_i \tag{10}$$

where $n_{\epsilon i}$ are the corresponding coefficients in (8), as can easily be seen if we consider the terms:

$$V(\begin{matrix} \vec{k}_1 \ldots \vec{k}_1, \ldots\ldots, \vec{k}_d \ldots \vec{k}_d, & -\vec{k}_\epsilon) \\ j_1 \ldots j_{n_{\epsilon 1}}, \ldots\ldots, j_{r+1} \ldots j_{r+n_{\epsilon d}}, j_m \end{matrix}$$

for which equation (7) reduces to (10). It can be easily proved that this last equation is also sufficient for (7) to be satisfied in any other case. Equation (10) can be rewritten as:

$$2\pi\alpha_\epsilon = \vec{q}_\epsilon \cdot \vec{\alpha} \tag{11}$$

where we consider $\vec{q}_\epsilon$ as a vector given by the set of components $2\pi(n_{\epsilon 1},\ldots,n_{\epsilon d})$ with respect to the reciprocal basis of the d-dimensional "phase space" in which the "vector" $\vec{\alpha} \equiv (\alpha_1,\ldots,\alpha_d)$ can be considered. This phase space corresponds obviously to the so-called internal space in the supercrystal formalism.

According to all these considerations, we can state that, in a displacive incommensurate phase resulting from an irreducible distortion, the most general transformation keeping the free energy invariant is a space group operation followed by a phase translation so that:

$$Q'_{\eta i} = \exp(i\vec{q}_\eta \cdot \vec{\alpha}) \sum_{\epsilon j} D_0(k_1,\{R|\vec{t}\})_{\eta \epsilon i j} Q_{\epsilon j} \tag{12}$$

where all the magnitudes have been defined above. We can represent this transformation by the symbol $\{R|\vec{t},\vec{\alpha}\}$, and using the properties of irreducible representations, it is straightforward to prove that it satisfies the following product rule:

$$\{R_2|\vec{t}_2,\vec{\alpha}_2\}\{R_1|\vec{t}_1,\vec{\alpha}_1\} = \{R_2 R_1 | R_2\vec{t}_1 + \vec{t}_2, \, \tilde{R}_I(R_2)\vec{\alpha}_1 + \vec{\alpha}_2\} \tag{13}$$

where $\tilde{R}_I(R_2)$ is the transpose of dxd matrix $R_I(R_2)$ determined by R_2 and defined by its action on the reciprocal basis $\{\vec{q}_1 \ldots \vec{q}_d\}$:

$$R_I(R)\vec{q}_i = \vec{k}_\eta \quad \text{if} \quad R\vec{k}_i = \vec{k}_\eta \tag{14}$$

It is important to point out that all this development strongly depends on the incommensurability of the $\{\vec{k_i^*}\}$ star. If not, terms would exist in the free energy expansion, $V(\vec{k}_{\varepsilon_1},\ldots,\vec{k}_{\varepsilon_m})$ for which $\Sigma\vec{k}_\varepsilon = \vec{g} \neq 0$, in particular for instance $\vec{k}_{\varepsilon_i} = \vec{k}_\varepsilon$ $\forall i$, so that $m\vec{k}_\varepsilon = \vec{g}$, and Equation (7) will result in $\alpha_\varepsilon = n/m$, for any ε, and therefore no kind of continuous group of phase translations will keep Φ invariant.

The important fact is that in a displacive incommensurate phase some distortion amplitude phases (not all in general) can be varied arbitrarily maintaining the free energy invariant. While in a commensurate case, two distortions described by amplitudes differing in some of their phases correspond to situations energetically and therefore structurally different, in an incommensurate structure may not be the case. The theory connects directly in this manner with the existence in these types of structures of the so-called phasons (10).

The symmetry group (superspace group) of the distorted structure appears now in this approach as the subgroup of those transformations $\{R|\vec{\varepsilon}\}$, defined by (12), which keep the whole structure invariant, that is:

$$Q'_{\eta i} = Q_{\eta i} \quad \eta=1\ldots n,\ i=1\ldots p \tag{15}$$

where $Q_{\eta i}$ are given by (12). From Equations (15), (12), (13) and the knowledge of the relevant IR $D_0(\vec{k}_1)$, the associated superspace group can be determined in a straightforward manner. The vector $\vec{\alpha}$ and the matrix $\tilde{R}_I(R)$ defined above correspond directly to the internal space translation and rotation respectively of the superspace symmetry element for the "supercrystal", which is associated in the usual formalism (3).

As an example, let us obtain the translation lattice $\{E|\vec{T},\vec{\alpha}(\vec{T})\}$ of the superspace group. From (15) and (12), we have in this sample case:

$$Q_{\varepsilon j} = \exp(i\vec{q}_\varepsilon \cdot \vec{\alpha}(\vec{T}))\exp(i\vec{k}_\varepsilon \cdot \vec{T})Q_{\varepsilon j} \quad \varepsilon=1\ldots n,\ j=1\ldots p \tag{16}$$

so that

$$-\vec{q}_\varepsilon \cdot \vec{\alpha}(\vec{T}) = \vec{k}_\varepsilon \cdot \vec{T} \tag{17}$$

and from this equation we obtain that:

$$\alpha_j = \sum_{i=1}^{3} \sigma_{ij} m_i \quad j=1\ldots d \tag{18}$$

where $_{ji}$ are defined by:

$$\vec{k}_j = \sum_{i=1}^{3} \sigma_{ji} \vec{a}_i^* \qquad\qquad j=1...d \qquad\qquad (19)$$

and $\vec{T} = \sum_{i=1}^{3} m_i \vec{a}_i \cdot \vec{a}_i$ and $\vec{a}_i^*$ the direct and reciprocal lattice basis vectors respectively for space group G. We have then that the generators of the translation lattice should be:

$$\{E|100,-\sigma_{11}\cdots\cdots-\sigma_{d1}\}, \quad \{E|010,-\sigma_{12}\cdots\cdots-\sigma_{d2}\},$$

$$\{E|001,-\sigma_{13}\cdots\cdots-\sigma_{d3}\}$$

which coincides with the superspace lattice introduced in the usual formalism (3).

Equations (15) and (12) are the simplest method to be used to determine the superspace group corresponding to an incommensurate structure described by the onset of an irreducible distortion. This symmetry group will depend strongly on the "direction" taken by the displacement amplitude set $\{Q_{\epsilon j}\}$ in the IR space.

Until now we have restricted our analysis to irreducible distortions. The situation is however essentially analogous in a more general case, as a general distortion can be considered a linear superposition of irreducible distorting modes. It is important to note that in the case that the incommensurate structure arises in a phase transition from the onset of a soft mode amplitude or order parameter, the secondary modes contributing to the total distortion result from their linear coupling with this primary mode (11), and as a consequence, they cannot diminish the superspace symmetry which the soft mode determines. A general demonstration of this fact will be published elsewhere. Here we limit ourselves to put a clarifying example.

The incommensurate phase in K_2SeO_4 (11) is due to a distortion Σ_2, from the basic structure of space group Pnam, for a certain wave vector $\vec{k}_o = \sigma\vec{a}^*$. The resulting superspace group for the distorted phase is p_{1ss}^{nam} (3,5), one of its elements being $\{I|000,-\beta/\tau\}$, where β is the amplitude phase of the Σ_2 distortion.

As a secondary mode, we can consider for instance, a mode Σ_4 for $\vec{k}_c = \vec{a}^*-2\vec{k}_o$. Then it can be shown that the superspace group for this distortion is p_{111}^{nam} with the same lattice as the formerly mentioned group, except for the internal space lattice, which is generated by $\{E|000,\frac{1}{2}\}$, if we express the element with respect to lattice basis of the first group. The inversion is also displaced along the fourth dimension in this case, $\{I|000,\beta'/2\pi\}$, where β'

is the amplitude phase of the Σ_4 mode.

The secondly mentioned group will be a supergroup of the former one only in the case that the inversion is centered at the same point for both of them; that is $\beta' = -2\beta + n\pi$, and only in this case, therefore, the mode $\Sigma_4(\vec{k}_c)$ will not diminish the superspace symmetry resulting from the order parameter onset. It can be easily shown that this relationship between the amplitude phases of both modes becomes enforced when we consider, in the frame of Landau theory, the material free energy expansion with the coupling term between the modes, and set the minimizing conditions to search for the equilibrium configuration.

Summarizing, we have shown that superspace symmetry in incommensurate displacive structures can be introduced from the invariance properties of their thermodynamic free energy without having to refer to an imagined supercrystal. This approach stresses the physical cause of this symmetry and its intimate relationship with the existence of the so-called phason excitations in these materials.

ACKNOWLEDGEMENTS

This work was supported in part by the Department of Education of the Basque Government (GM).

REFERENCES

(1)   de Wolff, P.M., Acta Cryst. A30 (1974), 777.
(2)   de Wolff, P.M., Acta Cryst. A33 (1977), 493.
[3]   Janner, A. and T. Janssen, Acta Cryst. A36 (1980), 399.
(4)   Janner, A. and T. Janssen, Physica 99A (1974), 47.
(5)   de Wolff, P.M., T. Janssen and A. Janner, Acta Cryst. A37 (1981), 625.

(6)   Yamamoto, A., Acta Cryst. A38 (1982), 87 and 79.
(7)   Yamamoto, A., Acta Cryst. B38 (1982), 1446.
(8)   Yamamoto, A., Acta Cryst. B38 (1982), 1451; Phys. Rev. B22 (1980), 373.
(9)   Rasing, Th., P. Wyder, A. Janner, T. Janssen, Phys. Rev. B25 (1982), 7504.
(10) Cowley, R.A. and A.D. Bruce, Adv. Phys. 29 (1980), 1.
(11) Iizumi, M., J.D. Axe, G. Shirane and K. Shimakoa, Phys. Rev. B15 (1977), 4392.

CRYSTAL STRUCTURES OF COMPLEX SULFIDES:
FROM MODULATED TO MODULAR STRUCTURES

Emil Makovicky

Institute of Mineralogy, University of Copenhagen

ABSTRACT

Crystal structures of complex sulfides of As, Sb and Bi with
Pb, Ag, Cu, Fe and other metals are often based on a simple
archetype, PbS or SnS, modified by a number of modulation
operators. They range from long-range modulated structures to
small-scale modular structures. Individual homologous series
(examples given) often span this range without a change in
building principles.

1 INTRODUCTION

Crystal structures of complex sulfides of As, Sb or Bi
combined, often in several distinct atomic ratios, with one or
more of the following cations: Pb, Ag, Cu, Fe, Sn, Hg, Tl or Mn,
often fall into the border-line area between compositionally and
geometrically modulated layer structures and structures built on
the same crystal chemical principles but containing two-
dimensionally finite building blocks, moduli. Increase in
modulation intensity and progressive disintegration of layers
lead from the former category to the latter.

The number of structural types involved not only confirms
the importance of these transitional structures for the complex
sulfide phases but also precludes any simple nomenclature for
transitional stages. Understanding and nomenclature of structure
modulation phenomena (structure building principles) will follow
those of Amelinckx (1) and Makovicky (2). Both of these articles
refer to original sources. Generally the structures in question
represent interface- and composition-modulated structures.

Superstructure phenomena in sulfides were briefly summarized
by Wuensch (3). For the Pb-Bi-(Ag, Cu) sulfides the structure
building principles were analysed in detail by Makovicky (2),
whereas the modulated noncommensurate layer structures in complex
sulfides and other compounds were reviewed by Makovicky and
Hyde (4,5).

2 PRINCIPAL STRUCTURAL FEATURES OF COMPLEX SULFIDES OF As, Sb and Bi.

The complex sulfides of As, Sb and Bi with Pb, Ag, Cu, Fe,
etc. are typically covalent structures with at least some
electropositive atoms having large and irregular coordination
polyhedra. In the majority of cases As, Sb and Bi display p^3
character with three short bonds nearly perpendicular to each
other. These are often supplemented by two longer, weaker bonds,
leading to the irregular square pyramidal coordination $[MS_{3+2}]$
with the metal atom at the pyramidal base. Below the base
lone electron pairs of As, Sb and Bi are accommodated. They require
additional space and lead to long, very weak M-S interactions
that involve one to three additional S atoms. Progressive
hybridization from the p^3 configuration towards a regular
octahedral one is very weak for As, weak for Sb and weak to
complete for Bi. Often several hybridization states (coordination
types) coexist in one structure.

For these structures collective accommodation of lone electron
pairs in extended channels or voids is typical. These spaces
represent, together with the surrounding metalloid atoms, lone
electron pair micelles regularly distributed over the structure.
Lead forms a number of distinct, large coordination polyhedra
which span all the stages from octahedral to three-capped
trigonal coordination prisms. The other metals also display
several types of coordination polyhedra but with the exception
of Tl the latter are less plastic than those of Pb. The small
size of the Cu and Fe polyhedra often renders these metals
efficient valency balancing agents in these complex structures.

Size and shape misfit of coordination polyhedra play an
important role in these covalent structures (primarily) composed
of irregular polyhedra. Therefore, these structures often
differentiate into layers, rods or blocks, based on a rather
simple structural principle (archetype) and composed of polyhedra
which easily fit together.

Two distinct types of simple, archetypal structures were
recognized: (1) the PbS-like, ccp structure with all octahedral
voids filled which occurs in the units with (usually) indistinct
lone electron pair character; (2) the SnS-like structure for the

units in which some elements possess distinct lone electron pair character. Blocks or layers based on the archetypes are mutually related by means of various structure building principles, primarily by the full-set or contracted-set unit cell twinning (either on mirror planes or on glide-reflection planes), by the noncommensurability interfaces, crystallographic shear planes or by the out-of-phase boundaries. On the composition interfaces, additional cations different from the majority as well as further lone electron pairs, can reside. Generally, the successful combination of basic elements leads to several homologous compounds or to entire homologous series in which a certain structure building operator is applied with varying frequency to the given archetypal structure.

3 GEOMETRICALLY MODULATED NON-COMMENSURATE LAYER STRUCTURES – THEIR MODIFICATION AND DISINTEGRATION

Two families of compounds with geometrically and compositionally long-range modulated layer structures are known in detail: the cylindrite-franckeite homologous series and the cannizzarite homologues. Both consist of two types of layers which can be described respectively as the (100) and (111) slabs of the archetypal PbS structure. Along the direction of the complex, non-commensurate match of the two layer sets, sometimes over 100Å long, the two layer sets undergo simultaneous compositional and geometrical modulation. In cannizzarite (6) the sinusoidal modulation clearly follows the approximate repetition period of interlayer match, equal to about 5 primitive subcells in the $(100)_{PbS}$ layer (Fig. 1). In cylindrite and franckeite (5, 7) the two layer types are rotated against each other by 45° and sinusoidal modulation has the wavelength of 6 or 7 centered subcells of the $(100)_{PbS}$ layer. Cylindrite and franckeite are complex sulfides of Pb, Sn^{2+}, Sn^{4+}, Sb and Fe; cannizzarite is a Pb-Bi sulfide. In both of them small compositional variations cause variation in the modulation periodicity and yield different homologues of the series. Larger variations may result in altered thicknesses of the two component layers in either structural family.

With increasing Bi/Pb ratio, the layer misfit and valency problems cannot be accomodated by the sinusoidally modulated structure and interface modulation is introduced. The first family of structures is created by application of crystallographic shear planes or out-of-phase boundaries to the cannizzarite-like structure. It can be typified by the structure of junoite, $Cu_2Pb_3Bi_8(S,Se)_{16}$ (8) with kinked $(111)_{PbS}$ layers and interrupted $(100)_{PbS}$ Layers (Fig. 2). In a number of these structures further adjustment of interlayer match by substitution of Se for S takes place. With the width of uninterrupted layer structure between adjacent shear planes reduced to a minimum, as in the structure

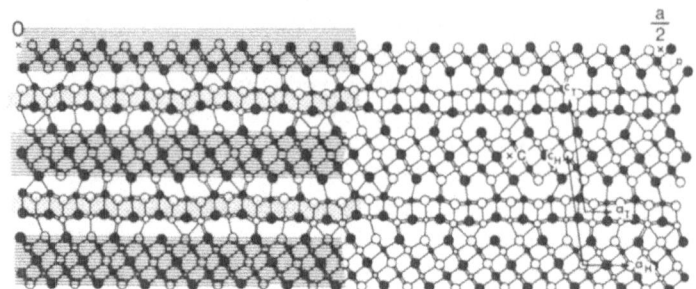

Fig. 1. The crystal structure of cannizzarite $Pb_{46}Bi_{54}S_{127}$ (6). In order of decreasing size the circles indicate $S,Pb,(Pb,Bi)$ and Bi. Here and in all the other figures (projections along the 4Å period), empty and full circles indicate atoms at two discrete levels, 2Å apart. Slabs $(100)_{PbS}$ are stippled, slabs $(111)_{PbS}$ are ruled. Modulation is parallel to $\underline{a}$.

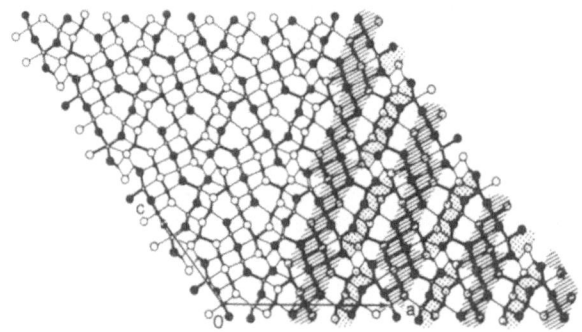

Fig. 2. The crystal structure of junoite, $Cu_2Pb_3Bi_8(S,Se)_{16}$ (8). In order of decreasing size circles represent (S,Se), Pb or Bi, Cu. For slab description see Fig. 1.

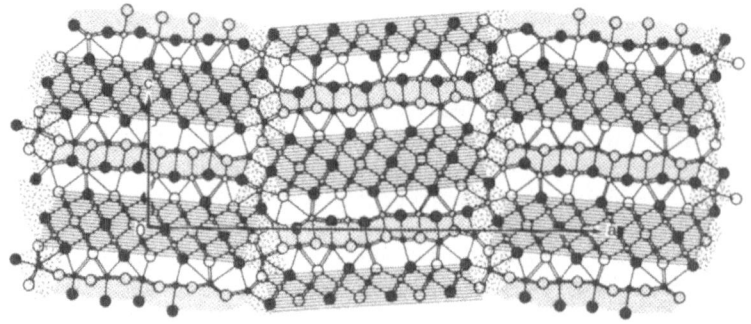

Fig. 3. The crystal structure of weibullite, $Ag_{0.33}Pb_{5.33}Bi_{8.33}(S,Se)_{18}$ (11). Small circles: Me atoms, large circles: (S,Se) atoms. Slab description as in Fig. 1.

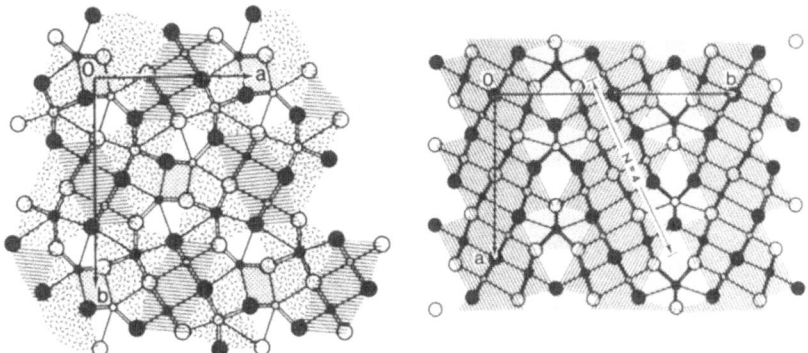

Fig. 4. (left). The crystal structure of galenobismuthite, $PbBi_2S_4$ (12). Circles in order of decreasing size: S, Pb, Bi. Slabs denoted as in Fig. 1.

Fig. 5. (right). The crystal structure of lillianite, $Pb_3Bi_2S_6$ (13). Circles in order of decreasing size: S, Pb and (Pb,Bi), (Bi,Pb). PbS like layers are ruled, trigonal coordination prisms stippled.

of $Cu_{1.57}Bi_{4.57}S_8$ (9, 10), the concept of the interface-modulated layer structure reaches the limits of its applicability.

Another type of interface modulation is represented by introduction of glide-reflection planes. These planes breaks up both types of continuous layers of the cannizzarite-like structure into strips of finite width. Introduction of interfaces, and of Se into the $(111)_{PbS}$ layers, solves the problems of mismatch between the two layer types. While the highest known member of this series, weibullite $Ag_{0.33}Pb_{5.33}Bi_{8.33}(S,Se)_{18}$ (11) represents a long-range modulated structure (Fig. 3) similar to cannizzarite, the lowest member, galenobismutite, $PbBi_2S_4$ (12), although based on exactly the same principles, can only be described as a structure composed of small-scale moduli of two different types in a glide-reflection arrangement (Fig. 4).

4. INTERFACE MODULATED, CHEMICALLY TWINNED STRUCTURES BASED ON PbS ARCHETYPE

Only a few complex sulfides have structures which can be described as (short-range) compositionally modulated structures based on an untwinned PbS archetype. As soon as the average metal valency departs from two, the structure motif has to be modified to accept the increased number of S atoms. For a large number of complex sulfides rich in either Pb or Bi (octahedral hybridization!) this adjustment proceeds by means of chemical twinning of PbS-like arrays on $(311)_{PbS}$ (Fig. 5). It may be

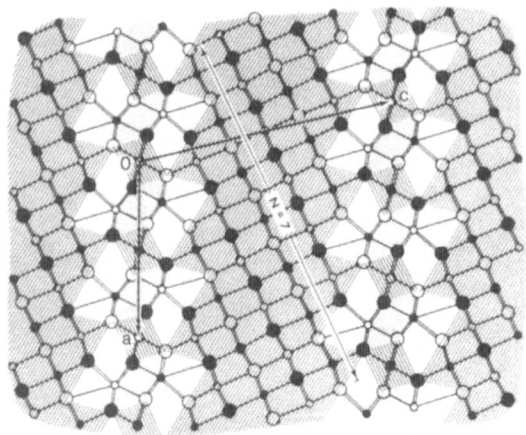

Fig. 6. The crystal structure of benjaminite, $Cu_{0.5}Pb_{0.4}Ag_{2.3}$ $Bi_{6.8}S_{12}$ (16). Circles in order of decreasing size: S,Ag and (Ag;Cu),Bi. PbS like layers of two distinct thicknesses are ruled, square coordination pyramids of Bi are stippled.

a full-set twinning in which the mirror-related portions are perfectly symmetrical or a contracted-set twinning (14) with only partial correspondence of the two portions, mainly due to their unequal thicknesses. Composition planes contain trigonal coordination prisms occupied by Pb or, asymmetrically, by Bi.

While in the Pb-Ag-Bi sulphides with high Pb contents (the lillianite homologous series, 15) the PbS-like slabs can achieve impressive thicknesses of up to 11 octahedra running diagonally through the slab, the compounds with low Pb (high Bi) contents display much thinner slabs for at least one slab orientation. With the width of only two octahedra per slab, the latter start to disintegrate and turn into isolated octahedral rods for the lowest member of the series. With all the principles of the homologous series still valid, at this stage the octahedra obtain additional degrees of freedom not available for those in thicker slabs. They can be distorted and occupied by elements different from those which occur in thicker slabs. This borderline case forms the basis for an entirely new series of asymetrically built, chemically twinned structures, the pavonite homologues (Fig. 6) which accommodate Bi, Ag, Cu and Pb in complicated proportions (16). Both the symmetrically and asymmetrically developed reflection-twinned PbS arrays display an appreciable number of periodicity errors (diffuse streaking in x-ray photographs, intimate twinning where applicable).

5. INTERFACE MODULATED, CHEMICALLY TWINNED STRUCTURES BASED ON SnS ARCHETYPE.

Structures with a pronounced lone electron pair character of metalloid atoms can often be derived by chemical twinning or other interface modulation of SnS-like atomic arrays. Here twinning proceeds by means of glide reflection planes. Besides the already mentioned disintegration of modulated structures into modular ones with decreasing order number in each homologous series, they display yet another type of departure from the usual image of a modulated structure. As the twin and composition planes $(h.0.1)_{SnS}$ swing from one family of structures to another towards higher and higher $\underline{h}$ values, the SnS-like motifs in the adjacent slabs finally meet at quite acute angles. In this type of structure, disintegration of the modulated structure for the lower homologues is very intense. It can be appreciated when comparison is made of the crystal structure of meneghinite, $CuPb_{13}Sb_7S_{24}$ (17), the fifth member of the series (Fig. 7), with that of aikinite, $CuPbBiS_3$ (18, 19), the second member of the same series (Fig. 8), that was always considered a typical modular structure.

6. MODULAR STRUCTURES WITHOUT MODULATED ANALOGUES

Among the structures based on glide reflection twinning on (hk0) planes of the SnS archetype, no higher homologues are known at present in which the slabs of SnS-like structure display sufficient continuity. The principal structures of this category, the Pb-Sb sulfides, represent typical modular structures. The

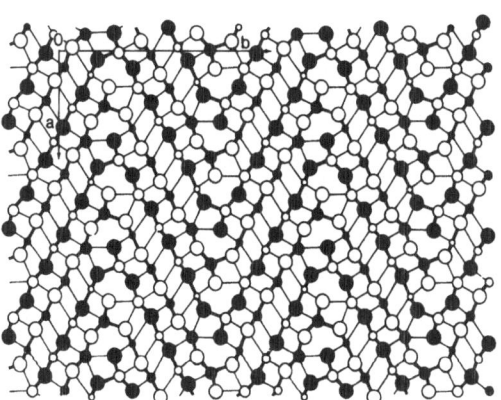

Fig. 7. The crystal structure of meneghinite, $CuPb_{13}Sb_7S_{24}$ (17). Circles in order of decreasing size: S, Pb, (Sb,Pb).

166

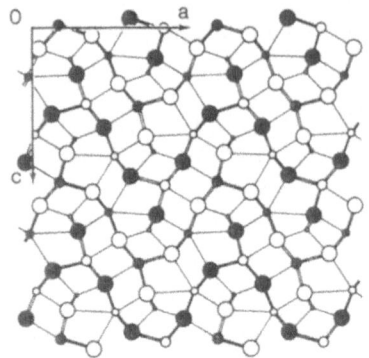

Fig. 8. The crystal structure of aikinite, $CuPbBiS_3$ (18, 19).
Circles in order of decreasing size: S,Pb,Bi,Cu.

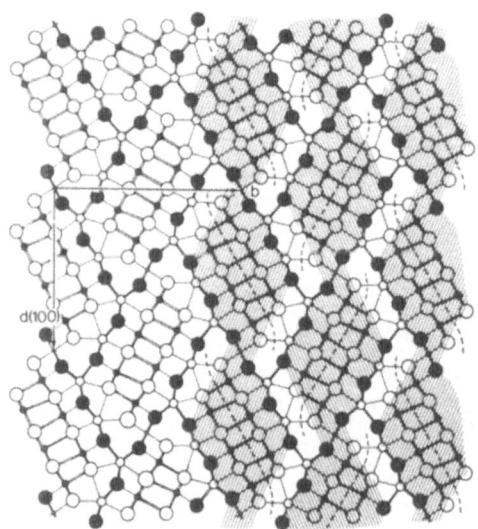

Fig. 9. The crystal structure of jamesonite, $FePb_4Sb_6S_{14}$ (20).
Circles in order of decreasing size: S,Pb,Sb,Fe.

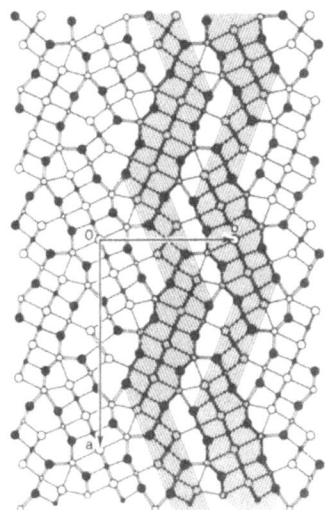

Fig. 10. The crystal structure of $Pb_4In_9S_{17}$ (21). Circles in order of decreasing size: S,Pb,In. Layers are emphasized by ruling.

moduli, rods with lozenge-like cross-sections, are formed by periodic and usually complete constriction of (hk0) slabs of SnS-like structure (Fig. 9). However, the Pb-Bi and Pb-In relatives, based on the PbS archetype, represent a special type of modulated layer structure (Fig. 10).

Another family without modulated analogues are the kobellite homologues, the structures of which are composed of alternating moduli respectively based on PbS- and SnS archetypes, in a locally pseudohexagonal arrangement.

7. CONCLUSIONS

Description of crystal structures of sulfides of As, Sb or Bi with other metals, primarily Pb,Ag,Cu and Fe, can often be based on a simple archetype (PbS or SnS) and a number of modulation (structure-building) operators. These structures range from typical long-range modulated structures towards those based on two-dimensionally finite moduli. The misfit and valency problems are first solved by geometrical and compositional modulation, followed by interface modulations of various kinds and intensity. Many homologous series built according to these principles have higher members as typical interface modulated structures whereas, with increasing frequency of interfaces, the structure of the lower members disintegrate into modular structures. The building

principles characteristic for the given homologous series remain
valid in this process.

1. Amelinckx, S. Survey of Modulated Structure Phenomena. AIP
   Conf. Proceedings 53 (1979) 102-113.
2. Makovicky, E. The Building Principles and Classification of
   Bismuth-Lead Sulphosalts and Related Compounds. Fortschritte
   der Mineralogie 59 (1981) 137-190.
3. Wuensch, B.J. Superstructures in Sulphide Minerals. AIP Conf.
   Proceedings 53 (1979) 337-354.
4. Makovicky, E. and B.G. Hyde. On Modulated, Non-Commensurate
   Layer Structures. AIP Conf. Proceedings 53 (1979) 99-101
5. Makovicky, E. and B.G. Hyde. Non-Commensurate (Misfit) Layer
   Structures. Structure and Bonding 46 (1981) 103-170.
6. Matzat, E. Cannizzarite. Acta Crystallographica. B35 (1979)
   133-136.
7. Makovicky, E. Crystallography of Cylindrite. Part I. Crystal
   Lattices of Cylindrite and Incaite. Neues Jahrbuch für
   Mineralogie, Abhandlungen 126 (1976) 304-326.
8. Mumme, W.G. Junoite, $Cu_2Pb_3Bi_8(S,Se)_{16}$, A New Sulphosalt from
   Tennant Creek, Australia: Its Crystal Structure and
   Relationship with Other Bismuth Sulfosalts. Amer.Mineralogist
   60 (1975) 548-558.
9. Ohmasa, M. and W. Nowacki. The Crystal Structure of Synthetic
   $CuBi_5S_8$. Zeitschrift für Kristallographie 137 (1973) 422-432.
10. Tomeoka, K., M. Ohmasa and R.Sadanaga. Crystal Chemical
    Studies on Some Compounds in the Copper-Bismuth Sulfide
    $(Cu_2S-Bi_2S_3)$ System. Miner. Journal 10 (1980) 57-70.
11. Mumme, W.G. Weilbullite $Ag_{0.32}Pb_{5.02}Bi_{8.55}Se_{6.08}S_{11.92}$ from
    Falun, Sweden: A Higher Homologue of Galenobismutite. Canad.
    Mineralogist 18 (1980) 1-18.
12. Iitaka, Y. and W. Nowacki.A redetermination of the Crystal
    Structure of Galenobismutite, $PbBi_2S_4$. Acta Crystallographica
    15 (1962) 691-698.
13. Takagi, J. and Y.Takéuchi. The Crystal Structure of Lillianite.
    Acta Crystallographica B28 (1972) 649-651.
14. Takéuchi, Y. "Tropochemical Twinning": a Mechanism of Building
    Complex Structures. Recent Progress of National Science in
    Japan 3 (1978) 153-181.
15. Makovicky, E. and S. Karup-Møller. Chemistry and
    Crystallography of the Lillianite Homologous Series. Part I.
    General. Properties and Definitions. Neues Jahrbuch für
    Mineralogie, Abhandlungen 130 (1977) 264-287.
16. Makovicky, E. and W.G. Mumme. The Crystal Structure of
    Benjaminite $Cu_{0.50}Pb_{0.40}Ag_{2.30}B_{6.80}S_{12}$. Canad. Mineralogist
    17 (1979) 607-618.

17. Euler, R. and E. Hellner. Über Komplex Zusammengesetzte Sulfidische Erze VI. Zur Kristallstruktur des Meneghinits, $CuPb_{13}Sb_7S_{24}$. Zeitschrift für Kristallographie 113 (1960) 345-372.

18. Omasa, M. and W. Nowacki. A Redetermination of the Crystal Structure of Aikinite [$BiS_2/S/Cu^{IV}Pb^{VIII}$]. Zeitschrift für Kristallographie 132 (1970) 71-86.

19. Kohatsu, I. and B.J. Wuensch. The Crystal Structure of Aikinite, $PbCuBiS_3$. Acta Crystallographica B27 (1971) 1245-1252.

20. Niizeki, W. and M.J. Buerger. The Crystal Structure of Jamesonite, $FePb_4Sb_6S_{14}$. Zeitschrift für Kristallographie 109 (1957) 161-183.

21. Ginderow, D. Structures Cristallines de $Pb_4In_9S_{17}$ et $Pb_3In_{6.67}S_{13}$. Acta Crystallographica B34 (1978) 1804-1811.

CHAPTER 3: DIFFRACTION METHODS

HIGH RESOLUTION ELECTRON MICROSCOPY STUDY OF MANGANESE SILICIDES

H.Q. Ye* and S. Amelinckx**

*On leave from: Institute of Metal Research, Wenhua Road, Shenyang (People's Republic of China); **RUCA, B 2020-Antwerpen (Belgium) and SCK/CEN, B2400-MOL (Belgium)

ABSTRACT

High resolution electron microscopy confirms the one-dimensional doubly periodic nature of the so-called "chimney ladder" structures of Nowotny-phases. The manganese and the silicon c-period can be observed separately. The c-parameter of the structure is the smallest common multiple of the c-periods. The orientation anomaly in the diffraction pattern is related to systematic variations of phase of the silicon helices. The spatial variations of the c-parameter and of the orientation anomaly can be imaged.

1. INTRODUCTION

High resolution electron microscopy of modulated crystals has already contributed significantly to our understanding, although the application of this technique to such problems started recently only. In particular it has allowed to determine unambigeously the origin of incommensurate diffraction patterns. In this paper we shall discuss its application to a particular type of modulated crystals : one-dimensional doubly periodic crystals i.e. crystals in which the sublattices of two constituents have different periods along one common direction the other two lattice parameters being equal. Examples are $Ba_{1+x}Fe_2S_4$ (1) $MnSi_{2-x}$ (2) (4), $MnGe_{2-x}$ (3)(4). We present in particular some new observations on $MnSi_{2-x}$ ($x \approx 0,25$); a so called "chimney ladder" structure or Nowotny phase (1).

2. CRYSTAL STRUCTURES

The manganese silicides $MnSi_{2-x}$ ($x \approx 0,25$) can be considered as derivatives of the $TiSi_2$ structure (3). The manganese atoms form a β-tin like arrangement with a tetragonal subcell

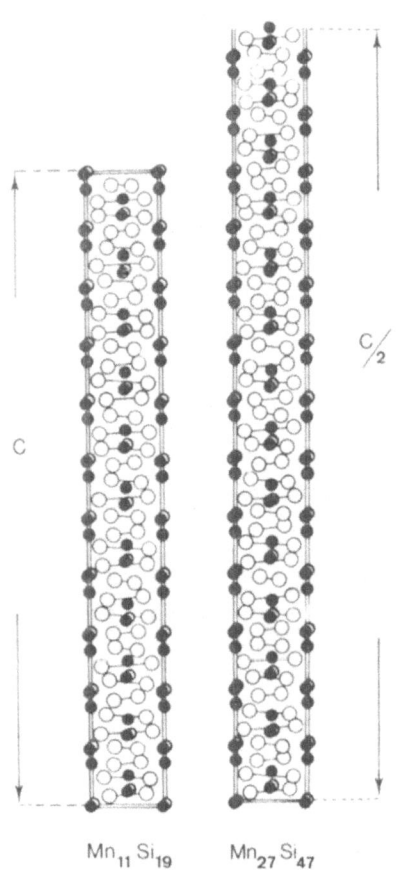

$Mn_{11}Si_{19}$ $Mn_{27}Si_{47}$

FIG.1. Chimney ladder structures or Nowotny phases; manganese atoms are represented as full dots; silicon atoms as open dots (after ref. (4)).

(a=0,552 nm; c_{Mn} = 0,437 nm). The silicon atoms adopt a double helical arrangement, with a period equal to the pitch of this helix $c_{Si} \simeq 4c_{Mn}$, filling the interstices left in the manganese sublattice. The resulting, approximately rationalized, c-period is the smallest common multiple of c_{Mn} and c_{Si}; it may become quite large (up to 20 nm). Strictly speaking there will often be no periodicity. Two of these structures are represented in fig.1 (after ref.(4)).

3. DIFFRACTION EFFECTS (5)(6)(7)

The diffraction patterns of such doubly periodic crystals consist of spots due to both sublattices, the spot positions being given by the diffraction vectors $\bar{g} = \bar{H}+\bar{h}$, where $\bar{H}$ is a reciprocal lattice vector of one of the sublattices and $\bar{h}$

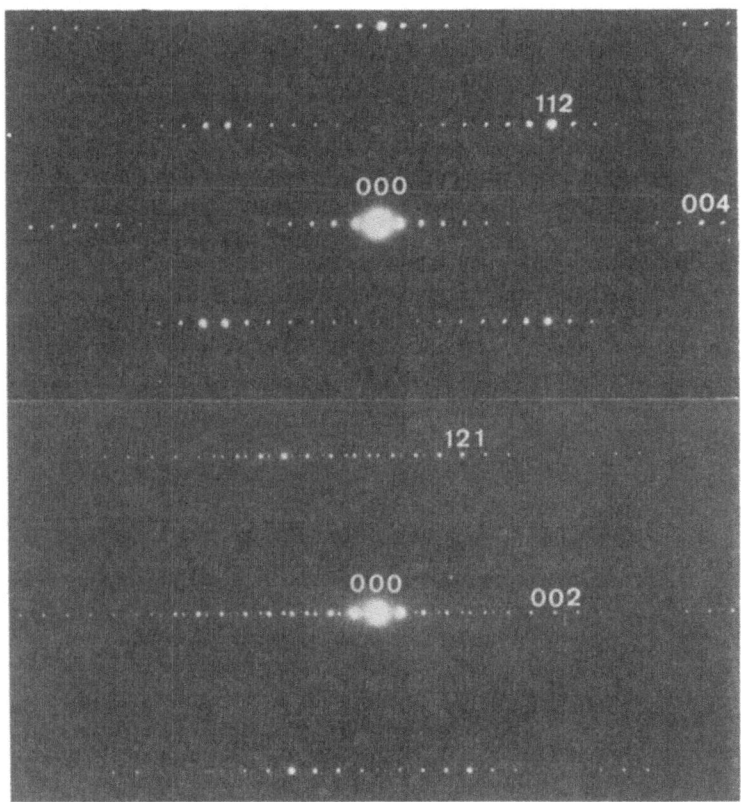

FIG.2. Diffraction patterns of $MnSi_{2-x}$ along two different zones. Manganese spots are encircled.
a. $[110]$ zone
b. $[120]$ zone.

176

a reciprocal lattice vector of the other sublattice. The c-parameter of the manganese sublattice c_{Mn} being much smaller than c_{Si}, in the manganese silicides, one observes a set of intense spots due to the manganese sublattice (usually, but not always the most intense spots) and linear arrays of closely spaced

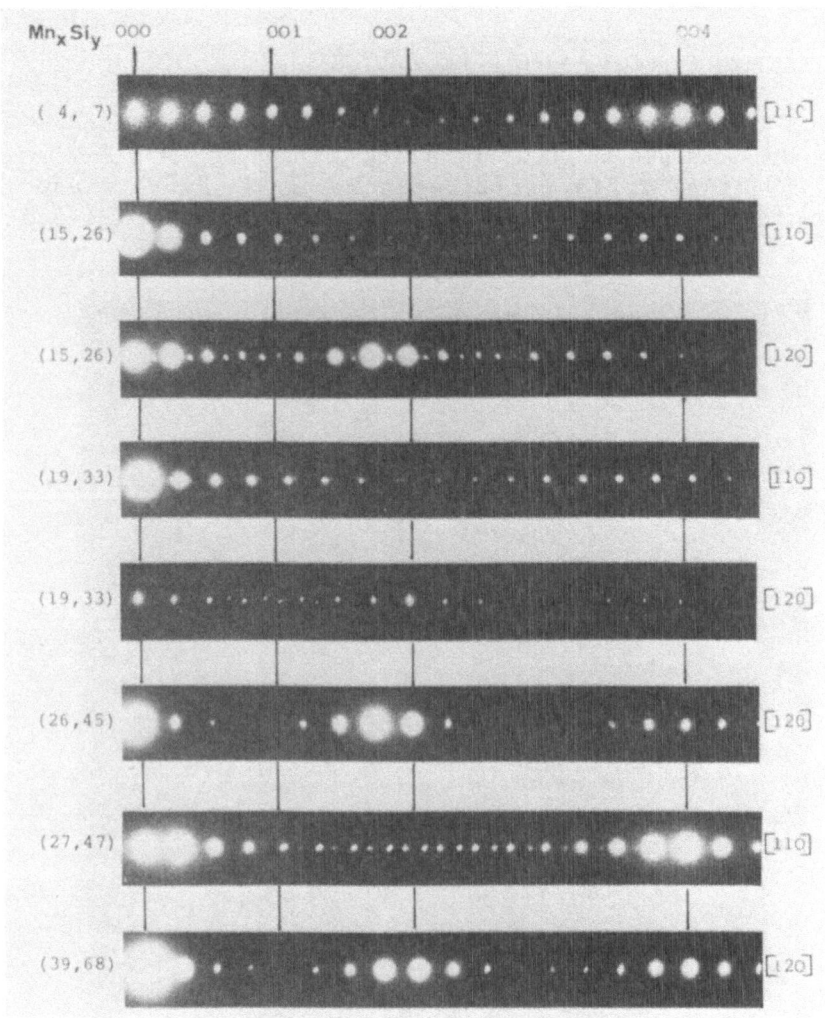

FIG.3. A number of rows of "superlattice" spots as observed along the two zones $[110]$ and $[120]$, for different phases, illustrating commensurate as well as incommensurate diffraction patterns.

spots (spacing $1/c_{Si}$) which are obviously associated with the silicon sublattice. The intensity of the spots in these linear arrays decreases towards both ends. Where different linear arrays of such spots meet, spacing and orientation anomalies often occur; this is not always the case however.

The spacing anomaly is obviously due to the fact that the ratio $c_{Si}/c_{Mn} \simeq 4$ is in general not an integer and may even be irrational. The orientation anomaly is related to the fact that although the silicon arrangements in neighbouring "chimney ladders" may have the same period, their "phases" may be different. The normal to the planes of equal phase may then enclose a small angle with the c-axis. The "surfaces" of equal phase need not be planar, leading to an orientation anomaly, which varies continuously over the specimen area.

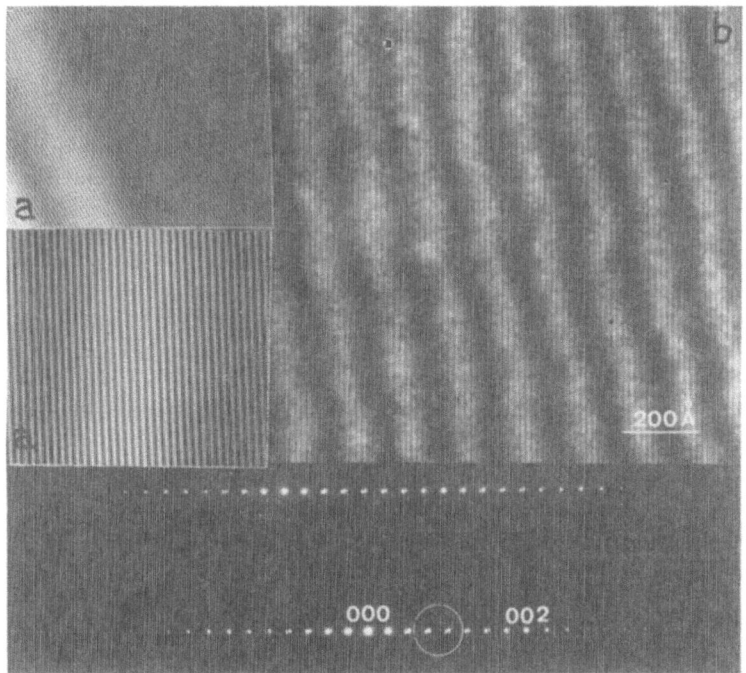

FIG.4. One dimensional lattice fringes in manganese silicides.
  (a) Collecting of spots belonging to a single sequence of superlattice spots. The fringes represent the silicon periodicity.
  (b) Collecting spots belonging to two overlapping sequences of superlattice spots. The image represents the lattice fringes due to the silicon arrangement as well as moiré-like fringes, which map the spatial variation of Δg.

The diffraction effects are clearly similar to those associated with interface modulated structures (6) except that there are no fractional shifts of the linear arrays of "superlattice" spots with respect to basic spots. In a structure of this type the relationship between the Mn and the Si sublattice spots is reciprocal i.e. either one can be considered as basic and the other as satellites.

Diffraction patterns along two different relevant zones $[110]$ and $[120]$, exhibiting the described characteristics are reproduced in fig.2. The long period is quite visible : the manganese sublattice spots are surrounded by small circles.

Rows of superlattice spots from different phases and taken from diffraction patterns along $[110]$ and along $[120]$ are reproduced in fig.3. Some of these patterns are commensurate; others contain anomalies.

4. IMAGES

4.1. One - Dimensional Lattice Fringes (6)(7)

One-dimensional lattice fringes are obtained when images are made in the superlattice spots (or satellites) belonging to one sequence only. (fig.4a.); they only reveal the silicon spacing c_{Si}. Moiré-like fringes are obtained when collecting beams belonging to two overlapping linear sequences (fig.4b.). Such images show the lattice fringes due to the silicon arrangement, together with and superposed on them, the moiré-like fringes with wave-vector $\Delta \bar{g}$. They display a map of the spatial variation of $\bar{\Delta} g$, in length and in orientation. They thus demonstrate the variability of the orientation and spacing anomaly in the diffraction pattern and hence also of the phase relationship between neighbouring helices.

4.2. High Resolution Images

We have obtained high resolution images from a number of different phases along the $[110]$ and $[120]$ zones. We shall discuss only three examples in order to illustrate the possibilities; a detailed paper is being published elsewhere. (8).

We first discuss fig. 5 which shows the bright and dark field structure images of $Mn_{27} Si_{47}$ taken along the $[110]$ zone. The bright field (B.F.) image (a) was taken by collecting the direct beam and two shells of manganese reflections around the origin, together with their satellites. The dark field (D.F.) image (b) was taken by including only four manganese spots situated at the summits of a lozenge, together with their satellites.

The bright dots in both images have the configuration and scale of the manganese sublattice. On the other hand the broad

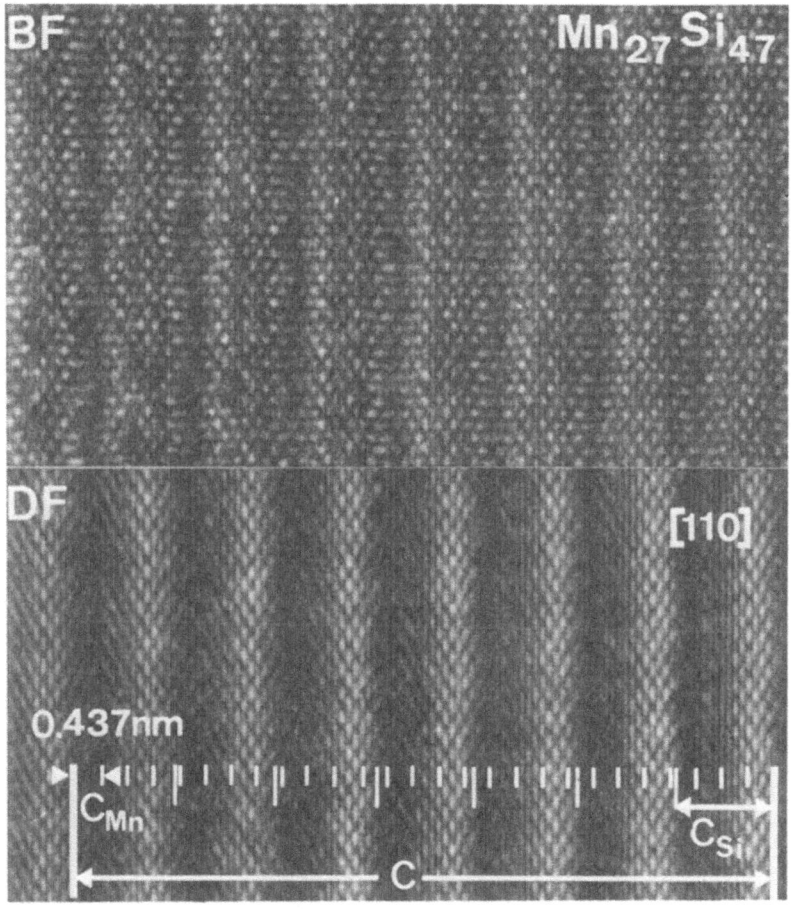

Fig.5. Structure images of $Mn_{27}Si_{47}$ viewed along the $[110]$ zone;
c_{Si} and c_{Mn} are indicated on the image.
a) Bright field image made by collecting the direct beam
   and two shells of manganese reflections around the
   origin together with their satellites.
b) Dark field image made by collecting four manganese spots
   situated at the summits of a lozenge, as well as the
   associated satellites.

one-dimensional fringes have a spacing, which is slightly less
than four times the c_{Mn} spacing; they represent the silicon
periodicity. In the case shown the diffraction pattern was incommensurate without orientation anomaly. As a result fringes are
parallel with the manganese fringes. We have indicated the
c_{Mn} period as well as the c_{Si} period on the photographs.

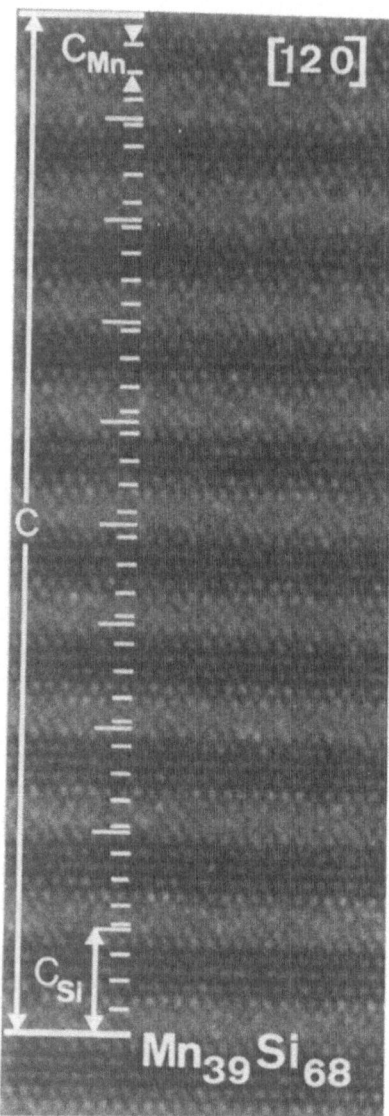

FIG.6. Dark field structure image of $Mn_{39}Si_{68}$ viewed along the [120] zone; c_{Si} is very approximately equal to $4c_{Mn}$ leading to a large parameter $c \simeq 17$ nm.

The latter is best visible in the D.F. image. It is clear that the c-parameter of the structure must be about $c \simeq 7c_{Si}$.

Note, in the B.F. image, the wavy nature of the rows of bright dots representing manganese columns, the period of the wave being c_{Si}. Probably the manganese atoms adapt their positions somewhat to the silicon arrangement; in this way the manganese sublattice becomes modulated with a period equal to c_{Si}.

Fig.6 shows a D.F. image of $Mn_{39} Si_{68}$ taken along the $[120]$ zone, made by collecting one Mn-beam and its six nearest neighbours, together with their satellites. The bright dots have again the same configuration and scale as the manganese columns viewed along this zone. The broad dark fringes have a spacing equal to c_{Si}. We have indicated c_{Si} and c_{Mn} on the photograph; c_{Si} is close to $4c_{Mn}$ leading to a very long unit cell: $c \approx 10c_{Si} = 39 c_{Mn} = 17$ nm.
This can be deduced directly from the indications on the photograph of fig.6.

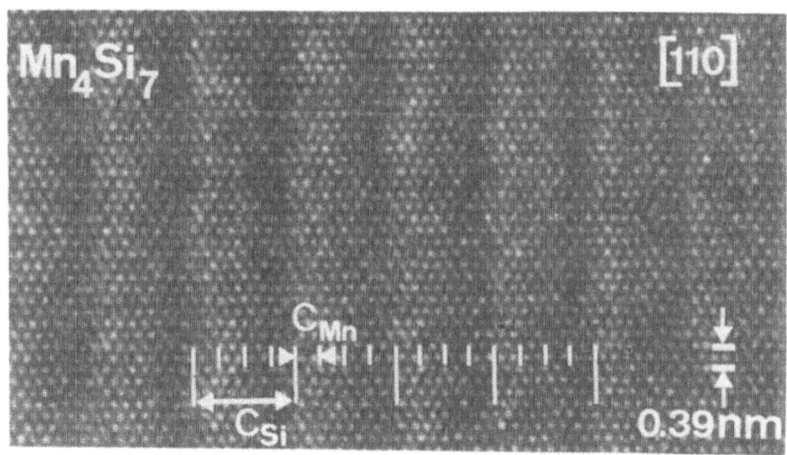

Figure 7 shows a B.F. image along the [110] zone of
simplest structure of the series.

Fig.7 finally shows a B.F. image along the $[110]$ zone of Mn_4Si_7, one of the simplest commensurate structures of this series, where $c_{Si} = 4 c_{Mn} = c = 1.78$ nm.
Areas exhibiting a diffraction pattern with large orientation anomalies produce complicated images which shall not be discussed here. (8).

5. CONCLUSIONS

All structure images obtained from manganese silicide crystals are consistent with the building principle of the "Chimney ladder" structures i.e. with the presence of two periods in the c-direction, one for the Mn-arrangement and another one for the silicon arrangement, which is about four times larger. The high resolution images allow to visualize clearly these two distinct periods.
The long unit cells introduced in ref.(2) are somewhat artificial since many more phases can be found with c-parameters of

almost any magnitude, depending on the exact ratio c_{Si}/c_{Mn}, which is moreover variable over small areas. This variability can be displayed directly by means of moiré-like fringes.

The high resolution images allow to determine directly any phase, given the building principle.

REFERENCES

1. Gray, I.E., J. Sol. State Chem 11 (1974) 128; Acta Cryst. B31

   Nakayama, N., K. Kosuge and S. Kachi. J. Sol. State Chem. 33 (1980) 267; ibid 36 (1981) 9.

2. Flicker, G.H., H. Völlenkle and H. Nowotny. Mh. Chem. 98 (1967)

   Knott, K.W., M.H. Mueller and L. Heaton. Acta Cryst. 23 (1967

3. Nowotny, H. The Chemistry of Extended Defects in Non-Metallic

4. Pearson, W.B. The Crystal Chemistry and Physics of Metals and

5. De Ridder, R., J. Van Landuyt and S. Amelinckx. Phys. Stat. Sol.

6. De Ridder, R., G. Van Tendeloo and S. Amelinckx. Phys. Stat. Sol

7. De Ridder, R., G. Van Tendeloo and S. Amelinckx. Phys. Stat. Sol

8. Ye, Hq. et al. (to be published).

THE USE OF HIGH RESOLUTION ELECTRON MICROSCOPY IN THE STUDY OF
MODULATED STRUCTURES IN ALLOY SYSTEMS

S. AMELINCKX, J. VAN LANDUYT and G. VAN TENDELOO

University of Antwerp, RUCA, Groenenborgerlaan 171,
B-2020 Antwerp, Belgium.

ABSTRACT

Electron microscopy and electron diffraction have become powerful tools in studying long period antiphase boundary structures in
alloy systems; they can provide information about a very small volume
of material,in direct space as well as in reciprocal space. In particular if the basic structure is known a detailed structure determination can be made using geometrical features of the diffraction
pattern, complemented by high resolution images.

Evidence is presented to support the idea that in a number of
systems the formation of such structures is composition driven, at
least in part.

Some conclusions are presented as to the building principle of
such periodic antiphase boundary structures.

1. INTRODUCTION

"Single crystals" of ordered alloys are invariably fragmented
in a number of symmetry related translation and orientation variants
derived from a common basic disordered matrix, as a result of the
loss of translation and rotation symmetry on ordering (1,2,3,4).
With X-rays or neutrons such a finely textured "single crystal" produces at best a diffraction pattern which is the superposition of
the diffraction patterns due to all orientation variants present.
It is often an impossible task to unscramble such patterns. It is
of course possible to obtain powder diffraction patterns from such
materials, however their interpretation is subject to the inherent

limitations. Selected area electron diffraction on the other hand
allows in general to obtain monodomain diffraction patterns, i.e.
true "single crystal" patterns. Moreover in the same instrument, one
can produce high resolution images which are under suitable condition
direct representations of the projected crystal potential. Even in
cases where no single domain can be selected, high resolution images
can be used as diffraction gratings in an optical diffraction experiment. In such a way it is often possible to obtain optical diffraction patterns from regions only containing a small number of
unit cells. With this knowledge the superposed electron diffraction
patterns can then usually be unscrambled. In this paper we shall
describe the application of these methods to the study of various
alloy systems. This study has allowed to determine a number of
structures corresponding with not simple compositions; these structures have been identified from electron diffraction patterns combined with high resolution images; they are often interface modulated
structures, where the role of the interfaces is at least in part to
incorporate the deviations from a simple composition.

2. TYPES OF MODULATED STRUCTURES

Modulated structures may result from the periodic introduction
of a number of different types of planar defects such as twin interfaces, antiphase boundaries, inversion boundaries, or stacking faults
from the periodic deformation of the structure in one, two or three
directions or from periodic changes in composition (5). Examples,
largely taken from work in our laboratory, are referred to the survey
of table I.

Modulated structures often produce incommensurate diffraction
patterns. In interface modulated structures this is almost invariably due to the fact that the spacing of the "sharp" interfaces, as
deduced from diffraction, is due to a not strictly periodic mixture
of two or more integral spacings. We shall call this "pseudo-incommensurate" because the incommensurability in the diffraction pattern
only results from the averaging effect of diffraction. On an atomic
scale, as revealed by high resolution electron microscopy, the separation of two successive interfaces is "quantized" i.e. always an
integral number of units (and thus commensurate).

Deformation modulated structures are often truly incommensurate,
the period of the displacement pattern being to some extent independent of the repeat distance of the basic structure. Such incommensurate modulated structures often occur as a transition phase between
two commensurate phases, which transform reversibly one into the
other at a well defined temperature (a part from possible hysteresis
(examples : quartz (6)), VSe_2 (7), TaS_2 (8,9), K_2SeO_4 (10), perov-
skites (11).

TABLE I. Types of modulated structures.

Interface modulated :	Examples :
- Antiphase boundaries	Au-Mn (15) Au-Mg (16) Au-Zn (17) Pt-Ti (18)
- Twin interfaces	YSeF (19) nBi_2S_3mPbS (20) $Au_{0.8}Ag_{0.2}Te_2$ (Krennerite) (21) SiO_2 (quartz) (6)
- Inversion boundaries	χ-phase (22) γ-brass (23)
- Stacking faults	SiC (24) ZnS (25) Au-Mn (26) (Au-Cd)
- Crystallographic shear planes	TiO_{2-x} (27) WO_{3-x} (28)

Composition modulated :	
- Vacancy ordering in planes	$Cu_{2-x}S$ (digenite) (34) $Ni_{2-x}Te_3$ (35)
- Mixed layer compounds	Ba-ferrites (36) Bastnaesite-synchisite series (37)
- Chimney ladder structures	$Mn Si_{2-x}$ (12,13)

The incommensurability may also be related to the fact that the two sublattices out of which the crystal consists have two different, mutually incommensurate periods. This is for instance the case in the "chimney-ladder" structures (12,13). In $MnSi_{2-x}$ the manganese atoms form a simple tetragonal framework; the silicon atoms occupy interstices in this framework following a helical pattern with a period which is not simply related to the period of the manganese arrangement (12).

In many instances the modulation will have mixed character and a clear cut classification becomes difficult (13). A composition modulated structure will also be somewhat deformation modulated. Interface modulated structures may also be composition modulated if the interfaces are non-conservative and in general they will also be deformation modulated as a result of relaxation along the interfaces (14).

In alloy systems modulation usually occurs by the introduction of periodic antiphase boundaries, in one or two dimensions, often also by the introduction of periodic stacking faults thus generating polytypes. The modulation period can often be influenced by the addition of impurities that change the electron/atom ratio, or by changing the composition in binary systems (43). In a number of alloy systems long period antiphase boundary structures are formed at temperatures below, but close to the order-disorder temperature.

3. DIFFRACTION EFFECTS DUE TO MODULATED STRUCTURES

3.1. Interface Modulated Structures

3.1.1. Translation interfaces. We shall discuss in the first place
the one-dimensional modulation resulting from the periodic introduction of parallel planar translation interfaces (antiphase boundaries,
shear planes or stacking faults) with a displacement vector $\bar{R}$. We
shall show that if the basic structure is known, the modulated
structure can be deduced from geometrical features of the diffraction pattern, which consists of arrays of "satellite" spots, associated with the spots due to the basic structure. The satellite
positions are defined with respect to those of the basic spots.
The direction of the arrays of satellites is parallel with the unit
normal $\bar{e}_n$ on the interface. The spacing between satellite spots is
determined by the average spacing d between interfaces. The diffraction vectors $\bar{g}$ of the superstructure are given in terms of the diffraction vectors $\bar{H}$ of the basic structure by the relation :

$$\bar{g} = \bar{H} + \frac{1}{d} (m - \bar{H}.\bar{R}) \bar{e}_n$$

where m is an integer, determining the order of the satellite. Two
sections of reciprocal space are sufficient to determine $\bar{R}$ from the
fractional shift $\bar{H}.\bar{R}$ of the satellite sequence with respect to the
basic spots (38). The intensity of the satellite spots decreases
with increasing m i.e. with increasing distance away from the basic
spots with which they are associated. Although the theory was made
in terms of the kinematical approximation, the positions of the spots
remain unaltered in dynamical conditions. The procedure can easily
be generalized to two and even to three dimensions. We have made
extensive use of this method to determine long period antiphase
boundary structures in ordered alloys.

Orientation and spacing anomalies in incommensurate diffraction
patterns are often difficult to interpret in terms of structures.
However with the help of the corresponding high resolution images an
unambigeous interpretation is in general possible. Orientation
anomalies resulting from systematic ledging of periodic APB's are
directly visible in the image. Spacing anomalies may result from
a systematic mixing of different spacings between APB's as in Au_4Zn
(17); also this can be concluded from the images.

If the required information is the average spacing of APB's the
use of the diffraction pattern is indicated; if also the frequency
distribution of spacings is needed, lattice fringes are required.
The degree of resolution to be used should be adapted to the level
of information searched for.

3.1.2. Periodic reflection - twin interfaces. We consider a crystal periodically and coherently twinned on the (hkℓ) planes, without a contraction or a dilatation along the twin interface in such a way that the distance between interfaces is an integral number n of the $d_{hkℓ}$ value. A polysynthetically twinned crystal can then be characterized by labeling the widths of successive twin lamella of type I and II by the number sequence n_1, n_2, n_3, ... n_N in $d_{hkℓ}$ units. Let the lattice potential of a block with thickness b, limited by two twin planes, be given by $V_B(\bar{r}) = V_0(\bar{r})B(x)$ where $V_0(\bar{r})$ is the potential of the considered unperturbed twin variant, I or II, the origin taken in the center of the block, and B(x) is a factor describing the width of the block; it is defined as B(x) = 1 for $-\frac{b}{2} < x < \frac{b}{2}$ and B(x) = 0 outside of this interval. The x-axis is taken along $\bar{e}$, normal to the twin plane. Calling $\bar{A}_3$ the translation vector, corresponding to one superperiod (fig. 1a & b) of which the magnitude is $\bar{A}_3.\bar{e}$, the potential of the periodic assembly can be described by the convolution product :

$$V(\bar{r}) = V_B(\bar{r}) * p(\bar{r}) = [V_0(\bar{r}) B(x)] * p(\bar{r})$$

with $p(\bar{r}) = \sum_n \delta(\bar{r}-n\bar{A}_3)$ (δ = Dirac's functional; n is an integer). Fourier transformation of $V(\bar{r})$ leads to :

$$\tilde{V}(g) = [\tilde{V}_0(g) * \tilde{B}(g_x)] \tilde{p}(\bar{g})$$

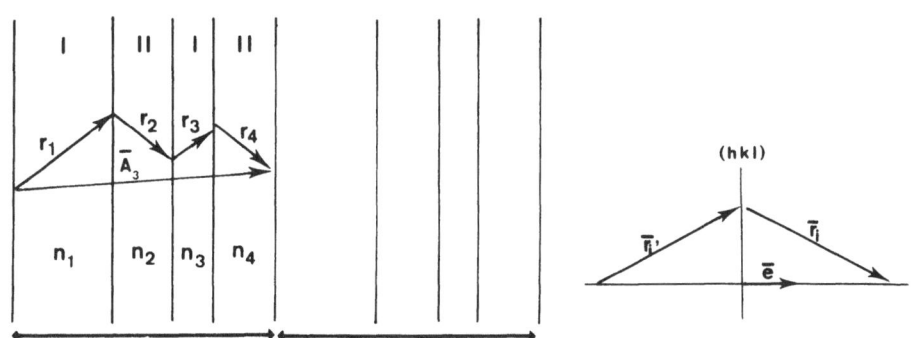

FIG. 1. (a) Geometry of the unit cell. $\bar{r}_i$ (i=odd) are parallel lattice vectors of variant I, $\bar{r}_i$ (i=even) are parallel lattice vectors of variant II. It is clear that the period is given by
$$\bar{A}_3 = \sum_{i=1}^{N} \bar{r}_i.$$
(b) Relation between a lattice vector $\bar{r}_i$ and its twin related vector $\bar{r}'_i$, $\bar{e}$ is the unit vector normal to the twin planes.

where $\tilde{V}_0(\bar{g})$ represents the diffraction pattern of the perfect crystal consisting of Bragg spots at positions $\bar{h}$. From the convolution product :

$$\tilde{V}_0(\bar{g}) * \tilde{B}(\bar{g}_x) = \tilde{V}_0(\bar{g}) * (sin\pi\, g_x b/\pi g_x)$$

it follows that the diffracted intensity is located on lines parallel to $\bar{e}$ through each of the Bragg spots of the basic structure. The intensity decreases with increasing distance g_x away from the Bragg spot position $\bar{h}$.

The positions g_x at which diffraction spots occur are determined by the factor :

$$p(\bar{g}) = \int e^{-2\pi\bar{g}.\bar{r}} \quad \sum_n \delta(\bar{x} - n\bar{A}_3) = \sum_n e^{-2\pi i\bar{g}.n\bar{A}_3}$$

which is a sharply peaked function around the positions $\bar{g}.\bar{A}_3 = m$ (m = integer) from which we deduce :

$$g_x = \frac{1}{d}\,(m - \bar{h}.\bar{A}_3) \tag{1}$$

i.e. the spot sequence associated with the Bragg reflection $\bar{h}$ is shifted over the fraction $-\bar{h}.\bar{A}_3/d$ (mod 1/d) of the distance between satellite spots which are regularly spaced over intervals 1/d such that :

$$d = d_{hk\ell}\,\sum_{i=1}^{N} n_i$$

The shift of the satellite reflections of $\bar{h}_I$ results from the fact that $\sum_{i=even} \bar{r}_i$ is not necessarily a lattice vector of I. The vector $\bar{r}_i^{-1} = 2n_i d_{hk\ell}\,\bar{e}-\bar{r}_i$ is twin related to $\bar{r}_i$ with respect to the (h k ℓ) twin plane ;it is therefore a lattice vector of the variant I. Hence the satellite shift at the reflection h_I is equal to :

$$\Delta_I = -2(\bar{h}_I.\bar{e})\,\sum_{even} n_i \,/\, \sum_{i=1}^{N} n_i$$

so that :

$$(h_x^I + \Delta_I)\,/\,h_x^I = [\sum_{i=odd} n_i - \sum_{i=even} n_i]\,/\,\sum_{i=1}^{N} n_i$$

This result suggests a simple procedure to deduce the number of blocks $\sum_{i=odd} n_i = N^I$ and $\sum_{i=even} n_i = N^{II}$ within one cell, and hence to find the structure.

We shall illustrate the procedure by means of fig. 2b, which is an optically simulated diffraction pattern of the grating shown in fig. 2a. Take the a-axis parallel with the satellite rows through the origin and the b-axis perpendicular to it. Take the c-axis along a systematic line of satellites through the origin. The first major reflection $\bar{h}$ along the a-axis is the (hkℓ) Bragg reflection corresponding with the (hkℓ) plane in the untwinned matrix. Construct

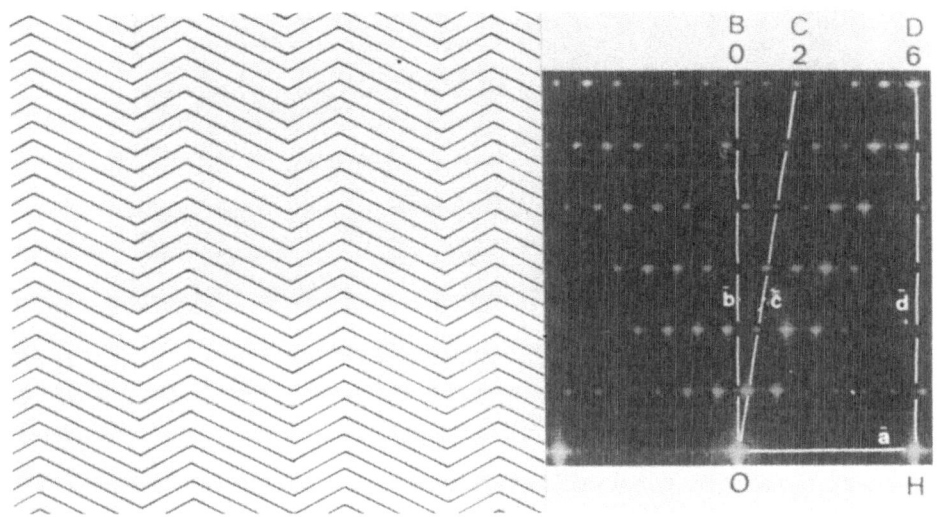

FIG.2. (a) Model of twinned system used as a grating for the simulation of an optical diffraction pattern (see fig. 2b).
(b) Optical diffraction pattern obtained from the grating of fig.2a.

a line d through the spot $\bar{h}$ parallel to $\bar{b}$. This line will pass through the strongest satellite of say the mth row. In the case shown $m = 6$. Let us call B, C and D respectively the satellites of the mth row situated on the lines b, c and d respectively. If one starts counting at B, each satellite is assigned a number $B \rightarrow n_B$, $C \rightarrow n_C$, $D \rightarrow n_D$, etc. in the concrete case $n_B = 0$, $n_C = 2$ and $n_D = 6$. According to the results deduced above $N^I + N^{II} = 6$; $N^I - N^{II} = n_C = 2$ so that $(N^I, N^{II}) = (4,2)$ in agreement with the grating used (fig.2b). Applications of this method can be found in ref. (39).

3.1.3. Inversion boundaries. A non-centrosymmetric crystal consisting of strips of equal width containing + and - structures related by an inversion operation also acts as a diffraction grating with a lattice constant equal to the distance between homologeous interfaces. Domain contrast is produced in multiple beam situations along zones which do not produce a center of symmetry in projection. This is a result of the violation of Friedel's law under multiple beam conditions. One-dimensional as well as three-dimensional inversion boundary modulated structures have been observed first in the χ-phase $Fe_{36}Cr_{22-x}$ (Ti, Mo)$_x$ (22) with $8 < x < 10$ and subsequently in γ-brass (23) by means of electron microscopy as well as by electron diffraction.

FIG. 3. Phase transition between the α-phase and β-phase in quartz. The phase front is seen to consist of a regular array of prismatic twins of which the mesh is reducing as the β-region is approached.

3.1.4. Periodic rotation-twin interfaces. Dauphiné twins in non-centrosymmetric α-quartz are related by a 180° rotation about the threefold axis. Dauphiné twin interfaces separating α_1 and α_2 can be revealed by structure factor contrast, by imaging in a common reflection such as $30\bar{3}1$ for which the structure factor is quite different in the two twin related parts, or by a reflection which is absent in one of the two parts.

It was found that close to the transition temperature (573°C) the α-crystal phase breaks up in triangular prismatic domains with the prismatic axis parallel with the c-axis. These domains form regular arrays of which the mesh size is a decreasing function of th temperature (fig. 3). Close to the transition temperature an incommensurate intermediate structure consisting of a regular arrangement of α_1 and α_2 domains is formed with a lattice parameter of ∿150 Å and less. This incommensurate superstructure can be observed direct in electron microscopy but can also be detected by small angle neutr Bragg scattering (see Dolino et al. this conference).

It is found by heating in the electron microscope close to the transition temperature that isolated Dauphiné twin interfaces as wel as those in the coarse networks constantly vibrate, creating in this way a transition region along the interface which transforms con-stantly from α_1 into α_2 and vice versa. Fine networks with a mesh size in the range down to about 20 nm remain relatively stable and can be photographed as long as the vibration amplitude remains below the mesh size.

3.2. Deformation Modulated Structures

In the simplest case the periodic deformation can be described by a displacement wave of the type :

$$\bar{R}(\bar{r}) = \bar{A} \cos (\bar{q}.\bar{r} + \phi)$$

where $\bar{A}$ is the polarization vector of the wave, $\bar{q}$ the wave vector and ϕ the phase. The diffraction pattern then consists of the Bragg spots $\bar{H}$ due to the undeformed structure and each Bragg spot has usually only two visible satellites at positions given by $+\bar{q}$ and $-\bar{q}$. If the deformation wave is not purely sinusoidal, higher harmonics with wave vector $2\bar{q}$, $3\bar{q}$, ... may become important. This is reflected in the diffraction pattern by the presence of higher order satellites corresponding with these wave vectors. In electron diffraction it is not always possible to decide whether these higher order satellites are due to double diffraction or to genuine higher harmonics. The polarization vector of different harmonics may be different. If the polarization vector is perpendicular to a diffraction vector the corresponding satellite disappears unless it can be produced by double diffraction. The symmetry of the deformation pattern may also give rise to systematic extinctions in the satellite pattern. Deformation waves may be present simultaneously along two or more directions related by symmetry elements of the basic structure; this is reflected in the diffraction patterns by sequences of satellites along each of these directions.

The intensity of the satellites is proportional to the intensity of the Bragg spot $\bar{H}$ with which they can be associated and moreover with the quantity $|(\bar{H}+\bar{q}).\bar{A}|^2$ (40). This means that the satellites associated with high order Bragg reflections of the basic lattice, i.e. with large $\bar{H}$ vectors will be relatively more intense than those associated with low order Bragg reflections, i.e. with small $\bar{H}$. This is a characteristic feature of the diffraction pattern of deformation modulated structure.

Typical incommensurate deformation modulated structures occur in many transition metal dichalcogenides such as TaS_2 (8,9), VSe_2(7). At sufficiently low temperatures they may become commensurate; e.g. in TaS_2, in $NbTe_2$ and in VSe_2 respectively the $\sqrt{13}$, $\sqrt{19}$ and the $\sqrt{7}$ commensurate structures lock-in.

4. HIGH RESOLUTION IMAGES

Different imaging modes have been used in the study of ordered alloys. A first mode consists in using dark field DF images made in superstructure spots, excluding the spots due to the basic structure (often the FCC structure). In this way irrelevant and redundant information is rejected intentionally. One should make sure that the intensities of the selected superstructure spots are as large as possible as compared to those due to the basic spots and that their intensity distribution reflects the symmetry of the crystal zone, which is being imaged. In this way, all beams producing spots located on a circle centered on the optical axis undergo the same phase shift

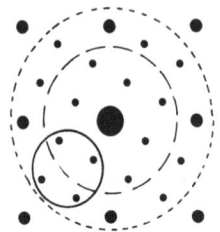

FIG. 4. Schematic view of positions and sizes of the aperture with respect to the diffraction pattern of an fcc based alloy. In the case shown the alloy is Au_4Mn.
Mode 1 :Dark field superlattice mode. Only superlattice reflections within one mesh of the fcc reciprocal lattice are admitted (full circle).
Mode 2 :Bright field superlattice mode. The direct beam as well as all superlattice reflections are admitted (dashed circles).
Mode 3 : Basic reflection mode. At least one shell of fcc basic reflections is admitted together with the superlattice reflections (dotted circle).

as a result of instrumental factors, since they enclose the same angle with the optical axis of the microscope; they then interfere with the correct phase relationship.

A second mode (bright field) combines the direct beam with all surrounding superlattice reflections, again excluding all basic spots. The specimen thickness should preferentially be such as to maximize the intensities of the superlattice reflections, as compared to that of the direct beam, in order to obtain optimum contrast. Whereas the images of antiphase boundaries are emphasized in the DF mode, the atomic positions are better defined in this BF image.

One can alternatively produce BF images, including also the basic spots for such a thickness as to minimize the intensities of the latter. In the thinner parts the resulting image is complicated by the simultaneous presence of dots representing the superstructure and the basic structure. In the very thin parts the basic structure is revealed, whereas in the thickest parts, the superstructure is emphasized. This behaviour is essentially a consequence of the large difference in effective extinction distance between beams due to the basic structure and those due to the superstructure.

The different imaging modes are represented schematically in fig.4.
For "column structures" it was shown semi-analytically that under suitable conditions, i.e. for the correct thickness and for the appropriate defocus value, images can be obtained which are directly interpretable in terms of structural features (41) e.g. the minority atoms being represented as bright dots. These conditions are fortunately often fulfilled for images which exhibit the highest contrast and are thus intuitively looked for by a good observer.
In order to confirm the intuitive interpretation of high resolution images it is always necessary to compare the observed images

with computer simulated ones. The computation of images of long period superstructures is time consuming in view of the large size of the unit cell and the large number of atoms involved. It was therefore of interest to investigate the effect of the presence of antiphase boundaries on structure images (42). It was found that the presence of an antiphase boundary only affects the images of the rows of atom columns immediately adjacent to the antiphase boundary, in sufficiently thin crystals. The thinner the crystal the more the effect of the antiphase boundary is restricted to its immediate vicinity. The image of the structure within the domains is the same as in the perfect crystal. As a result one can interpret the images of long period interface modulated structures

     by comparing the structure image within the domains with computed images of the perfect structure without worrying about the effect of antiphase boundaries on the image. The shift of the structure image in the domains on either side of the boundary is a good representation of the displacement caused by the APB.

     The intensities of the dots representing columns of a different chemical nature are different under the appropriate conditions i.e. in sufficient thick crystals. Zones along which the structure is a "column structure" will therefore allow to image columns containing atoms of a different chemical nature as dots of different intensities. In some non-column zones of multilayer crystals it was found that the image only reveals the structure in one layer for one focus and of another layer for a different defocus. The phenomenon is not well understood as yet. It clearly carries the promise of obtaining images of sections of the crystal rather than imaging the projection only. A striking example of this phenomenon was obtained in the ABCB stacked alloy Au-Mg (16).

5. CASE STUDIES

     We shall now present a number of typical case studies based on work in our laboratory, which will make it possible to illustrate a number of aspects and allow some general conclusions.

5.1. The Cu₃Sn Alloy (43)

     The Cu_3Sn alloy structure, as described in the literature, is derived from the Cu_3Ti type orthorhombic basic structure, which itself is a superstructure of the hexagonally close-packed structure. Our electron diffraction data and structure images confirm the structure proposal by Schubert (44) and its description as a long period antiphase boundary structure with a period of 10 times the b-spacing of the orthorhombic cell (fig.5). A diffraction pattern of the [001] zone is shown in fig. 6, where the superlattice spots are clearly visible along the (oko) rows (k=odd) the spots along k=even rows are systematically weaker, they are due to double diffraction. The frac-

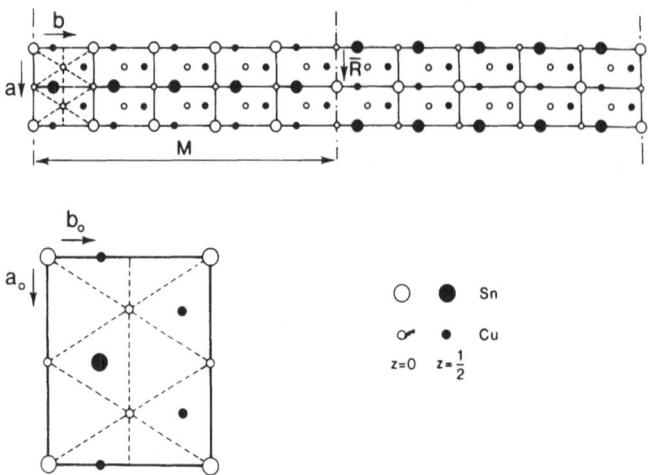

FIG. 5. Structure model for the LPAPB structure with b = 10b$_o$ of
the basic orthorhombic cell shown below. Also the displacement
vector $\bar{R} = \frac{1}{2}[100]$ of the APB's is indicated. , ● Sn, • Cu
(O z = 0, ● z = 1/2).

tional shifts are 1/2 for k : odd and 0 or 1 for k : even. This is
consistent with an LPAPB structure consisting of APB's with a displacement vector R = 1/2[100]. The structure images of fig. 7 show
the increasing degree of information obtained if more reflections
can be used for imaging (see § 4). In fig. 7a only superstructure
reflections and in 7b also the basic reflections are included in the
aperture whereas the structure image of fig.7c was obtained with the
spots shown in the inset; it reveals the antiphase boundary structure
in an alloy containing some Zn and which, as a result exhibits a
period smaller than 10b. The configuration and scale of the bright
dots is the same as that of the minority atoms in the structure model of fig. 5.

In order to investigate the effect of aliovalent impurities on
the long period alloys with small additions of Zn and Ni were investigated. $Cu_{3-x}SnZn_x$ alloys with x = 0,04; 0,07 and 0,10 and 0.25
were prepared. Dark field images made by using superlattice spots
of one row k = odd are sufficient to reveal the long period and
study also the frequency distribution of the spacing. A series of
such images is shown in fig. 8 and the relevant rows of superlattice
spots are reproduced along side. The diffraction pattern yields the
average period only. It is found that for an addition x = 0,10 of
zinc the period is approximately only half of that in pure Cu_3Sn.
The addition of nickel also causes a reduction of the period but in
a much less pronounced way (43).

The presence of three orientation variants, differing 120° in
orientation, consistent with the fact that the disordered basic

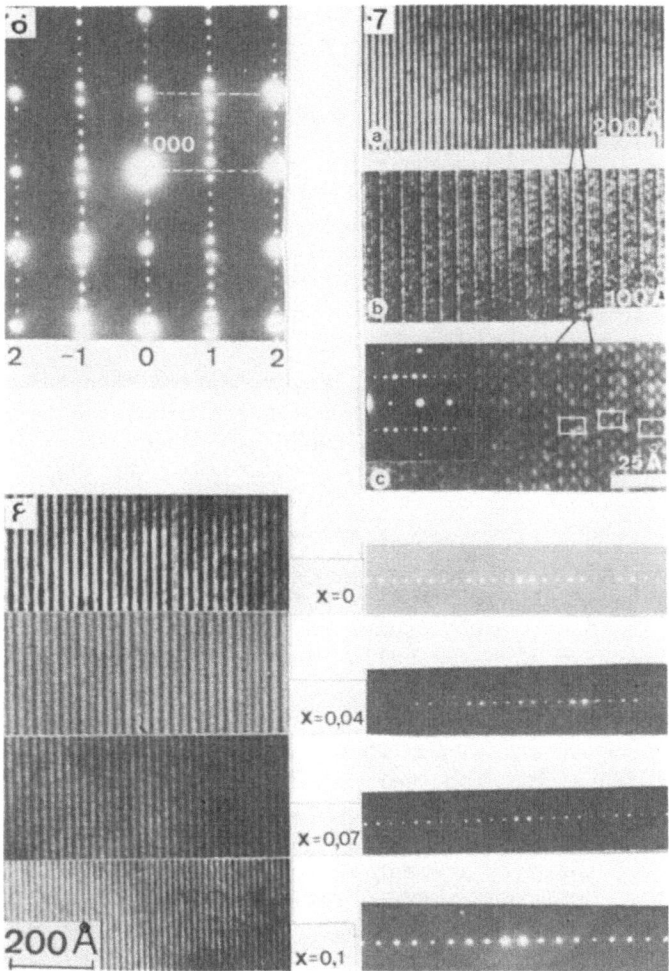

FIG. 6. Diffraction pattern of the long periodic antiphase boundary structure observed in Cu$_3$Sn. [001]-zone pattern.

FIG. 7. Line images revealing the periodic antiphase boundaries :
(a) only superlattice spots are used for imaging;
(b) by using also basic spots the (100) planes corresponding with the basic orthorhombic structure can also be observed in between the APB's. LPAPB strips with M = 3, 4 and 5 are observed;
(c) high resolution structure image of the Cu$_{3-x}$Zn$_x$Sn (x=0.14) alloy where the configuration of white dots is found to be that of the Sn atoms in the Cu$_3$Sn structure projection. The inset show the spots used for imaging.

FIG. 8. Long period antiphase boundary arrangements in alloys of the type Cu$_{3-x}$Zn$_x$Sn, with x-values 0, 0.04, 0.07 and 0.1. The corresponding diffraction patterns are also shown.

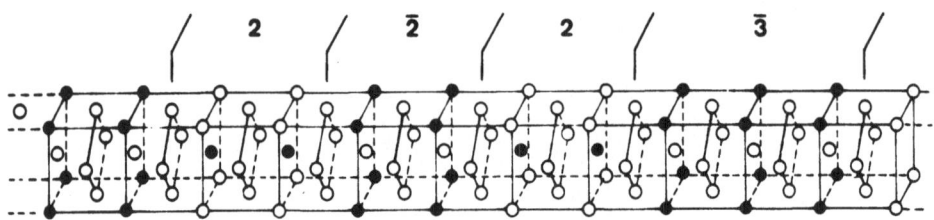

FIG. 9. Long period antiphase boundary modulated structure derived
from the Cu$_3$Au-structure. FCC unit cells are displaced over
$\bar{R} = \frac{1}{2}[110]$. The number of cells between subsequent APB's is indicated.

structure is hexagonally close packed,was also observed.

Whereas the effect of impurities on the long period in FCC
based alloys is rather well documented, Cu$_3$Sn was the first example
of such an effect in an hexagonally close-packed alloy.

5.2. The Au$_{4-x}$Zn Alloy System (17)

The one-dimensional long period structure of this alloy is
derived from the Cu$_3$Au type structure which itself is based on the
FCC structure. It is a long period, antiphase-boundary-modulated
LPAPB structure, the displacement vector of the APB's being of the
type $\bar{R} = \frac{1}{2}[110]$. This superstructure is shown in fig. 9. For this
particular case we can designate the numbers of FCC unit cells along
the direction of the long period by a symbol such as 2 $\bar{2}$ 2 $\bar{3}$ ...;
where the -sign designates a block of displaced unit cells.

The average number of FCC unit cells in one block is called the
M number; the case shown thus corresponds to M = 2.25. The long
period structure of this alloy is somewhat dependent on the exact
heat treatment; it is even possible to obtain two-dimensional
long period superstructures under suitable conditions.

Diffraction patterns such as the one shown in fig.10(b,d,f) are
often obtained; they contain spacing anomalies and sometimes also
orientation anomalies, suggesting that the structures are incommensurate. We shall show by means of electron microscopy, that the
structures are in fact pseudo-incommensurate.

The pattern of fig. 10b is the simplest. The positions of the
Cu$_3$Au spots, which are absent, are marked by crosses. It is clear
that the fractional shift is $\frac{1}{2}$ for the spots with k = odd and 0 or 1
for the spots with k = even. This is consistent with the displace-

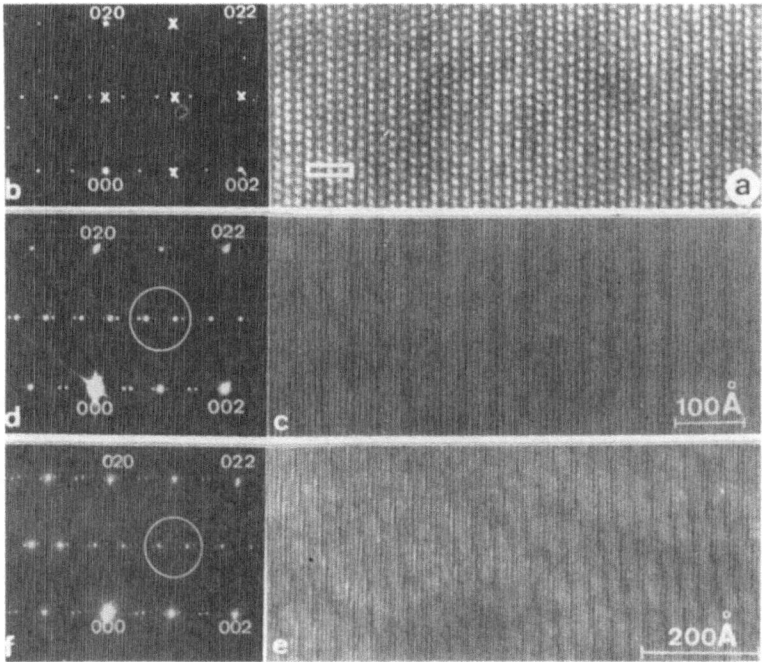

FIG. 10. Diffraction patterns and corresponding images of Au$_{4-x}$Zn
LPAPB structure. (a) and (b) belong to the commensurate 2$\bar{2}$2$\bar{2}$-struc-
ture; (c)(d) and (e) (f) are pseudo incommensurate structures. For
the images (c) and (e) the reflections encircled in the corresponding diffraction patterns have been used. For fig. (a) a larger
objective aperture has been used; white dots correspond to the Zn-
configuration.

ment vector R = ½[110]. The period (which is 2M) is clearly equal
to four times the lattice parameter of the FCC unit cell. The diffraction pattern thus corresponds to M=2, i.e. to the sequence
2 $\bar{2}$ 2 $\bar{2}$. A high resolution image of such a structure is shown in
fig. 10a and confirms the model.

The incommensurate pattern of fig. 10d can be analysed in a
similar manner. The satellites belonging to the same basic spot
are connected by brackets in fig. 11. It is clear that now the M
value is 2.25, since the distance between superlattice spots is 4.5
times smaller than the distance between FCC spots. The fringes obtained by collecting the spots encircled in fig. 10(d,f) are shown
in fig. 10 (c,e). The sequence deduced from the spacing of the
fringes is 2 $\bar{2}$ 2 $\bar{3}$, i.e. this is the sequence represented in the
model of fig. 9. The incommensurate value is thus due to the regular mixture of spacings, leading to an average value of 2.25 (17).
This is a typical example of a pseudo-incommensurate superstructure.

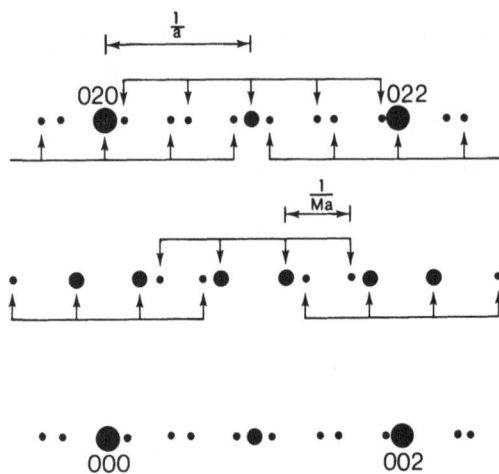

FIG. 11. Schematic representation of a diffraction pattern exhibiting a spacing anomaly; satellites belonging to the same group are
indicated.

In some cases also an <u>orientation anomaly</u> is present in the
diffraction pattern (fig. 10f). The interpretation can be deduced
from the corresponding image of fig. 10e. The lattice fringes
corresponding with the antiphase boundaries, and obtained by collecting the spots encircled in fig. 10f are shown in fig. 10e. Although
the antiphase boundaries are still in cube planes the "singular" strip
domains, which are three FCC unit cells wide, have an <u>average</u> orientation which encloses a small angle with the cube direction. This
is a consequence of the presence of systematic ledges in the antiphase boundaries, in the manner shown schematically in fig.12. In
this way the singular strips are in fact uniformly dispersed
throughout the configuration of antiphase boundaries.Such orientation
anomalies frequently occur in alloys which have only been annealed
for a short period of time; after a sufficiently long annealing
time the ledges are eliminated by migration along the APB's. The
interpretation of the orientation anomaly was confirmed by optical
simulation experiments, whereby a line pattern, as the one shown in
fig. 12, was used as an optical grating.

5.3. The Au-Mn System (15)

This alloy system is extraordinarily rich in ordered phases.
We shall limit our discussion to phases with compositions in the
intervals $Au_{3+x}Mn$ (0.3 < x<0.5) and $Au_{4-x}Mn$ with o < x < 0.5. We
shall first discuss the basic alloy structures corresponding with
the simple compositions Au_4Mn and Au_3Mn. Afterwards we discuss the
structure of alloys with deviating compositions and we shall show
that one or two-dimensional long period superstructures of these
simple structures are formed, which incorporate these composition

$$\bar{3}\ \ 2\ \ \bar{2}\ \ 2\ \ \bar{3}\ \ 2\ \ \bar{2}\ \ 2\ \ \bar{3}\ \ 2\ \ \bar{2}\ \ 2$$

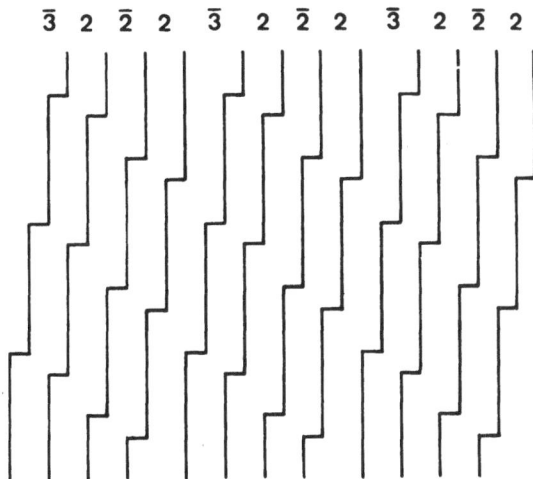

FIG. 12.   Schematic illustration of the APB jogging giving rise to
inclined average APB planes.   The concentration of ledges has been
exaggerated.

changes along non conservative APB's.
This phenomenon is rather similar with the formation of crystallographic shear planes in non-stoichiometric oxides.   Still smaller
changes can be accommodated in periodic APB structures by orientation changes of the APB's similar to the "swinging" CS-planes in
shear structures.

5.3.1. Ordering in Au_4Mn.   The alloy Au_4Mn has the tetragonal Ni_4Mo
structure, which is a superstructure of the FCC structure (fig. 13a).
It can be formed in six orientation variants within the same   FCC
matrix, the variant generating group being 32.   There are two variants for each direction of the tetragonal axis, which is parallel
to one of the cube directions of the FCC structure (fig. 14).   The
reciprocal lattice of one variant is represented in fig. 13b and the
diffraction pattern for one and two variants is shown as inset in
fig. 15.   Dark field high resolution images have been made using
the eight superlattice reflections on a circle within the reciprocal
unit cell of the FCC lattice, and which are produced by two coaxial
variants.   This imaging mode is exceptional in that all beams can be
made to suffer the same phase shift as a result of instrumental
abberations since they then all enclose the same angle with the optical axis. Both variants are then imaged equally well.   This imaging
mode makes use of the minimum information needed to image the superstructure; no information is included concerning the basic structure.

The arrays of bright dots in the images have the scale and
configuration of the manganese columns in the structure when viewed
along the tetragonal axis.   Image calculations have confirmed that
under the right imaging conditions and at an appropriate thickness

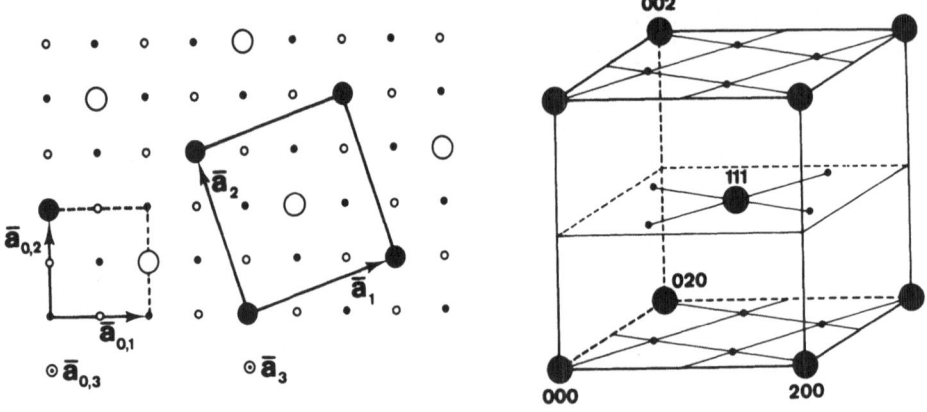

FIG.13. (a) Structure of Au_4 Mn. The large circles represent Mn at
two different levels $(0, \frac{1}{2})$ whereas the small circles represent gold
atoms at two levels $(0, \frac{1}{2})$. The base vectors of the FCC basic struc-
ture are $\bar{a}_{0,1}$, $\bar{a}_{0,2}$ and $\bar{a}_{0,3}$; the base vectors of the Au_4Mn structure
are $\bar{a}_1$, $\bar{a}_2$ and $\bar{a}_3$.
(b) Reciprocal lattice of the Au_4Mn-structure.

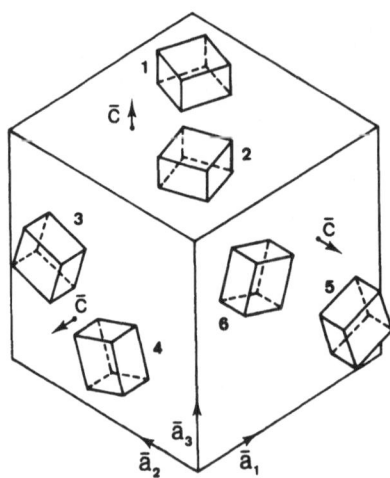

FIG.14. Six orientation variants of the Au_4Mn structure represented
with respect to the FCC cube.

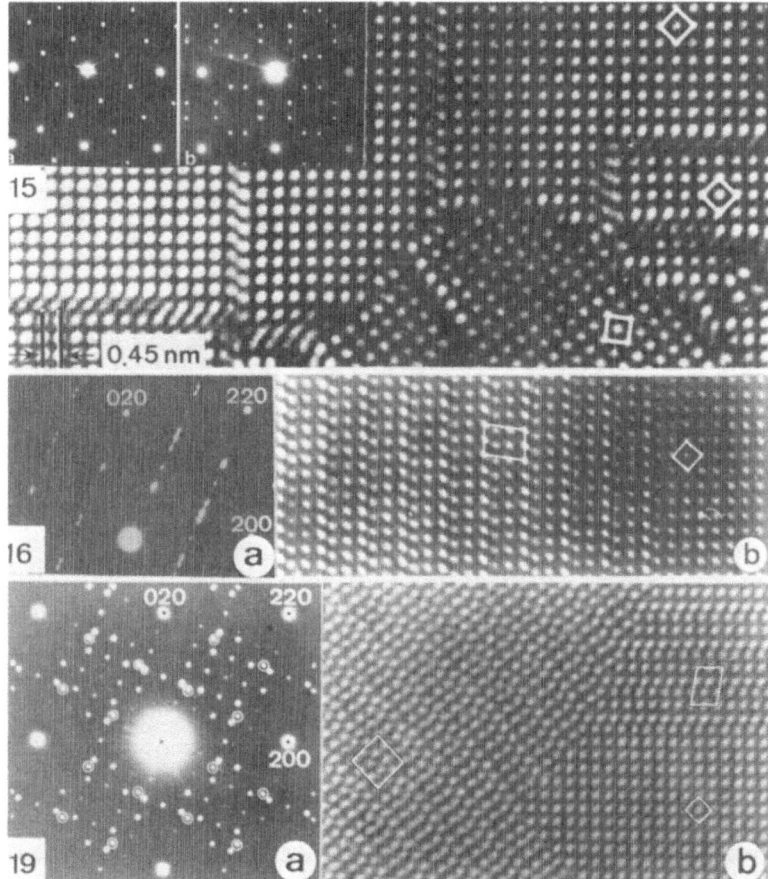

FIG. 15. Dark field structure image of multidomain structure of Au$_4$Mn APB's as well as a twin interface can be recognized. The corresponding pattern is shown as an inset.

FIG. 16. (a) Mono-domain diffraction pattern of the one-dimensional long period superstructure and from the Au$_4$Mn structure from which it is derived. Note the relative spot shifts which are represented schematically in fig. 17.
(b) Bright field image in superstructure spots of the monoclinic one-dimensional long period superstructure derived from the Au$_4$Mn structure. The projection of the unit cells are indicated. The white dots represent manganese columns.

FIG. 19. (a) Diffraction pattern of the square shaped island structure, derived from the Au$_4$Mn structure. Also the spots due to the variant of the Au$_4$Mn structure from which the island structure is derived are visible. Note the relative shifts of the spot sequences which are represented schematically in fig. 20.
(b) Multiphase area containing the one-dimensional structure (upper right), the square island structure (left) and the Au$_4$Mn structure (bottom right). The unit cells are outlined.

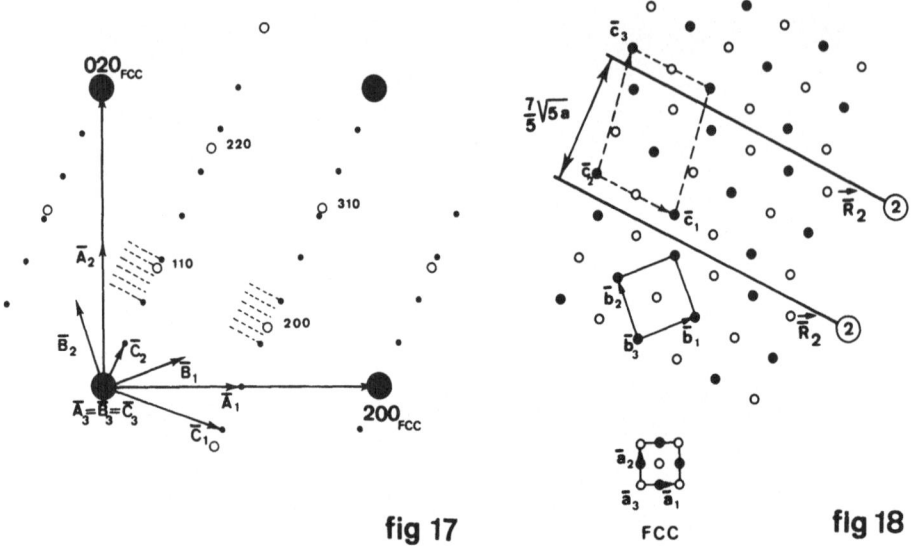

fig 17

FCC   **fig 18**

FIG. 17. Schematic representation of the mono-domain pattern of the monoclinic one-dimensional long period superstructure (full dots). The open circles represent the diffraction spots due to the variant of the Au_4Mn structure from which the long period superstructure is derived. Note the spot shifts which are multiples of 1/5 of the interspot distance.

FIG. 18. Model of the monoclinic one-dimensional long period superstructure of the Au_4Mn structure. The unit cells of the Au_4Mn and of the basic f.c.c. structure are indicated as well. Only manganese atoms are represented.

the manganese columns are imaged as bright dots. Antiphase boundaries and twin interfaces are clearly revealed under these conditions (fig. 15).

5.3.2. One-dimensional superstructures. In certain parts of the specimen with composition $Au_{4-x}Mn$ one finds a diffraction pattern as reproduced in fig. 16a and represented schematically in fig. 17. From the geometry of this pattern one can deduce the structure model represented in fig. 18 and in which only the manganese positions are shown; it is evidently a one-dimensional long period superstructure derived from the Au_4Mn structure. From the shifts of the satellites (table 2) the displacement vectors of the antiphase boundaries can be derived to be $\bar{R} = \frac{1}{10}[3\bar{1}5]$ as referred to the Au_4Mn basic lattice. The separation between successive APB's is 1,25 nm. The monoclinic unit cell has base vectors $\bar{c}_1$, $\bar{c}_2$ and $\bar{c}_3$ which are related to those of the tetragonal unit cell of Au_4Mn, $\bar{b}_1$, $\bar{b}_2$, $\bar{b}_3$ by means of the relation :

TABLE II. One-dimensional superstructure

$\bar{g}(Au_4Mn)$	calculated shift in array 2 $\bar{R}_2 = \frac{1}{10}[3\bar{1}5]$ $\bar{g}.\bar{R}$	observed fractional shifts
(1) 110	1/5	1/5 (or 4/5)
(2) 220	2/5	2/5 (or 3/5)
(3) 200	3/5	3/5 (or 2/5)
(4) 310	4/5	4/5 (or 1/5)
(5) 3$\bar{1}$0	0	0 (or 1)

$$\begin{pmatrix} \bar{c}_1 \\ \bar{c}_2 \\ \bar{c}_3 \end{pmatrix} = \begin{pmatrix} 1 & -1 & 0 \\ 0 & 0 & 1 \\ 5/5 & 8/5 & 0 \end{pmatrix} \begin{pmatrix} \bar{b}_1 \\ \bar{b}_2 \\ \bar{b}_2 \end{pmatrix}$$

The pointgroup is 2/m and the structure can thus be formed in twelve orientation variants starting from a disordered FCC matrix. The ideal composition is $Au_{11}Mn_3$. This model can be compared with the high resolution image of fig. 16b. Note that along the antiphase boundaries the configuration of atoms is similar to that in the DO_{22}-structure.

5.3.3. Two-dimensional long period structures. In high resolution images of an alloy with composition $Au_{4-x}Mn$ one finds areas such as reproduced in fig. 19b (left part). A monodomain diffraction pattern from such an area is reproduced in fig. 19a and represented schematically in fig.20; the Au_4Mn spots coming from an adjacent area are also present. The most intense superlattice spots occur close to Au_4Mn spots (surrounded by small circles) the intensity of these spots decreases with their distance from the Au_4Mn spots. This suggests that the long period superstructure is based on the Au_4Mn basic structure. This could of course also be concluded even in the absence of the Au_4Mn spots, of which the positions can be computed with respect to the FCC spots. The long period superstructure spots form a square array which is shifted, with respect to the Au_4Mn spots, over 1/5 (or 4/5) and 2/5 (or 3/5) of an interspot distance along the densest rows of spots. These shifts are represented schematically in fig. 20. The interfaces are perpendicular to the densest rows of spots and their distance is inversely proportional to the interspot distance. The high resolution images thus suggest the presence of two mutually perpendicular families of APB's indicated as (1) and (2) in fig. 21. Let the corresponding displacement

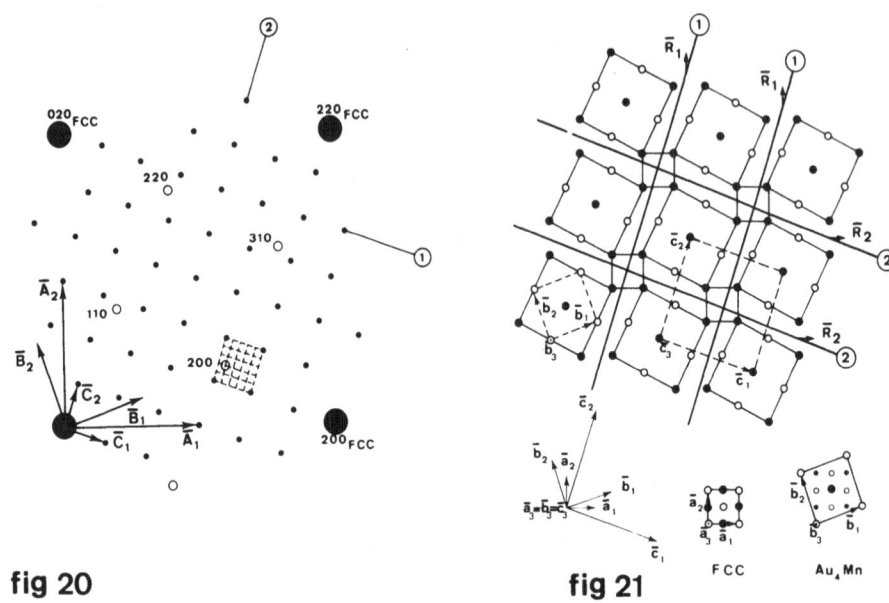

fig 20

fig 21

FIG.20. Schematic representation of the mono-domain diffraction patte of the square island superstructure of the Au_4Mn structure (full dots Open dots represent the diffraction spots due to the variant of the Au_4Mn structure from which the island structure is derived. Note the spot shifts which are multiples of 1/5 in both directions.

FIG. 21. Model of the square island structure and its relation with the Au_4Mn and the f.c.c. structure. The family of APB's 1 (2) is determined by the row of superlattice spots 1 (2) in fig. 20.

vectors be respectively $\bar{R}_1$ and $\bar{R}_2$. If $\bar{R}_1 = [u\ v\ w]$ we can conclude from the fractional displacements, summarized in table 3, that $2u = = 1/5$ (mod.1) or $u = 1/10$; whereas $u + v = 2/5$ (mod 1), i.e. $v = 3/10$ The w component is 5/10 since the vector $\bar{R}_1$ has to be of the type $\frac{1}{2}<110>_{FCC}$; finally $\bar{R}_1 = \frac{1}{10}[135]$. Similarly one finds $\bar{R}_2 = \frac{1}{10}[3\bar{1}5]$. Both displacement vectors are consistent with the observed spot shifts. The structure model suggested by these considerations is represented in fig. 21; it can be compared with the high resolution image of fig. 19a. The size of the unit cell in the direction perpendicular to the plane of the drawing was deduced from a diffraction pattern along a zone perpendicular to the normal on this plane. The ideal composition is $Au_{31}Mn_9$. This structure can best be described as being a "square island" superstructure of Au_4Mn. It was completely determined by a combination of electron diffraction and high resolution electron microscopy.

5.3.4. Ordering in Au_3Mn. The Au_3Mn phase adopts the DO_{22} structure represented in fig. 22a and of which the reciprocal lattice is shown in figure 22b. This ordered structure can occur in three orientation variants within the same FCC disordered matrix; the tetragonal

TABLE 3. Two-dimensional superstructure

$\bar{g}(Au_4Mn)$	$\bar{g}.\bar{R}$ values			
	observed(mod 1)		calculated $\bar{R}_1=1/10\ [135]$ $\bar{R}_2=1/10[3\bar{1}5]$	
	array 1	array 2	array 1	array 2
(1) 200	1/5 or 4/5	2/5 or 3/5	1/5	3/5
(2) 110	2/5 or 3/5	4/5 or 1/5	2/5	1/5
(3) 220	4/5 or 1/5	3/5 or 2/5	4/5	2/5
(4) 310	3/5 or 2/5	1/5 or 4/5	3/5	4/5

axis being parallel to one of the cube directions. The variant generating pointgroup is 3 (1). Two orientation variants are visible in
the high resolution image of fig. 23. From computed images it can
be concluded that under suitable conditions the bright dots represent the manganese columns. A number of long period superstructures
derived from this structure were found in alloys with nominal
composition $Au_{3+x}Mn$.

5.3.5. One-dimensional long period superstructures. We shall first
discuss the simplest one-dimensional orthorhombic long period
superstructure of which the diffraction pattern is reproduced in the
inset of fig.24. It is clear that sequences of satellites are

DO₂₂

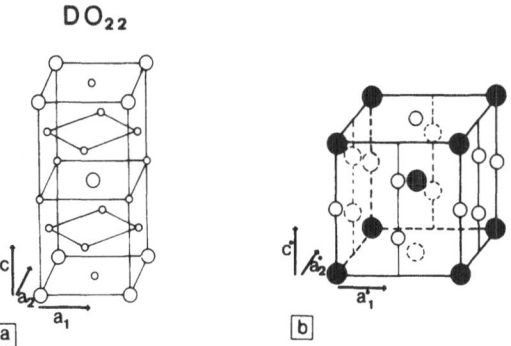

FIG. 22. Schematic representation of the DO_{22}-structure.
(a) Direct structure
(b) Reciprocal lattice
Black dots are fcc reflections, open circles represent superlattice
reflections.

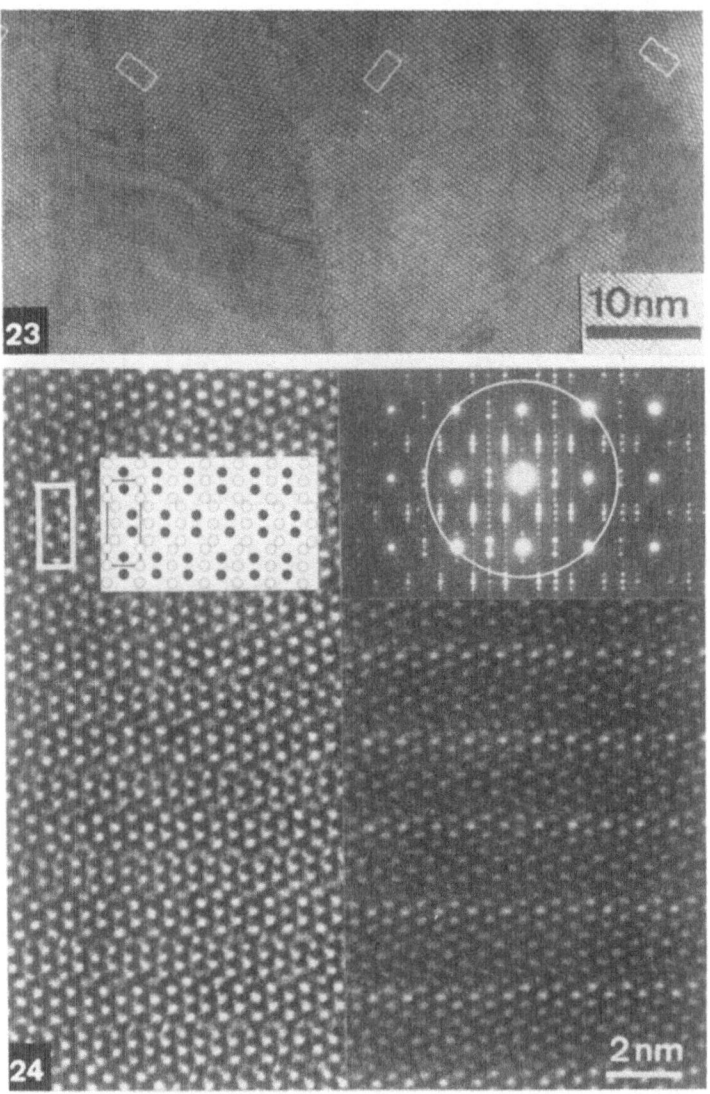

FIG. 23. Area of DO_{22} structure viewed along [100] exhibiting two orientation variants with roughly mutually perpendicular c-axis in the foil plane. The area also contains antiphase boundaries.

FIG. 24. Orthorhombic I structure viewed along the [100] zone. The insets show the corresponding diffraction pattern as well as a model of the structure at the same magnification as the image.
(a) The 1 nm spacing between APB's is prominent. (b) The 2 nm period is prominent.

centered on the DO_{22} spots. From the fact that the intensity of the superlattice spots decreases with their distance away from the DO_{22} spots one can conclude that the superstructure must be derived from the DO_{22} structure. Making use of the geometrical features of the diffraction pattern in particular from the fractional shifts one can deduce a model for the structure. Such a model is shown as an inset of fig.24 at the same scale as the high resolution images, which are reproduced, as obtained under two different contrast conditions. The structure clearly consists of strips of DO_{22} structure, 2 1/2 unit cells wide, limited by (010) planes and which are shifted one with respect to the other over the displacement vectors $\frac{1}{4}$ [021] and $\frac{1}{4}$ [02$\bar{1}$] alternatively. The long period is ∿1.0 nm. The anti-phase boundaries are of the conservative type since the layers which are removed or inserted as a result of the displacements $\frac{1}{4}$ [02$\bar{1}$] (or $\frac{1}{4}$ [021]) have the stoichiometric composition Au_3Mn. The structure can also be described as being the result of periodic twinning.

If the interfaces are no longer in the (100) or (010) plane but for instance in the (021) planes a different one-dimensional long period superstructure can be formed. The diffraction pattern in the inset of fig. 25 shows again satellite sequences associated with DO_{22} spots and it is therefore clear that also this superstructure is derived from the basic DO_{22} structure. The satellite sequences are now perpendicular to (021) planes; from the fractional shifts we can deduce the same displacement vector as obtained above i.e. of the type 1/4 [021]. The model deduced from the diffraction pattern is shown as an inset in fig.25 at the same magnification as the corresponding high resolution image. Associating again the bright dots with manganese columns perfect correspondence is found between model and image. In this case the antiphase boundaries are non-conservative : the ideal composition of the monoclinic structure of fig. 25 is $Au_{31}Mn_{11}$.

A second type of orthorhombic superstructure was found in which the domain strips are again 2 1/2 DO_{22} unit cells wide and limited by (010) or (100) planes, but the displacement vector is now the same for all interfaces and equal to $\bar{R}$ = 1/2 [110] (fig.26). The image of two different orientation variants of this structure, as viewed along the common [001] axis, is reproduced in fig. 26.

Also the Au_5Mn_2 structure can be considered as a long period antiphase boundary superstructure of the DO_{22} structure; this is represented in fig. 27. The antiphase boundary planes are now on the (101) planes and the displacement vector is again 1/4 [201]. A high resolution image of this phase is reproduced in fig. 28.

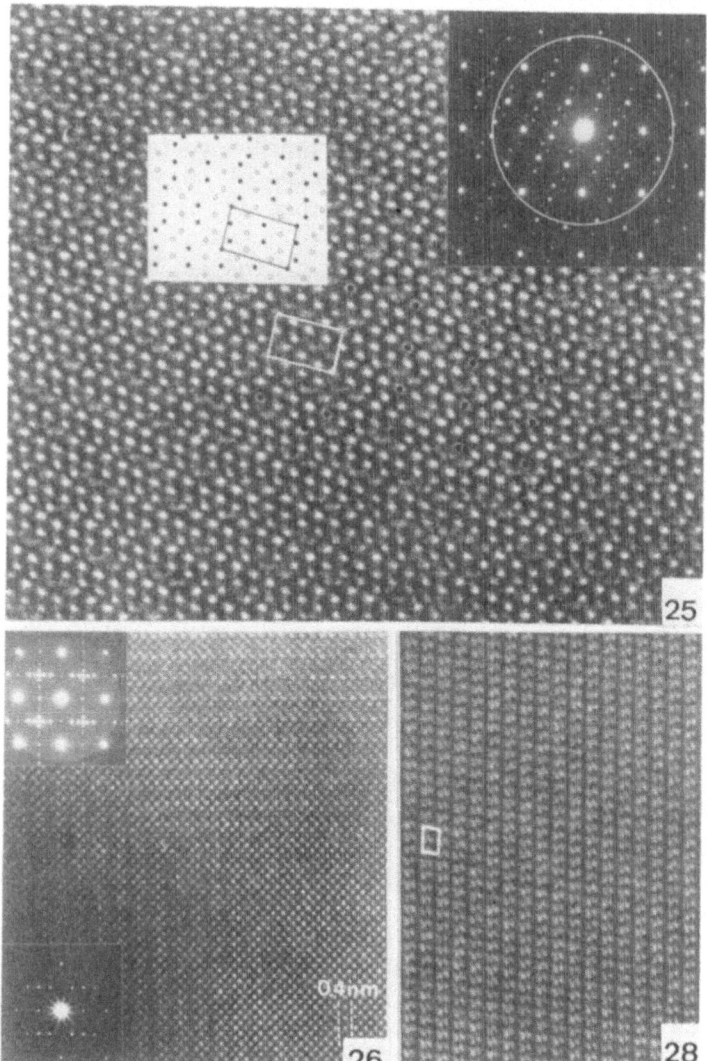

FIG. 25. The monoclinic II structure as observed by HREM. The insets show the corresponding diffraction pattern and a detailed structure model on the same scale as the micrograph.

FIG. 26. Two families of antiphase boundaries as viewed along the [001]zone; across each set all rows of manganese columns are continuous. The electron diffraction pattern over the whole region and the optical diffraction over one family are shown as insets.

FIG. 28. High resolution image of the Au_5Mn_2-structure. This structure as shown in fig. 27 can also be considered as an LPAPB superstructure of DO_{22}.

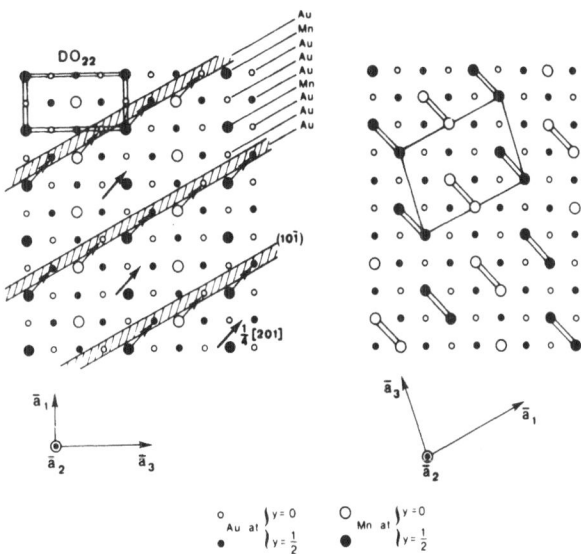

FIG. 27. Removing every sixth pure gold layer from the DO_{22} structure generates the Au_5Mn_2 structure.
(a) Removing periodically pure gold layers along $(10\bar{1})$ from the
    DO_{22} structure.
(b) The Au_5Mn_2 structure contains strings of manganese pairs along
    the $(0\bar{1}0)$ direction.

5.3.6. Two-dimensional long period superstructures derived from the
DO_{22} structure. In fig. 29 we have reproduced the diffraction patterns of three different areas of a crystal with nominal composition $Au_{3+x}Mn$. The first one (fig. 29a) is pseudo incommensurate, as
can be deduced from its schematic representation in fig. 30. The
satellites associated with different basic spots are represented
as open or closed dots. Sequences associated with different basic
spots clearly do not match when they meet and a spacing anomaly
results. The pattern of fig. 29b on the other hand is commensurate. The most intense satellites occur close to the positions of
DO_{22} basic structure. The fractional displacements are $(\frac{1}{4}, \frac{1}{4})$
(for the two directions) at the spots 101 and 103; they are $(\frac{1}{2}, \frac{1}{2})$
for the spots 202 and 002. The average orientations of the two
families of antiphase boundaries have clearly to be roughly along
(101) and $(10\bar{1})$ and their spacing has to be four times d_{101}.

The pattern of fig. 29c finally only exhibits very few satellites suggesting considerable variability of the antiphase boundary spacing.

This observation leads to the lozenge shaped island model of
fig. 31 which corresponds with a diffraction pattern as fig.29b.

The high resolution images of fig. 29d, e confirm this model, but show at the same time that in some areas there is a considerable variability in size and shape of the islands. The islands consist of DO_{22} structure, whereas along the antiphase boundaries the configuration of columns is based on elements of the Au_4Mn structure. These interfaces are non-conservative and incorporate an excess of gold. Some islands are lozenge shaped and contain 3x3 or 4x4 manganese columns; others are parallelogram shaped and contain 4x3 columns. Regular mixtures of these islands constitute complicated two-dimensional structures (fig. 29e).

Fig. 32 gives a survey of all the "local" structures which we have been able to identify in the high resolution images. Note that the "average" orientations of the two families of APB's are

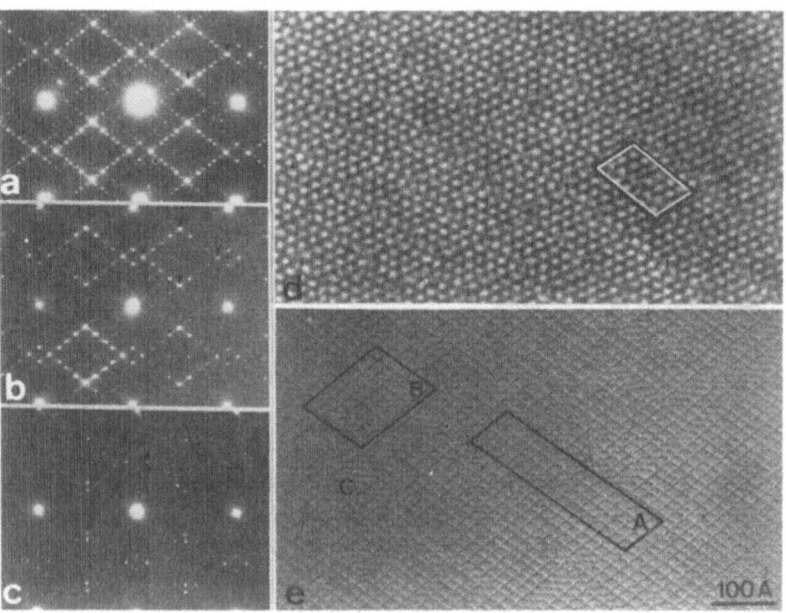

FIG. 29. Mono-domain diffraction patterns of the "lozenge" shaped island superstructure of the DO_{22} structure. Note that the most intense spots or groups of spots are situated in the immediate vicinity of DO_{22} spots.
(a) Well ordered structure containing a mixture of 3x3 and 3x4 islands.
(b) Less well ordered structure containing predominantly 4x4 islands.
(c) Structure with variable island sizes.
(d) Bright field image of an area containing a structure of the type of fig. 31, i.e. containing only 3x4 islands.
(e) Specimen area containing different types of island structures derived from DO_{22}. The structure in (A) is the one represented in fig.32(c) the structure in (B) is the one in fig. 32(e).

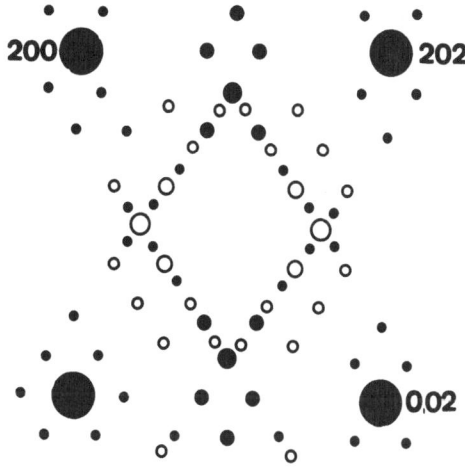

FIG. 30. Schematic representation of a diffraction pattern of type (a) of fig. 29. Superstructure spots derived from different basic spots are represented by a different symbol (full or open dot).

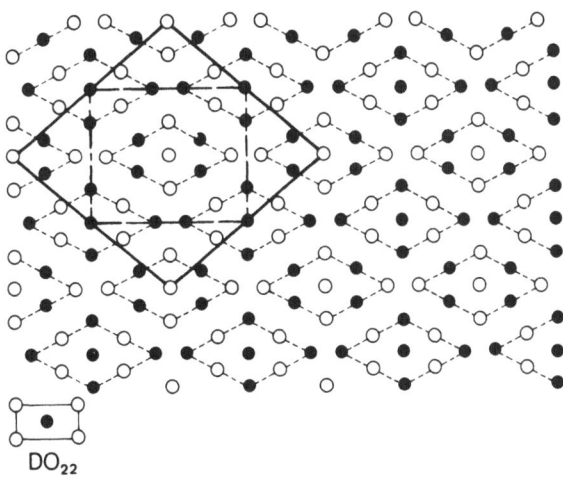

FIG. 31. "Lozenge" shaped island superstructure as derived from the diffraction pattern as in fig.29b. The directions of the APB's are perpendicular to the rows of superstructure spots, whereas their spacing follows from the interspot distance. The displacement vectors are derived from the fractional shifts in the diffraction pattern.

TABLE 4. Lozenge shaped island structures derived from the DO_{22} structure.

model (fig.32)	unit cell size*	Bravais lattice	pointgroup	angle**	ideal composition	
					chemical formula	at% Mn
(a)	$\begin{pmatrix} 4 & 0 & 0 \\ 0 & 5 & 0 \\ 0 & 0 & 1 \end{pmatrix}$ $c_1 = 4a$ $c_2 = 5a$ $c_3 = a$	orthorhombic body centered	$\frac{2}{m}\frac{2}{m}\frac{2}{m}$	77°	$Au_{31}Mn_9$	22.50
(c)	$\begin{pmatrix} 5 & 0 & 0 \\ 0 & 7 & 0 \\ 0 & 0 & 1 \end{pmatrix}$ $c_1 = 5a$ $c_2 = 7a$ $c_3 = a$	orthorhombic one face centered	$\frac{2}{m}\frac{2}{m}\frac{2}{m}$	71°	$Au_{27}Mn_8$	22.85
(e)	$\begin{pmatrix} 9 & 12 & 0 \\ -5/2 & 7/2 & 0 \\ 0 & 0 & 1 \end{pmatrix}$ $c_1 = 15a$ $c_2 = \frac{a}{2}\sqrt{74}$ $c_3 = a$	monoclinic one face centered	$\frac{2}{m}$	74°	$Au_{95}Mn_{28}$	22.76
(f)	$\begin{pmatrix} 13/2 & 7/2 & 0 \\ 5/2 & 17/2 & 0 \\ 0 & 0 & 1 \end{pmatrix}$ $c_1 = \frac{a}{2}\sqrt{218}$ $c_2 = \frac{a}{2}\sqrt{314}$ $c_3 = a$	monoclinic primitive	$\frac{2}{m}$	77°	$Au_{72}Mn_{21}$	22.58
(g)	$\begin{pmatrix} 9 & 9 & 9 \\ 0 & 12 & 0 \\ 0 & 0 & 0 \end{pmatrix}$ $c_1 = 9a$ $c_2 = 12a$ $c_3 = a$	orthorhombic body centered	$\frac{2}{m}\frac{2}{m}\frac{2}{m}$	74°	$Au_{167}Mn_{49}$	22.68
(h)	$\begin{pmatrix} 5/2 & 7/2 & 0 \\ -4 & 5 & 0 \\ 0 & 0 & 1 \end{pmatrix}$ $c_1 = \frac{a}{2}\sqrt{74}$ $c_2 = a\sqrt{31}$ $c_3 = a$	monoclinic one face centered	$\frac{2}{m}$	74°	$Au_{41}Mn_{12}$	22.64

* The transformation matrix relating the base vectors $\bar{c}_1$, $\bar{c}_2$ and $\bar{c}_3$ of the superstructure with those of the f.c.c. basic lattice $\bar{a}_1$, $\bar{a}_2$ and $\bar{a}_3$. The lengths of the vectors are expressed in units $a = |\bar{a}_1| = |\bar{a}_2| = |\bar{a}_3|$.

** Angle between average directions of the APB's.

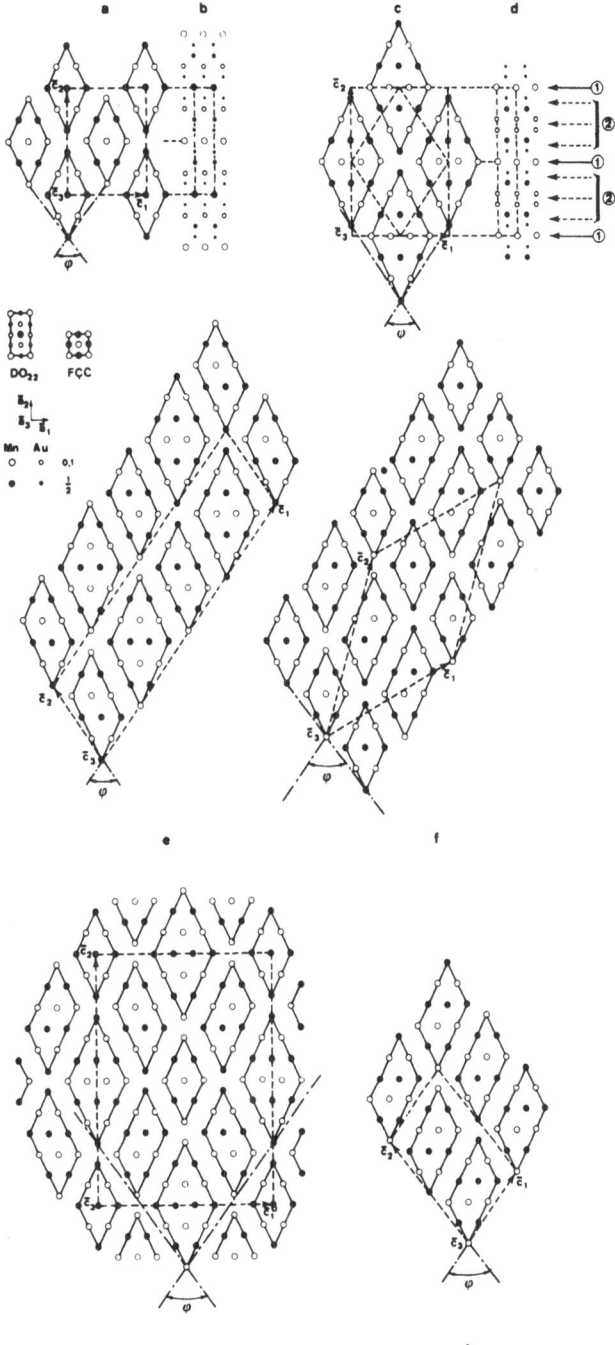

FIG. 32. Different "lozenge" shaped island superstructures of the DO$_{22}$ structure a) and b) structure consisting of 3x3 islands only; a) is a view along the normal to the island plane whereas b) is a side view along another cube direction, c) and d) structure consisting of 4x4 islands only; c) is again a "top" view and d) a side view. e) Mixture of 3x4 and 4x4 islands. f) Mixture of 3x4 and 3x3 islands. g) Mixture of three types of islands : 3x3, 4x4, and 3x4. h) Structure consisting of 4x3 islands only.

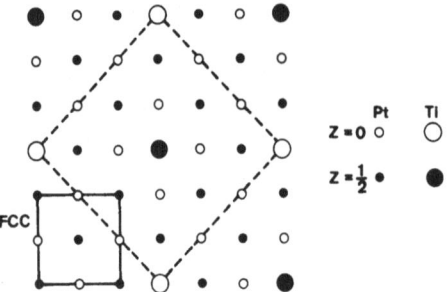

FIG. 33. [001] projection of the model for the Pt_8Ti phase; the primitive body centered tetragonal unit cell is indicated with dashed lines and the FCC unit cell with full lines.

slightly different from (101) and (10$\bar{1}$), although the islands them-selves are limited by such planes. These different local structures correspond with slightly different compositions,which are indicated in table 4.

5.4. The Pt-Ti System (18)

5.4.1. The Pt_8Ti structure. We shall discuss only structures corres-ponding with compositions in the range Pt_4Ti to Pt_8Ti and which are derivatives from the Pt_8Ti structure. The structure of Pt_8Ti is a tetragonal superstructure based on the FCC basic lattice; it is shown in fig. 33. It can be formed in three orientation variants the tetra gonal c-axis being parallel with one of the cube directions of the FCC structure; the variant generating pointgroup is 3. Figure 34a shows two orientation variants.

5.4.2. One-dimensional superstructure. The diffraction pattern of the one-dimensional long period superstructure is shown schematically in fig. 35, in which the positions of the Pt_8Ti spots are indicated by crosses. The most intense superlattice spots are those which are closest to the Pt_8Ti spots. The structure is thus a derivative of the Pt_8Ti structure. The fractional displacements lead to a dis-placement vector of the type 1/6[103]. A model for the structure is shown in fig. 36b it can be compared with the high resolution image of fig. 34b, in which the bright dots should be associated with titanium-columns. The diffraction pattern is shown as an inset.

5.4.3. Two-dimensional superstructure. A two-dimensional derivative of the Pt_8Ti structure is visible in fig. 34c. The corresponding diffraction pattern is reproduced in the inset of fig. 34c while a schematic representation is shown in fig. 37; the positions of the Pt_8Ti spots are indicated by crosses. There are now two mutually perpendicular families of periodic antiphase boundaries. From the fractional displacements we can conclude that both families have displacement vectors of the same type i.e. $\frac{1}{6}[103]$ and $\frac{1}{6}[013]$. This

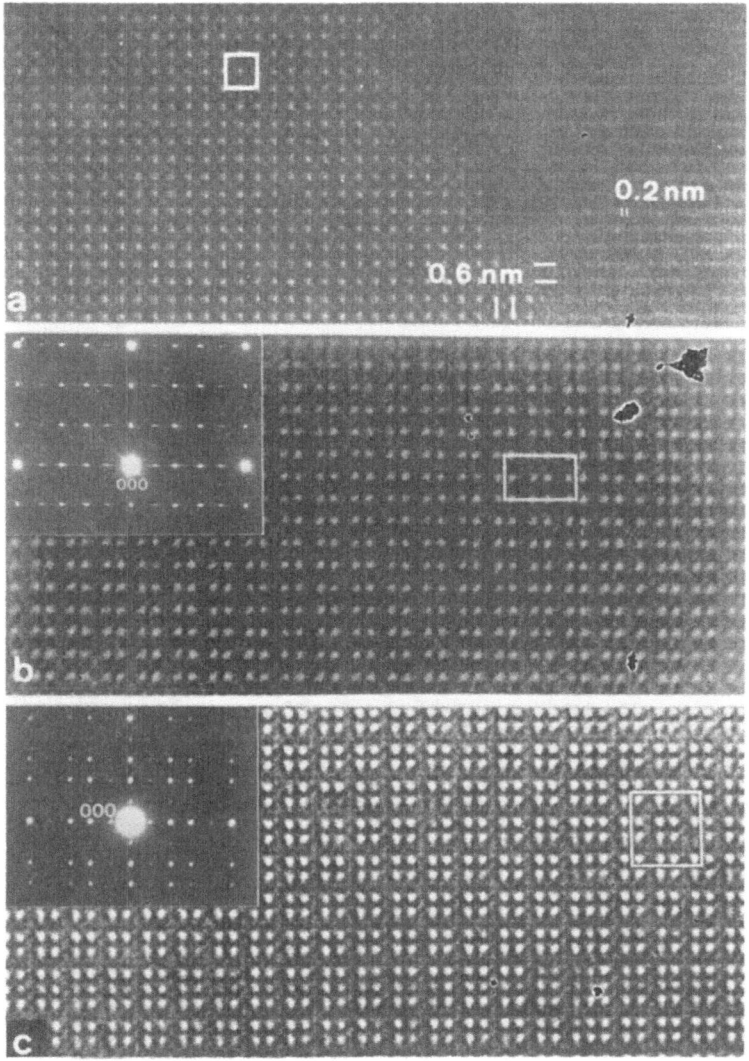

FIG. 34. (a) High resolution image of two orientation variants of the Pt_8Ti phase. In the variant imaged by a square array of dots (left) the electron beam lies parallel with the tetragonal axis and the drawn square indicated the face centered base of the unit cell. In the line region (right) the electron beam is perpendicular to the tetragonal axis.

(b) High resolution image of the one-dimensional $Pt_{13}Ti_2$ structure with the electron beam parallel with the tetragonal axis. The rectangle indicates the unit cell.

(c) High resolution image of the two-dimensional $Pt_{21}Ti_4$ structure, viewed along the tetragonal axis. The square indicates the unit cell. Diffraction patterns of both structures are shown as insets.

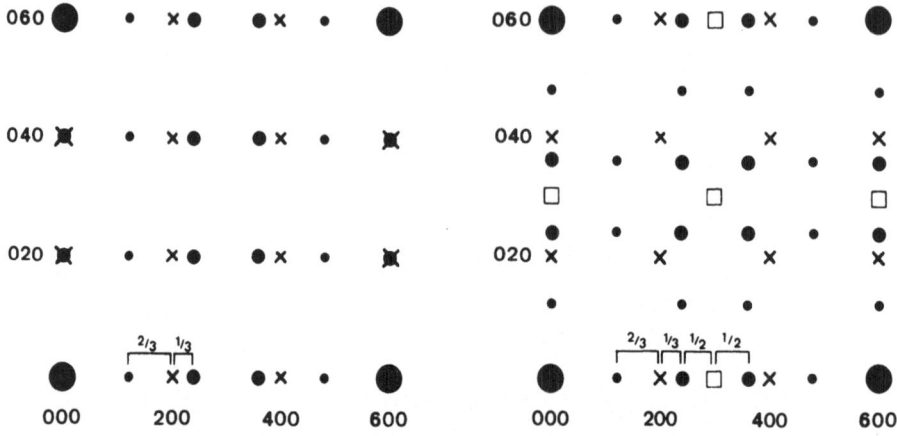

FIG. 35. Schematic representation of the cube zone diffraction pattern observed in Pt_6Ti specimens showing the basic FCC reflections (large dots) and superreflections (two kinds of small dots). The crosses indicate the positions of the Pt_8Ti spots and the indices are given according to the Pt_8Ti basic structure. The fractional shifts are indicated at one reflection.

FIG. 37. Schematic view of a cube zone pattern in Pt_5Ti where the same symbols are used as in fig. 35.
Positions of reflections due to Pt_8Ti and Pt_3Ti are added as follows:
□: Pt_3Ti (Ll_2) reflections;   x : Pt_8Ti reflections.
The indices refer to the Pt_8Ti basic lattice.

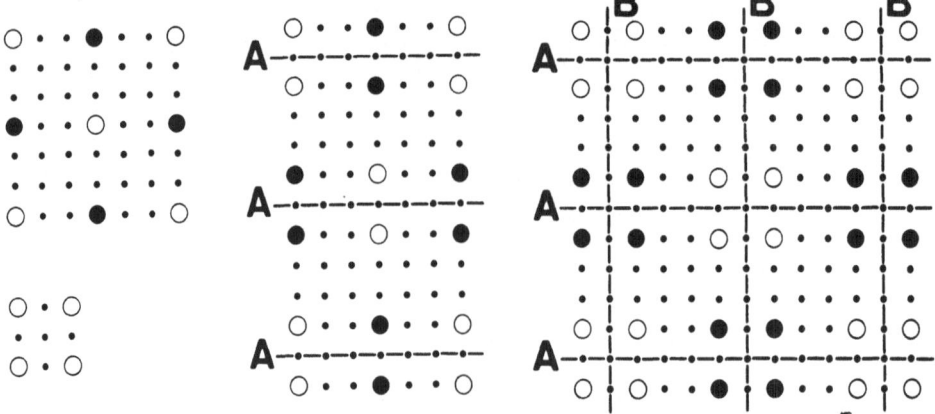

FIG. 36 : Schematic representation of a) the Pt_8Ti; b) the one-dimensional LPAPB structure; c) the two-dimensional LPAPB-structure. The two families of APB's are indicated A and B respectively.

leads to the model fo fig. 36c which can be compared with the high
resolution image of fig. 34c. It was verified that in the third
direction this unit cell was one FCC unit cell thick. The structure
can thus be considered as consisting of a regular mixture of islands
having the Pt_8Ti structure and islands having the Pt_3Ti (i.e. the
Cu_3Au) structure.

6. DISCUSSION

The results of the study of long period structures in a number
of alloy systems suggests that deviations from a simple chemical
composition, with which a simple structure is associated, result in
the appearance of non-conservative periodic antiphase boundaries
within this simple structure. The larger the deviation the larger
the concentration of such interfaces. Sometimes, if the simple
structure has tetragonal symmetry for instance two crystallographically equivalent families of antiphase boundaries are introduced
avoiding in this way a too high concentration of interfaces in a
single family.

If the deviation exceeds a certain value, the concentration of
interfaces would have to become too large. It may then become more
favourable to form a long period structure derived from a different
simple structure, having a composition which approximates better the
actual composition of the alloy. The density of antiphase boundaries
can then be reduced. It should be noted that in such an alloy system
the configuration of atoms along the antiphase boundaries in one
structure consists of structural elements to be found in the other
structure and vice versa. Let for instance A be a basic superstructure with a composition C_A and B the basic superstructure with
composition C_B, both structures A and B being based on the same
Bravais lattice e.g. FCC. For compositions close to C_A the structure will be LPAPB structure derived from A, the antiphase boundary
region consisting of structural elements of B. On the other hand
for compositions close to C_B the structure will be a LPAPB structure
derived from B with antiphase boundaries exhibiting structural
elements of A. This empirical finding is documented with examples
in the Au-Mn system and in the Pt-Ti system. In the Au-Mn system
the simple structures A and B are for instance Au_4Mn (Ni_4Mo-type)
and Au_3Mn (DO_{22}) whereas in the Pt-Ti system A and B are Pt_8Ti and
Pt_3Ti (Cu_3Au structure). This adaptation of the composition can
be achieved by the introduction of the appropriate concentration
of antiphase boundaries with a suitable orientation and displacement
vector.

The antiphase boundary energy being often anisotropic certain
planes may be strongly preferred e.g. the cube planes in many alloys
derived from FCC. (In passing it is noted that in alloys where the
antiphase boundary energy is isotropic no periodic antiphase boundary structures have been found as yet). The energy depends further-

more on the displacement vector. In ordered structures derived from
the FCC structure the displacement vector is usually of the type
1/2<110>.

A large change in orientation with respect to the displacement
vector may transform   a conservative APB into a non-conservative
one and as a result increase the energy and lead to the dissociation
of a single antiphase boundary into two (possibly more) components,
lowering in this way the total energy (fig. 39). In any given structure usually only one type of APB occurs after long anneal; the one
which leads to the smallest energy. The other geometrically possible
APB's dissociate into APB's of this type. Small changes in orientation with respect to the displacement vector may be associated
with small changes in composition. The orientation with respect to
the displacement vector and the spacing of the interfaces are therefore the main parameters which determine the composition change
caused by a periodic array. There is a strong indication that the
appearance of LPAPB structures is at least partly composition driven.
However, in some cases the interfaces in periodic arrays are conservative and hence do not change the composition. In such cases
their presence can of course not be composition induced, but must
be related to either strain energy minimization (45) or to electronic
effects such as proposed by Sato and Toth (46) or possibly to both.

In many periodic antiphase boundary structures the spacing
between the interfaces is not strictly constant, especially when
the period is very long. Also in the latter case the average
spacing depends sensitively on the addition of impurities changing
the e/a ratio, and on deviations from the simple composition. In
alloys with a small APB spacing, the spacing has a strong tendency
to be uniform; a perturbation introduced by the insertion of an
additional interface is rapidly smoothened out so as to ensure again
a uniform spacing (fig. 38). This suggests a repulsive interaction
between antiphase boundaries. On the other hand fig. 39 which illustrates the dissociation into two components of a single antiphase
boundary in Au_4Mn on changing 90° in orientation, suggests that an
equilibrium distance is established, which in the case shown is
equal to the distance between APB's in the one-dimensional superstructure of Au_4Mn (fig. 16). The interaction presumably has thus
also an attractive component which leads to an energy minimum at
a certain distance.
The observations on a few alloy systems (Au-Zn, Au-Mn) suggest a
mechanism whereby periodic APB's can be generated in certain alloy
systems, starting from a single antiphase boundary.

The main observations can be summarized as follows (fig.40abc).
(a) Zig-zag shaped APB's are often found to have formed around a
    single boundary adopting a "fish bone" configuration.
(b) Hairpin shaped APB's are usually present. It has not been
    possible as yet to observe the motion of such configuration

"in situ" therefore it cannot be concluded unambigeously whether these features are formed during the formation process or in the elimination process of the APB's.

(c) Periodic boundaries in "irrational" orientations frequently occur in the initial stages of the formation process. They consist in fact of "rational" parts separated by periodic ledges.

These observations suggest the following possible formation mechanism. Let us assume that a single non-conservative APB is formed by chance

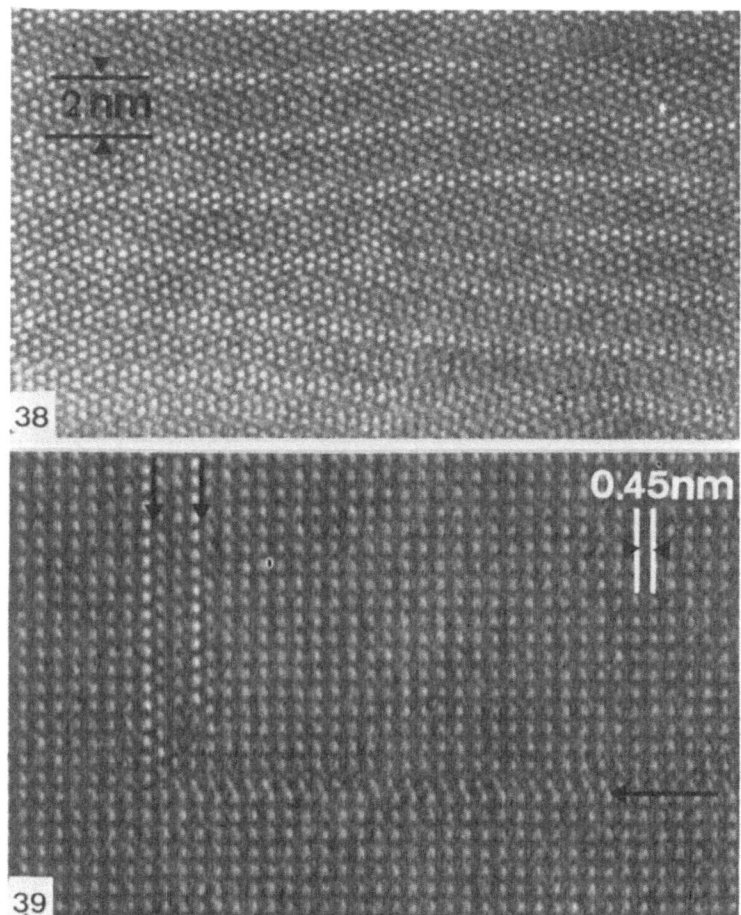

FIG. 38. Dislocation-like configuration in the orthorhombic I structure derived from DO_{22}. The insertion of an additional interface is rapidly smoothened out to reinsure a uniform spacing.

FIG. 39. Antiphase boundary in Au_4Mn changing 90° in orientation whereby a dissociation occurs in two APB's with a spacing corresponding with the equilibrium spacing in the one-dimensional PAPB-structure.

and let the alloy be in the temperature range where the long period
APB structure is the stable one. The non-conservative APB may then
lower its energy by transforming into a zig-zag shaped APB consisting of mainly conservative parts, only keeping some non-conservative
segments at the tips of the "fingers" or "hairpins' ;lengthening(or
shortening) of the APB's occurs by diffusion at the tip i.e. along
the non-conservative parts. Sidewise motion on the other hand is
produced by the lengthwise motion of ledges. In this way the equilibrium arrangement can be adopted by atom movements preferentially
along APB's. Such a mechanism does not require long range diffusion but only local rearrangements along planar defects.

Acknowledgements :

    Thanks are due to D. Van Dyck, M. Van Sande and D. Schryvers
for the use of illustrations from joint papers.

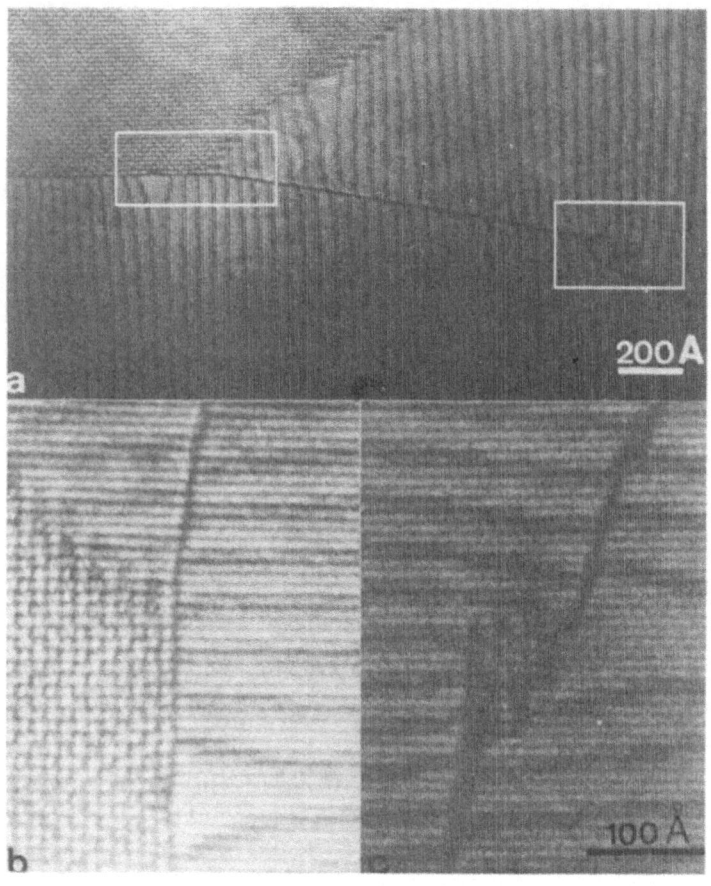

FIG. 40. Antiphase boundary configurations in the early stages
of formation of the two-dimensional LPAPB-structure in Au_4Zn.
Two particular configurations are shown enlarged in b) and c).

References

1. Van Tendeloo, G. and S. Amelinckx, Acta Cryst. A30, 431 (1974).
2. Wondraschek, H. and W. Jeitschko, Acta Cryst. A32, 664 (1976).
3. Bärnighausen, H., Comm. Math. Chem. 9, 139 (1980).
4. Portier, R. and D. Gratias, J. de Physique 43, C4-17 (1982).
5. Amelinckx, S., AIP Conference Proceedings nr. 53, p. 102 (1979).
6.-Van Tendeloo, G., Van Landuyt, J., and S. Amelinckx, Phys.stat.
    sol. (a) 33, 723 (1976).
   -Liebau, F. and H. Böhm, Acta Cryst. A38, 252 (1982).
7. Van Landuyt, J., Wiegers, G.A. and S. Amelinckx, Phys.stat.sol.
    (a) 46, 479 (1978).
8.-Van Landuyt, J., Van Tendeloo, G. and S. Amelinckx, Phys.stat.
    sol. (a) 26, 359 (1974).
   -Van Landuyt, J., Van Tendeloo, G. and S. Amelinckx, Phys.stat.
    sol. (a) 26, 585 (1974).
   -Van Landuyt, J., Van Tendeloo, G. and S. Amelinckx, Phys.stat.
    sol. (a) 36, 767 (1976).
   -Van Landuyt, J., Van Tendeloo, G. and S. Amelinckx, Phys.stat.
    sol. (a) 42, 565 (1977).
9.-Wilson, J.A., di Salvo, F.J. and S. Mahajan, Adv. Phys. 24,117
    (1975).
   -Williams, P.M., Parry, G.S. and C.B. Scruby, Phil. Mag. 29, 695
    (1974).
10.Iizumi, M., Axe, J.D., Shirane, G. and K. Shimaoka, Phys. Rev.
    B15, 4392 (1977).
11.-Tanaka, M., Saito, R. and K. Tsuzuki, Jap. J. Appl. Phys. 21,
    291 (1982).
   -Yamamoto, N., Acta Cryst. A38 (1982).
   -Hutchison, J.L., Chem. Scripta 14, 181 (1979).
   -Smith, D.J. and J.L. Hutchison, J. Microscopy 129, 285 (1983)
12.-De Ridder, R., Van Tendeloo, G. and S. Amelinckx, Phys.stat.
    sol. (a) 33, 383 (1976).
   -Nowotny H. in : The Chemistry of Extended Defects in Non-Metallic
    Solids, Ed. L. Eyring and M.O.'Keeffe, North Holland PUblishing
    Co., Amsterdam, 1970, p.223.
13.-Van Tendeloo, G., Den Broeder, F.J.A., Amelinckx, S., De Ridder,
    R., Van Landuyt, J. and H.J. van Daal, Phys. stat. sol. (a)
    67 , 217 (1981).
   -Den Broeder, F.J.A., Van Tendeloo, G., Amelinckx, S., Hornstra,
    J., De Ridder, R., Van Landuyt, J. and H.J. van Daal, Phys.
    stat. sol. (a) 67, 233 (1981).
14.-Van Tendeloo, G. and S. Amelinckx, Phys. stat. sol. (a) 22,
    621 (1974).
   -Van Tendeloo, G. and S. Amelinckx, Phys. stat. sol. (a) 27,
    93 (1975).
15.-Van Tendeloo, G., Wolf, R., Van Landuyt, J. and S. Amelinckx,
    Phys. stat. sol. 47, 539 (1978).
   -Wolf, R., Van Tendeloo, G., Van Landuyt, J. and S. Amelinckx,
    Phys.stat.sol. (a) 49, 337 (1978).

222

-Van Tendeloo, G. and S. Amelinckx, Phys. stat. sol.(a) 65, 73 (1981).

-Van Tendeloo, G. and S. Amelinckx, Phys. stat. sol.(a) 65, 431 (1981).

-Van Tendeloo, G., Van Landuyt,J. and S. Amelinckx, Phys. stat. sol. (a), 70, 145 (1982).

-Van Landuyt, J., Van Tendeloo, G. and S. Amelinckx, Phys.stat. sol. (a), 70, 407 (1982).

-Van Tendeloo, G. and S. Amelinckx, Phys. stat. sol. (a) 71, 185 (1982).

16.-Van Tendeloo, G., Van Sande, M., Amelinckx, S. and P. Airo, Electron Microscopy 1980, volume 1, p.276.

-Van Tendeloo, G. and S. Amelinckx, Phys. stat. sol. (a) 69, 103 (1982).

17.-Van Tendeloo, G. and S. Amelinckx, Phys. stat. sol. (a) 43, 553 (1977).

-Van Tendeloo, G. and S. Amelinckx, Phys. stat. sol. (a)50, 53 (1978).

-Schryvers, D., Van Tendeloo, G., Van Landuyt, J. and S.Amelinckx Phys. stat. sol. (a) 75, 607 (1983).

18. Schryvers, D., Van Landuyt, J.,Van Tendeloo, G. and S.Amelinckx, Phys. stat. sol. (a) 76, 575 (1983).

19. Van Dyck, D., Van Landuyt, J., Amelinckx, S., Nguyen Hiu-Dung and C. Dagron, Journal of Solid State Chemistry 19, 179 (1976).

20. Colaitis, D., Van Dyck, D. and S. Amelinckx, Phys. stat. sol. (a) 68, 419 (1981).

21. Van Tendeloo, G., Gregoriades, P. and S. Amelinckx, submitted to J. Sol. St. Chem.

22.-Meulemans, M., Delavignette, P., Garcia-Gonzales, F. and S. Amelinckx, Mat. Res. Bull. 5, 1025 (1970).

-Snykers, M., Serneels, R., Delavignette, P., Gevers, R. and S. Amelinckx, Crystal Lattice Defects 3, 99 (1972).

-Snykers, M., Delavignette, P. and S. Amelinckx, Phys. stat. sol. (a) 48, K1 (1971).

23.-Van Sande, M., Van Landuyt, J. and S. Amelinckx, Phys. stat. sol. (a) 55, 41 (1979).

-Morton, A.J., Acta Met. 27, 863 (1979).

24.-Van Landuyt, J. and S. Amelinckx, Mat. Res. Bull. vol. 6, 613 (1971).

-Dubey, M., Singh, G. and G. Van Tendeloo, Acta Cryst. A33, 276 (1977).

-Gai, P.L., Anderson, J.S. and C.N.R. Rao, J. Phys. D8, 1157 (1975).

-Jepps, R.W. and T.F. Page, J. of Microscopy 116, 159 (1979).

25. Akizuki, M., Am. Mineralogist, 66, 1006 (1981).

26.-Van Tendeloo, G., Van Landuyt, J. and S. Amelinckx, "40th Ann. Proc. EM Soc. Amer.", Washington D.C., 1982, G.W. Bailey (ed.) p. 540.

-Van Tendeloo, G., Wolf, R., Van Landuyt, J. and S. Amelinckx, Phys. stat. sol. (a) 60, 581 (1980).

ELECTRON MICROSCOPIC STUDY OF MODULATED STRUCTURES IN (Au,Ag)Te$_2$

G. Van Tendeloo[1], P. Gregoriades[2], S. Amelinckx[1,3]

[1] Rijksuniversitair Centrum Antwerpen, B 2000-Antwerpen (Belgium)
[2] University of Thessaloniki, Thessaloniki (Greece)
[3] Also at S.C.K., B 2400-Mol (Belgium)

ABSTRACT

The compounds (Au,Ag)Te$_2$ exhibit remarkable modulated structures of different types, depending on the exact ratio of the concentrations of gold and silver. Calaverite (AuTe$_2$) has a deformation modulated structure, which in the case of non-stoichiometric Sylvanite (Au$_{1+x}$Ag$_{1-x}$Te$_4$) is moreover coupled with a long period anti-phase boundary superstructure. Finally in Krennerite (Au$_{0.8}$Ag$_{0.2}$Te$_2$) the structure is modulated by poly-synthetic twinning. In all cases the modulation planes are approximately along the (101) planes of calaverite. The modulation periods are also approximately of the same magnitude in the different cases.

1. INTRODUCTION

The pseudo binary system (Au,Ag) Te$_2$ is of interest since it is possible to substitute up to 50 % of the gold atoms in calaverite (AuTe$_2$) by silver atoms, without changing fundamentally the basic structure. We have studied this system by means of electron diffraction and high resolution electron microscopy for different Au/Ag ratios (1). From X-ray diffraction there were already indications that a modulated structure might be present in AuTe$_2$ (2)(3). The other phases in the system (Au,Ag)Te$_2$ have not yet been studied from this point of view however.

2. MATERIALS AND SPECIMEN PREPARATION

The materials were prepared by melting together the constituent elements in evacuated quartz tubes at 550°C followed by slow cooling over several days. Specimens for electron microscopy were prepared by crushing. The described observations were all performed at room temperature unless otherwise stated.

3. CRYSTAL STRUCTURES

The "average" or "idealized" structures of calaverite ($AuTe_2$), and sylvanite ($Au_{0.5}Ag_{0.5}Te_2$) as well as the structure of Krennerite have been determined by Tunnell and Pauling (2).These structure are represented in fig.1.

Calaverite has a structure which is in fact a somewhat deformed version of the CdI_2 structure; alternatively it can also be considered as a superstructure based on a slightly deformed primitive cubic lattice with a $\cong 0.3$ nm. The schematic representation of fig.1,a is a projection on a $\{110\}$ type plane of this quasi-cubic lattice or along the $[010]$ direction of the monoclinic structure of calaverite.

The structure of sylvanite as determined by Tunell and Pauling (2) is closely related to that of calaverite, it is in fact a superstructure of the latter structure in which gold and silver atoms are ordered within the same tellurium sublattice as in calaverite.

Along the $\bar{c}$-axis gold and silver atoms alternate in projection; as a result gold and silver containing layers alternate along the b-direction. Moreover it was found that along the $\bar{c}$-direction successive silver atoms are displaced along the $\bar{b}$-direction alternatively above and below the plane y = 0.5. As a result the unit cell of sylvanite has twice the size of that of calaverite. The structure of sylvanite is thus in a sense commensurately modulated; along the $\bar{c}$-direction the level of the heavy atoms (Ag and Au) can be represented by a sinusoid describing the deviation from the y = 0.5 level for silver or from the o-level for gold. This modulation causes the unit cell to be doubled along the modulation direction. Also the tellurium atoms are periodically displaced with respect to the positions which they would occupy in ideal calaverite; the period of this displacement is only half that of the silver atoms. The planes of equal phase are parallel with the $(101)_{cal}$ planes for both atomic species.

Also the krennerite structure is closely related to that of calaverite. It results from the latter structure by repeated subunit cell twinning leading to an orthorhombic structure (fig.1c). It is not known from the literature whether or not silver and gold are ordered in this compound.

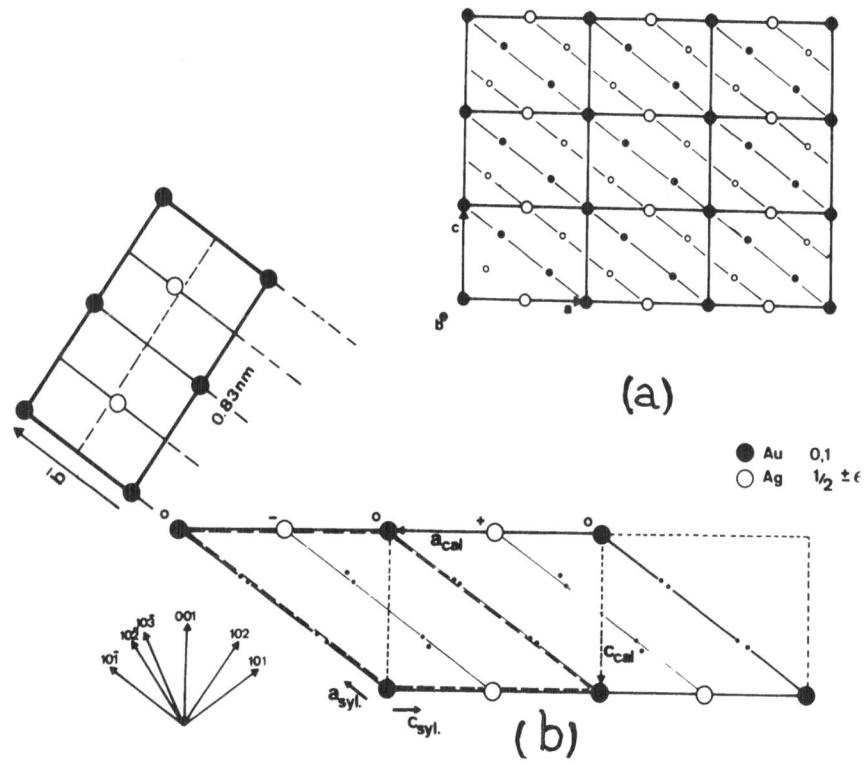

FIG. 1a. Structure of calaverite projected along [010]. Large circles represent Au-atoms. Open circles are at $y=1/2$ while filled circles are at $y=0$. The unit cell is outlined and the same convention as in ref.2 is used. Fine dotted lines indicate (202) planes which are nearly perpendicular to the modulation vector.

FIG. 1b. Crystal structure of sylvanite projected on the (010) plane. The conventional unit cell is shown in full lines. The dotted line shows an alternative unit cell which consists of two juxtaposed calaverite unit cells. The small full dots represent the actual Te-position, as compared to the ideal ones, shown as open dots. The silver atoms along lines marked + are displaced upwards and those along lines marked − are displaced downwards. A side view of the sylvanite structure along the $\bar{a}_{sylv.}$ direction is shown at the left. The displacements of the silver atoms are exaggerated.

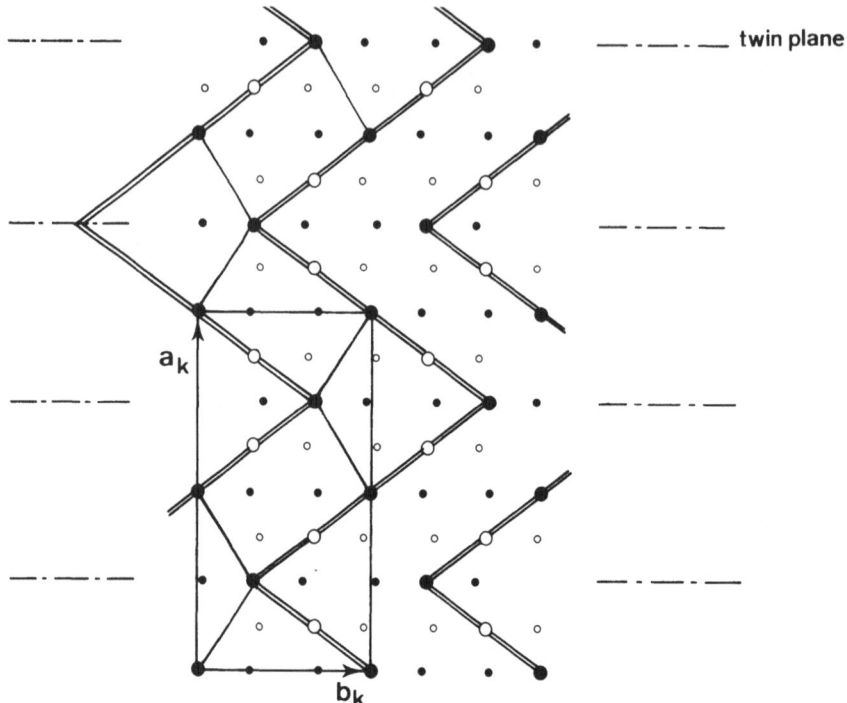

twin plane

FIG.1,c. Structure of krennerite projected along the (001) zone.
Note that the structure can be considered as a polysynthetically twinned version of calaverite (see
fig.1,a).

The crystallographic data of these three compounds are
summarized in table I, according to Tunnell and Pauling (2).

Table I

	spacegroup	a_o(nm)	b_o(nm)	c_o(nm)	β
Calaverite	C 2/m	0.719	0.440	0.507	90°13'
Krennerite	Pma	1.654	0.882	0.446	
Sylvanite	P 2/c	0.896	0.449	1.462	145°26'

4. CALAVERITE

4.1. Diffraction Patterns

The electron diffraction patterns as well as the high resolution images confirm to a first approximation the known average
structure. However in all specimens satellite reflections occur
in a number of sections of reciprocal space. The ratio of the
intensity of the satellite spots to that of the basic spots, from

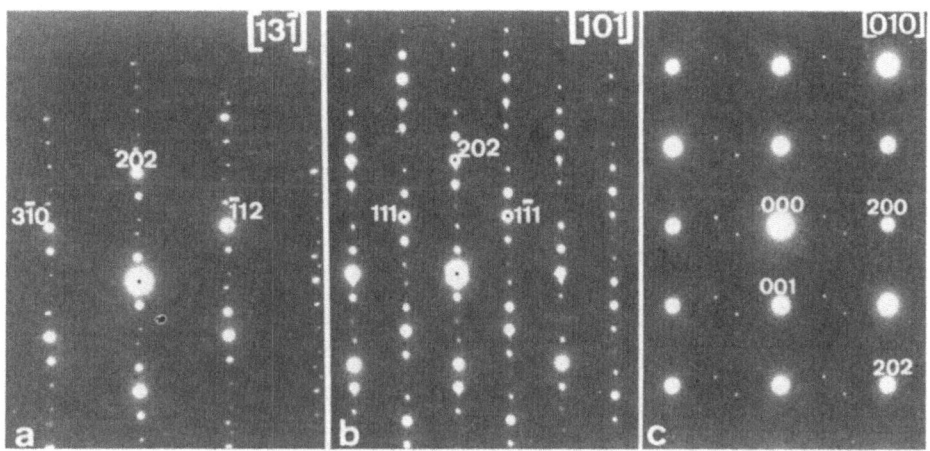

FIG.2,a) $[13\bar{1}]$ diffraction pattern (a pseudo hexagonal section) clearly showing satellite sequences forming an incommensurate diffraction pattern.

b) $[10\bar{1}]$ zone (a pseudo cubic section) exhibiting the same sequences of satellites.

c) $[010]$ diffraction pattern showing quite clearly the satellite pairs around the positions hol with h=odd.

which they are derived, increases with increasing length of the diffraction vector of the basic spot. This behaviour suggests that the satellites are probably due to a deformation modulated structure. The idealized structure does not occur apparently.

Such sequences of satellites are for instance prominently visible in the zone patterns $[13\bar{1}]$ and $[10\bar{1}]$ which are respectively a pseudo-hexagonal and a pseudo cubic zone of the underlying primitive cubic lattice (fig.2a,b). The linear sequences of satellites are roughly oriented along the $[202]^*_{cal}$ direction; their spacing is approximately $(1/4.5 \, d_{202})$ (i.e. $1/q = 0.94$ nm). There is an orientation as well as a spacing anomaly.

The $[202]^*$ direction also belongs to the $[010]$ zone. However in the diffraction pattern along this zone (fig.2c) only pairs of weak satellite spots are visible. One could think of these as being derived from the hol spots with h odd, by splitting. From a comparison with the satellite sequence along $[202]^*$ in other sections we can conclude that the observed spots in the $[010]$ zone are in fact the second order satellites of the same $[202]^*$ sequences; in other words in the $[010]$ zone, i.e. for hol basic reflections, the satellites of odd order are extinguished. All observed spots can thus be explained as being due to a single linear set of satellites associated with each basic spot, enclosing a small angle with the $[202]^*$ direction and with a q-vector approximately equal to 1.06 nm^{-1}.

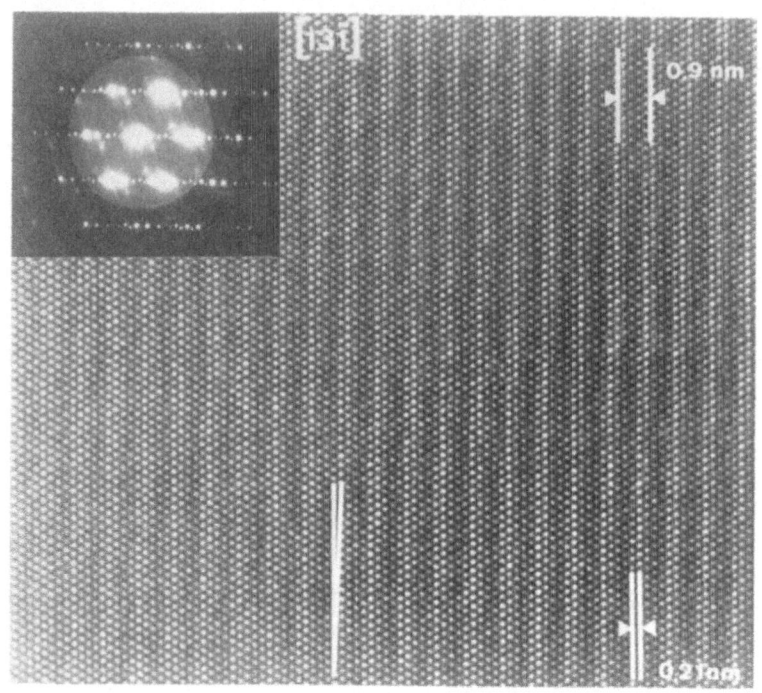

FIG.3. High resolution image in the $[13\bar{1}]$ section. The 0.94 nm modulation is very prominent. The diffraction pattern is shown as an inset. The rows of bright dots enclose a small angle with the close packed rows.

The satellites close to the basic Bragg spots in the $[010]$ zone are either due to double diffraction, or they may be the fourth order reflections of the $[202]*$ sequences. Since the spot positions according to both assumptions, coincide it is difficult to distinguish between them.

4.2. High Resolution Images

High resolution images were obtained from a number of zones, which could all be interpreted in terms of a modulation of the described "average" or "ideal" structure. We shall discuss two relevant zones.

In the pseudo-hexagonal zone $[13\bar{1}]$ (fig.3) the atom sites clearly form a hexagonal close-packed array. The intensity of the dots is modulated however. The average direction of the

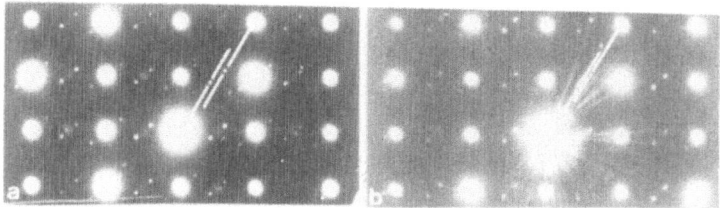

FIG.4. Comparison of $\bar{q}$-vectors in (a) pure $AuTe_2$ ($1/q = 0,19$ nm) and in (b) $AuTe_{1.75}Se_{0.25}$ ($1/q = 0.925$ nm).

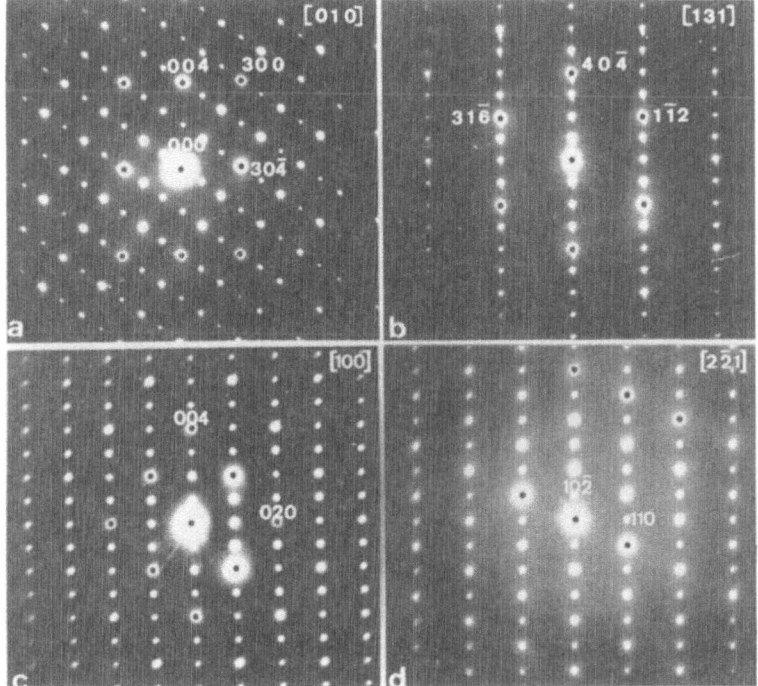

FIG.5. Four sections of reciprocal space of commensurate sylvanite indexed with respect to the sylvanite lattice.
(a) $[110]$ zone pattern ; note the alternation of weak and strong spot along the $[101]$ direction; the weak spots are due to ordering. Note also the absence of 001 and 003 which is related to the Ag-displacements parallel with $[010]$.
(b) $[131]$ zone pattern; pseudo hexagonal section of the quasi-cubic lattice. Note the fourfold periodicity.
(c) $[100]$ zone. Note the presence of 001 and 003 in this zone due to double diffraction.
(d) $[221]$ zone. Note the alternation of weak and strong reflections in the $[102]$ direction.

lines of maximum brightness, which reveal the modulation, encloses a small angle with the close-packed direction of the array of
bright dots. This is in accordance with the orientation anomaly
observed in the diffraction pattern. Also the spacing between
modulations is not an integral number of times the spacing between the close packed rows of bright dots, which is consistent
with the spacing anomaly in the diffraction pattern.

4.3. Effect of Temperature and Substitution

We have heated specimens up to their melting point whilst
observing the diffraction pattern. No directly observable changes
have been found, i.e. the satellites are still present close to
the melting point. The same applies to specimens which had been
cooled in the specimen holder of the microscope down to liquid
nitrogen temperature. However, it is not excluded that small
changes of the q-vector occur, which are not directly observable
in the microscope. More careful experiments are under way, to
determine possible small changes.

We have also studied specimens in which part of the
tellurium was replaced by selenium. It has been shown that this
does not change the structure to any great extent (1)(4). It was
found that such a substitution changes the $\bar{q}$-vector in orientation as well as in length. This is evident from fig.4 which
refers to a specimen with nominal composition $AuTe_{1.75}$
$Se_{0.25}$ and in which $1/q = 0.925$ nm as compared to $1/q =$
0.940 nm in pure $AuTe_2$.

5. SYLVANITE

5.1. Diffraction Patterns

5.1.1. Ideal structure. Whereas the "average" structure of
calaverite does not seem to exist even in a wide temperature
range, we did find evidence at room temperature for the "ideal"
sylvanite structure, as determined by Tunell and Pauling (2).
Next to this ideal structure we discovered, also at room
temperature, a "modulated" structure, which produces a
diffraction pattern which is practically undistinghuisable from
that of calaverite. The analysis of the images reveals
significant differences with calaverite however.

Different sections of reciprocal space of "ideal" sylvanite
are reproduced in fig.5. In the $[010]$ zone the pairs of satellites which occur in calaverite are now replaced by a single superstructure spot, which can be attributed to the ordering of gold
and silver. Four sections of reciprocal space are reproduced in
fig.5 where the spots are indexed with respect to the sylvanite
lattice.

The distance between the origin and the $[004]_{sylv.}$ spot which corresponds to the 202_{cal} spot is now divided in four equal parts, whereas in calaverite the satellites divide this distance in approximately 4.5 equal parts. In the $[010]$ section this distance is divided in two parts i.e. the reflections ooℓ are only present when ℓ = even. This is a consequence of the fact that the periodic displacements of the silver atoms are along the $[010]$ direction and do not cause a doubling of the unit mesh in this zone.

5.1.2. Modulated structure. The diffraction patterns in modulated sylvanite and those of the corresponding zones in calaverite are not essentially different. However the satellites seem to be more intense compared to the basic spots, than in calaverite, but their geometry is the same. Their positions depend somewhat on the exact silver content but we have not yet made a systematic study of this. These remarks suggest that the modulated structures of calaverite and sylvanite must be closely related. The high resolution images will prove to be essential in proposing a model.

The increased intensity of the satellites as compared to that in calaverite is striking in the $[010]$ zone (fig.6). Their satellite intensity is now relative to that of the basic spots largest for the low order reflections, which seems to suggest that the satellites are not exclusively due to deformation modulation.

5.2. High Resolution Images

5.2.1. Ideal structure. The high resolution images confirm the "ideal" structure. We shall briefly discuss a few relevant zones. In this section the indices refer to the sylvanite lattice.

In the pseudo-hexagonal zone $[001]$ of fig.7 the structure is imaged as an hexagonal array of bright dots of two different intensities. One third of the dots are extra bright ones; they form an hexagonal array by themselves in such a way that each extra bright dot is surrounded by six less bright dots. From the structure as viewed along the $[001]$ zone, and from the chemical composition it is evident that the extra bright dots must be associated with gold or silver columns, whereas the less bright dots must be associated with Te-columns. This is confirmed by computer simulated images.

The $[101]$ zone produces a pseudo square array of dots; although all columns have the same composition there is some variation in the intensity, of the bright dots revealing in this way the true periodicity. The corresponding zone image of the modulated structure is very informative as we shall see below.

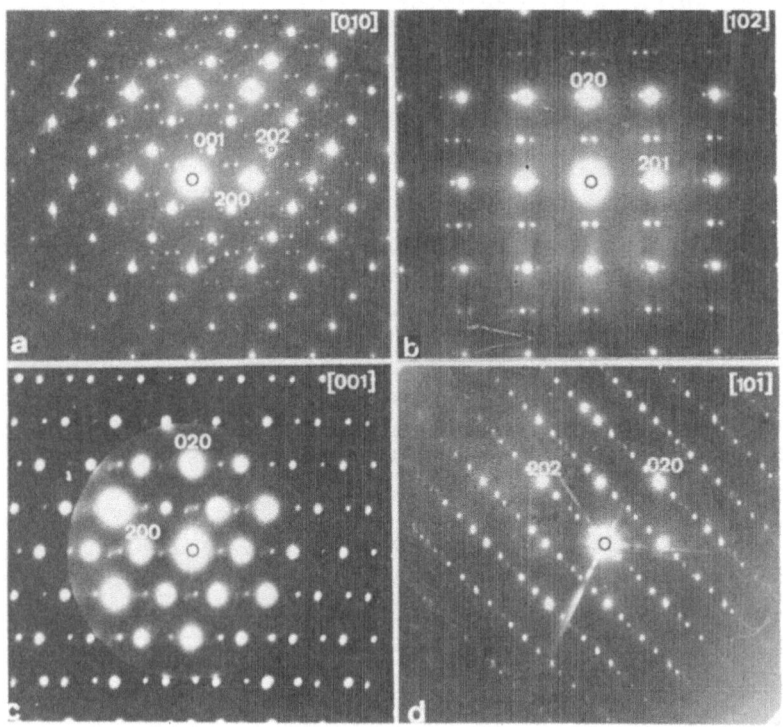

FIG.6. Four sections of the reciprocal lattice of the incommensurate sylvanite structure (calaverite indices).
a) [010] zone pattern. The pairs of weak spots are second order satellites (see fig.3.).
b) [102] zone pattern. Note the pairs of spots in the 201 direction.
c) [001] zone pattern perpendicular to the close packed planes. Note that the satellites are relatively more intense at high order basic reflections.
d) [101] zone pattern. The basic spots form a pseudo square arrangement. One diagonal of the squares is divided into ~ 4.5 equal parts. The orientation anomaly is small in this section.

The [010] zone is very similar in aspect with the corresponding zone of calaverite. Under certain conditions successive bright dots along the c-direction have slightly different intensities, revealing in this way the alternation of gold and silver columns. The Te-columns only produce visible dots under well defined conditions. As opposed to calaverite no modulations are visible in the [010] zone.

The pseudo hexagonal zone $[131]$ gives rise to an image consisting of an hexagonal array of bright dots. However the intensity of the dots is modulated revealing in this way the true periodicity; there is no orientation or spacing anomaly, the spacing between extra bright dots being exactly equal to four times the spacing between close-packed rows of bright dots, the extra bright rows being exactly parallel with the close packed rows of dots.

We have produced many more zone images of the structure; they are all consistent with the structure proposed in (2).

5.2.2. Modulated structure. We have produced high resolution images of the modulated structure along different zones. The most informative ones are those along the $[010]_{cal}$ zone, and along directions normal to the $[010]_{cal}$ direction such as $[10\bar{2}]_{cal}$ and $[10\bar{1}]_{cal}$. In this section we use indices referred to the calaverite lattice.

In fig.8 we have reproduced the images along the $[010]$ zone. The quasi-rectangular arrangement of dots represents under these conditions the heavy metal columns, as could be deduced from computer simulated images. Along horizontal lines in fig.8 the

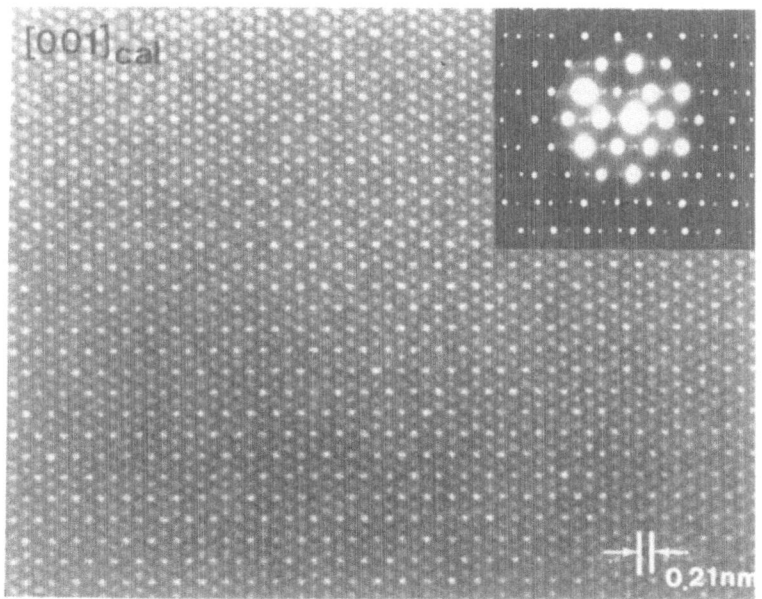

FIG.7. Ideal sylvanite : view along the direction $[001]_{cal}$ perpendicular to the close packed planes. According to the calculations the bright dots represent heavy metal columns and the weaker dots tellurium columns.

234

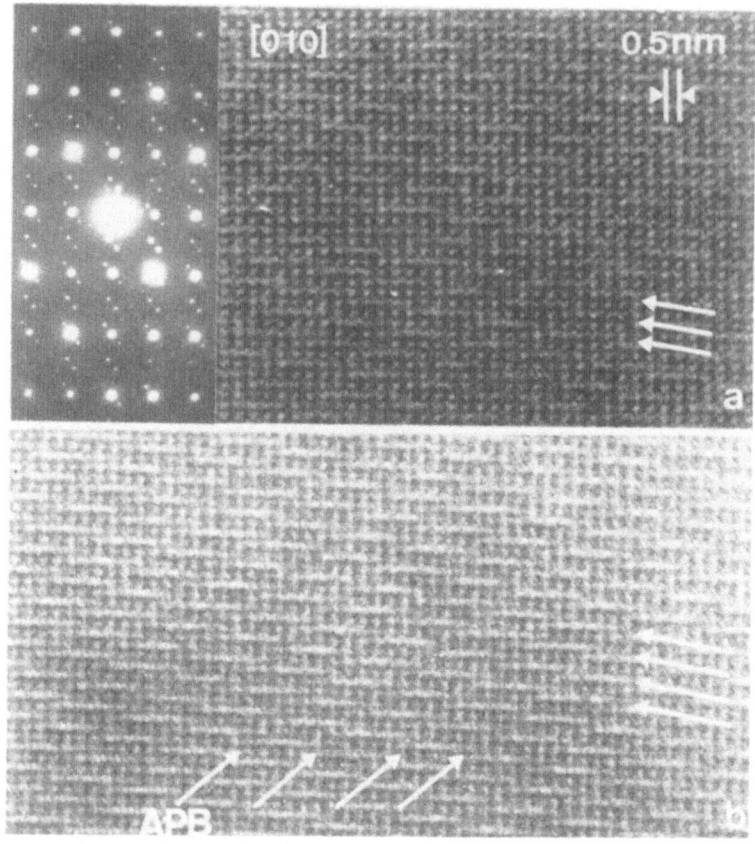

FIG.8. High resolution image along the 010 zone of incommensurate
sylvanite. The modulations are due to the presence of
anti-phase boundaries along $[101]_{cal}$ planes. Along such
planes rows of silver columns interchange with gold columns (compare with fig.14).

intensity of the bright dots changes periodically. The positions
of maximum (and of minimum) intensity are gradually shifted in
successive rows giving in this way rise to two visible wave
fronts of modulations, indicated by arrows in fig.8. The most
pronounced modulation corresponds in wavelength and orientation
with the vector $\bar{q}_1$ (in fig.9), connecting the two weak satellites
in a pair. The second family of modulations corresponds with the
vector indicated as $\bar{q}_2$ in fig.9. The vector $\bar{q}'$ does not produce a
visible modulation in this zone because the corresponding wave
length is too small. The distinction between the different
vectors is not essential however since they are related : e.g.
$\bar{q}' + \bar{q}_1 + \bar{q}_2$ = lattice vector and also $\bar{q}_1 + \bar{q}_2 = \bar{q}_3$.

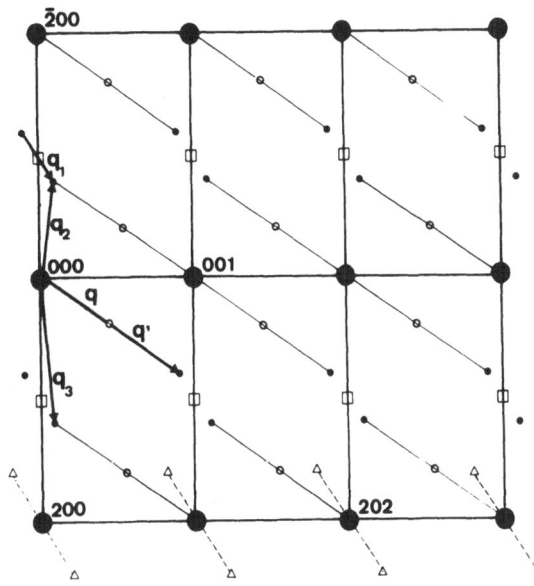

FIG.9. Schematic $[010]$ diffraction pattern. Basic calaverite reflections are represented as large dots while the observed satellites are indicated by small black dots. The open circles represent structurally forbidden first order satellites while the open squares are at the structurally forbidden but commensurate positions with h=odd. The open triangles (only indicated at the h=2 row) shows the double diffraction spots observed in fig.5,b. They could also be the fourth order spots of q.

The $[10\bar{1}]$ zone is of interest since it provides a view edge-on along the wave fronts of the modulations which are perpendicular to $[202]*$. A high resolution image along this zone is reproduced in fig.10b; the atom columns form a square grid. This image should be compared with fig.10a which is the corresponding image in the ideal structure. It is clear that the rows of bright dots are out of phase every second bright row and that moreover the directions of the rows of bright dots enclose a small angle with the b-direction, which leads to an orientation anomaly in the corresponding diffraction pattern (fig.6d).

The period i.e. the distance between rows of bright dots, which in de ideal structure (fig.10a) is equal to $4d_{202}$ is now equal to $\sim 4.5\ d_{202}$. Moreover the configuration of bright dots is shifted over a vector $R = 1/4\ [021]_{syl} = 1/2\ [110]_{cal}$ along the lines of brightest dots. These obser- vations suggest the presence of antiphase boundaries with R

236

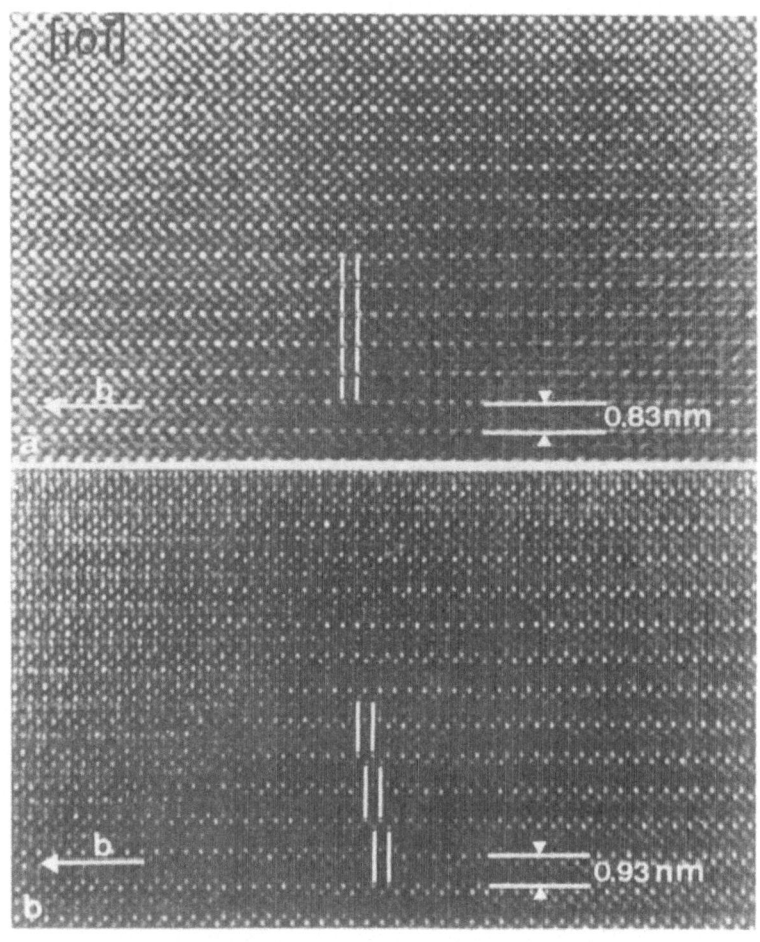

FIG.10. View of the sylvanite structure along the $[10\bar{1}]_{cal}$ direction.

(a) Commensurate structure. The intensity modulation in the thick part reveals the true period. Separate $AuTe_2$ and $AgTe_2$ columns can clearly be recognized following a pseudo square pattern. A few unit cells are outlined.

(b) Incommensurate phase. The intensity modulation reveals again the true period, but also shows that antiphase boundaries occur. A few unit cells are outlined to illustrate the side ways shift. The orientation anomaly is also apparent when viewing the photograph along horizontal lines.

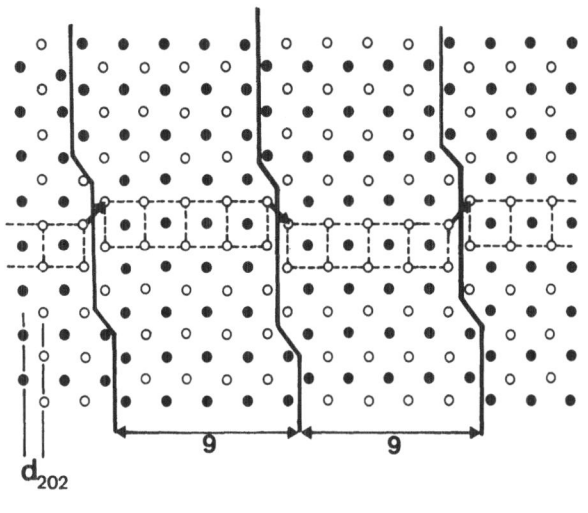

(a)

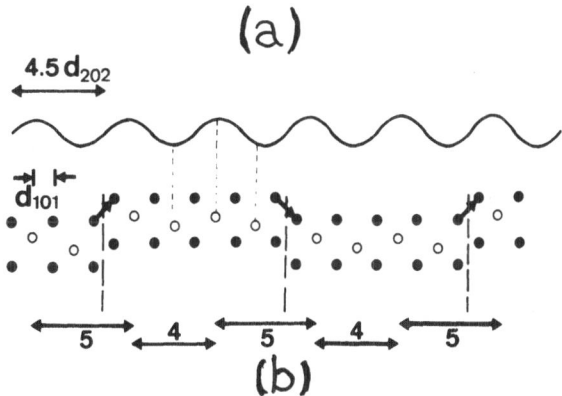

(b)

FIG.11. (a) View of the periodic anti-phase boundary structure as viewed along the $[101]$ direction. The full dots represent $AuTe_2$ columns, whereas the open circles represent $AgTe_2$ columns. The vector is $1/2$ $[110]_{cal}$. Systematic ledging of the anti-phase boundaries causes an orientation anomaly.

(b) The coupling between a displacement wave with a period $4.5\ d_{202}$ in the superstructure and the periodic anti-phase boundary arrangement. The displacements of the silver containing columns (open circles) are exagerated. The average period of the deformation wave is 4.5 whereas the distance between APB's is twice as large in this case.

as a displacement vector, and which are situated in planes normal to $[202]^*$. A model for such a boundary is shown in fig.11,a in which the small displacements of the silver atoms have been ignored and in fig.11,b.

238

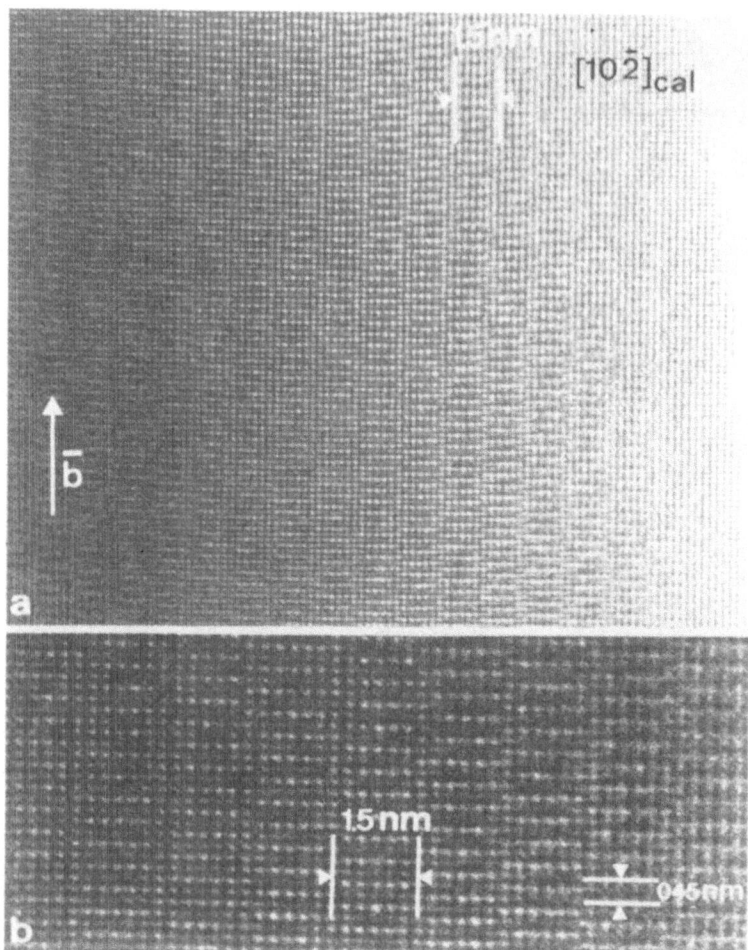

FIG.12. View along the $[10\bar{2}]_{cal}$ zone. Note the presence of periodic APB's roughly parallel with the b-direction. The
boundaries are not seen edge-on and have an apparent
width.

The image along the $[10\bar{2}]$ zone is represented in fig.12.
Along the lines perpendicular to b, the rows of bright dots
vary periodically in intensity; the variations are in anti-phase
in successive rows leading to lines having a certain width along
which the switching-over bright-dark occurs. These lines can be
considered as oblique views of the anti-phase boundaries introduced above. The variations in intensity can then be attributed
to the changing over from a silver-containing (010) layer into a
gold-containing one and vice versa. A model for the anti-phase
boundary as viewed along the $[10\bar{2}]$ direction is represented
in fig.13a. The apparent width of the lines is due to the obli-
queness of the interfaces with respect to the viewing direction;

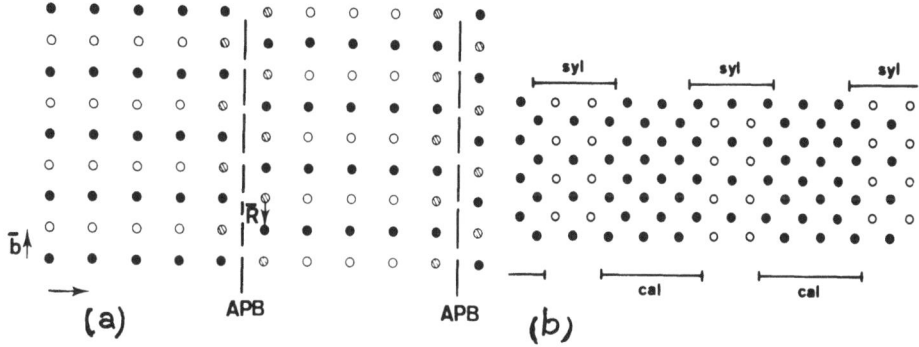

FIG.13 (a). Model for the periodic APB structure revealed in fig.12 as viewed along the $[10\bar{2}]_{cal}$ zone. The cross hatched columns may contain either silver or gold.

(b). View along the $[10\bar{1}]_{cal}$ zone. Along the APB's "interleaving" may occur in order to incorporate an excess of gold without changing the period.

as a result some columns adjacent to the APB may contain silver as well as gold; they are represented as shaded circles in fig.13. It is also possible that some "interleaving" occurs; by this we mean that for instance gold containing planes extend sowewhat beyond the interface, generating locally, along the interface, a narrow slab of calaverite. In this way an appreciable excess of gold can be taken up in the crystal without changing the structure (fig.13b).

5.3. Discussion

A model for the periodic antiphase boundary structure which is consistent with all the observations is represented in a somewhat idealized fashion in fig.14 as projected on the $(010)_{cal}$ - plane. It consists of antiphase boundaries all with the same displacement vector, situated in planes roughly perpendicular to the vector $\bar{q}_1$ i.e. in planes which are roughly parallel with the $(101)_{cal}$ planes. The traces of these APB's are shown as double lines (full + dotted). Along these planes a vertical row of silver atoms changes into a row of gold atoms and vice versa. The separation of the APB's is equal to $\sim 9d_{202}$; this separation is consistent with the vector $\bar{q}_1$ (fig.9).

The question then arises how this can be understood in terms of the modulation vector $|q_1|$. For simplicity we ignore the difference in orientation of the vectors q and

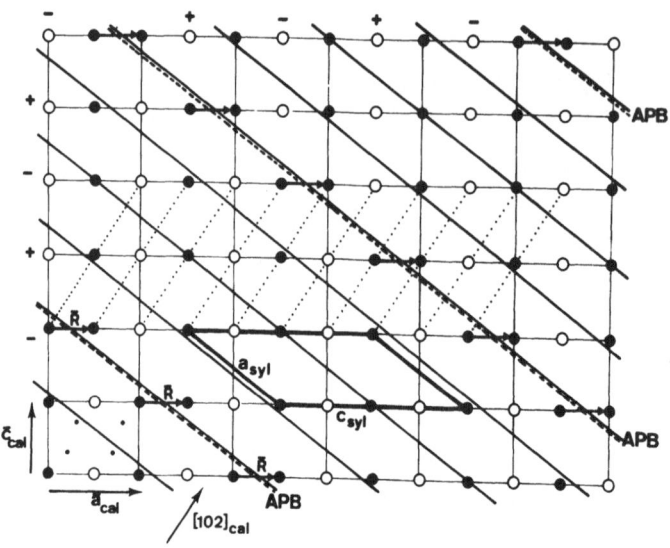

FIG.14. Model for the incommensurate sylvanite structure as viewed along the [010] zone. Zeros of the modulation wave are marked by full lines. The APB's are marked by dashed lines. Along the APB's vertical rows of silver columns (open circles) interchange with rows of gold columns (full circles); tellurium atoms are not shown.

$\bar{q}_1$; periodic ledging of the APB's can take care of this small orientation difference. The generation of the antiphase boundaries as a result of the presence of a modulation wave with wavelength equal to 4,5 d_{202} can be explained by assuming the following ordering principle in non-stoichiometric incommensurate sylvanite. We refer to fig.14 to assist the reasoning. In this fig. we have represented on the one hand the atomic positions and on the other hand the displacement wave with polarization vector along the [010] direction and with wavelength equal to ~ 4.5 d_{202} and which describes the silver displacements pattern. An atomic position which is closer to an extremum than to a zero will be occupied preferentially by a silver atom and be displaced in the sense of the elogation vector. The atomic positions closest to zeros will be occupied by gold. Applying this simple rule to the wave represented in fig.14 leads to a structure in which the gold and silver atoms occupy the position represented in the view along the $[10\bar{1}]_{cal}$ direction (fig.11) and along [010] in fig.13. It is clear that the structure now contains interfaces separated by a distance ~ 9 d_{202} and which can to a good approximation be considered as antiphase boundaries with a displacement vector $1/2[110]_{cal}$. The high resolution image of

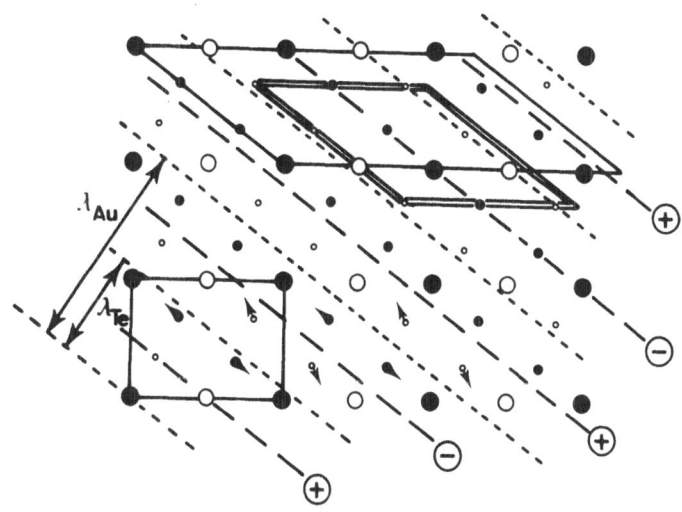

FIG.15. Qualitative model for the modulated structure of calaverite. The unit mesh of the displacement pattern of gold
is outlined in full lines; the corresponding unit mesh
for tellurium is outlined by double lines. This model is
suggested by the commensurate structure of sylvanite (2)
(+ means displaced upwards; − downwards of gold; arrows
indicate displacements in the plane of Te).

fig.12 provides in fact direct evidence for the occurence of
interfaces with this displacement vector. This periodic
anti⁻phase boundary structure conserves the same average period
of 4.5 d_{202} of the modulation wave.
     If only the silver atoms were displaced along the $\begin{bmatrix}010\end{bmatrix}$
direction one should observe no satellites at all in the $\begin{bmatrix}010\end{bmatrix}$
zone pattern since $\bar{g}\cdot\bar{R}=0$ for all reflections of this
zone. The fact that relatively weak second order satellites are
observed in this zone can be explained by noting that the
positions of the Te-atoms are periodically displaced as well, in
a direction parallel with the (010) plane and with a period which
is half of that of the silver displacement pattern. This is
represented in fig.15.
     We have described here a remarkable coupling between
displacement modulation and antiphase boundary modulation. It is
believed that in calaverite a similar displacement modulation
occurs, but of course no interface modulation (fig.15).

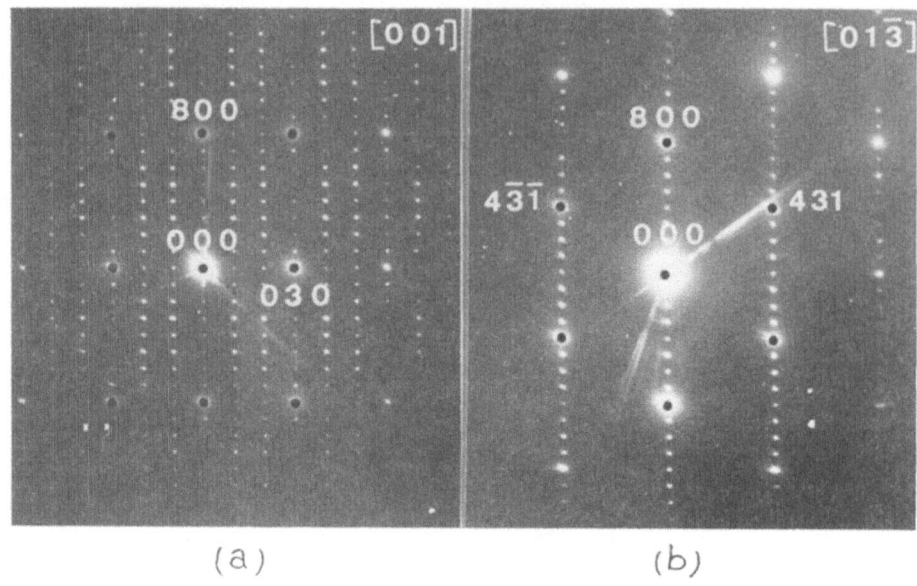

FIG.16. (a) $\begin{bmatrix} 00\underline{1} \end{bmatrix}$ diffraction pattern of krennerite
       (b) $\begin{bmatrix} 013 \end{bmatrix}$ pseudo hexagonal diffraction pattern.

6. KRENNERITE

We have examined specimens obtained from melts with compositions in the vicinity of $Au_{0.8}Ag_{0.2}Te_2$. Crystal fragments exhibiting the Krennerite structure were found, mixed with fragments exhibiting the other structures. The material was clearly heterogeneous. It is believed that the Krennerite phase only occurs in a narrow composition range.

6.1. Diffraction Patterns

The most relevant diffraction pattern is that along the $\begin{bmatrix} 001 \end{bmatrix}$ zone; it is reproduced in fig.16; the superperiod is clearly visible. The distance between the 202_{cal} spot and the origin is now devided in 8 intervals by superstructure spots in such a way that $200_{cal} = 800_{Kr}$; similarly $\overline{2}01_{cal} = 030_{Kr}$. The spots hko with k = threefold are very weak, although they are not prohibited by the spacegroup; it is possible that they are due to double diffraction. Also the pseudo-hexagonal zone $\begin{bmatrix} 013 \end{bmatrix}$ exhibits the eightfold periodicity (fig.16).The diffraction patterns are in agreement with the X-ray results in ref.$\begin{pmatrix} 2 \end{pmatrix}$.

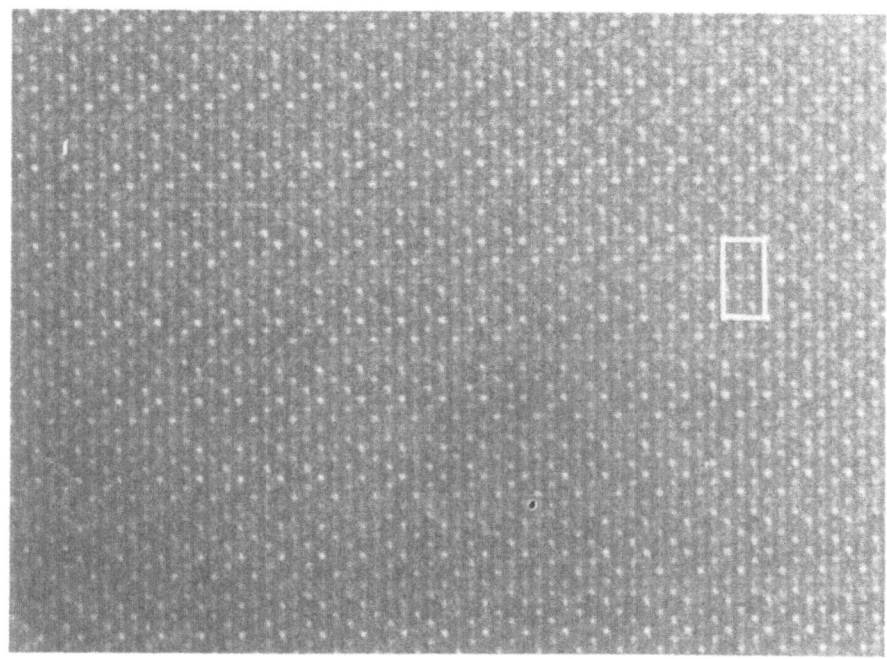

FIG.17. High resolution image of krennerite with the electron
beam along $[001]$; the 1.65 x 0.88 nm orthorhombic unit
cell has been outlined. The twin planes are horizontal
(compare with fig.1,c.).

The pattern of fig.16 can be analysed in terms of repeat
twinning using the method described in ref.(55). The structure
deduced in this way from the geometry of the pattern can be noted
as (4.4); this is in accordance with (2).

We have not found any variability of the diffraction
pattern; in particular the widths of the twins were the same in
all cases; no incommensurability was noted. However sometimes
isolated untwinned strips of the calaverite phase bound by
$[101]_{cal} = [010]_{Kr}$ planes have been observed in high
resolution images.

6.2. High Resolution Images

The most relevant images were obtained along the $[001]$ zone.
(fig.17). Image simulations have indicated that the bright dots
in this image mark the heavy metal columns. The sub-unit cell
polysynthetic twins are clearly revealed in accordance with (2).

In certain parts of the image the atoms located in the twin interface are marked by somewhat brighter dots; we believe that this may be significant. We have simulated images where all atom colums located along the interface were assumed to contain silver. Under suitable conditions these are indeed imaged as especially bright dots; this asumption is thus consistent with the observed images.

This observation may shed some light on the question whether or not silver and gold are ordered. This problem remained unsettled in (2). If silver atoms should adopt an ordered configuration which changes neither the unit cell nor the space group it may well go unnoticed in an X-ray diffraction experiment; this would be the case if silver atoms are assumed to occupy sites along the twin interfaces. This assumption would moreover lead to the correct chemical composition. It is possible that the silver atoms may play a role in inducing twinning and hence in determining the period.

7. CONCLUSIONS

The system $(Au,Ag)Te_2$ has proved to be of particular interest. Depending on the silver/gold ratio this system exhibits modulated structures of three different types, all derived from

the same basic structure. The possibility to change the silver/gold ratio over such a wide range without changing fundamentally the basic structure is presumably related to the fact that the substitution of gold by silver does not change the e/a ratio and is also a consequence of the almost identity in atom radius of silver and gold.

In the calaverite phase the structure is non-commensurate deformation modulated with a wavelength of about 4.5 d_{202}. In the ideal sylvanite phase the same type of deformation modulation is present but it is commensurate its wavelength being $4d_{202}$ (2). For compositions close to the sylvanite phase $(Au,Ag\ Te_4)$ the structure is simultaneously deformation and interface modulated as a result of the coupling between these two modes of modulation. The interface modulation also allows to accomodate an excess of gold. The wavelength of the deformation wave is again $\sim 4.5\ d_{202}$.

In Krennerite the modulation results from the introduction of sub-unit cell twin interfaces; the distance between interfaces has again a comparable value, it is 4 d_{202}, the crystallographic repeat distance being 8 d_{202} however. It is quite remarkable to find that the wavelength for the different types of modulations is always in the same range 4 - $4.5d_{202}$ for compositions covering a wide range.

The defect structure of incommensurate sylvanite, as proposed here, i.e. the coupling between deformation modulation and interface modulation can also be described as a periodic array of parallel discommensurations (D.C.). Whether one or the other description is more adequate depends on the width of the defect. If it is abrupt or narrow it would rather have to be called an anti-phase boundary; if it has a certain width it would perhaps more adequately be called a discommensuration. The same relation for the defect spacing is valid in both cases. One has $2m\bar{q}^- + \bar{q}_1$ = reciprocal lattice vector, with m = integer which in the present case is $m=2$. The vector $\bar{q}$ is the wavevector of the array of anti-phase boundaries or discommensurations, i.e. the array is perpendicular to $\bar{q}_1$ and it has a spacing $1/q_1$. The displacement vector $\bar{R}$ of the APB or of the D.C. has to be consistent with the fractional displacement with respect to the spot $\bar{g}$ corresponding with the commensurate sylvanite structure and which is midway the spot pair in the $[010]$ zone. The fractional displacement is thus 1/2 for these spots; hence $\bar{g} \cdot \bar{R} = 1/2$ (mod.1). This is consistent with $\bar{R}_{cal}= 1/2 \, [110]$ and $g_{cal}=$ hol with h= odd, where $\bar{R}$ as well as $\bar{q}$ were referred to the calaverite lattice.

We believe that in calaverite the defects can be described as DC's whereas they have more the nature of APB's in sylvanite, because of the two kinds of atoms involved in the latter case.

REFERENCES

1. G. Van Tendeloo, P. Gregoriades, S. Amelinckx. J. Sol. State Chem.(in the press).
2. G. Tunell and L. Pauling. Acta Cryst. 5 (1952) 375.
3. S. Suneo, M. Kimata and M. Ohmasa. Modulated Structures 1979. (AIP-Conf. Proc. n°53). 333.
4. G.E. Cranton and R.D. Meyding. J.Chem. 46 (1968) 2637.
5. D. Van Dyck, D. Colaitis and S. Amelinckx. Phys. Stat. Sol. (a) 68 (1981) 385.

INCOMMENSURATE STRUCTURES IN THE LONG-PERIOD ORDERED ALLOYS
STUDIED BY HIGH-RESOLUTION ELECTRON MICROSCOPY

D. Watanabe and O. Terasaki

Department of Physics, Faculty of Science,
Tohoku University, Sendai 980, Japan

ABSTRACT

The Au-Zn and Au-Mg alloys with the incommensurate two-dimensional antiphase structures, containing 16-19 at.% Zn and 16-22 at.% Mg respectively, were investigated by high-resolution structure imaging method with the 1000 kV electron microscopy, and it was revealed that the actual domain configurations change delicately with the incommensurate periods along the two directions. "Domain-like" atomic arrangement was observed in the Au-Zn alloys: two types of "domains" with commensurate structures of the same unit cell size, $6a \times 5c$, are arranged in such a characteristic way that the incommensurate periods are produced in the [100] and [001] directions. In the Au-Mg alloys, however, the "domain-like" structures were not observed.

1 INTRODUCTION

Long-period antiphase structures in ordered alloys often show incommensurate character; antiphase domain sizes M's estimated from the diffraction patterns are in general non-integral multiples of the fundamental cell. The fractional property of M value in the one-dimensional antiphase structure has been interpreted by Fujiwara (1) in terms of the uniform mixture of domains with different sizes. In the two-dimensional cases (2d-APS), however, it is not easy to understand the actual domain configurations in real space from the diffraction study alone.

The Cu_3Pd-type 2d-APS based on the $L1_2$ structure has two kinds of antiphase boundary; one is of the first kind parallel to

the (001) plane, existing at every M_3 cells of the Ll_2, and the
other is of the second kind parallel to the (100) at every M_1
cells. This structure model was originally proposed for the Cu-
Pd by Watanabe *et al.* (2), assuming the stoichiometric composition
A_3B and the integer values for M_1 and M_3. The model for $M_1=3$ and
$M_3=2$ is illustrated in Fig. 1(a), where A and B atoms are shown
by open and solid circles, respectively, and the expected diffraction pattern in Fig. 1(b). There are two types of the second-
kind antiphase boundary. Type 1 is characterized by the subtrac-
tion of an atom plane of the major component, forming the nearest-neighbour like-atom pairs of the minor component across the
boundary, and type 2 by the interpolation of an extra plane of
major component. In the ideal model shown in Fig. 1(a), composition of the structure is conserved to A_3B by the alternate occurrence of these two types of second-kind boundaries.

The Cu_3Pd-type 2d-APS exists also in the Au-Zn and Au-Mg alloys (3,4,5,6). However, there is no necessity to conserve the
composition to A_3B, since the structure appears in the off-stoi-
chiometric composition; 16-19 at.% Zn in the Au-Zn and 16-22 at.%
Mg in the Au-Mg. A model in which excess Au atoms replace Zn or
Mg atoms statistically throughout the unit cell or preferentially
at the second-kind boundaries may be considered. However, there
are other important problems. Firstly, the structure is incommensurate: the values of M_1 and M_3 measured from the diffraction
patterns are non-integers and vary with composition. Secondly,
weak superlattice reflections, which can not be explained by the
idealized models as shown in Fig. 1(a), have been observed recently on the electron diffraction patterns of well-annealed specimens (6,7). For the Au-Zn, Iwasaki (8) performed a single crystal X-ray study on the 19 at.% Zn alloy and derived the structure
model which contains displacement modulation as well as concentration modulation, assuming M_1 and M_3 to be 3 and 2, instead of
the observed values, 2.9 and 2.4. However, the weak reflections

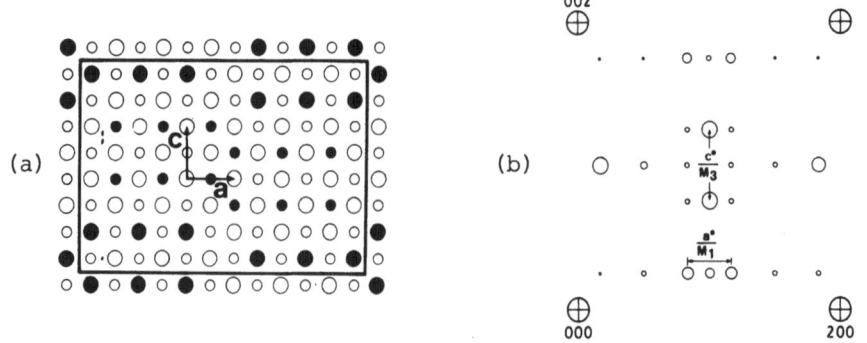

Fig. 1. (a) Model of the Cu_3Pd-type 2d-APS ($M_1=3$, $M_3=2$).
(b) Illustration of diffraction pattern expected from (a).

observed on the electron diffraction patterns can not be explained, even if such lattice modulations as derived by Iwasaki are
taken into account.

High-resolution electron microscopy and diffraction studies
of these two alloy systems were recently made by use of the 1 MV
electron microscope, and the incommensurate characters were
clearly revealed by observing the atomic arrangements for different compositions and different M values through the high-resolution superstructure images (9,10,11).

2 EXPERIMENTAL

Specimens for electron microscopy were obtained from bulk
alloys by jet-electropolishing after proper heat treatment for
homogenization and ordering. Zn contents of the specimens were
determined by chemical analysis with EDTA Titration method and Mg
contents by atomic absorption spectrochemical analysis. High-
resolution images were taken with the Tohoku University 1000 kV
electron microscope (JEM-1000) having the theoretical resolution
of 1.9 Å, under [010] axial illumination, allowing beams of up to
four 200 fundamental reflections to pass through the objective
aperture. Direct magnification was 3×10^5 and exposure time was
about 10 sec. In the micrographs, minor atom rows along the beam
direction are seen as bright or dark dots depending on specimen
thickness and defocus condition, and the ordered arrangement of
minor atoms in the superstructure is determined directly from the
image in an atomic scale.

3 Au-Zn ALLOYS

3.1 Results from Electron Diffraction

Diffraction patterns of the 2d-APS were observed from the alloys containing 15.3 to 18.9 at.% Zn annealed at 250°C for 7-47
days. Fig. 2(a) shows an example obtained from the 17.2 at.% Zn
alloy annealed for 21 days and Fig. 2(b) its schematic illustration. Indices refer to the fundamental cell. Strong superlattice
spots are explained by the structure model such as shown in Fig.
1(a), and mean antiphase domain sizes, M_1 and M_3, can be estimated
from the separations of the split spots, for example, from the
separation of spots 1 and 2, and that of 1 and 3, respectively.
The results are plotted against alloy composition in Fig. 3. The
value of M_1 increases slowly towards 3, whereas M_3 decreases from
about 2.5, as the composition of Zn increases.

Since the experimental values of M_1 and M_3 are close to 3 and
2.5, respectively, the structure consisting of a regular mixture

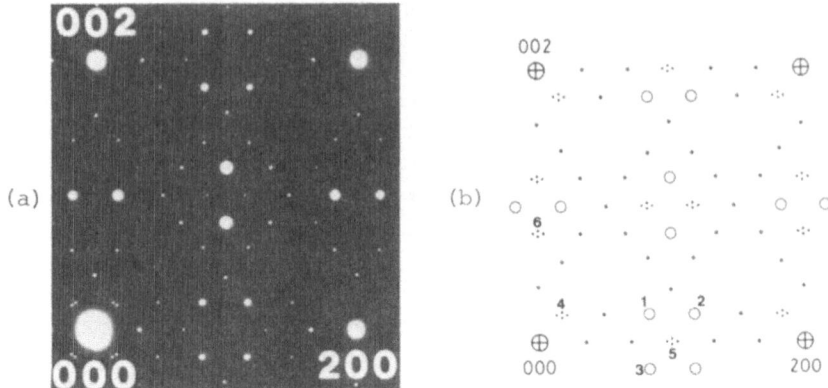

Fig. 2. (a) Diffraction pattern of 2d-APS obtained from Au-17.2
at.% Zn alloy. (b) Schematic illustration of (a).

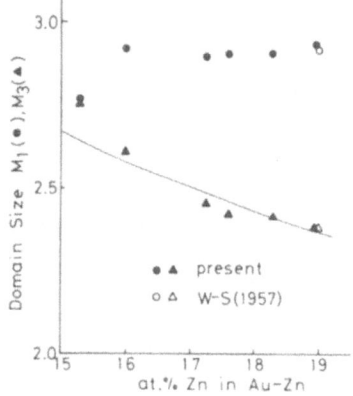

Fig. 3. Antiphase domain sizes M_1 and M_3 vs. Zn content.

of domains with $M_3=2$ and 3, such as $\cdots 2\bar{3}2\bar{3}\cdots$, may be considered
as an idealized model. However, it is to be noted that many
weak, extra spots are observed in the diffraction patterns (Fig.
2(a)) and they are not explained by such idealized model. On
the other hand, other models such as shown in Fig. 4(a), (b) and
(c) give diffraction patterns similar to the observed ones. In
Fig. 4, Au and Zn atoms are represented by open and solid circles, respectively, and large and small circles are at two different levels along the [010] direction. Thin solid and dotted
lines indicate the first- and second-kind boundaries, respectively, and heavy lines the superlattice unit cell with the size,
$A \times C = 6a \times 5c$, where a and c are the lattice constants of the
fundamental $L1_2$. The first-kind boundaries exist at every 2.5
cells of the $L1_2$ in these models. In Fig. 4(a), type 1 and type 2

of the second-kind boundaries exist alternately and composition
of the structure is conserved to 25 at.% Zn. However, all the
second-kind boundaries are of the type 2 in Fig. 4(b) and (c): the
structure of Fig. 4(b) is derived by the substitution of Zn atoms
at the boundaries indicated by thick arrows in Fig. 4(a) with
excess Au atoms and the structure of Fig. 4(c) by the substitution
of Zn atoms indicated by thin arrows, and consequently Zn contents
are reduced to 21.7 and 20.0 at.%, respectively. Calculated
kinematical intensity distributions corresponding to the models,
Fig. 4(a), (b) and (c), are shown in Fig. 5(a), (b) and (c), respectively. Comparison of the observed diffraction patterns
(Fig. 2) with Fig. 5 shows that most of the fine weak spots are
not explained by the model (a), but by the models (b) and (c).
However, it is difficult to distinguish these two models by diffraction experiment alone, mainly because the structure is incommensurate.

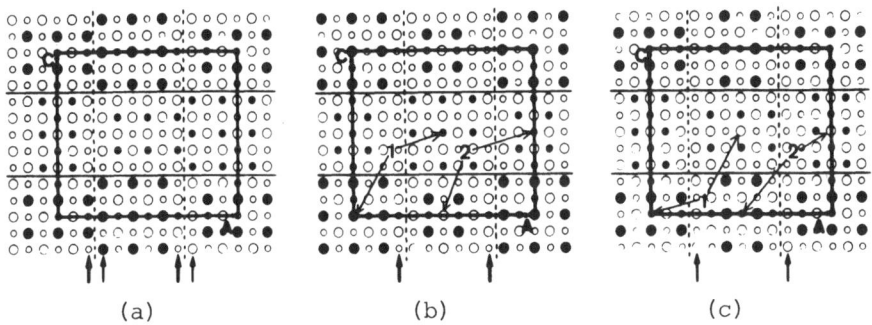

Fig. 4. Structure models of the 2d-APS with $M_1=3$ and $M_3=2.5$.

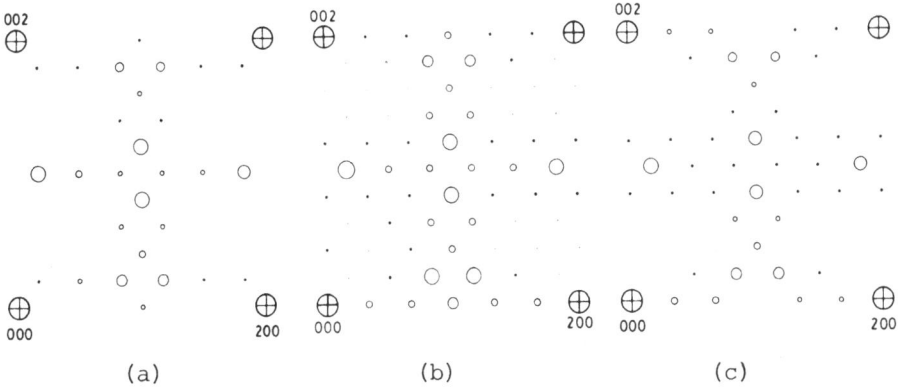

Fig. 5. (a), (b) and (c) are the intensity distributions calculated from the models, Fig. 4(a), (b) and (c), respectively.

3.2 High-Resolution Images and Structure Models

    Fig. 6 shows a high-resolution image obtained from the 17.2
at.% Zn alloy annealed at 250°C. The corresponding diffraction
pattern is Fig. 2(a). The dark fringes parallel to the (100) and
(001) planes, with spacings of about 12 and 10 Å respectively,
correspond to the second- and first-kind boundaries. In addition,
dark stripes parallel to the {560} are seen, as indicated by thin
white lines. These lines divide the crystal into two parts, i.e.,
the "domain A" enclosed by them and "domain B" between the A's.
An enlarged image of Fig. 6 is shown in Fig. 7, where the centers
of A and B "domains" are indicated by small and large arrows,
respectively. The closest distance between bright dots is 4 Å
which corresponds to the lattice constant of the fundamental $L1_2$.

    Plotting the bright dots, i.e. Zn atom positions, from Fig.
7, atomic arrangements in A and B "domains" are determined. It
is remarkable that the structures in A and B are exactly the same
as those of Fig. 4(b) and (c), respectively: the structure A consists of domains containing 3 x 3 and 2 x 2 columns of Zn and the
structure B those of 3 x 2 and 2 x 3 columns, and these domains
are so distributed that Zn-Zn nearest-neighbour pairs (NNP's) are
not formed across the second-kind boundaries. Both "domains" are
observed also in the images obtained from the 17.6 and 18.3 at.%
Zn alloys.

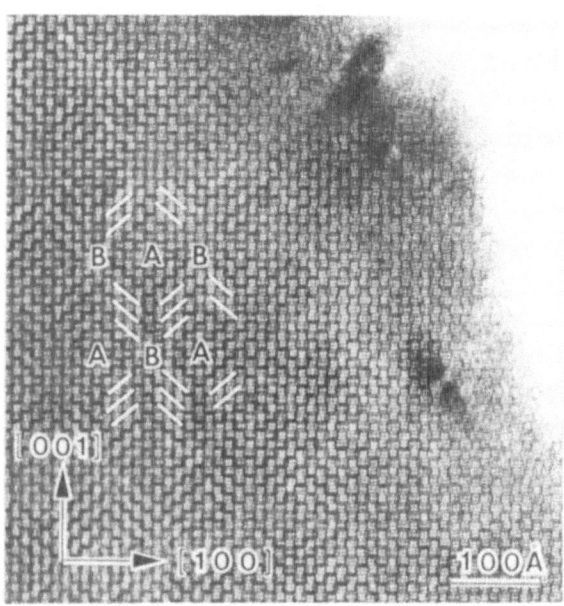

Fig. 6. High-resolution image of the Au-17.2 at.% Zn alloy.
    White lines are parallel to {560}.

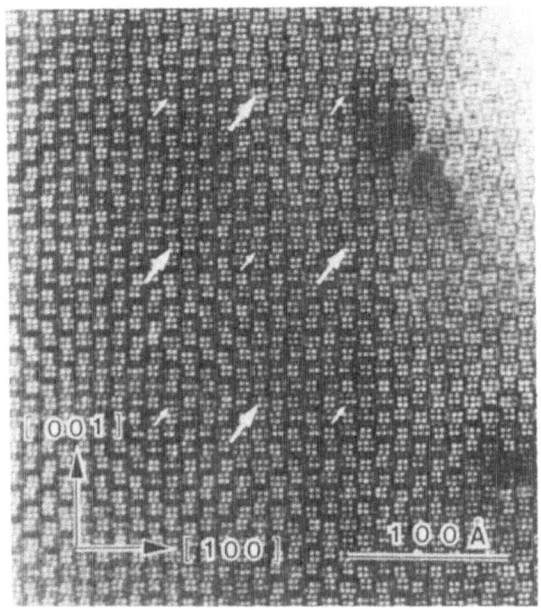

Fig. 7. Enlarged image of a part of Fig. 6.

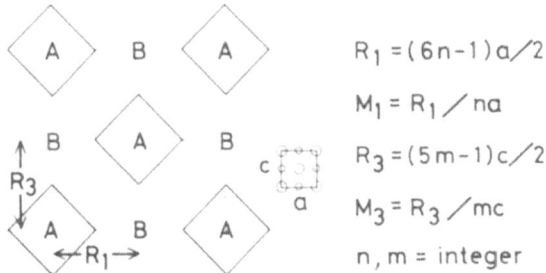

$$R_1 = (6n-1)a/2$$

$$M_1 = R_1/na$$

$$R_3 = (5m-1)c/2$$

$$M_3 = R_3/mc$$

$$n,m = integer$$

Fig. 8. Sketch of the geometrical arrangement of A and B "domains".

The geometrical arrangement of the "domains" can be sketched from the images, as shown in Fig. 8. The distance between the centers of both "domains" is given by $R_1=(6n-1)a/2$ and $R_3=(5m-1)c/2$ along the [100] and [001] directions respectively, where n and m are integers. The average values of M's are then given by $M_1=R_1/na$ and $M_3=R_3/mc$. n and m measured on the images vary with specimen composition, as shown in Table 1. The domain sizes, M_1 and M_3, calculated from these relations are in good agreement with those obtained from the diffraction patterns (see Table 1).

The structures in A and B "domains" shown in Fig. 4(b) and (c) possess the concentration modulations along both the [100] and

Table 1. Measured values of M_1, M_3, n and m

Alloy (at.% Zn)	E.D. Pattern		E.M. Image			
	M_1	M_3	n	M_1	m	M_3
17.2	2.89	2.46	5	2.90	10	2.45
17.6	2.91	2.42	6	2.92	7	2.43
18.3	2.91	2.42	5	2.90	7	2.43
			6	2.92	6	2.42
			7	2.93		
18.9	2.93	2.38	(8)		(4)	

[001] directions. But, there are phase differences in the modulation waves for the two structures, arising from the fact that positions of the planes where Zn atoms are replaced by Au atoms are different by $(1/2)a$ in the [100]. The role of the phase difference can be seen clearly in the structure factors $F(hk\ell)$ of both structures, which are given by

$$F(hk\ell) = \delta_{Au}[\sum_{j=Au} \exp 2\pi i(hx_j+ky_j+\ell z_j) + \sum_{j=Zn} \exp 2\pi i(hx_j+ky_j+\ell z_j)]$$
$$+ (\delta_{Zn}-\delta_{Au}) \sum_{j=Zn} \exp 2\pi i(hx_j+ky_j+\ell z_j).$$

Only the second term contributes to the superlattice reflections, and the values for some representative superlattice reflections within the first Brillouin zone of the fundamental lattice are given in Table 2, where coordinates of the reflections are given as well in terms of M_1, M_3, $a*=1/a$ and $c*=1/c$. A(1) and A(2) or B(1) and B(2) mean that the origin of A or B "domains" is positioned at the center of symmetry, (1) and (2) respectively, as indicated in Fig. 4(b) and (c). There is the phase difference $2\pi\vec{k}\cdot\vec{R}$ in the $\vec{k}$ reflections from the two "domains" separated by the distance $\vec{R}$. The phase differences divided by 2π for the two

Table 2. Values of $\sum_{j=Zn} \exp 2\pi i(hx_j+ky_j+\ell z_j)$

$hk\ell$	Coordinates of $hk\ell$			A(1)	A(2)	B(1)	B(2)
501	($a*(2M_1-1)/2M_1$, 0, $c*/2M_3$		)	12.1	-12.1	12.1	-12.1
105	($a*/2M_1$, 0, $c*$		)	18.9	-18.9	-18.4	18.4
604	($a*$, 0, $c*(2M_3-1)/2M_3$)			16.2	16.2	-16.2	-16.2

Table 3. Values of $\vec{k}\cdot\vec{R}$ for hkl reflections

hkl	$\vec{k}\cdot\vec{R}_1$	$\vec{k}\cdot\vec{R}_3$
$50\underline{1}$, $\overline{50}\overline{1}$ $50\overline{1}$, $\overline{5}01$	$\dfrac{\pm(2M_1-1)\,nM_1}{2M_1} = \pm(5n-1)/2$	$\pm mM_3/2M_3 = \pm m/2$
105, $\overline{1}05$	$\pm nM_1/2M_1 = \pm n/2$	$mM_3 = (5m-1)/2$
604, $60\overline{4}$	$nM_1 = (6n-1)/2$	$\pm mM_3(2M_3-1)/2M_3 = \pm(4m-1)/2$

cases, $R=R_1$ and R_3, are given in Table 3. They are integers or half-integers depending on n and m values, odd or even. This result means, for example, that the signs of the structure factors change for the 105 and 604 reflections but not for the 501, if n is odd. In other words, the phases of structure factors for the "domains" separated by R_1 in the [100] from the A(1) "domain" coincide with those for B(1) "domain" if n is odd, and coincide with those for B(2) "domain" if n is even. Similar situation occurs in the [001]. These facts mean that structure factors of all the superlattice reflections from both the A and B "domains" existing at intervals of R_1 and R_3 are in phase, and the

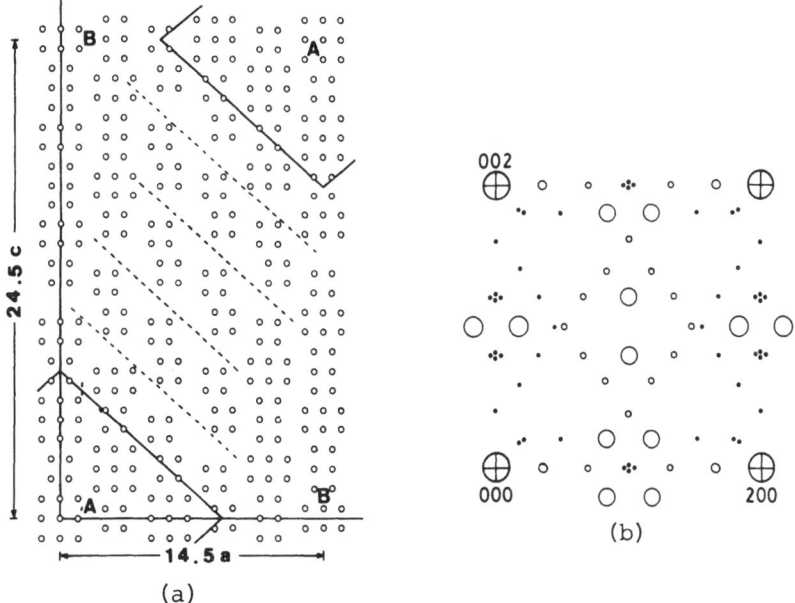

Fig. 9. (a) Positions of Zn atoms plotted from Fig. 7.
        (b) Intensity distributions calculated from the atomic
            arrangement shown in (a).

"domain-like" configurations can exist, as observed in the 17.2, 17.6 and 18.3 at.% Zn alloys.

The structures shown in Fig. 4(b) and (c) are the idealized models for the A and B "domains" in the sense that they have compositions of 21.7 and 20.0 at.% Zn, respectively. The deficiency of Zn atoms in the alloys is accommodated mostly at the interfaces parallel to the {560}, where the A "domains" are neighbours to each other. The positions of Zn atoms plotted from the image of Fig. 7 are shown in Fig. 9(a), which covers only a quarter of the area shown in Fig. 8. It is seen that Zn atoms at the antiphase boundaries are preferentially substituted by Au atoms and antiphase domains consisting of 2 x 2 columns of Zn are predominant at the interfaces of A "domains", and consequently the Zn deficient regions parallel to the {560} become noticeable, as indicated by the dotted lines. Zn content of this structure is 17.7 at.%, which can be compared with the alloy composition, 17.2 at.% Zn. Fig. 9(b) shows the calculated kinematical intensity distribution, which explains qualitatively all the spots observed in the diffraction pattern shown in Fig. 2(a).

4 Au-Mg ALLOYS

4.1 Results from Electron Diffraction

Diffraction patterns of the 2d-APS were observed from the alloys containing 16-22 at.% Mg. Fig. 10(a) shows an example obtained from the 16.1 at.% Mg alloy annealed at 300°C for 28 days, and (b) its schematic illustration. Mean antiphase domain sizes, M_1 and M_3, estimated from the diffraction patterns are plotted against alloy composition in Fig. 11. The value of M_1 is fixed to about 2.5 independently of Mg content, but the value M_3 decreases from about 2.3, as Mg content increases.

Although most of the superlattice reflections are interpreted by the Cu_3Pd-type 2d-APS, some of them cannot be explained, unless a multiple diffraction effect or periodic lattice modulation associated with the antiphase structure is assumed. For example, the spots assigned the numbers 1 and 2 and their higher-order spots in Fig. 10(b) cannot be expected from the idealized structure model, as shown in Fig. 1(a). When the crystal is rotated on the [100] or [001] axis, these spots weaken but do not disappear. This fact means that part of the intensity is caused by multiple diffraction but the rest is due to some kind of lattice modulation. Other spots represented by open circles cannot be explained either by the Cu_3Pd-type 2d-APS.

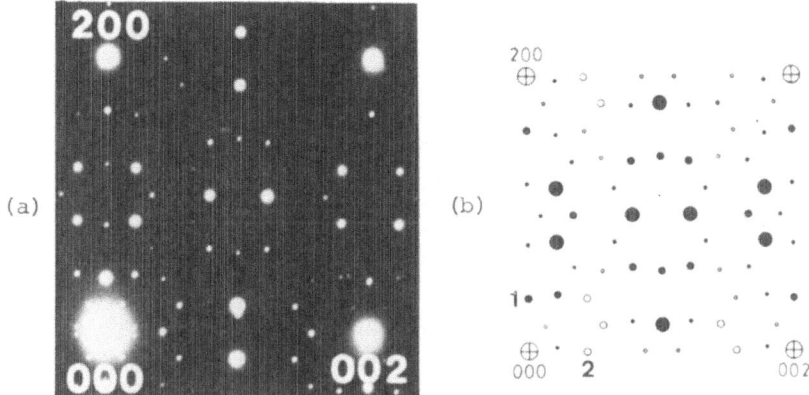

Fig. 10. (a) Diffraction pattern of 2d-APS obtained from Au-16.1 at.% Mg alloy. (b) Schematic illustration of (a).

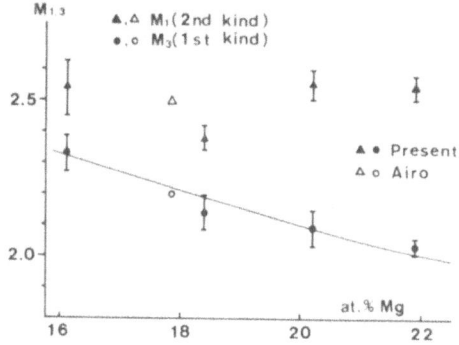

Fig. 11. Antiphase domain sizes M_1 and M_3 vs. Mg content.

4.2 High-Resolution Images and Structure Models

An example of high-resolution images taken from the 16.1 at.% Mg alloy with the [010] incidence is shown in Fig. 12. Dark fringes parallel to the (100) and (001) planes, corresponding to the second- and first-kind boundaries, are seen, but domain configuration is different from that of the Au-Zn alloys: the image consists mainly of the square-shaped domains containing 2 x 2 dots with the separation of 4.1 Å, which corresponds to a unit-cell size of the fundamental structure.

The idealized structure model can be derived, based on the 2 x 2 domains, as shown in Fig. 13, where thin lines indicate the fundamental unit cell and thick lines the superlattice unit cell.

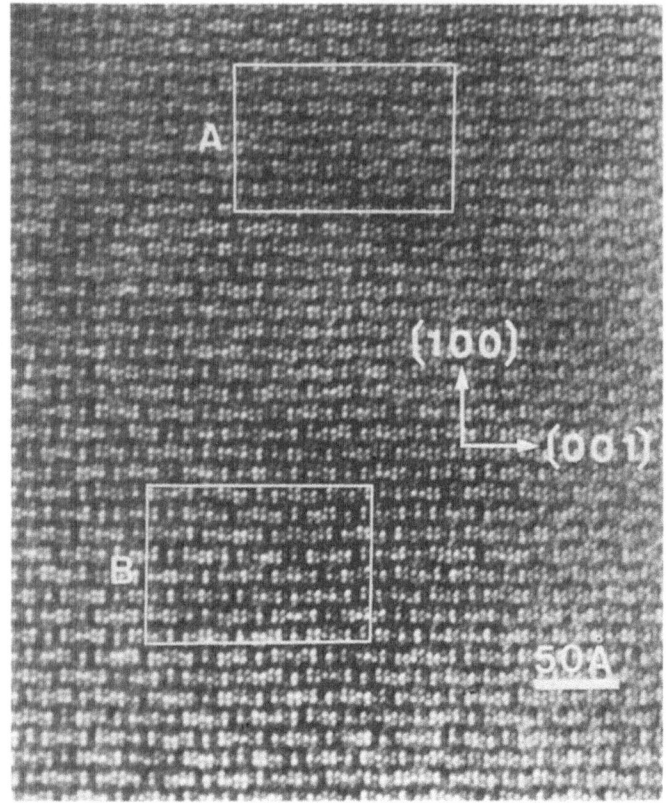

Fig. 12. High-resolution image of the Au-16.1 at.% Mg alloy.

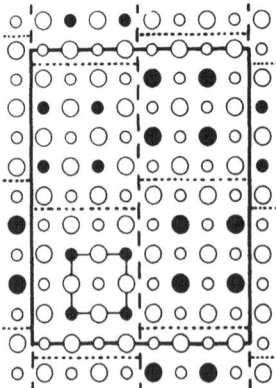

Fig. 13. Idealized structure model, Au$_4$Mg, for the 2d-APS.
Dashed and dotted lines indicate the first- and
second-kind boundaries, respectively.

Au and Mg atoms are shown by open and solid circles, and large and small circles are at two different levels. The structure has the first- and second-kind boundaries parallel to (001) and (100), as indicated by dashed lines and dotted lines, and the domain size is $M_1a \times M_3c = 2.5a \times 2c$. The second-kind boundaries are of the type 2 and Mg content is reduced to 20 at.%. Agreement between the observed diffraction pattern and the calculated kinematical intensity distribution is generally good. The spot 1 and its higher-order spots in Fig. 10(b) appears, reflecting the fact that a density modulation exists in the [100] direction. However, the observed intensity of the spot 1 is still stronger than the calculated intensity, and the spots shown by open circles in Fig. 10 (b) are not explained.

A close examination of the image shows, however, that arrangement of the 2 x 2 domains is not perfectly regular, reflecting the incommensurate character of the structure, i.e. M_1=2.57 M_3=2.29. A schematic drawing of the framed region A of Fig. 12 is shown in Fig. 14, where the unit cell of the idealized model is outlined by thin lines. The rectangle-shaped domains with 2 x 3 dots are discernible at some places if the dots with low contrast (represented by small open circles in Fig. 14) are taken into account: the domains with three dots along the [001] are mixed uniformly with those of 2 x 2 dots, giving the average domain size of M_3=2.3, which corresponds to the value measured on the diffraction pattern. In addition, some of the domains have three dots along the [100] direction.

It is considered that contrast of the dots is low or invisible, if Mg atoms in the column along the incident electron beam are substituted partially or completely by Au atoms. Assuming

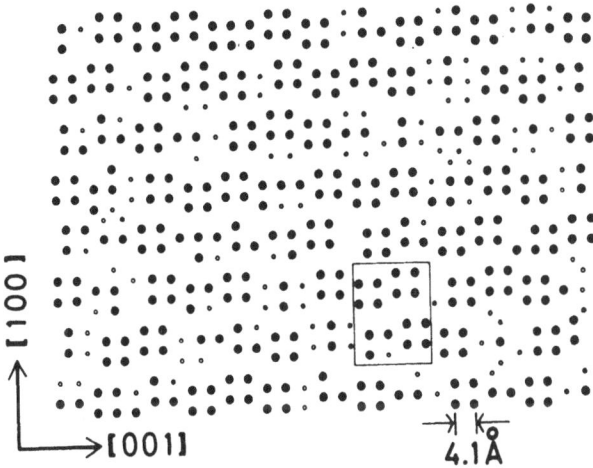

Fig. 14. Schematic drawing of the framed region A of Fig. 12.

the occupation probabilities of Mg atoms to be x and y for large
solid circles and small open circles, respectively, in Fig. 14
and counting the number of the circles, the Mg concentration in
at.% is given approximately by $15.0x + 4.7y$. If it is assumed
that the region A has the composition of 16.1 at.% Mg and x is
simply taken as 1, the value of 0.24 is obtained for y. A similar
estimation for the region B of Fig. 12 gives 0.46. Although the
estimation is very crude, it may be said that about a half or more
of the Mg atoms are replaced by Au atoms along the columns corresponding to the dots with low contrast. It is to be noticed that
such dots are seen mostly on the domain boundaries, especially on
the boundaries of 2 x 3 domains, meaning the existence of density
modulation along the [001] as well, and that these 2 x 3 domains
are not neighbours. A possible model exhibiting these features
is shown in Fig. 15(a), where M_1 and M_3 correspond to 2.5 and
2.33, respectively. It has a density modulation with the period
M_3 along the [001] direction: occupation probability of Mg atoms
at the Mg sites is given by the sine curve, and the composition

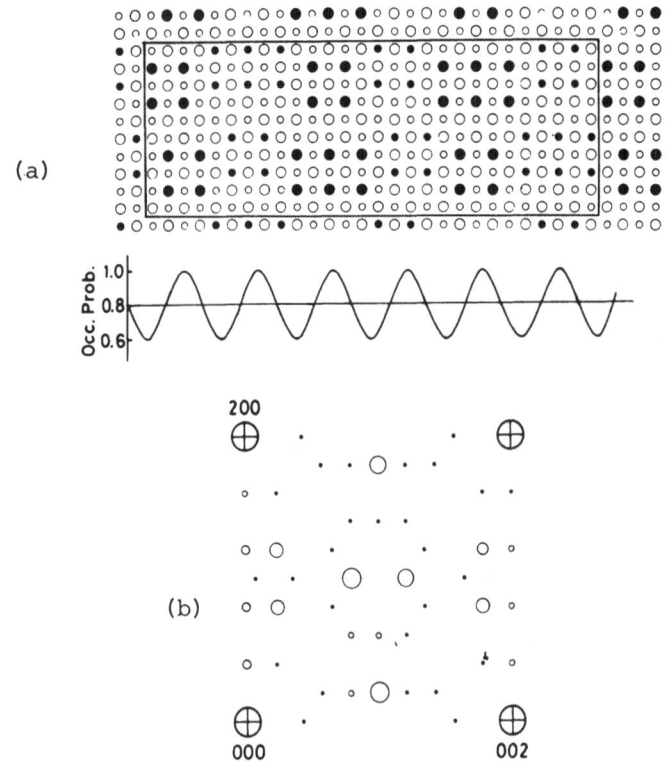

Fig. 15. (a) Structure model with sinusoidal density modulation
along the [001] direction. Open and solid circles
represent the Au and Mg sites, respectively.
(b) Intensity distribution calculated from model (a).

of Mg is reduced from 20 at.% to 16 at.%. Relative intensity distribution calculated kinematically from this model is shown in Fig. 15(b). Agreement between observation and calculation is improved remarkably; appearance of the spot 2 and some of the spots shown by open circles in Fig. 10(b) is now explained. If the higher-order modulations are included, then all the spots will be explained.

5 DISCUSSION

The mean domain sizes, M_1 and M_3, of the incommensurate 2d-APS's described in the preceding sections vary with alloy system and composition, and the domain structures change with temperature. As seen in Figs. 3 and 11, the values of M_1 are fixed to about 2.9 and 2.5 for the Au-Zn and Au-Mg alloys, respectively, but M_3 decreases continuously. The composition dependence of M_3 can be compared with the following relation derived by Sato and Toth (12), based on the concept of the stabilization of long-period superlattice at the Brillouin zone boundary:

$$(e/a) = (\pi/12t^3)[2+(1/M_3)+(1/4M_3^2)]^{3/2},$$

where (e/a) is the number of free electrons per atom and t the truncation factor for the Fermi surface. This relation explains well the experimental results, if the number of electron is assumed to be 1 for Au and 2 for Zn and Mg and t is taken to be 0.948 and 0.958 for Au-Zn and Au-Mg, respectively, as shown by solid curves in Figs. 3 and 11. The truncation factor t for the Au-Zn is comparable with the value, 0.945, obtained for the disordered Au-Zn system from the analysis of short-range order diffuse scattering, based on the Fermi-surface imaging concept (13).

In the Au-Zn system, the "domain-like" structure was observed in the 17.2, 17.6 and 18.3 at.% Zn alloys. It consists of the A and B "domains" with commensurate structures and they are arranged in such a way that the incommensurate periods are produced along the [100] and [001] directions. In the 18.9 at.% Zn alloy, the 2d-APS coexists with the 1d-APS with $M=2$, i.e. DO_{23} structure, as seen in the image shown in Fig. 16. Although the A and B "domains" are discernible at some places, the 2d-APS does not show the "domain-like" configuration clearly. The reason for this is in the M_3 value, which is related to the (e/a), i.e. the dimension of the Fermi surface: since M_3 is measured to be 2.38 (Fig. 3), R_3 is estimated to be $4M_3c = 9.5c$, according to the idea used in the preceding section. This distance is too short to give a distinct "domain" structure. The "domain-like" structure was not observed in the 16.0 at.% Zn alloy too: M_3 is measured to be 2.61, which is larger than 2.5 inherent in the commensurate structure of "domains".

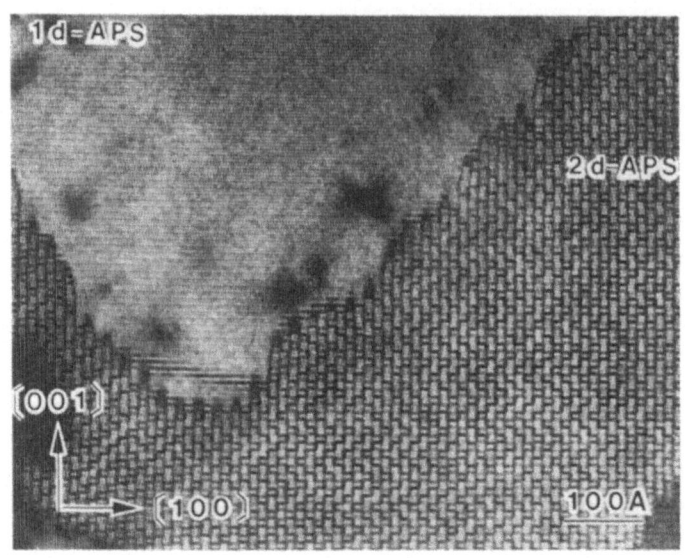

Fig. 16. High-resolution image of the Au-18.9 at.% Zn alloy
annealed at 250°C for 8 days.

The "domain-like" structure was not recognized also in the
Au-16.1 at.% Mg alloy for which $M_1=2.57$ and $M_3=2.29$. It was un-
successful to obtain the superstructure images from the alloys
containing 18-22 at.% Mg: the crystal regions having a single domain orientation were very small, because of the low ordering
temperature below 300°C. However, such "domain" structures as observed in the Au-Zn alloys may not be expected for the Au-Mg alloys, if the values of M_1 and M_3 are taken into account (Fig. 11).

The 2d-APS in the Au-Zn system transforms into the disordered
structure at about 300°C. On the other hand, when the Au-Mg alloys containing 18-22 at.% Mg are annealed at 300-400°C, the
structure transforms into another 2d-APS of $Au_{30}Mg_8$ (6,14,15).
Fig. 17 shows the structure model derived from the diffraction
pattern and superstructure image of the 21.9 at.% Mg alloy annealed at 300°C for 21 days. The superlattice unit cell is outlined
with solid lines. The relation between the unit cell parameters
of superlattice and fundamental lattice is as follows; $A=4a-b$,
$B=(-a+5b)/2$ and $C=c$. The unit cell contains 30 Au and 8 Mg atoms,
corresponding to the composition of 21.1 at.% Mg, which agrees
with the specimen composition, 21.9 at.%.

The structure of $Au_{30}Mg_8$ also can be described in terms of
the periodic antiphase structure. It consists of the 2 x 2 domains of $L1_2$ structure and the domains align along the two directions simultaneously, forming the first- and second-kind

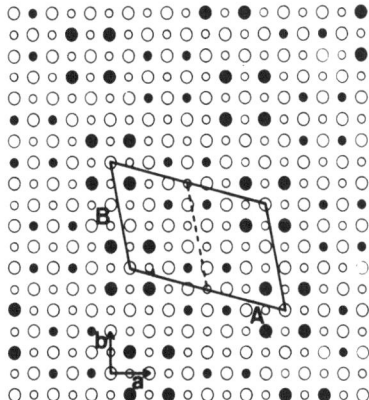

Fig. 17. Structure model of $Au_{30}Mg_8$. Open and solid circles represent Au and Mg atoms, respectively.

boundaries. The latter boundaries are of the type 2. However, in contrast with the 2d-APS appearing at lower temperatures (Fig. 13), the domains align along the A and B axes, instead of the a and b axes. Because of this difference, the 2d-APS shown in Fig. 17 can accommodate a bit more Mg atoms than the 2d-APS given in Fig. 13, whereas the nature of the boundaries is the same for the two structures.

REFERENCES

1. Fujiwara, K. On the Period of Out-of-step of Ordered Alloys with Antiphase Domain Structure. J. Phys. Soc. Jpn. 12 (1957) 7-13.
2. Watanabe, D., Hirabayashi, M. and Ogawa, S. On the super-structure of the alloy Cu_3Pd. Acta Cryst. 8 (1955) 510-511.
3. Wilkens, M. and Schubert, K. Über einige metallische Ordnungsphasen mit grosser Periode. Z. Metallkde. 48 (1957) 550-557.
4. Iwasaki, H., Hirabayashi, M., Fujiwara, K., Watanabe, D. and Ogawa, S. Study on the Ordered Phases with Long Period in the Au-Zn Alloy System. I. J. Phys. Soc. Jpn. 15 (1960) 1771-1783.
5. Airo, P. On the Superstructure of the Alloy Au-18 at.% Mg. phys. status solidi (a), 47 (1978) K107-110.
6. Terasaki, O. and Watanabe, D. Long-Period Superstructures of the Au-Rich Au-Mg Alloys. *Modulated Structures*-1979, AIP Conf. Proc. No.53, (1979) pp.253-255.

7. Terasaki, O. and Watanabe, D. Two-Dimensional Antiphase Structures of the 2d-Cu_3Pd Type Studied by High Voltage, High Resolution Electron Microscope. Japan. J. Appl. Phys. 20 (1980) L381-384.

8. Iwasaki, H. Study on the Ordered Phases with Long Period in the Au-Zn Alloy System. II. J. Phys. Soc. Jpn. 17 (1962) 1620-1633.

9. Terasaki, O. Direct observation of "domain like" atomic arrangement of incommensurate 2d-$Au_{3+}Zn$ by high-voltage, high-resolution electron microscope. J. Phys. C: Solid State Phys. 14 (1981) L933-938.

10. Terasaki, O. Study of the Incommensurate Two-Dimensional Antiphase Structure of $Au_{3+}Zn$ by High Voltage, High Resolution Electron Microscopy. J. Phys. Soc. Jpn. 51 (1982) 2159-2167.

11. Terasaki, O., Mikata, Y., Watanabe, D., Hiraga, K., Shindo, D. and Hirabayashi, M. Two-Dimensional Antiphase Structure of Au_4Mg Studied by High-Voltage, High-Resolution Electron Microscopy. J. Appl. Cryst. 15 (1982) 65-71.

12. Sato, H. and Toth, R.S. Long-Period Superlattices in Alloys. II. Phys. Rev. 127 (1962) 469-484.

13. Watanabe, D. Short-range order diffuse scattering from disordered alloys. *Electron Diffraction* 1927-1977, Inst. Phys. Conf. Ser. No.41, (1978) pp.88-97.

14. Terasaki, O., Watanabe, D., Hiraga, K., Shindo, D. and Hirabayashi, M. Structure Analyses of Long-Range Ordered Alloys by High Voltage, High Resolution Electron Microscopy. Micron 11 (1980) 235-240.

15. Sande, M.V., Van Tendeloo, G., Amelinckx, S. and Airo, P. High Resolution Electron Microscopic and Electron Diffraction Study of the Au-Mg System. I. The Monoclinic $Au_{15}Mg_4$ Phase. phys. status solidi (a), 54 (1979) 499-512.

ANALYSIS OF DIFFUSE SCATTERING FROM COMPOSITION MODULATIONS IN CONCENTRATED ALLOYS

P. Georgopoulos and J.B. Cohen

Department of Materials Science and Engineering, The Technological Institute, Northwestern University, Evanston, IL 60201

ABSTRACT

Theoretical and experimental methods for the measurement and analysis of diffuse x-ray and neutron scattering have progressed greatly in the last few years. It is now possible to obtain very detailed information on local atomic arrangements in concentrated and, in some cases, fairly dilute metallic alloys, ceramics and intermetallic compounds. The advent of synchrotron radiation promises substantial savings in data collection times and hence makes possible a whole new class of experiments, too time consuming to be carried out with laboratory apparatus. In this paper, the most recent developments in data analysis will be reviewed. Additionally, experimental procedures for both laboratory and synchrotron environments will be discussed.

1. INTRODUCTION

The analysis of x-ray and neutron diffuse scattering is to this date the most accurate and direct method to obtain information about local atomic arrangements in crystalline solid solutions (1). By the term "local atomic arrangements", we refer to atomic groupings present in the solid solution, exhibiting no periodicity (hence the term "local"), or, in some cases, periodicities incommensurate to that of the average lattice. It is these local atomic configurations, i.e. the aperiodic component of the electron density, that gives rise to diffuse scattering, whereas the periodic component produces scattering in well-defined directions, giving Bragg peaks.

Before we proceed with the formal x-ray scattering theory, certain philosophical points must be addressed first: Just as every other physical theory, the theory of x-ray scattering attempts to describe real crystals in terms of idealized models, that is ideal atoms arranged on a perfect infinite lattice with no defects, no breaks in periodicity and all the atoms precisely on lattice sites. Real crystals obviously defy such exact theoretical straightjackets and so it becomes a matter of perspective to regard certain crystals as "perfect" and certain others as "imperfect". In fact, one and the same crystal can be considered as perfect, when one is after gross features of the crystalline state, and as imperfect, when more subtle features are investigated.

Let us consider an example to illustrate this point: A solid solution of gold and nickel, exhibiting very weak clustering tendencies, that is, there is a slight preference for like atoms to occupy near neighbor positions. Because of the different atomic size of gold and nickel, the atoms are not expected to be at their precise lattice positions, but they may be slightly displaced. The scattering pattern of such a crystal will contain sharp Bragg reflections, as well as weak continuous scattering between the Bragg reflections. If we choose to regard this weak diffuse scattering as "background" and ignore it, then a simple scattering theory involving average "goldnickel" atoms on an ideal lattice with only random thermal disorder is sufficient to account very satisfactorily for the scattering pattern. Such a theory cannot, however, account for the weak diffuse intensity away from Bragg reflections. A more sophisticated theory, which can accomplish this will be reviewed in the following chapter.

2. SCATTERING THEORY

In this chapter we present a kinematical (single scattering) theory, applicable to a binary solid solution with cubic symmetry and with one atom per lattice point (2). This theory has been generalized in a straightforward manner to cover multicomponent solutions with more than one atom per lattice point (3) and non-cubic symmetry but, as we will see, the equations for the scattered intensity quickly become very cumbersome, even for minor extensions.

The general equation for the total scattered intensity in electron units (one electron unit is the amount of radiation scattered by a free electron) is:

$$I_T = \sum_m \sum_n f_m f_n \exp[ik.(r_m - r_n)] \tag{1}$$

where:

    m, n are lattice indices. Each one is understood to represent a triplet of coordinates in space.

    f_m, f_n are atomic scattering factors of the atoms located at sites m and n.

    r_m, r_n are vectors from an arbitrary origin to the atoms located at m, n.

Allowing the atoms to be displaced from their ideal lattice positions by small amounts, the vectors r_m and r_n can be expressed as:

$$r_m = R_m + \delta_m, \qquad r_n = R_n + \delta_n, \tag{2}$$

where R_m, R_n are ideal lattice vectors and δ_m, δ_n are small displacement vectors from the ideal lattice sites. Then, equation (1) can be written as:

$$I_T = \sum_m \sum_n f_m f_n \exp[ik.R_{mn}] \exp[ik.\delta_{mn}], \tag{3}$$

where:

$$R_{mn} = R_m - R_n \quad \text{and} \quad \delta_{mn} = \delta_m - \delta_n. \tag{4}$$

Let us now call the two components of the solid solution A and B and define a conditional pair probability P_{mn}^{AB} as the probability of finding a B atom at site n when there is an A atom at site m. Then the spatially averaged intensity given by equation (3) can be expressed as:

$$
\begin{aligned}
\langle I_T \rangle = \sum_m \sum_n \big(& c_A f_A^2 \, P_{mn}^{AA} \, \langle \exp[ik.\delta_{mn}^{AA}] \rangle \\
& + c_A f_A f_B \, P_{mn}^{AB} \, \langle \exp[ik.\delta_{mn}^{AB}] \rangle \\
& + c_B f_A f_B \, P_{mn}^{BA} \, \langle \exp[ik.\delta_{mn}^{BA}] \rangle \\
& + c_B f_B^2 \, P_{mn}^{BB} \, \langle \exp[ik.\delta_{mn}^{BB}] \rangle \, \exp[ik.R_{mn}],
\end{aligned}
\tag{5}
$$

where the carats imply an average over either AA, or AB or BB pairs. c_A and c_B are the atom fractions of components A and B respectively.

If the displacements of atoms from their ideal lattice sites are assumed small compared with the interatomic distance, the exponentials in equation (5) can be expanded as follows, denoting by i, j the species on the m or n site (i and j stand for A or B):

$$\langle \exp[ik.\delta_{mn}^{ij}]\rangle = \langle 1 + ik.\delta_{mn}^{ij} - \tfrac{1}{2}(k.\delta_{mn}^{ij})^2$$
$$- \tfrac{i}{3!}(k.\delta_{mn}^{ij})^3 + \tfrac{1}{4!}(k.\delta_{mn}^{ij})^4 + \ldots \tag{6}$$

Grouping all odd-powered terms together and factoring out $ik.\delta_{mn}^{ij}$:

$$\langle \exp[ik.\delta_{mn}^{ij}]\rangle = \langle [1 - \tfrac{1}{2}(k.\delta_{mn}^{ij})^2 + \tfrac{1}{4!}(k.\delta_{mn}^{ij})^4 + \ldots]$$
$$+ ik.\delta_{mn}^{ij} [1 - \tfrac{1}{3!}(k.\delta_{mn}^{ij})^2 + \ldots]\rangle . \tag{7}$$

Now both terms in square brackets are approximately Taylor expansions of $\exp[-\tfrac{1}{2}(k.\delta_{mn}^{ij})^2]$. Hence:

$$\langle \exp[ik.\delta_{mn}^{ij}]\rangle \approx \exp[-\tfrac{1}{2}\langle(k.\delta_{mn}^{ij})^2\rangle].(1+i\langle k.\delta_{mn}^{ij}\rangle)$$
$$= \exp[-\tfrac{1}{2}\langle(k.\delta_{m}^{i})^2\rangle] \exp[-\tfrac{1}{2}\langle(k.\delta_{n}^{j})^2\rangle] \tag{8}$$
$$\exp[\langle(k.\delta_{m}^{i})(k.\delta_{n}^{j})\rangle] (1+i\langle k.\delta_{mn}^{ij}\rangle).$$

The first two exponentials on the right-hand side of equation (8) involve lattice averages of squared displacements of atoms and depend only on the atomic species involved. Hence, they are nothing but Debye-Waller factors, M_i. Equation (8) can then be rewritten as:

$$\langle \exp[ik.\delta_{mn}^{ij}]\rangle = (1+i\langle k.\delta_{mn}^{ij}\rangle$$
$$+ \langle(k.\delta_{m}^{i})(k.\delta_{n}^{j})\rangle) \exp(-M_i) \exp(-M_j), \tag{9}$$

expanding $\exp\langle(k.\delta_{m}^{i})(k.\delta_{n}^{j})\rangle$, multiplying and keeping terms up to the second order in displacements.

Substituting (9) in (5) and grouping together terms with the same powers of δ, one obtains:

$$\langle I_T\rangle = \sum_{m}\sum_{n} [c_A f_A^2 P_{mn}^{AA} + c_A f_A f_B P_{mn}^{AB} + c_B f_B f_A P_{mn}^{BA} + c_B f_B^2 P_{mn}^{BB}$$
$$+ ic_A f_A^2\langle k.\delta_{mn}^{AA}\rangle P_{mn}^{AA} + ic_A f_A f_B\langle k.\delta_{mn}^{AB}\rangle P_{mn}^{AB}$$
$$+ ic_B f_A f_B\langle k.\delta_{mn}^{BA}\rangle P_{mn}^{BA} + ic_B f_B^2\langle k.\delta_{mn}^{BB}\rangle P_{mn}^{BB}$$
$$+ c_A f_A^2\langle(k.\delta_{m}^{A})(k.\delta_{n}^{A})\rangle P_{mn}^{AA}$$
$$+ c_A f_A f_B\langle(k.\delta_{m}^{A})(k.\delta_{n}^{B})\rangle P_{mn}^{AB} \tag{10}$$
$$+ c_B f_A f_B\langle(k.\delta_{m}^{B})(k.\delta_{n}^{A})\rangle P_{mn}^{BA}$$
$$+ c_B f_B^2\langle(k.\delta_{m}^{B})(k.\delta_{n}^{B})\rangle P_{mn}^{BB}] \exp[ik.R_{mn}] .$$

Using the following relations for a binary solid solution:

$$P_{mn}^{AA} + P_{mn}^{AB} = 1,$$

$$P_{mn}^{BA} + P_{mn}^{BB} = 1, \tag{11}$$

$$c_A P_{mn}^{AA} \langle \delta_{mn}^{AA} \rangle + c_A P_{mn}^{BA} \langle \delta_{mn}^{BA} \rangle + c_B P_{mn}^{AB} \langle \delta_{mn}^{AB} \rangle + c_B P_{mn}^{BB} \langle \delta_{mn}^{BB} \rangle = 0,$$

equation (10) becomes (see also Gragg (4), Hayakawa (5) and Borie and Sparks (6)):

$$
\begin{aligned}
\langle I_T \rangle = {} & (c_A f_A + c_B f_B)^2 \sum_m \sum_n \exp[ik.R_{mn}] \\
& + c_A c_B (f_A - f_B)^2 \sum_m \sum_n \alpha_{mn} \exp[ik.R_{mn}] \\
& + i c_A c_B (f_A - f_B)^2 \sum_m \sum_n [\ \eta\ (c_A/c_B + \alpha_{mn})\ \langle k.\delta_{mn}^{AA} \rangle \\
& - \zeta\ (c_B/c_A + \alpha_{mn})\ \langle k.\delta_{mn}^{BB} \rangle]\ \exp[ik.R_{mn}] \qquad (12) \\
& + c_A c_B (f_A - f_B)^2 \sum_m \sum_n [\eta^2\ (c_A/c_B + \alpha_{mn})\ \langle (k.\delta_m^A)(k.\delta_n^A) \rangle \\
& + 2\eta\zeta\ (1 - \alpha_{mn})\ \langle (k.\delta_m^A)(k.\delta_n^B) \rangle \\
& + \zeta^2\ (c_B/c_A + \alpha_{mn})\ \langle (k.\delta_m^B)(k.\delta_n^B) \rangle]\exp[ik.R_{mn}],
\end{aligned}
$$

where $\eta = f_A/(f_A - f_B)$, $\zeta = f_B/(f_A - f_B)$ and $\alpha_{mn} = 1 - P_{mn}^{AB}/c_B$ are the familiar Warren short range order parameters.

The first term in equation (12) represents the Bragg reflections and will not concern us any further. In addition we note that, in the case of short range order, coefficients like α_{mn}, $\langle k.\delta_{mn} \rangle$ and $\langle (k.\delta_m)(k.\delta_n) \rangle$ vanish when the sites m and n are distant. Hence, the sums over m and n in equation (12) can be replaced by N times a single sum over all interatomic vectors, N being the total number of atoms. This expresses the fact that lattice averages are independent of the choice of origin of coordinates.

Finally, for the case of m3m cubic symmetry and substituting:

$$R_{mn} = la_1 + ma_2 + na_3,$$

$$\delta_{lmn}^{AA} = x_{lmn}^{AA} a_1 + y_{lmn}^{AA} a_2 + z_{lmn}^{AA} a_3, \tag{13}$$

$$k = 2\pi (h_1 b_1 + h_2 b_2 + h_3 b_3),$$

where the a_i and the b_i are the real and reciprocal basis vectors respectively, equation (12) becomes (see also refs. 7 and 10):

$$I_D = c_A c_B (f_A - f_B)^2 [I_{SRO}$$

$$+ \eta h_1 Q_x^{AA} + \zeta h_1 Q_x^{BB} + \eta h_2 Q_y^{AA} + \zeta h_2 Q_y^{BB} + \eta h_3 Q_z^{AA} + \zeta h_3 Q_z^{BB}$$

$$+ \eta^2 h_1^2 R_x^{AA} + 2\eta\zeta h_1^2 R_x^{AB} + \zeta^2 h_1^2 R_x^{BB}$$

$$+ \eta^2 h_2^2 R_y^{AA} + 2\eta\zeta h_2^2 R_y^{AB} + \zeta^2 h_2^2 R_y^{BB}$$

$$+ \eta^2 h_3^2 R_z^{AA} + 2\eta\zeta h_3^2 R_z^{AB} + \zeta^2 h_3^2 R_z^{BB} \qquad (14)$$

$$+ \eta^2 h_1 h_2 S_{xy}^{AA} + 2\eta\zeta h_1 h_2 S_{xy}^{AB} + \zeta^2 h_1 h_2 S_{xy}^{BB}$$

$$+ \eta^2 h_2 h_3 S_{yz}^{AA} + 2\eta\zeta h_2 h_3 S_{yz}^{AB} + \zeta^2 h_2 h_3 S_{yz}^{BB}$$

$$+ \eta^2 h_3 h_1 S_{zx}^{AA} + 2\eta\zeta h_3 h_1 S_{zx}^{AB} + \zeta^2 h_3 h_1 S_{zx}^{BB}],$$

where I_D is the diffuse intensity in electron units per atom and:

$$I_{SRO} = \sum_l \sum_m \sum_n \alpha_{lmn} \cos2\pi l h_1 \cos2\pi m h_2 \cos2\pi n h_3,$$

$$Q_x^{AA} = -2\pi \sum_l \sum_m \sum_n (c_A/c_B + \alpha_{lmn}) \langle x_{lmn}^{AA} \rangle$$

$$\sin2\pi l h_1 \cos2\pi m h_2 \cos2\pi n h_3, \qquad (15)$$

$$R_x^{AA} = 4\pi^2 \sum_l \sum_m \sum_n (c_A/c_B + \alpha_{lmn}) \langle x_0^A x_{lmn}^A \rangle$$

$$\cos2\pi l h_1 \cos2\pi m h_2 \cos2\pi n h_3,$$

etc. for the rest of the terms in (14).

What have we accomplished so far? Quite a lot! We have managed to write the diffuse intensity as a sum of various components, each of which contains a particular piece of information about the local arrangements of the atoms in the solid solution. The first term is a function of the short range order parameters, which tell us about the chemical environment about each atomic species, i.e. on the average how many atoms of each kind and at what distance from an A or B atom. The remaining twenty four terms in the intensity expansion contain very detailed information about the local displacements. The first order terms give us the preferred interatomic distance of AA or BB pairs, separated by a given lattice vector. The second order terms give us correlations of atomic displacements: $\langle x_0^A y_{lmn}^B \rangle$, for instance, gives the probable displacement of a B atom at the tip of the vector [lmn] in the y direction when an A atom at the origin has been displaced by x in the x direction. In the next chapter we will see how we can extract this information from measurements of the diffuse scattering.

3. DATA ANALYSIS

Right at this point, we note that the various terms in equation (14) are Fourier series, whose coefficients are the short range order parameters and the displacement averages. So, it seems that, if we managed to break down the total diffuse intensity into its components contained in equation (14), a three-dimensional Fourier transform of each component in reciprocal space should yield the corresponding parameters. But how can we achieve this decomposition of the diffuse intensity?

Let us note one more thing: Each of the intensity components is periodic in reciprocal space and they all have the same periodicity. It should be possible, then, to find a set of reciprocal space points, at which each of the components has the same value. After measuring the total diffuse intensity at these reciprocal space points, we can form a system of linear equations like (14), in which the twenty five components are the constants to be determined. It is obvious that we need twenty five measurements or more for the solution of the linear system to be feasible.

Now we will systematically construct the reciprocal space point set referred to in the previous paragraph, identify it by its member that lies in the first Brillouin zone (i.e. the one closest to the origin of reciprocal space) and call it the "associated set" of this point (8). It is clearer to illustrate this procedure with a concrete example: Take 0.4,0.3,0.6 as the base point and assume BCC symmetry. By examining equation (14), we can spot some obvious candidates for the associated set: All reciprocal space points which lie integral reciprocal lattice vectors away from 0.4,0.3,0.6, such as 2.4,0.3,0.6, 1.4,1.3,0.6 etc. All such points accessible experimentally are members of the associated set of 0.4,0.3,0.6. But we can find many more, not so obvious. Besides translational symmetry, reciprocal space possesses additional symmetry elements, such as mirror planes, rotational axes and inversion centers. These additional symmetries and their effects of the various intensity components in equation (14) are listed in tables 1, 2 and 3 for simple cubic, BCC and FCC symmetry respectively. Using these tables, we can find more points belonging to the associated set of 0.4,0.3,0.6. Table 4 lists part of this set.

From tables 1,2 and 3 we notice also that the various symmetry elements present in each case can impose certain constraints on the intensity components. Let us again illustrate this with an example: From the definition of Q_x in equation (15), the following can be easily verified:

$$Q_x(h_1h_2h_3) = Q_x(h_1h_3h_2) = Q_z(h_3h_2h_1) \qquad (16)$$

TABLE 1

Symmetry relations among the intensity components for simple cubic symmetry.

Mirror planes : $\{\frac{n}{2},0,0\}$, $n = 0,\pm1,\pm2,\ldots$

$$I_{SRO}\ (\tfrac{1}{2}+h_1h_2h_3) = I_{SRO}\ (\tfrac{1}{2}-h_1h_2h_3)$$

$$Q_x(\tfrac{1}{2}+h_1h_2h_3) = -Q_x(\tfrac{1}{2}-h_1h_2h_3)$$

$$R_x(\tfrac{1}{2}+h_1h_2h_3) = R_x(\tfrac{1}{2}-h_1h_2h_3)$$

$$S_{xy}(\tfrac{1}{2}+h_1h_2h_3) = -S_{xy}(\tfrac{1}{2}-h_1h_2h_3)$$

Relations among components

$$I_{SRO}(h_1h_2h_3) = I_{SRO}(h_2h_3h_1) = I_{SRO}(h_3h_1h_2)$$

$$I_{SRO}(h_1h_3h_2) = I_{SRO}(h_3h_2h_1) = I_{SRO}(h_2h_1h_3)$$

$$Q_x(h_1h_2h_3) = Q_x(h_1h_3h_2) = Q_y(h_2h_1h_3) = Q_z(h_3h_2h_1)$$

$$R_x(h_1h_2h_3) = R_x(h_1h_3h_2) = R_y(h_2h_1h_3) = R_z(h_3h_2h_1)$$

$$S_{xy}(h_1h_2h_3) = S_{xy}(h_2h_1h_3) = S_{yz}(h_3h_2h_1) = S_{zx}(h_1h_3h_2)$$

TABLE 2

Symmetry relations among the intensity components for body centered cubic symmetry.

Mirror planes : $\{n,0,0\}$, $n = 0,\pm 1,\pm 2,...$

$$I_{SRO}(n+h_1h_2h_3) = I_{SRO}(n-h_1h_2h_3)$$

$$Q_x(n+h_1h_2h_3) = -Q_x(n-h_1h_2h_3)$$

$$R_x(n+h_1h_2h_3) = R_x(n-h_1h_2h_3)$$

$$S_{xy}(n+h_1h_2h_3) = -S_{xy}(n-h_1h_2h_3)$$

Rotation axes :   2-fold $\langle\frac{1}{2},\frac{1}{2},0\rangle$, $n = 0,\pm 1,\pm 2,...$

$$I_{SRO}(\tfrac{1}{2}+h_1,\tfrac{1}{2}+h_2h_3) = I_{SRO}(\tfrac{1}{2}-h_1,\tfrac{1}{2}-h_2h_3)$$

$$Q_x(\tfrac{1}{2}+h_1,\tfrac{1}{2}+h_2h_3) = -Q_x(\tfrac{1}{2}-h_1,\tfrac{1}{2}-h_2h_3)$$

$$R_x(\tfrac{1}{2}+h_1,\tfrac{1}{2}+h_2h_3) = R_x(\tfrac{1}{2}-h_1,\tfrac{1}{2}-h_2h_3)$$

$$S_{xy}(\tfrac{1}{2}+h_1,\tfrac{1}{2}+h_2h_3) = -S_{xy}(\tfrac{1}{2}-h_1,\tfrac{1}{2}-h_2h_3)$$

Relations among components

Same as for simple cubic symmetry

TABLE 3

Symmetry relations among the intensity components for face centered cubic symmetry.

Mirror planes : Same as for BCC symmetry

Inversion centers : $(n+\frac{1}{2}, n+\frac{1}{2}, n+\frac{1}{2})$, $n = 0, \pm1, \pm2, \ldots$

$$I_{SRO}(\tfrac{1}{2}+h_1, \tfrac{1}{2}+h_2, \tfrac{1}{2}+h_3) = I_{SRO}(\tfrac{1}{2}-h_1, \tfrac{1}{2}-h_2, \tfrac{1}{2}-h_3)$$

$$Q_x(\tfrac{1}{2}+h_1, \tfrac{1}{2}+h_2, \tfrac{1}{2}+h_3) = -Q_x(\tfrac{1}{2}-h_1, \tfrac{1}{2}-h_2, \tfrac{1}{2}-h_3)$$

$$R_x(\tfrac{1}{2}+h_1, \tfrac{1}{2}+h_2, \tfrac{1}{2}+h_3) = R_x(\tfrac{1}{2}-h_1, \tfrac{1}{2}-h_2, \tfrac{1}{2}-h_3)$$

$$S_{xy}(\tfrac{1}{2}+h_1, \tfrac{1}{2}+h_2, \tfrac{1}{2}+h_3) = -S_{xy}(\tfrac{1}{2}-h_1, \tfrac{1}{2}-h_2, \tfrac{1}{2}-h_3)$$

Relations among components

Same as for simple cubic symmetry

TABLE 4

The associated set of 0.4,0.3,0.6 and the constraints due to body centered cubic symmetry.

CONSTRAINTS	ASSOCIATED SET		
	h_1	h_2	h_3
$Q_x^{AA} = -Q_z^{AA}$	0.4	2.3	1.4
	0.4	0.7	0.4
$Q_x^{BB} = -Q_z^{BB}$	0.6	0.7	0.6
	0.6	0.3	1.6
$R_x^{AA} = R_z^{AA}$	0.6	0.7	1.4
	0.6	1.3	0.6
$R_x^{AB} = R_z^{AB}$	0.4	0.7	1.6
	0.4	1.7	0.6
$R_x^{BB} = R_z^{BB}$	0.4	1.3	1.6
	0.6	1.3	1.4
$S_{xy}^{AA} = -S_{yz}^{AA}$	0.4	1.7	1.4
	1.4	0.3	1.6
$S_{xy}^{AB} = -S_{yz}^{AB}$	1.4	0.7	1.4
	1.6	0.7	1.6
$S_{xy}^{BB} = -S_{yz}^{BB}$	1.4	1.3	1.4
	1.4	1.7	1.6
	1.6	1.3	1.6
	0.4	0.3	2.6
	0.4	2.3	0.6
	2.4	0.3	0.6
	0.4	0.7	2.4
	0.4	2.7	0.4
	0.6	1.7	1.6
	2.4	0.3	1.4
	0.4	1.3	2.4
	0.6	2.3	1.6
	.	.	.
	.	.	.
	.	.	.

which, for our base point 0.4,0.3,0.6 implies:

$$Q_z(0.4,0.3,0.6) = Q_x(0.6,0.3,0.4) = Q_x(0.6,0.4,0.3) \qquad (17)$$

From table 2, using the two-fold axial symmetry:

$$Q_x(0.6,0.4,0.3) = Q_x(0.5+0.1,0.5-0.1,0.3)$$

$$= -Q_x(0.5-0.1,0.5+0.1,0.3) = -Q_x(0.4,0.6,0.3) \qquad (18)$$

which gives:

$$Q_z(0.4,0.3,0.6) = -Q_x(0.4,0.6,0.3) = -Q_x(0.4,0.3,0.6) \qquad (19)$$

Similar constraints can be found for other intensity components. For our base point 0.4,0.3,0.6, they are all listed in the right column of table 4. Some of these constraints can result from a combination of the symmetry elements shown in table 2. Note that each constraint reduces the number of unknowns in equation (14) by one.

Now the strategy for the diffuse scattering measurement has been laid out: For each base point in reciprocal space that we wish to separate the diffuse intensity into its components, we construct the associated set, carry out the measurements and form a linear system of equations like (14). This system is then solved by a linear least squares procedure. Care should be exercised, because the matrix of this system is often ill-conditioned, i.e. the coefficients of the intensity components are almost multiples of one another, since and do not vary much as a function of $\sin\theta/\lambda$. The procedure in use in our laboratory is a Householder least squares reduction with iterative optimization, described by Golub (9). Additional tricks to improve the stability of the solution have been employed, such as weighting and a modified Marquardt algorithm. These are described in detail in ref. 10.

In the case of diffuse neutron scattering (for example, see ref. 11), equation (14) is of course still valid, but the various components are no longer linearly independent. Since the neutron scattering length is independent of the scattering angle, the quantities and defined by equation (15) are constant and the matrix of the linear system of equations is singular. Equation (14) must be modified to remove this singularity, by combining all the AA, AB and BB terms of each component into one. The number of unknowns is thus reduced to ten. The short range order intensity component is still the same, but the new Q, R and S terms contain less information, since they are now unseparable combinations of the AA, AB and BB pair contributions.

This idea is not new. In fact, it is the basis of the original method of diffuse scattering analysis, known as the Borie-Sparks separation procedure (6),(12). It is valid, as mentioned above, for neutron scattering and, for x-ray scattering, only if the variation of η and ζ can be assumed negligible. This is generally not a good assumption, as we have found out after extensive tests in our laboratory.

In order to be able to Fourier invert each intensity component to recover the short range order coefficients and the various displacement moments, the proper separation procedure must be carried out for base points in a uniform net, covering one repeat volume of reciprocal space. These minimum volumes for simple cubic, BCC and FCC symmetry are shown in figure 1. Note that they are much smaller than one reciprocal unit cell in every case.

How well do these methods of analysis work? How reliable are the calculated values for the short range order parameters and the displacement moments? Of course, the error in intensity measurements is well-known, since it follows Poisson statistics. The least squares procedure will propagate these errors through to give error estimates for the intensity components and Wu et al. (10) have shown how to further propagate these errors, through the Fourier transform, to produce error estimates for the short range order parameters and the displacement moments. But this is only part of the problem. Many other sources of errors exist besides counting statistics. First and foremost is the fact that equation (14) is an approximation, valid only if the local atomic displacements are small and third and higher order displacements negligible. The high order displacement terms are, in fact, included in the second order terms, R and S, in an approximate manner (13,14), but the small displacement assumption may not be a legitimate one in certain alloy systems.

Other errors of experimental nature also enter. The net diffuse intensity is only part of the measured value. Other contributions, such as Compton and parasitic scattering, sample fluorescence etc. must be measured or estimated and subtracted. Uncertainties in the composition of the specimen can also introduce errors in the analysis. Finally, the diffuse intensity must be placed on an absolute scale, that is the intensity of the incident beam (which is orders of magnitude higher than the measured intensities) must be accurately measured. It is very difficult to estimate the contribution of all the possible errors to the uncertainty in the final parameters, even if each can be accurately assessed.

278

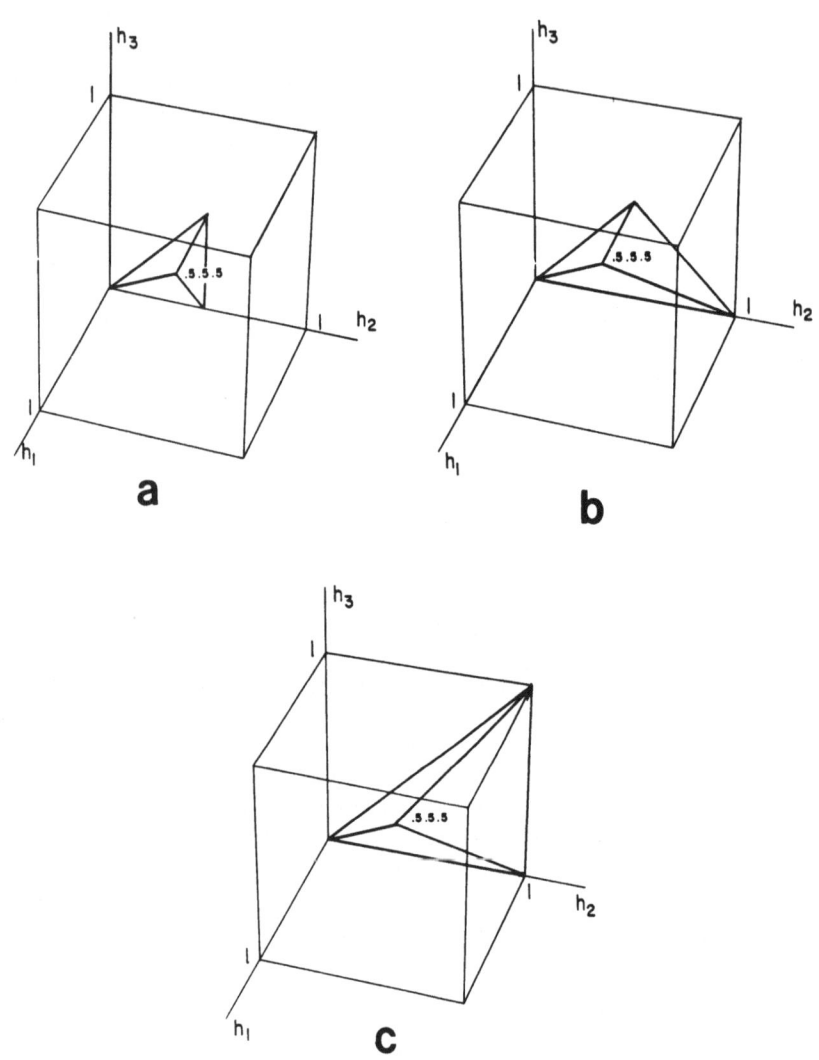

FIGURE 1

Minimum reciprocal space volume required for Fourier inversion of
the intensity components: a. Simple cubic symmetry, b. Body
centered cubic symmetry and c. Face centered cubic symmetry.

Fortunately, there are some independent, global checks that one can apply after the data analysis is completed. They are all based on physical arguments, which the numerical analysis does not have to obey automatically.

First, since the short range order intensity is just a redistribution of the Laue monotonic scattering (the uniform diffuse scattering from a random solid solution), it cannot be negative anywhere in reciprocal space. The same holds true for the second order size effect scattering (the combination of the R and S terms of equation (14)), since it is merely the redistribution of the intensity lost from the Bragg reflections through the Debye-Waller factor. Any successful experiment should produce positive values for these components, and the few occasional negative values should be within the statistical error.

From the definition of the short range order parameters, it is easy to verify that the first coefficient, α_{000} must have the value of unity, since the probability of having a B atom on top of an A atom is obviously zero. The value of α_{000} is a very sensitive check on the quality of a diffuse scattering measurement and the validity of the data analysis. Typically now, it is possible to obtain α_{000} values within 5 pct. of unity. Other values have errors typically of 0.02. Also, one can establish ranges of values that the α's can take, by setting the pair probabilities involved equal to zero or unity. Any value far outside these physical bounds (more than twice the standard deviation) again indicates that the results are not reliable. The displacements must be realistic as well.

Once the short range order parameters have been obtained, they can be employed in computer simulations, involving up to 108,000 atoms on a three-dimensional lattice (15). The "atoms" are rearranged to satisfy some of the observed short range order parameters. The resulting local atomic configurations can then be examined in detail. Thermodynamic information can also be obtained by calculating pair interaction potentials from the observed α's (16-18).

Before leaving this chapter to discuss experimental procedures, it is worth mentioning that a diffuse scattering measurement is a long one: The diffuse intensity must be sampled with reasonable statistical accuracy (say 1-2%) at a large number of reciprocal lattice points, typically 3000-5000. Given the fact that diffuse scattering is fairly weak, an experiment in the laboratory can easily last one to two weeks, sometimes longer. Even with synchrotron radiation, a high quality measurement takes 24-36 hours. In view of this and the complexity of the data analysis procedures, it is hardly worth emphasizing the necessity of using a computer for automation of the experiment, as well as the generation of the associated sets and the final data analysis. Over the past few years, extensive software development has been undertaken in our laboratory and we are now able to perform data collection and analysis almost completely automatically, with minimal operator intervention.

4. EXPERIMENTAL PROCEDURES

The experimental apparatus for diffuse scattering measurements is centered around a single crystal four-circle diffractometer, preceeded by an incident beam monochromator. Figure 2 shows a typical laboratory setup. Figure 3 shows a diffuse scattering station that our group has installed at CHESS. Note that in the latter case, the diffractometer has been mounted such that diffraction is measured in the vertical plane, due to the polarization of the synchrotron beam.

There are no fundamental differences between a laboratory and a synchrotron setup, except for the incident beam monochromator. The same basic components are present, perhaps with small adaptations. Due to the long duration of the measurement and the variability of the incident beam flux (much more severe at the synchrotron), a beam flux monitor is always used. The sample is placed under an evacuated beryllium dome to eliminate air scattering.

We will now examine various procedures that must be followed in the measurement of diffuse scattering. These procedures are also discussed in ref. 2. They are enumerated below to make the presentation more systematic; it is not implied that they are followed in the indicated sequence.

1.   The crystallographic orientation of the sample with respect to the diffractometer axes must be measured accurately. Since this is a fairly time-consuming procedure to perform manually, it is highly desirable to have the computer search for Bragg reflections, determine their exact locations and compute the orientation matrix automatically.

2.   The set of reciprocal space points, at which measurement of the diffuse intensity is needed, must be generated and converted to diffractometer settings. In order to save time spent in moving the instrument from one setting to the next, the settings should be ordered in a sequence that minimizes diffractometer motion.

3.   The measurements should be performed by counting each data point until a fixed monitor count is reached, rather than for a fixed time. This is much more important at the synchro-

FIGURE 2

Laboratory diffuse scattering apparatus. a. Monochromator, b.
Monitor detector, c. Sample chamber and d. Diffracted intensity
detector.

FIGURE 3

Diffuse scattering station at CHESS. Components indicated the
same as figure 2.

tron, where the incident beam flux can vary by a factor of three or more.

4.   Parasitic scattering must be measured as a function of scattering angle, in order to be subtracted later from the measured diffuse intensity. Parasitic intensity may include contributions such as slit scattering, residual air scattering from the vicinity of the sample, scattering from sample supports and other items in the path of the incident beam. X-ray detector noise also makes a contribution. This extra intensity can be measured by replacing the sample with a beam trap, usually a lead box with a small hole for the beam to enter.

5.   Fluorescence from the major constituents of the sample, as well as impurities, must be measured. This is best done with a solid state detector, whose resolution is usually adequate to reveal and identify the various contributions to the measured intensity.

6.   Inelastic scattering must be estimated. This includes two contributions: Compton scattering, which can be calculated fairly accurately, and high order thermal diffuse scattering. This can be minimized by performing the measurements at low temperature, if possible. If not, it can be reasonably accurately calculated if the atomic force constants of the alloy are known.

7.   The incident beam polarization must be estimated, preferably measured directly, for the monochromator crystal employed.

8.   The intensity of the incident beam must be measured accurately, in order to be able to express the diffuse intensities in electron units. One can measure the incident beam intensity either directly, using an ion chamber, or indirectly, by measuring scattering from standard samples, such as aluminum powder or amorphous polystyrene, whose scattering power can be accurately calculated. It is best to use more than one of these methods in any experiment and compare the results for consistency.

5. SUMMARY

We have examined methods and techniques for diffuse scattering measurements and analysis. These methods have evolved steadily in recent years, to the point that it is now straightforward to obtain reliable, detailed information about local atomic arrangements in binary solid solutions. With the advent of synchrotron radiation, the duration of the experiment has been reduced to reasonable times. It is hoped that, with all these developments, the analysis of diffuse scattering can become a fairly standard tool for the study of condensed matter.

ACKNOWLEDGEMENTS

The authors would like to thank the U. S. National Science Foundation (Grant No. DMR-79-23825, and DMR-MRL-76-80847 through Northwestern University's Materials Research Center); it has been through their continuing support that we have been able to develop the methods and procedures summarized here. A large number of former and present graduate students, particularly Professors L.H. Schwartz, P.C. Gehlen, M. Hayakawa, T.B. Wu, Dr. J.E. Gragg, Jr and Mr. E. Matsubara, have been responsible for the bulk of this work. Finally, we would like to express our thanks to Prof. B.W. Batterman, director, D.H. Bilderback and the whole staff of the Cornell High Energy Synchrotron Source. Their assistance in setting up our diffraction station is deeply appreciated, as is their most gracious hospitality.

REFERENCES

1. Chen, H.D., R.J. Comstock, and J.B. Cohen. The Examination of Local Atomic Arrangements Associated with Ordering. Ann. Rev. Mater. Sci. 9 (1979) 51-86.

2. Schwartz, L.H. and J.B. Cohen. Diffraction from Materials. Academic Press, New York, 1977, PP.403-424.

3. Hayakawa, M. and J.B. Cohen. Equations for Diffuse Scattering from Materials with Multiple Sublattices. Acta Cryst. A31 (1975) 635-645.

4. Gragg, J.E.,Jr. PhD Thesis (1970). Northwestern Univ., Evanston, IL.

5. Hayakawa, M. PhD Thesis (1973). Northwestern Univ., Evanston, IL.

6. Borie, B. and C.J. Sparks, Jr. The Interpretation of Intensity Distributions from Disordered Binary Alloys. Acta Cryst. A27 (1971) 198-201.

7. Georgopoulos, P. and J.B. Cohen. The Determination of Short Range Order and Local Atomic Arrangements in Disordered Binary Solid Solutions. J. de Physique, 38C7 (1977) 191-196.

8. Georgopoulos, P. PhD Thesis (1979). Northwestern Univ., Evanston, IL.

9. Golub, G. Numerical Methods for Solving Linear Least Squares Problems. Num. Mathematik 7 (1965) 206-216.

10. Wu, T.B., E. Matsubara, and J.B. Cohen. Estimating Errors in Quantitative Studies of Diffuse X-ray Scattering. J. Appl. Cryst. In Press.

11. Lefebvre, S., F. Bley, M. Bessiere, M. Fayard, M. Roth, and J.B. Cohen. Short Range Order in Ni_3Fe. Acta Cryst. A36 (1980) 1-7.

12. Sparks, C.J. and B. Borie. Methods of Analysis for Diffuse X-ray Scattering Modulated by Local Order and Atomic Displacements. In "Local Atomic Arrangements Studied by X-ray Diffraction. Gordon & Breach, New York (1965) 5-46.

13. Hayakwa, M., P. Bardhan, and J.B. Cohen. Experimental Considerations in Measurements of Diffuse Scattering. J. Appl. Cryst. 8 (1975) 87-95.

14. Cohen, J.B. and P. Georgopoulos. The Structure of GP Zones in Al-Cu Alloys. Scripta Met. 16 (1982) 1107-1110.

15. Gehlen, P.C. and J.B. Cohen. Computer Simulation of the Structure Associated with Local Order in Alloys. Phys. Rev. 139 (1965) A844-A855.

16. Clapp, P.C. and S.C. Moss. Correlation Functions of Disordered Binary Alloys I. Phys. Rev. 142 (1966) 418-427.

17. Clapp, P.C. and S.C. Moss. Correlation Functions of Disordered Binary Alloys II. Phys. Rev. 171 (1968) 754-763.

18. Wilkins, S. Determination of Long-Range Interaction Energies from the Scattering of X-rays by Disordered Alloys. Phys. Rev. B 2 (1970) 3935-3942.

NEUTRON INVESTIGATION OF MODULATED STRUCTURES*

R. Currat

Institut Laue-Langevin, 156X, 38042 Grenoble Cedex, France

I. INTRODUCTION

1.1 Modulated Structure

Incommensurably-modulated solids are often described (1) as
long-range ordered structures which lack translational symmetry
due to the superposition within the same system of several mutually
incompatible periodicities. As such IC solids provide an intermediate step between the ordinary crystalline state and the
disordered non-periodic amorphous state.

In fact, part of the current work on IC structures is aimed
at testing the validity of the above definition. As shown by
de Wolff (2) and Janner and Janssen (3), spatial periodicity is
hidden rather than absent in the ideal IC solid and may be recovered
via a suitable generalization of the concept of symmetry operation.
On the other hand, experimental and theoretical evidence indicate
that real IC solids close to the C limit (4) or in the presence of
frozen defects (5,6) do not qualify as fully long-range ordered
structures.

A wide range of physical systems are known to present compositional or displacive modulations and the corresponding physical
situations are necessarily very diverse. For instance 2-dimensional
systems, such as adsorbed monolayers or intercalates exhibit
qualitatively unique features (see Nielsen (7) for an experimental
review). In this lecture the emphasis will be on displacive
modulations, i.e. modulations of the equilibrium atomic positions,

*Reprinted from "Multicritical Phenomena" Plenum NATO-ASI Series B
(Physics) (Plenum New York 1983) Ed. R. Pynn and A.T. Skjeltorp.

in ordinary 3-dimensional solids.

Perhaps the single feature common to all IC systems is the occurrence of irrational satellite reflections, in addition to the Bragg reflections associated with the unmodulated reference system. The observation of a continuous shift in the satellite diffraction pattern as a function of temperature or external field has been taken, perhaps too readily, as acceptable evidence for the true IC nature of the modulation. As suggested by theoretical models (see Bak (8) for a recent review), an apparent smooth change in modulation wavevector may correspond to an infinite sequence of higher-order C phases separated or not by true IC phases (incomplete or complete devil's staircase). In such models the concept of discommensuration (DC) or domain wall separating locally C regions is of central importance (9). In particular the value of the modulation wavevector in the IC state is determined by the average DC spacing, the C state corresponding to zero DC density.

Another characteristic feature of IC systems is the presence of global hysteresis affecting the temperature and pressure dependence of the modulation wavevector and of those macroscopic variables (e.g., dielectric constants, optical birefringence coefficients, etc.) which couple to the IC order parameter. As pointed out by Aubry (10), global hysteresis is not compatible with a smooth analytic behavior. Thus its observation lends support to the (complete) devil's staircase picture where irreversible effects arise as a result of pinning of the DC's by the underlying lattice. Recently, experimental evidence has shown that defects play an important role in the observed irreversibilities, giving rise to specimen-dependent behavior and memory effects (11-14). Thus in general, both types of pinning mechanisms, intrinsic and defect-induced, are expected to contribute to the observed irreversible behavior and incomplete ordering.

1.2 Phasons

IC solids are characterized by new types of excitations (15-16), corresponding to fluctuations in the phase and amplitude of the static modulation wave. In the long-wavelength limit the phase mode consists in an overall phase shift of the modulation with respect to the atomic lattice (sliding mode). For a well-behaved IC modulation, for which the atomic displacements can be described in terms of analytic functions of space, such a shift costs no energy and the corresponding phase branch is gapless. For a quasi-C modulation the phase branch becomes identical to the vibrational "phonon" branch of the DC lattice. Its gapless character is preserved as long as the DC's are not pinned.

Recently, it has been shown that phase modes are overdamped, at least at long wavelength (17). The damping arises from the fact that the atomic motions involved are inhomogeneous on a microscopic scale and are thus affected by ordinary anharmonic (i.e., dissipative) interactions.

Hence, unlike acoustic modes, phasons do not qualify as true Goldstone modes. In terms of conservation laws, the difference lies in the fact that the total particle-momentum operator commutes with the total crystal Hamiltonian, while the analogous phase-momentum operator commutes with the harmonic part only. The over-all phase shift is thus affected by friction.

Practically, the phase mode response is expected to be of the damped harmonic oscillator type:

$$S(q,\omega) = \frac{kT}{\Pi} \; \frac{\Gamma_o}{(\omega^2 - v^2 q^2)^2 + \omega^2 \Gamma_o^2}$$

where the velocity v and damping coefficient Γ_o are slowly-varying with q. The response is diffusive at low q:

$$\omega_{ph}(q) = i \frac{v^2 q^2}{\Gamma_o}(q < \frac{\Gamma_o}{v})$$

becoming progressively underdamped at large q:

$$\omega_{ph}(q) = vq + i\Gamma_o \; (q > \frac{\Gamma_o}{v}).$$

Rough estimates for v and Γ_o indicates that the two regimes may be probed by Brillouin and neutron scattering, respectively (18). Underdamped phase modes have been observed by neutron scattering in phase III of biphenyl (19) and in $ThBr_4$ (20) (see Fig. 1).

I.3 Experimental Techniques

Most of the experimental information on IC solids is based on X-ray neutron and electron diffraction data. The best q-resolutions ($\simeq 10^{-4}$ Å^{-1}) are obtained from X-ray sources. In $2H-TaSe_2$, minute deviations from commensurability have thus been revealed. In structurally-modulated $BaMnF_4$, satellite reflections of the type $(h+\xi, k+1/2, 1+1/2)$, with ξ ranging from 0.39 to 0.40 depending upon temperature (and specimen origin), have been shown to present a finite q-width along ξ, corresponding to a correlation length of a few hundred cells (22). In view of the absence of any lock-in transition at low temperatures, this result suggests a defect-induced pinning of the modulation.

Real space visualization of DC arrays in $2H-TaSe_2$ have been achieved by dark-field electron microscopy (23,24). In particular, dynamic studies as a function of temperature show that the basic mechanism through which the DC array readjusts its average spacing

288

is "stripple" nucleation (or evaporation) rather than homogeneous
expansion.

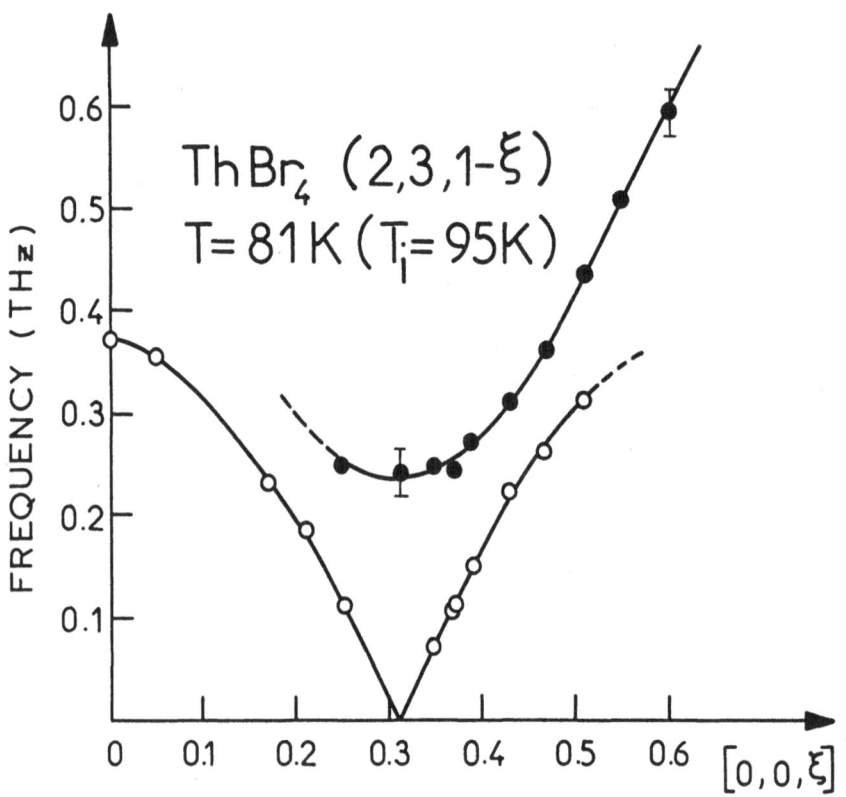

FIG. 1.   Phase mode (open circles) and amplitude mode (closed
circles) in ThBr$_4$ (from Ref. (20)).

Local information may be obtained from resonance techniques
(see Blinc (25) for a review), or at low temperatures, from optical
crystal-field spectroscopy (26).   Resonance lines are broadened
inhomogeneously due to the distribution of local environments
introduced by the modulation.   The line-shape of the resonance
spectrum mirrors the distribution of local fields, provided the
latter fluctuates slowly on the characteristic timescale of the
measurement (inverse linewidth).   This condition should be (and
appears to be) generally fulfilled except in the immediate vicinity
of the IC ordering temperature where motional narrowing should in
principle occur due to phase fluctuations.   In Rb$_2$ZnBr$_4$ a partial
crossover to the fast fluctuation regime has been reported (27) at
about 10K below the ordering temperature T$_i$ = 200K).   The [87]Rb NMR
linewidth at that temperature was ~20 KHz.

II. NOWOTNY PHASES: AN EXAMPLE OF UNIAXIAL INTERGROWTH COMPOUND

II.1 Structure

The structure of Nowotny phases (28) is based on two inter-penetrating tetragonal lattices with identical cell dimensions in the (a,b) plane, but different c parameters. Model systems of this type have been considered by Theodorou and Rice (29) and Axe and Bak (30). The chemical formula is TX_x where T is a transition metal element and X is an element of group III or IV. The observed variation of the atomic ratio x with the chemical nature of T and X suggests that the stability of the structure is largely determined by the value of the electron concentration per T-atom (31).

For each of the two sublattices one can define an idealized subcell (cf. Fig. 2) of dimensions (a,a,c_T) and (a,a,c_X) with 4 atoms/cell in each case (i.e., $c_T = xc_X$). The atomic positions on sublattice X are modulated in such a way as to avoid overlap with the surrounding T-atoms. The modulation consists in a rotation of the X-X dumbells in the (a,b) plane, together with a shift along c. The X-atomic positions along c are expressed as

$$z_X(n) = nc_X + u_X + g_T(nc_X+u_X-u_T) \tag{1}$$

where $u_{X,T}$ are appropriate origins and $g_T(z)$ is a periodic modulation function with period c_T:

$$g_T(z+c_T) = g_T(z) \tag{2}$$

An analogous expression can be written down for the T positions, although the observed distortions are much smaller.

In order to calculate the diffracted intensity from such a system let us write the density function of each sublattice as:

$$\tilde{\rho}_{X,T}(\vec{r}) = \rho_{X,T}(\vec{r}-\vec{g}_{T,X}(z)) \tag{3}$$

where $\vec{g}(\vec{g}_X,g_T)$ is a displacement vector satisfying (2) and $\rho_{X,T}(\vec{r})$ is the unmodulated sublattice density function:

$$\rho_{X,T}(\vec{r}) = \sum_{G,\lambda} \rho_{X,T}(\vec{G},\lambda)\ \exp\{i(\vec{G}\cdot\vec{r}+\lambda c^*_{X,T}z)\} \tag{4}$$

where $\vec{G}+\lambda c^*_{X,T}$ is a reciprocal lattice vector of sublattice X,T.

Inserting (4) into (3) yields:

$$\tilde{\rho}_X(\vec{r}) = \sum_{G,\lambda} \rho_X(G,\lambda)\ \exp\{i(\vec{G}\cdot\vec{r}+\lambda c^*_X z)$$
$$\times \exp\{-i[\vec{G}\cdot\vec{g}_T(z)+\lambda c^*_X g_T(z)]\} \tag{5}$$

290

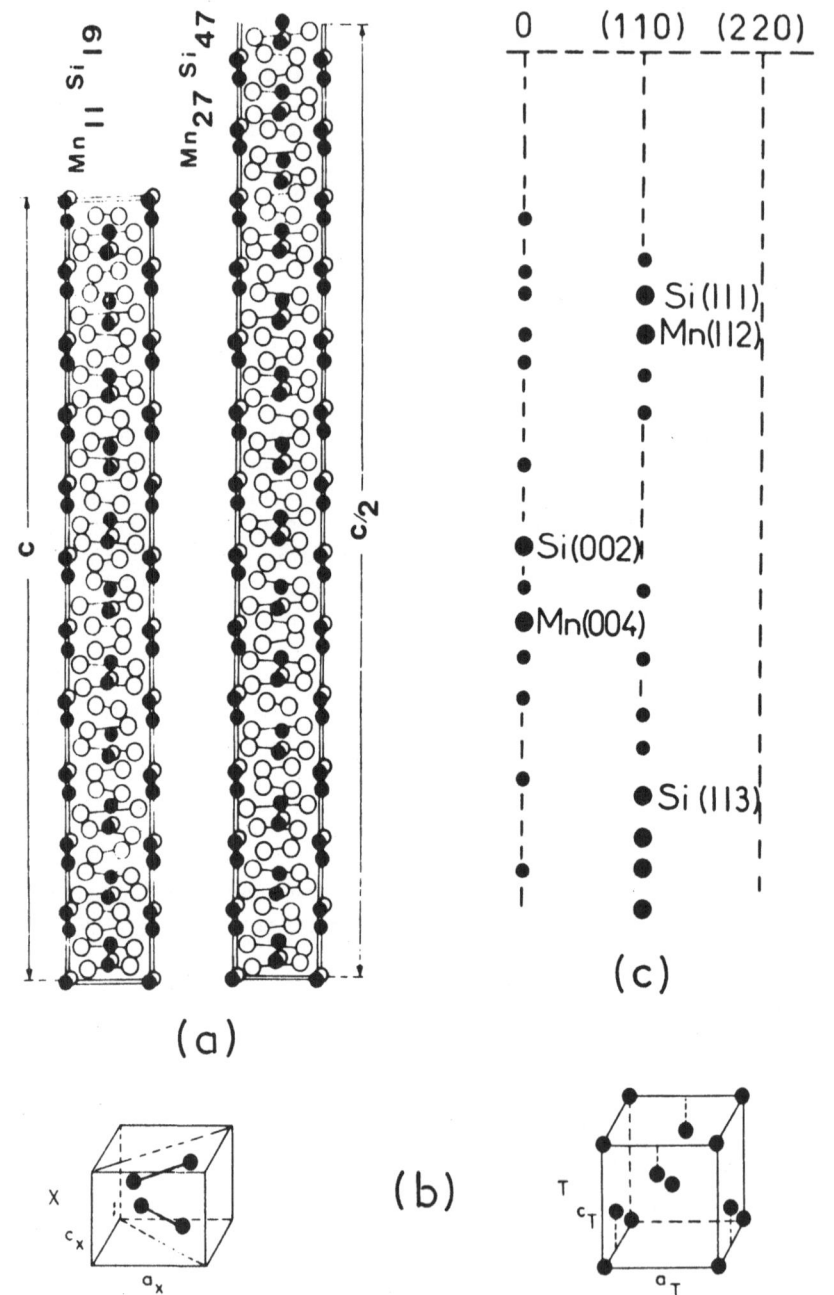

FIG. 2. Structure of Nowotny phases TX_x ($T = T_i, Zr, V, Mo, Cr, Mn, Rh$; $X = Si, Ge, Sn$). (a) Sketch of unit cell for two stoichiometric compounds in the $MnSi_x$ series ($x = 1.73 \pm 0.02$) (from Ref. (35)); (b) Idealized subcells ($a_T = a_X$; $C_T = xC_X$); (c) typical diffraction pattern. All spots can be indexed according to Eq. (6). Only the main reflections (λ or $\mu = 0$) are labelled in the figure.

The last exponential factor in (5) is periodic with period c_T, whence

$$e^{-i[\ ]} = \sum_\mu w_T(\mu;G\ ,\lambda)e^{i\mu c_T^* z}$$

and

$$\tilde{\rho}_x(\vec{r}) = \sum_{G\ ,\lambda,\mu} \rho_x(G\ ,\lambda)w_T(\mu;G\ ,\lambda)$$

$$\chi\ \exp\{i[\vec{G}\cdot\vec{r} +(\lambda c_X^* +\mu c_T^*)z]\}$$

Bragg peaks will be observed at:

$$\vec{Q} = \vec{G}\ + \lambda\vec{c}_X^* + \mu\vec{c}_T^* \quad (\lambda,\mu = \text{integers}) \tag{6}$$

with structure factors:

$$F_{G\ ,\lambda,\mu} = \rho_x(G\ ,\lambda)w_T(\mu;G\ ,\lambda) + \rho_T(G\ ,\mu)w_x(\lambda;G\ ,\mu) \tag{7}$$

Thus in general both sublattices contribute to each reflection. Because the T-sublattice is relatively rigid ($w_X(\lambda;G\ ,\mu)$ - 0 if $\lambda \neq 0$, the reflections corresponding to $\lambda \neq 0$ will essentially be pure X-reflections. The only pure T-reflections are obtained for $\lambda = G\ = 0$ ($w_T(\mu;0,0) = \delta_\mu$).

For rational values of x ($x = m/n$), the supercell dimension along c is $\bar{c} = nc_T = mc_X$. The space groups of the corresponding compounds are D_{2d}^6, D_{2d}^{12} or D_{2d}^8, depending upon the parities of m and n. The MnSi$_x$ system, the following compounds have been reported by X-ray diffraction (32,33):

Mn$_n$Si$_m$	x	a(Å)	$(\bar{c}/n)$(Å)
Mn$_{11}$Si$_{19}$	1.727	5.518	4.376
Mn$_{15}$Si$_{26}$	1.733	5.525	4.370
Mn$_{27}$Si$_{47}$	1.741	5.530	4.368

Commensurate structures have also been observed by de Ridder et al. (34) by electron diffraction. In addition, these authors observe compositional inhomogeneities across the volume of their specimen, giving rise to an alternance of C and IC diffraction patterns (35).

Since the observed concentration fluctuations cannot be eliminated by annealing, they indicate that the equilibrium Si concentration is inhomogeneous even at high temperature. This feature can be tentatively interpreted as resulting from a competition between the band structure energy, which is minimized for some IC value of x, and short-range lock-in terms, which favor simple rational fractions.

II.2 Dynamics

To the extent that the displacement functions $g_{T,x}$ (cf. Eq. (1)) are continuous, analytic function with mutually IC periodicities excitations polarized along c will propagate independently on each sublattice. Indeed inelastic neutron scattering data taken near a pair of strong Mn(Si) reflections reveal two longitudinal acoustic branches with different slopes (36).

In the long-wavelength limit, however, the microscopic incommensurability of the two sublattices becomes ireelevant and several coupling mechanisms arise (30,37). As a result, a single combined acoustic mode should emerge. In addition, one expects to observe quasielastic scattering corresponding to the (diffusive) phase mode response. For unscreened ionic crystals, however, the longitudinal phase branch should present a gap, corresponding to the two-sublattice plasmon frequency. All these predictions remain to be tested experimentally.

III. CHARGE DENSITY WAVES

Fermi surface instabilities in low-dimensional metals provide a specific mechanism leading to modulated structures. The modulation in that case, results from the screening of the electronic charge density wave by the nuclei (the "$2k_f$" instability). It is essentially through the lattice modulation that the instability is observed in a diffraction experiment.

The IC character of the modulation is connected to that of the Fermi wavevector, the latter arising from non-stoichiometry, as in KCP, charge transfer as in TTF-TCNQ, or just energy band structure as in the transition-metal dichalcogenides. In the latter case the lock-in potential from the lattice is sufficient to eventually drive the CDW periodicity to a commensurate value (38). This is not so in one-dimensional metals where commensurability can only be achieved by altering the value of the Fermi wavevector, i.e. by changing the conduction electron concentration.

The existence within the IC-state of a transition between the analytic, Peierls-conducting regime and the pinned insulating state has been predicted (39,40) and possibly observed in doped poly-acetylene (41).

IV. INSULATORS

IV.1 General

In insulators the IC-state results either from a soft mode-type of instability (displacive case) or from a collective ordering mode (order-disorder case), both of general wavevector.

Within the soft mode picture, the IC ordering transition appears as a mere extension of the conventional ferro- and anti-ferrodistorsive transitions. While the occurrence of a local minimum around an arbitrary wavevector on a phonon dispersion curve suggests a competition between interactions of different characteristic ranges, it does not in general provide any insight into the nature of these interactions.

It is likely that in ionic crystals the competition involves Coulomb vs. short-range forces but this applies equally well to most phase transitions in ionic crystals (42). Furthermore, microscopic calculations are generally too complex to yield a simple physical picture of the IC instability (43).

In a non-polar molecular crystal such as biphenyl (44) the IC modulation is believed to arise from the competition between the intermolecular potential favoring a finite torsional angle between the two molecular phenyl rings (as in the free molecule) and intermolecular interactions with favor a co-planar conformation.

On a phenomenological level, incommensurability may sometimes be associated with the non-fulfillment of Lifshitz criterion (45). For example, in non-symmorphic structures, most of the irreducible space-group representations corresponding to special zone-boundary wavevectors are pairwise degenerate and give rise to Lifshitz invariants. In that case a zone boundary phonon softening leads first to a modulated structure with a wavevector close to the high-symmetry zone-boundary point (46).

This type of argument, however, does not apply to IC phases with wavevectors close to the zone-center (and which eventually lock-in at $q = 0$), as encountered in $NaNO_2$ (47), quartz (48) and Thiourea (49), and alternative models, based on two-mode interactions have been proposed (50-52).

IV.2 Thiourea

Thiourea crystallizes in the orthorhombic space group Pnma with four molecules per unit cell (see Fig. 3). Each molecule is planar and possesses a permanent dipole moment oriented in the (a,c) plane. The net dipole moment per cell vanishes due to the (n) and (a) glide plane.

Below room temperature, the softening of a polar optic mode (B_{3u}) is observed by infra-red spectroscopy (53) and neutron inelastic scattering (54). The soft mode dispersion minimum lies in the $\vec{b}$*-direction where mixing with a TA branch of some symmetry (τ_4) occurs. For T < 220K, a heavily damped coupled mode response is observed.

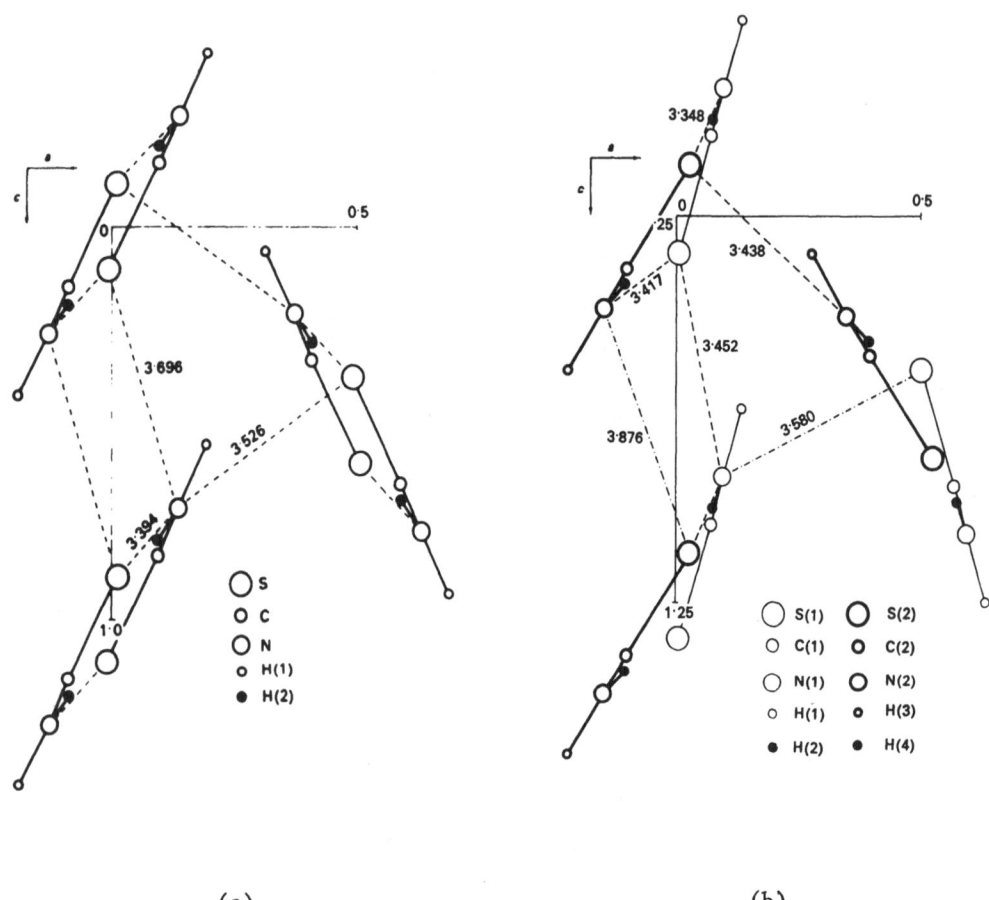

(a)   (b)

FIG. 3.   Structure of thiourea in the paraelectric (Pnma) phase (a) and ferroelectric ($P2_1ma$) phase (b). Molecules labelled (3) and (4) are displaced by b/2 with respect to (1) and (2) (after Ref. (57)).

Below T_i = 218K (all transition temperatures refer to the deuterated compound), satellite reflections appear, located at:

$$\vec{q}_\delta(T) = \delta(T)\vec{b}* \qquad\qquad (T < T_i)$$

$\delta(T)$ decreases continuously from $\delta(T_i)$ = 0.141 to $\delta(T_9)$ = 0.115. Between T9 = 193K and T_c = 191K, δ locks to the commensurate value 1/9. Below T_c, δ vanishes giving rise to a ferroelastic phase with spontaneous polarization along $\vec{a}(P2_1ma)$. Both lock-in transitions are distinctly first-order. A moderate (< 1 K) wave-vector hysteresis is observed in the IC phase.

On cooling from T_i, first, second and third order satellites appear successively. Only small intensity discontinuities are observed at T_9. Critical exponents for the first and second order satellite intensities are found to be in good agreement (55) with the critical behavior of the d = 3 XY-model, including a uniaxial perturbation (56).

Figure 4 shows the isothermal variation of δ with pressure for a few selected temperatures. Pressure hysteresis is within experimental error ($\pm$ 20 bars). Note the additional step at $\delta =$ 1/7 for isotherms below 207K, and the characteristic shape of the $\delta(P)$ curve just above and below the 1/7 plateau. Continuous lock-in transitions with logarithmic singularities have been predicted theoretically (10,8). Note also the absence of a well-defined plateau on the 157.6K isotherm.

Results are summarized in the (P,T) phase diagram shown in Fig. 5. The values of δ along the $T_i(P)$ and $T_c(P)$ lines are shown as δ_i and δ_c, respectively. The negative sign of dT_i/dP and dT_c/dP is typical of a ferroelectric-type instability (42).

Single crystal diffraction patterns (54) indicate that the high-pressure $\delta = 1/3$ phase belongs to the same class of lock-in phases as the $\delta = 1/7$ and 1/9 phases. Its persistence to temperatures much above the $T_i(P)$ line is however, not understood.

As far as the lower part of the diagram is concerned (P < 2 kbars), the results may be formulated as follows:

i)   a steady decrease of $\delta(T,P)$ with decreasing T and P is observed across the IC phase.

ii)   when the "natural" value of $\delta(T,P)$ approaches a simple rational value such as 1/7 or 1/9, small lock-in free energy terms stabilize the corresponding C-phase over a narrow T or P range. Note the absence of a lock-in phase at 1/8 and the presence of two distinct 1/7 phases.

Point (i) above may be accounted for in terms of a phenomeno-logical Landau-Ginzburg free-energy of the type (58-59):

$$F = \int \{\frac{1}{2} A_o P_x^2 + \frac{1}{4} BP_x^4 + \frac{1}{2} \alpha(\frac{dP_x}{dy})^2 + \frac{1}{4} \beta(\frac{d^2P_x}{dy^2})^2 + \gamma(P_x \frac{dP_x}{dy})^2\}dy$$

$$(8)$$

296

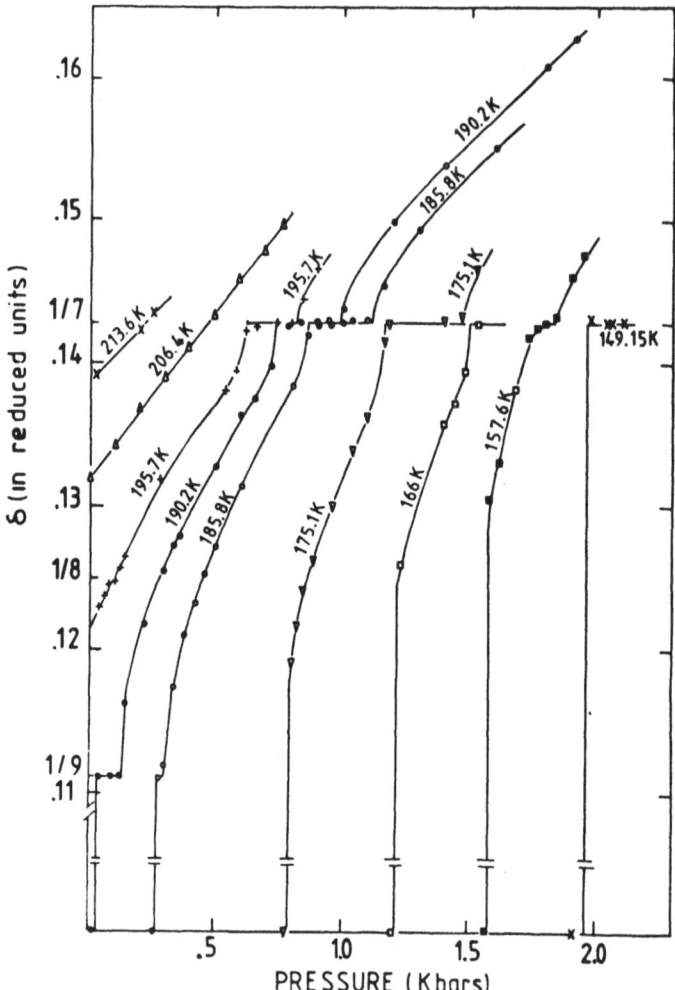

FIG. 4.   Thiourea: isothermal variation of modulation wavevector δ
with pressure (after Ref. (49)).

with: $A_o = a(T-T_o)$; $B > 0$; $\alpha < 0$; $\beta > 0$; $\gamma > 0$.

Approximating the polarization wave $P_x(y)$ by a sine-wave:

$$P_x(y) = P_o + 2P_\delta \cos(2\pi \; \delta y) \tag{9}$$

expression (8) yields:

$$F = \frac{1}{2} A_o P_o^2 + \frac{1}{4} BP_o^4 + A_\delta P_\delta^2 + \frac{3}{2} B_\delta P_\delta^4 + \frac{3}{2} (B+B_\delta)P_o^2 P_\delta^2 \tag{10}$$

where: $A_\delta = A_o + \alpha\delta^2 + \frac{1}{2} \beta\delta^4$

and: $B_\delta = B + \frac{4}{3} \gamma\delta^2$.

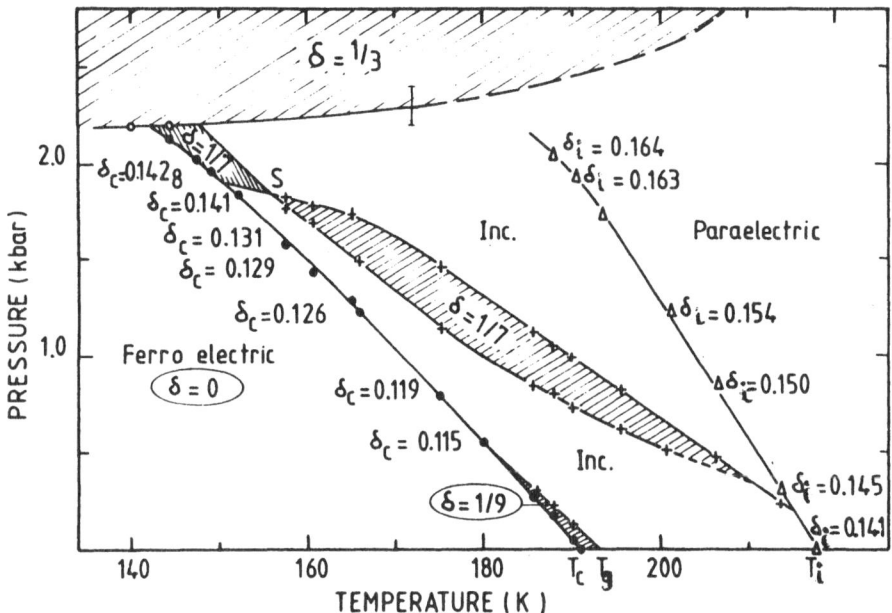

FIG. 5.   Thiourea: (P,T) phase diagram (after Ref. (49)).

Model (10) gives a second-order phase transition (at $T = T_i > T_o$),
from the paraelectric state ($P\delta = P_o = 0$) to the IC-state ($P_o = 0$,
$P_\delta \neq 0$) and a first-order transition (at $T = T_c < T_o$) from the
IC-state to the ferroelectric state ($P_o \neq 0$; $P_\delta = 0$). Minimization
of (10) with respect to δ, leads to the required variation of
across the IC-phase. Presumably the pressure dependence of the
various parameters in the model can be adjusted in such a way as
to reproduce the observed phase diagram. The inclusion of higher-
order harmonics in (9) will simply enhance the T-dependence of δ.

In order to discuss point (ii) it is necessary to introduce
microscopic order parameters and make use of their symmetry
properties. Let $Q(\vec{q}_\delta, \tau_4)$ denote the complex order parameter
corresponding to the first Fourier component of the modulation. A
commensurate phase with $\delta = 1/n$ is stabilized by an Umklapp free
energy term of the form:

$$\Delta F_n = V^{(n)} (\vec{q}_\delta, \tau_4) \, Q^n(\vec{q}_\delta, \tau_4) + c.c. \tag{11}$$

$V^{(n)}$ is an appropriate anharmonic coefficient whose symmetry
properties are determined by the space group of the paraelectric
phase. In particular $V^{(n)}$ vanishes unless:

$$n\vec{q}_\delta = m\vec{b}* \quad i.e. \quad \delta = \frac{m}{n} \quad (m,n = integers).$$

Furthermore point group selection rules require m and n to be of the same parity. Hence the 1/7 and 1/9 phases correspond to lock-in terms of order 7 and 9, while the 1/8 phase would require a lock-in term of order 16, which is too weak to be effective. It has been predicted (49), however, and confirmed experimentally (60,61) that the 1/8-phase could be stabilized by a uniform electric field applied along a. The corresponding lock-in term is of the form:

$$\infty \; Q^8(\vec{q}_\delta, \tau_4) \; Q(0, \tau_4),$$

where $Q(0, \tau_4)$ is induced by the applied field.

The stability range of the 1/7-phase appears to vanish in the vicinity of the point labelled S in Fig. 5. In fact, satellite intensity patterns recorded on either side of point S suggest two distinct 1/7-phases. This observation can be accounted for if one assumes that the pressure and temperature dependence of the $V^{(7)}$ Umklapp coefficient is such that it vanishes on a curve in the (P,T) plane. Point S would then be at the intersection of the two curves defined by:

$$V^{(7)}(P,T) = 0$$

$$\tilde{\delta}(P,T) - 1/7$$

where $\tilde{\delta}(P,T)$ is the natural value of δ, i.e. the value of δ which minimizes the free energy in the absence of Umklapp lock-in terms. Introducing real variables:

$$V^{(7)}(\vec{q}_{1/7}, \tau_4) = iv(T,P); \quad Q(\vec{q}_{1/7}, \tau_4) = \eta e^{i\phi}$$

(11) becomes:

$$\Delta F_7 = v\eta^7 \sin 7\phi.$$

Thus the two 1/7 phases are characterized by opposite values of the phase

$$\phi = \pm \frac{\pi}{14} \; (\text{mod} \; \frac{2\pi}{7})$$

or, equivalently, by opposite signs of the atomic displacements $\vec{u}_{\ell k}$:

$$\vec{u}_{\ell k} = (m_k)^{-1/2} \eta \vec{w}_k(\vec{q}_{1/7}, \tau_4) e^{i(\vec{q}_{1/7}\vec{r}_{\ell k} + \phi)} + \text{c.c.}$$

The two phases together with the IC phase would then coexist at S.

In fact, in a very narrow T and P range around S the contribution from the next Umklapp term:

$$V^{(14)}(\vec{q}_{1/7}, \tau_4) Q^{14}(\vec{q}_{1/7}, \tau_4) \equiv w \, \eta^{14} \cos 14\phi \qquad (12)$$

should be included into ΔF_7. Depending upon the sign of w, two different situations may arise, but in either case the isolated character of point S is removed.

Finally we wish to stress that although expressions such as (11) are written in terms of the primary Fourier component of the modulation alone, this does not preclude the existence of higher order distortion harmonics, which do contribute addition lock-in free-energy terms. Formally one assumes that the minimization of the free-energy with respect to high-order harmonics has already been performed and that the $V^{(n)}$'s have been renormalized accordingly.

This formulation, however, is not adequate for a quantitative discussion of the stability of the IC phase with respect to the higher-order C phases since it does not allow for the formation of phase distortions in the IC wave, corresponding to discommensurations with respect to the high-order C phase. These DC's, however, will presumably be broad on an atomic scale due to the small amplitude of the lock-in term and apart from aking the lock-in transitions continuous, no qualitatively new feature will be introduced.

In the same contenxt, it is of interest to determine the harmonic content of the modulation in the C phases. Hence a detailed structural study of the 1/9 phase has been initiated and is currently in progress (62). Preliminary results indicate that the simple ferroelectric-domain model (i.e. a square wave with 9 planes polarized up and 9 planes polarized down) is not realistic. The harmonic content of the modulation appears to be considerably lower than predicted by the above model. If domain walls are present some degree of disorder must also be present, in such a way as to smooth out the average wave profile. This result is not unexpected since the DC picture is not applicable to the case $\delta = 0$ (63).

IV.3 TMATC-Zn

Tetramethylammonium tetrachlorozincate (TMATC-Zn) belongs to a class of compounds of formula $(N(CH_3)_4)_2 \, XCl_4$ (X = Zn Co, Mn, Fe, etc.), isomorphous to Rb_2ZnCl_4 and K_2SeO_4 in their undistorted phase (Pnma). Below room temperature these compounds undergo various sequences of phase transitions involving IC, C, and quasi-C modulated structures, with modulation wavevector:

$$\vec{q}_\delta = \delta(T)\vec{a}*$$

300

Figure 6 shows the $\delta(T)$ curve measured on deuterated TMATC-Zn, using neutron diffraction (64). In addition to pronounced thermal hysteresis, the experimental curve exhibits a quasi-C plateau prior to the onset of the 3/7-phase. A similar effect has been observed in $Ba_2NaNb_5O_{15}$, where the C phase is in fact never reached (65). It has been ascribed to the stabilization by impurities of phase defects in the C wave (66).

In a later (cooling) run, using the same sample the 1/2 plateau was found to be much reduced (points marked * in Fig. 6). In other samples the same plateau appears only under stress (67).

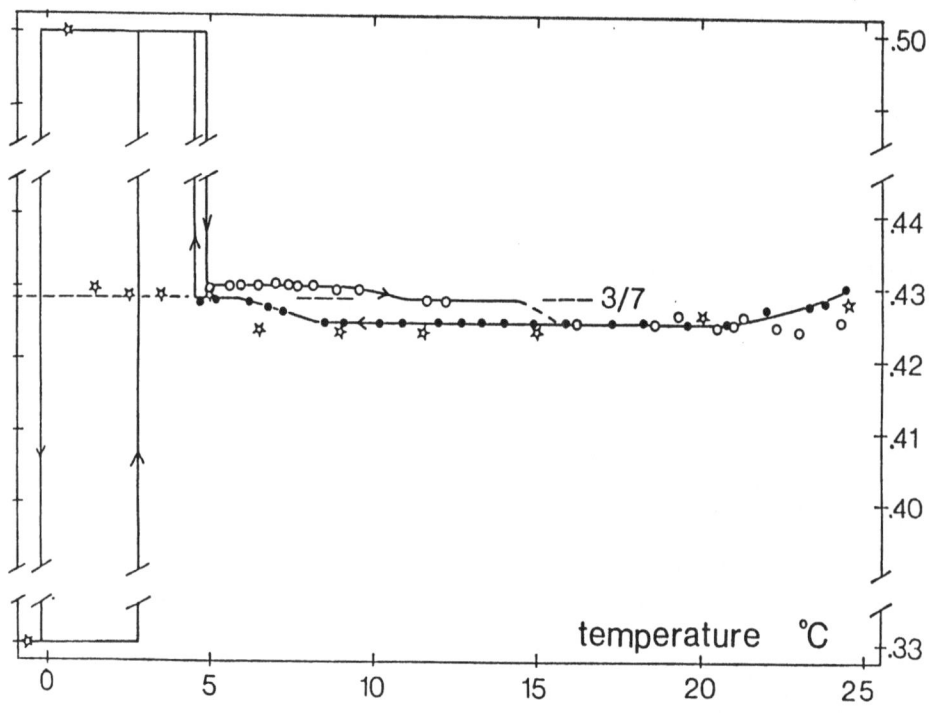

FIG. 6. TMATC-Zn: variation of modulation wavevector with decreasing (closed circles) and increasing temperature (open circles). The data points marks as * correspond to a later cooling run. (Deuterated compound from Ref. (64)).

Different behaviours are observed on deuterated and non-deuterated samples (68). In particular non-deuterated samples exhibit a plateau at 2/5, which is also observed on the deuterated compound under pressure (a few hundred bars).

Clearly the balance of interactions which determines the value of the modulation wavevector is affected by differences in crystal quality (residual strains, impurities) and by thermal

history. The corresponding effects in thiourea are either absent
or much attenuated.

Another qualitative difference between the two compounds is
the non-monotonic variation of $\delta(T)$ as shown in Fig. 6. Aubry (1)
has shown in the context of the discrete Frenkel-Kontorova model,
that the average lattice spacing misfit ℓ of the adsorbed chain
(here the modulation periodicity) needs not vary monotonically with
the substrate potential λ (here the temperature). He also points
out that very small changes in the chain tension μ, the other
parameter in the model, can produce large changes in the $\ell(\lambda)$
curve. Figure 7 shows two such curves calculated for slightly
different values of μ: one shows a monotonous decrease of $\ell/2a$
from 2/5 to zero while the other exhibits a plateau at a maximum
value of 1/2. The absence of an IC phase in the low-λ part of the
calculated curves is an artefact of the model, rated to the
particular form of substrate potential used.

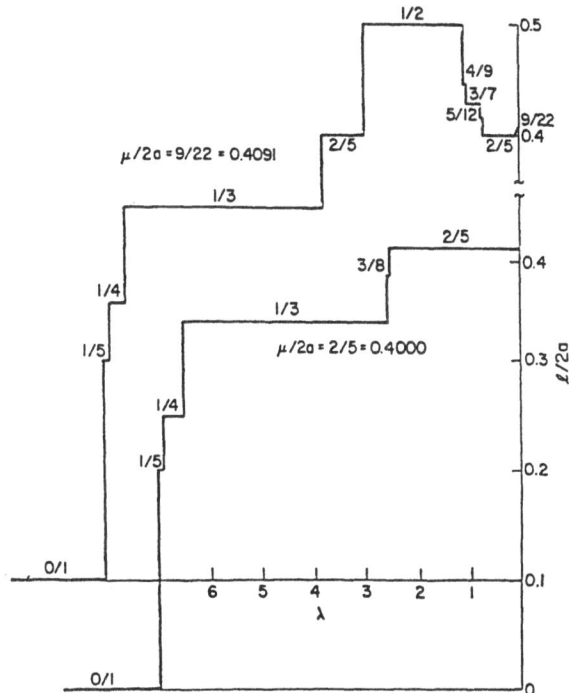

FIG. 7. Discrete Frenkel-Kontorowa model: average lattice-spacing
         misfit versus substrate potential for two values of the
         chain tension (from Ref. (74)).

In summary, it seems that the complex behaviour shown by the
present compound and by others in the same class, cannot be understood, even qualitatively, without taking explicit account of the

302

discrete nature of the modulated lattice.

V.   A MAGNETIC EXAMPLE: CERIUM ANTIMONIDE

    Below the magnetic ordering temperature (T_N = 16.2K) CeSb
displays a particularly complex behaviour (69), as evidenced by the
many specific heat anomalies observed (70) (cf. Fig. 8). The
complete (H,T) phase diagram is shown in Fig. 9 for H along [001].
The magnetic structures of the various phases observed are all
based on a periodic stacking sequence of non-magnetic and ferromagnetic (001) planes.

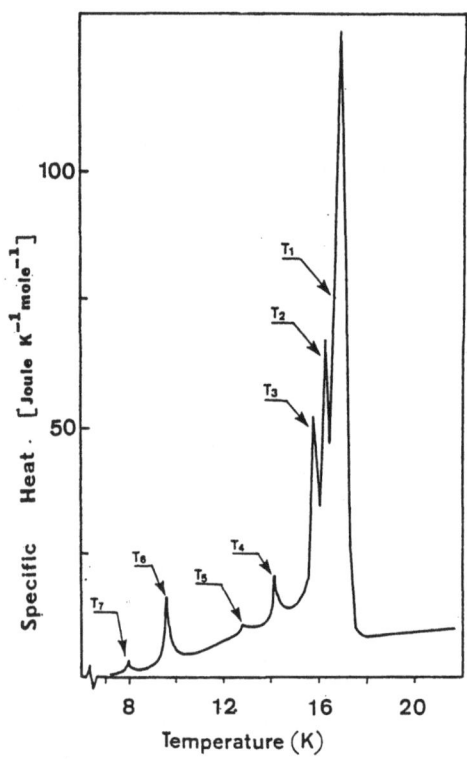

FIG. 8.   Specific heat of CeSb in zero applied magnetic field
          (from Ref. (70)).

    The three end-structures are:

        - the paramagnetic (P) phase at high T
        - the ferromagnetic (F) phase at high H and low T
        - the antiferromagnetic (AF) phase at low H and low T.

    The other phases can be regarded as intermediate between two
and the above three.

    The structure of the AF phase consists in a sequence of two
pairs of Ce^{3+} layers with opposite magnetization, i.e. (++--).

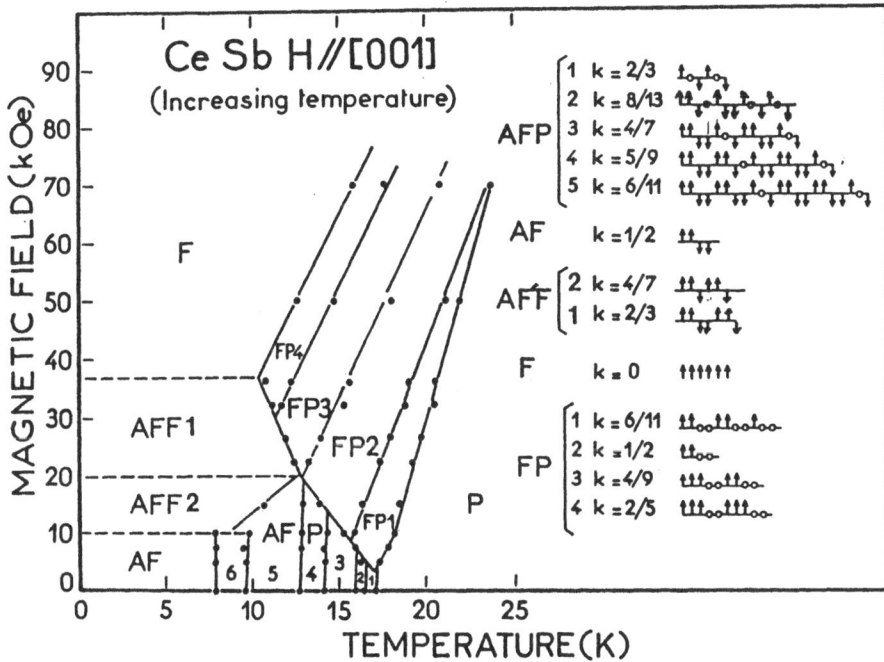

FIG. 9. (H,T) phase diagram of CeSb as deduced from specific heat anomalies (from Ref. (70)).

This sequence can be described by a square-wave of wavevector $k = (0,0,1/2) \, 2\pi/a$ in the f.c.c. Brillouin zone (interlayer spacing $= a/2$). The structures labelled AFF2 and Affl are obtained from AF by removing one down layer (−) every 7th and 3rd layer, respectively, i.e. (++−−++−) and (++−). The corresponding square-wave wavevectors are $k = 4/7$ and $2/3$.

Similarly the AFP phases are generated from the AF phase by introducing one non-magnetic plane every $2n-1$ layer. The resulting wavevector is of the form $k = n/2n-1$ with $n = 2,4,5,6,...\infty$ (an exception to the rule is the AFP2 phase of wavevector $k = 8/13$). In all cases no net magnetization is introduced.

Finally between the F and P phases the substitution involves non-magnetic and ferromagnetic up (+) planes. The corresponding structures can be visualized as resulting from the superposition of a ferromagnetic component and of a square-wave of wavevector $n/2n+1$ (FP3, FP4) or $n/2n-1$ (FP1).

The overall shape of the phase diagram is well accounted for in terms of a simple thermodynamic model in which each Ce^{3+} ion is assumed to contribute an entropy of $k\ln2$ in the non-magnetic state and a moment of 2.I μ_B in the magnetic state. For example the transitions AF → AFF and AFF → F involve a change in magnetization and no change in entropy ($\Delta M \neq 0$; $\Delta S = 0$). Hence from Clapeyron's relation:

$$\frac{\partial H}{\partial T} = -\frac{\partial S}{\partial M} = 0$$

in agreement with Fig. 9. Similarly the AF → AFP → P transitions correspond to $\Delta S = 0$ and $\Delta M = 0$ and thus should be field-independent while the P → FP → F ones involve changes in both S and M and thus should have a finite slope in the (H,T) diagram.

The physical origin of the non-magnetic planes has not been fully clarified, so far. The possibility of two different electronic ground states for the magnetic and non-magnetic Ce^{3+} ions has been considered. A related problem is the large anisotropy of the exchange interactions between in-plane (J_0) and out-of-plane (J_1,J_2) Ce^{3+} neighbours. The stability of the AF (++--) structure at T = H = 0, suggests $J_1 < 0$, $J_2 < 0$ and $J_0 \gg |J_1|$, $|J_2|$ which is somewhat surprising in view of the cubic symmetry of the structure for $T > T_N$.

On the other hand, the large exchange anisotropy makes CeSb an attractive model system and several attempts have been made at explaining the main features of the experimental phase diagram, particularly the H = 0 AF → P sequence, within the ANNI model (71,72,73). Noting that this sequence corresponds to introducing an increasing proportion of non-magnetic defects, Villain and Gordon (73) argue that the true devil's staircase behaviour will only be realized if the effective interaction between defects is repulsive at all distances. This condition imposes constraints on the relative values of J_1 and J_2 which are not met in the case of CeSb. Alternatively (8) one may argue that the missing intermediate phases are suppressed by the metastability of some of the adjacent ones. It has been noted that some samples miss more steps than others which implies that at least in some cases the number of observed phases is limited by (extrinsic) pinning mechanisms.

In any case, it is clear that the experimental limit has not yet been reached, in the sense that finer steps could still be resolved, if present. Until then, the search for the ideal devil's staircase system is likely to continue.

ACKNOWLEDGEMENTS

I am grateful to R. Almairac, G. Marion, J.D. Axe, R.A. Cowley and S. Aubry for communication of results prior to publication. Useful discussions with F. Fenoyer, L. Bernard, C. Vettier, J. Rossat-Mignod and J. Villain are acknowledged.

REFERENCES

1. S. Aubry, J. Physique 44, 147 (1983).
2. P.M de Wolff, Acta Cryst. A30:777 (1984).
3. A. Janner and T. Janssen, Phys. Rev. B15, 643 (1977): Acta Cryst. A36, 399 (1980); Acta Crystl. A36, 408 (1980).
4. P. Bak, Phys. Rev. Lett. 46, 791 (1981).
5. Y. Imry and S.K. Ma, Phys. Rev. Lett. 35, 1399 (1975).
6. L.J. Sham and B.R. Patton, Phys Rev. B13, 3151 (1976).
7. M. Nielsen, J. Bohr, K. Kjaer and J.P. McTague, Inst. Phys. Conf. Ser. 64, p. 289 (Schofield P., Editor) (1982).
8. P. Bak, Rep. Prog. Phys. 45, 587 (1982).
9. W.L. McMillan, Phys. Rev. B14, 1496 (1976).
10. S. Aubry, in "Solitons in Condensed Matter Physics," ed. by A.R. Bishop and T. Schneider (Springer, Berlin) (1978).
11. H. Mashiyama, S. Tanisaki and K. Hamano, J. Phys. Soc. Japan (1982), 51, 2538: K. Hamano, Y. Ikeda, F. Fujimoto, K. Ema and S. Hirotsu, J. Phys. Soc. Japan 49, 2278 (1980).
12. H.G. Unruh, to appear in J. Phys. C. (1983).
13. J.P. Jamet and P. Lederer, to appear in J. Phys. (Paris) (1983).
14. G. Errandonea, "Multicritical Phenomena," NATO-ASI B. Plenum, N.Y. (1983).
15. A.W. Overhauser, Phys. Rev. B3, 3173 (1971).
16. A.D. Bruce and R.A. Cowley, J. Phys. C. 11, 3609 (1978).
17. R. Zeyher and W. Finger, Phys. Rev. Lett. 49, 1833 (1982).
18. V.A. Golovko and A.P. Levanyuk, Sov. Phys. JETP 54, 1217 (1981).
19. H. Cailleau, F. Moussa, C.M.E. Zeyen and J. Bouillot, Solid State Comm. 33, 407 (1980).
20. L. Bernard, R. Currat, P. Delamoye, C.M.E. Zeyen, S. Hubert and R. de Kouchovsky, J. Phys. C 16, 433 (1983); L. Bernard, R. Currat and P. Delamoye, to be published.
21. R.M. Fleming, D.E. Moncton, D.B. McWhan and F.J. DiSalvo, Phys. Rev. Lett. 45:576 (1980).
22. D.E. Cox, S.M. Shapiro, R.J. Nelmes, T.W. Ryan, H.J. Bleif, R.A. Cowley, M. Eibschutz and H.J. Guggenheim, preprint (1983).
23. K.K. Fung, S. McKernan, J.W. Steeds and J.A. Wilson, J. Phys. C 14, 5417 (1981).
24. C.H. Chen, J.M. Gibson and R.M. Fleming, Phys. Rev. Lett. 47, 723 (1981).
25. R. Blinc, Phys. Repr. 79, 331 (1981).
26. P.D. Delamoye and R. Currat, J. Physique Lett. 43, L-655 (1982).
27. R. Blinc, D.C. Ailion, P. Prelovsek and V. Rutar, Phys. Rev. Lett. 50, 67 (1983).
28. H. Nowotny, The Chemistry of Extended Defects in Non-Metallic Solids, Ed. LeRoy, Eyring and M. O'Keefe, North-Holland Pub. Co., Amsterdam (1970).
29. G. Theodorou and T.M. Rice, Phys. Rev. B18, 2840 (1978).
30. J.D. Axe and P. Bak, Phys. Rev. B 26, 4963 (1982).
31. W. Jeitschko and E. Parthé, Acta Cryst. 22, 417 (1967).
32. G.H. Flicher, H. Vollenkle and H. Novotny, Mh. Chem. 98, 2173 (1967).

33. E.W. Knott, M.E. Meuller and L. Heaton, Acta Cryst. $\underline{23}$, 549 (1967).
34. R. de Ridder and S. Amelinckx, Mat. Res. Bull. $\underline{6}$, 1223 (1971).
35. R. de Ridder, G. Van Tendeloo and S. Amelinckx, Phys. St. Sol. (a) $\underline{33}$, 383 (1976).
36. J.D. Axe (private communication.
37. W. Finger and T.M. Rice, PEys. Rev. Lett. $\underline{49}$, 468 (1982).
38. For a review on 2H-TaSe$_2$ see J.D. Axe (1982).
39. P. Bak and V.L. Pokrovsky, Phys. Rev. Lett. $\underline{47}$, 958 (1981).
40. P.Y. Le Daeron and S. Aubry, Metal Insulator Transition in the Peierls Chain, submitted to J. Phys. C. (1982).
41. C.K. Chiang, C.R. Fincher, Y.W. Park, A.J. Heeger, H. Shirikawa, E.J. Louis, S.C. Gau and A.G. MacDiarmid, Phys. Rev. Lett $\underline{39}$, 1098 (1977).
42. G.A. Samara, Comments Sol. St. Phys. $\underline{8}$, 13 (1977).
43. M.S. Haque and J.R. Hardy, Phys. Rev. B $\underline{21}$, 245 (1980).
44. H. Cailleau, F. Moussa and J. Mons, Sol. St. Comm. $\underline{31}$, 521 (1979).
45. See A. Michelson, Phys. Rev. B $\underline{18}$, 459 (1978) for a discussion of the applicability of Lifshitz criterion to IC phase transitions.
46. M. Iizumi and K. Gesi, Sol. St. Comm. $\underline{22}$, 37 (1977).
47. D. Durand, F. Fenoyer, M. Lambert, L. Bernard and R. Currat, J. Physique $\underline{43}$, 149 (1982); and ref. therein.
48. G. Dolino, J.P. Bachheimer and C.M.E. Zeyen, Sol. St. Comm. $\underline{45}$, 295 (1983).
49. F. Denoyer, A.H. Moudden, R. Currat, C. Vettier, A. Bellamy and M. Lambert, Phys. Rev. B $\underline{25}$, 1697 (1982).
50. A.P. Levanyuk and D.G. Sannikov, Sov. Phys. Sol. St. $\underline{18}$, 1122 (1976).
51. T.A. Aslanian and A.P. Levanyuk, Sol. St. Comm. $\underline{31}$, 547 (1979).
52. V. Heine and J.D.C. McConnell, Phys. Rev. Lett. $\underline{46}$, 1092 (1981).
53. F. Brehat, J. Claudel, P. Strimer and A. Hadni, J. Physique Lettres, $\underline{37}$, L229 (1976).
54. A.H. Moudden, Thesis, Université Paris-Sud (1980) unpublished.
55. A.H. Moudden and J. Als-Nielsen (unpublished).
56. R.A. Cowley and A.D. Bruce, J. Phys. C $\underline{11}$, 3577 (1978).
57. M. Elcombe and J.C. Taylor, Acta Cryst. A $\underline{24}$, 410 (1968).
58. Y. Ishibashi and H. Shiba, J. Phys. Soc. Jap. $\underline{45}$, 409 (1978).
59. P. Lederer and C.M. Chaves, J. Physique Lett. $\underline{42}$, L127 (1981).
60. J.P. Jamet, P. Lederer and A.H. Moudden, Phys. Rev. Lett $\underline{48}$, 442 (1982).
61. A.H. Moudden, E.C. Svensson and G. Shirane, Phys. Rev. Lett. $\underline{49}$, 557 (1982).
62. Simonson, F. Denoyer, C. Vettier and R. Currat (to be published).
63. A.D. Bruce, R.A. Cowley and A.F. Murray, J. Phys. C $\underline{11}$, 3591 (1978).
64. G. Marion, R. Almairac, J. Lefebvre and M. Ribet, J. Phys. C $\underline{14}$, 3177 (1981).

65. J. Schneck, J.C. Toledano, C. Joffrin, J. Aubree, B. Joukoff and A. Gabelotaud, Phys. Rev. B $\underline{25}$, 1766 (1982).

66. K. Nakanishi, J. Phys. Soc. Jpn. $\underline{46}$, 1434 (1979).

67. H. Mashiyama, S. Tanisaki and K. Gesi, J. Phys. Soc. Jpn. $\underline{50}$, 1415 (1981).

68. G. Marion (private communication).

69. J. Rossat-Mignod, P. Burlet, J. Villain, H. Bartholin, W. Tcheng-Si, D. Florence and O. Vogt, Phys. Rev. B $\underline{16}$, 440 (1977).

70. J. Rossat-Mignod, P. Burlet, H. Bartholin, O. Vogt and R. Lagnier, J. Phys. C $\underline{13}$, 6381 (1980).

71. P. Bak and J. von Boehm, Phys. Rev. B $\underline{21}$, 5297 (1980).

72. W. Selke and M.E. Fisher, Phys. Rev. B $\underline{20}$, 257 (1979).

73. J. Villain and M. Gordon, J. Phys. C $\underline{13}$, 3117 (1980).

74. S. Aubry, to appear in J. Phys. C (1983).

ATOM PROBE FIELD-ION MICROSCOPY STUDIES OF MODULATED STRUCTURES

S.S. Brenner*, M.K. Miller* and W.A. Soffa[+]

*U.S. Steel Research Laboratory, Monroeville, PA 15146 USA
[+]University of Pittsburgh, Pittsburgh, PA 15261 USA

1. INTRODUCTION

Modulated microstructures can occur in metallic or non-metallic materials as a result of phase transformations or be synthesized by creating layered structures by atomic deposition techniques. In this review we consider only modulated microstructures forming during precipitation from solid solutions. These unique microstructures have both scientific and technical interest. Periodic precipitation processes generally are a manifestation of the fundamental aspects of solute clustering and continuous phase separation in solids (1). The formation of these periodic microstructures can produce marked changes in the physical and mechanical properties and may be used to tailor unique combinations of properties.

There have been several theoretical studies of the formation of modulated microstructures via continuous phase separation or spinodal decomposition. However, experimental confirmation of a number of important aspects of these models has been hampered by a lack of techniques that have sufficient resolution to measure the ultra-fine scale details of the transformations.

The atom probe field-ion microscope is an extremely effective tool in the investigation of these ultra-fine scale microstructures because it provides a means for atomic scale microstructural and microchemical analysis of these materials. This instrument provides a method for direct measurement of important microstructural and microchemical features of the transformation such as the wavelength of the dominant concentration waves, the spatial extent of the solute-enriched regions, and the three-dimensional morphology of the evolving phase mixture.

The technique and its general application to studies in material science have recently been reviewed by Wagner (2), Ralph et al. (3), and Brenner and Miller (4,5). Watts and Ralph (6) have discussed the use of the atom probe for the study of continuous transformations and several investigations into continuous transformation studies have been undertaken recently using the atom probe. Ralph and his coworkers have studied the early stages of decomposition in a Ni-12 at.% Ti alloy (7) and continuous phase separation in Ni-Al alloys (8). Biehl and Wagner have investigated the mechanism and kinetics of decomposition in supersaturated Cu-Ti alloys (9,10). The authors have also applied the atom probe technique to characterize the kinetics and morphology of the early stages of spinodal decomposition and coarsening in Fe-Cr (11,12,13), Fe-Cr-Co (14), and Fe-Be (15) alloys.

In this paper we review the atom probe technique and its application to the study of continuous transformation of modulated microstructures in spinodally decomposed alloys with examples of results from selected applications in which the authors have been involved.

2. THE ATOM PROBE FIELD-ION MICROSCOPE

The atom probe consists of two parts: an ultra-high resolution field-ion microscope and a mass spectrometer. Individual atoms can be resolved in the field-ion microscope (FIM). Coupled to the FIM is a time-of-flight mass spectrometer that can accurately identify these single atoms to provide a chemical analysis.

The principle of operation and the main components of the instrument are shown schematically in Figures 1 and 2, respectively. The needle-shaped specimen is inserted into the ultra-high vacuum chamber through a specimen exchange airlock and is attached to a rotatable platform. After cooling to cryogenic temperatures a small amount of imaging gas is admitted and a positive voltage is applied to the specimen. At a certain voltage, dependent on the sharpness of the specimen and the image gas used, an image of the specimen surface is formed by the projection of the ionized gas atoms onto the channel plate and phosphor screen assembly.

To make a chemical analysis the specimen is rotated until the image of the area of interest falls over a small hole in the channel plate assembly. This probe aperture serves as the entrance to

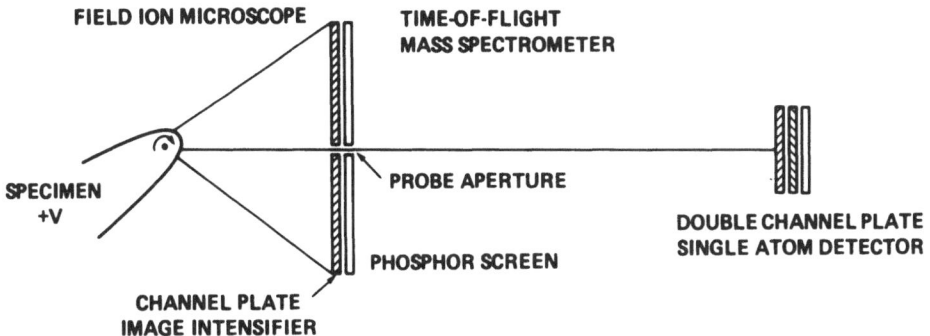

Fig. 1. Principle of operation of the atom probe.

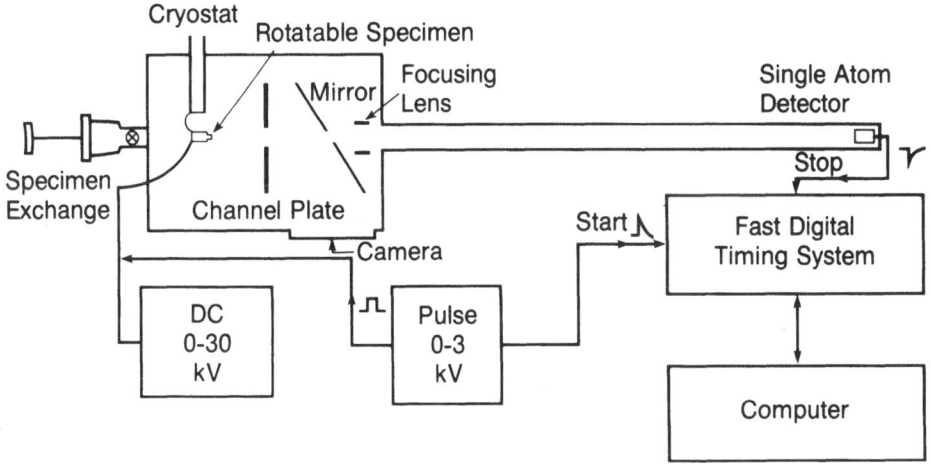

Fig. 2. Main components of the atom probe.

the time-of-flight mass spectrometer. The surface atoms are then
removed from the specimen, usually in the absence of the imaging
gas, by the superimposition of a short high-voltage positive pulse
(V_{pulse}) onto the standing voltage (V_{dc}). While atoms are removed
from the entire imaging area of the specimen only those whose tra-
jectories pass through the probe aperture are analyzed. The mass-
to-charge ratio (m/n) of the atoms analyzed are derived by equating
their potential energy and kinetic energy and is given by

$$m/n = 0.193 \, (V_{dc} + V_{pulse}) \, t^2/d^2,$$

where d is the distance traveled from the specimen to the detector

312

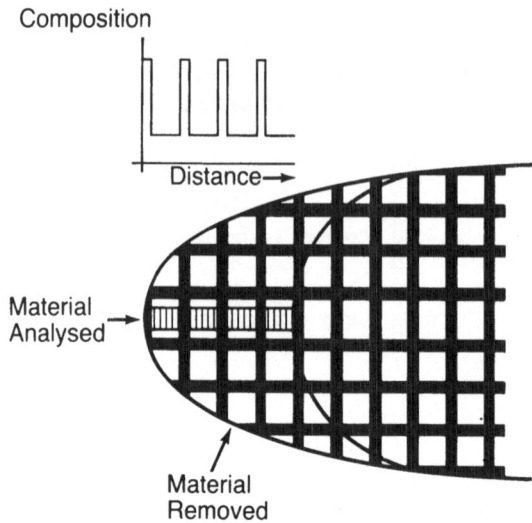

Fig. 3. Random area analysis of a modulated microstructure.

and t is the flight time of the atom. This experimental cycle is
repeated and the data analyzed under computer control at a rate of
up to 250 Hz in the current metallurgical atom probes.

The atom-by-atom sequence can be accumulated into spectra
from which the composition is directly measured by identifying the
peaks in the spectrum and then counting the number of atoms in each
peak. This type of analysis is called selected area analysis when
a region of the specimen is selected. The probe aperture may be
repositioned, if necessary, to ensure that only atoms from that
selected region are collected.

Another type of analysis that makes use of the atom-by-atom
sequence that is particularly suitable for the analysis of modulated microstructures is random area analysis. In this type of
analysis, shown schematically in Figure 3 for an aligned modulated
microstructure, a cylinder of material is examined, without re-
positioning the probe aperture, much in the same manner as a geo-
logical core sample. The diameter of the core of material is defined by the effective probe aperture but the length of the sample
is usually extended over many hundreds of angstroms. By visually
monitoring the amount of material removed from the specimen during
field evaporation the composition fluctuations can be measured as
a function of distance with atomic precision (5). It is important
that the diameter of the cylinder of analysis be smaller than the
microstructural feature under investigation but be large enough so
that a statistically significant number of atoms can be collected
to provide a composition over the small depth intervals.

These composition profiles, or time series, can be subjected to various statistical techniques such as autocorrelation or Fourier transform, to assist in the determination of the wavelength and dimensions of the phases present.

The early stages of decomposition can also be investigated by examining the frequency distribution of small blocks of atoms of the atom-by-atom sequence to detect deviations from the binomial distribution characteristic of the solid solution. This technique allows the detection of phase separation at times that do not produce noticeable changes in the field-ion micrograph.

3. ISOTROPIC SPINODAL DECOMPOSITION IN Fe-Cr AND Fe-Cr-Co ALLOYS

The phase changes occurring in iron-chromium alloys at low temperatures have been of metallurgical interest for a long time because of the embrittling effects that accompany the precipitation reaction. Although it was reasonably clear that high chromium alloys undergo spinodal decomposition within a very broad miscibility gap (16,17), it was not until recently that the microstructure was resolved micrographically for the first time by the authors using field-ion microscopy (11). Transmission electron microscopy and electron and X-ray diffraction have been of limited success in resolving the details of the reaction because of the similar sizes and scattering factors of iron and chromium.

A field-ion micrograph of an iron-32 at.% chromium alloy aged for 10,000 hours at 470°C is shown in Figure 4. The most striking feature of the highly interconnected network morphology is the complete absence of any crystallographic alignment of the phases. The microstructure bears a close resemblance to the computer simulated images of an isotropically decomposed spinodal microstructure by Cahn (17) and the morphology exhibited in phase-separated glasses. The isotropic morphology in the iron-chromium alloy is not surprising considering the small strain energy effects associated with the decomposition.

Atom probe selected area analysis revealed that the dark regions are the chromium-enriched phase and the brightly imaging regions the iron-rich phase. The contrast between the phases arises because of small differences in the field-evaporation behavior of the two phases which leads to slight differences in local radius of curvature. This in turn affects the local ionization probability of the image gas and hence produces a difference in contrast in the field-ion image. A specimen temperature of 78 K was used to give the maximum contrast between the two phases at the expense of some atomic resolution.

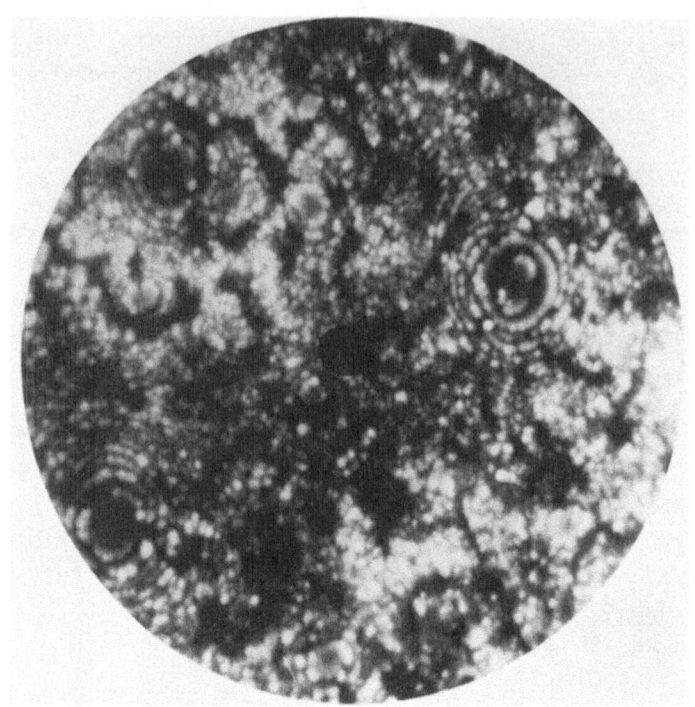

Fig. 4. Field-ion micrograph of Fe-32% Cr, aged 10,000 hrs, 470°C.

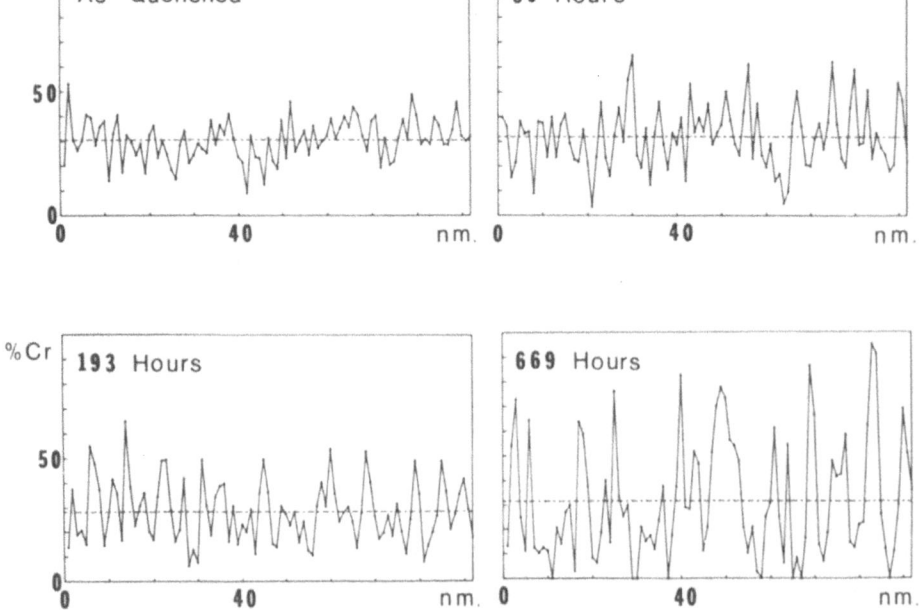

Fig. 5. Composition profile of Fe-32% Cr, aged at 470°C.

The evolution of the composition modulations in the iron-32% chromium alloy aged for times up to 669 hours at 470°C is shown in Figure 5. Each data point in these profiles is the composition of a volume element of material approximately 0.8 nm thick by 1 nm in diameter comprising of between 30 and 50 atoms. The amplitudes of the composition fluctuations continue to increase progressively with time from initially small values at least up to 200 hours, providing strong evidence that the alloy undergoes continuous phase separation.

At the earliest times it is difficult to detect deviations from the random solution in the composition profiles because of the statistical noise arising from the small sample size. However, the frequency distribution of the composition of small blocks of atoms can reveal changes in behavior at earlier times as shown in Figure 6. After 10 hours aging at 470°C, the distribution has become noticeably broader than a binomial distribution characteristic of a random solid solution. Eventually, a bimodal distribution appears corresponding to the development of a distinct two-phase mixture.

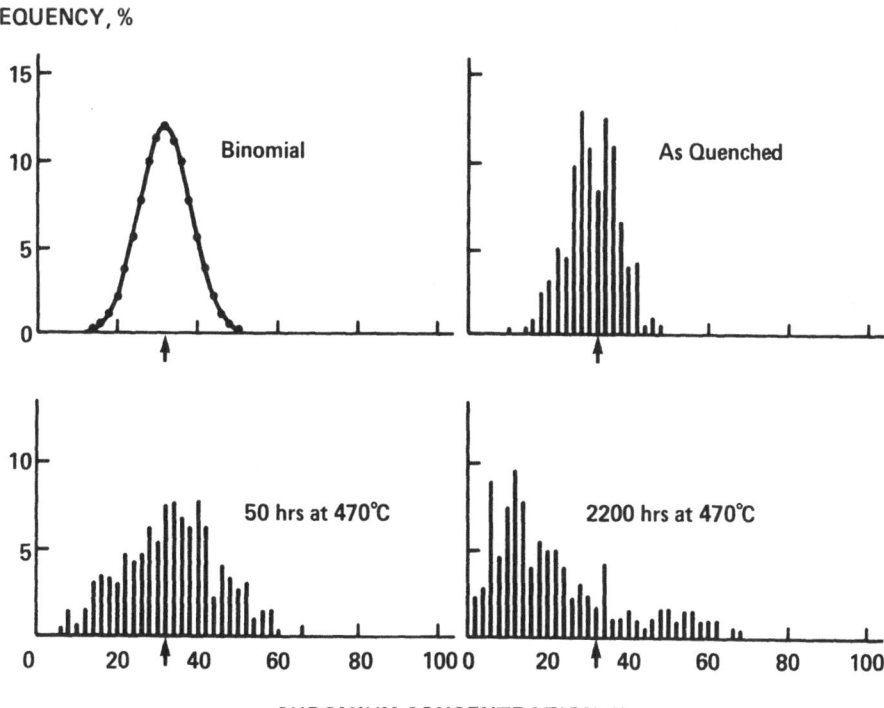

Fig. 6. Frequency distribution of blocks of atoms in Fe-32% Cr.

Iron-chromium-cobalt alloys undergo phase separation within a low temperature miscibility gap similar to the iron-chromium binary alloys. The addition of cobalt raises the critical temperature and produces an asymmetrical miscibility gap. The morphology and composition of the resulting phases also play an important role in determining the physical and magnetic properties of the alloy.

A field-ion micrograph of the spinodally decomposed microstructure of an iron-28.5 wt.% chromium-10.6 wt.% cobalt alloy, aged 8 hours at 600°C, is shown in Figure 7. The microstructure is essentially similar to the iron-chromium binary alloy and consists of highly interconnected networks of brightly imaging and darkly imaging regions. The contrast between the phases is not altered significantly by the addition of cobalt. Atom probe selected area analysis revealed that darkly imaging regions were the chromium-rich phase and the brightly imaging regions were the iron-rich phase. This irregular morphology showed no evidence of crystallographic alignment. The very high degree of interconnectivity of both phases is clearly revealed even in the planar section of the micrograph.

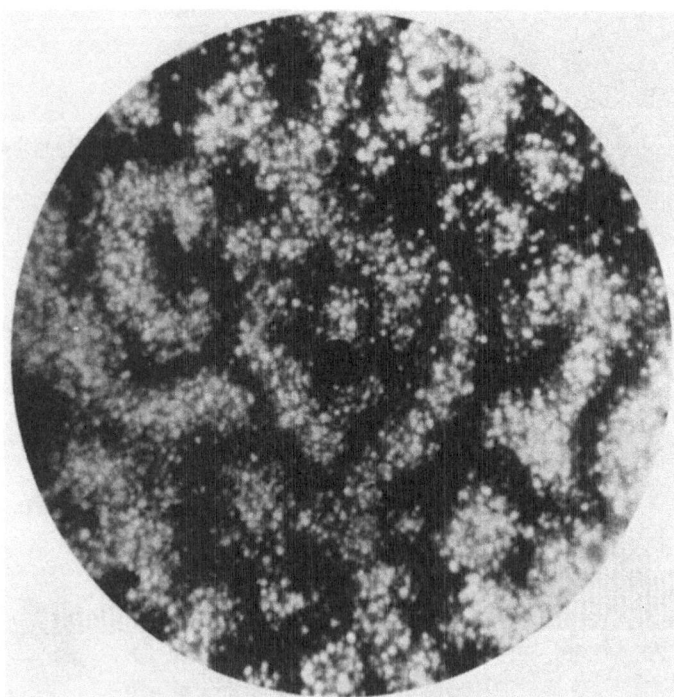

Fig. 7. Field-ion micrograph of Fe-28.5% Cr-10.6% Co; aged 8 hrs, 600°C.

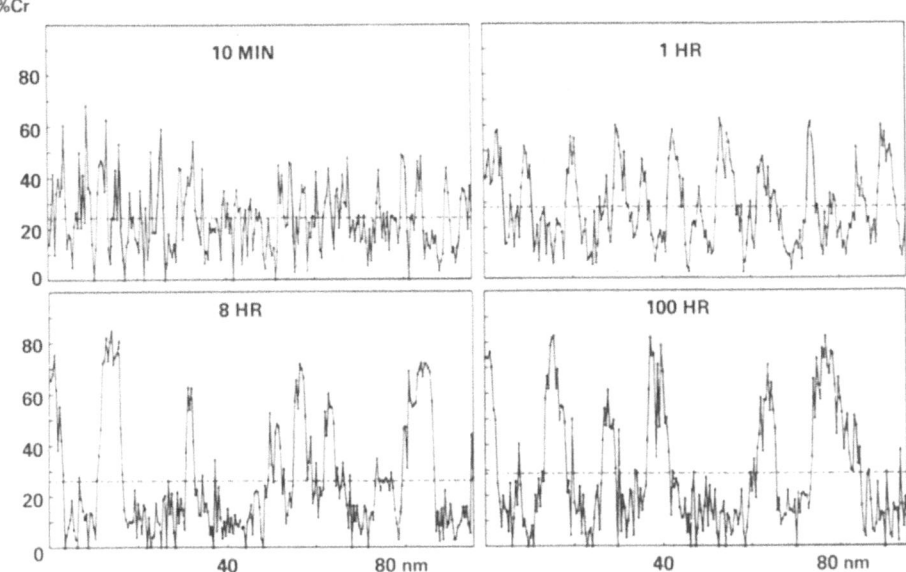

Fig. 8. Composition profile in Fe-28.5% Cr-10.6% Co; aged 560°C.

The evolution of the composition fluctuations that accompany the spinodal decomposition of this alloy aged at 560°C are shown in the series of composition profiles in Figure 8. The phase separation is clearly revealed in the composition differences of the two phases even at the earliest time of 10 minutes. The average composition of the chromium-rich regions increases with aging time consistent with the theory of spinodal decomposition.

Coarsening of the isotropic microstructure begins even while the composition of the two phases is still increasing. A series of field-ion micrographs of the material aged at 560°C is shown in Figure 9. The scale of the microstructure both in terms of the thickness of the chromium-rich regions and their mean spacing increases as aging proceeds. No spheroidization or marked degeneration of the microstructure was evident even up to 100 hours.

The thickness of the chromium-rich regions and the period between these regions was determined from the composition profiles and are shown in Figure 10. Both the thickness and the wavelength were found to increase with aging although at a very slow rate. This behavior was similar to the iron-chromium binary alloy. The data could be fitted to a t^a power law where the time exponent was approximately 0.15. No regime where a single dominant wavelength persisted was experimentally detected for the time and temperature ranges investigated.

318

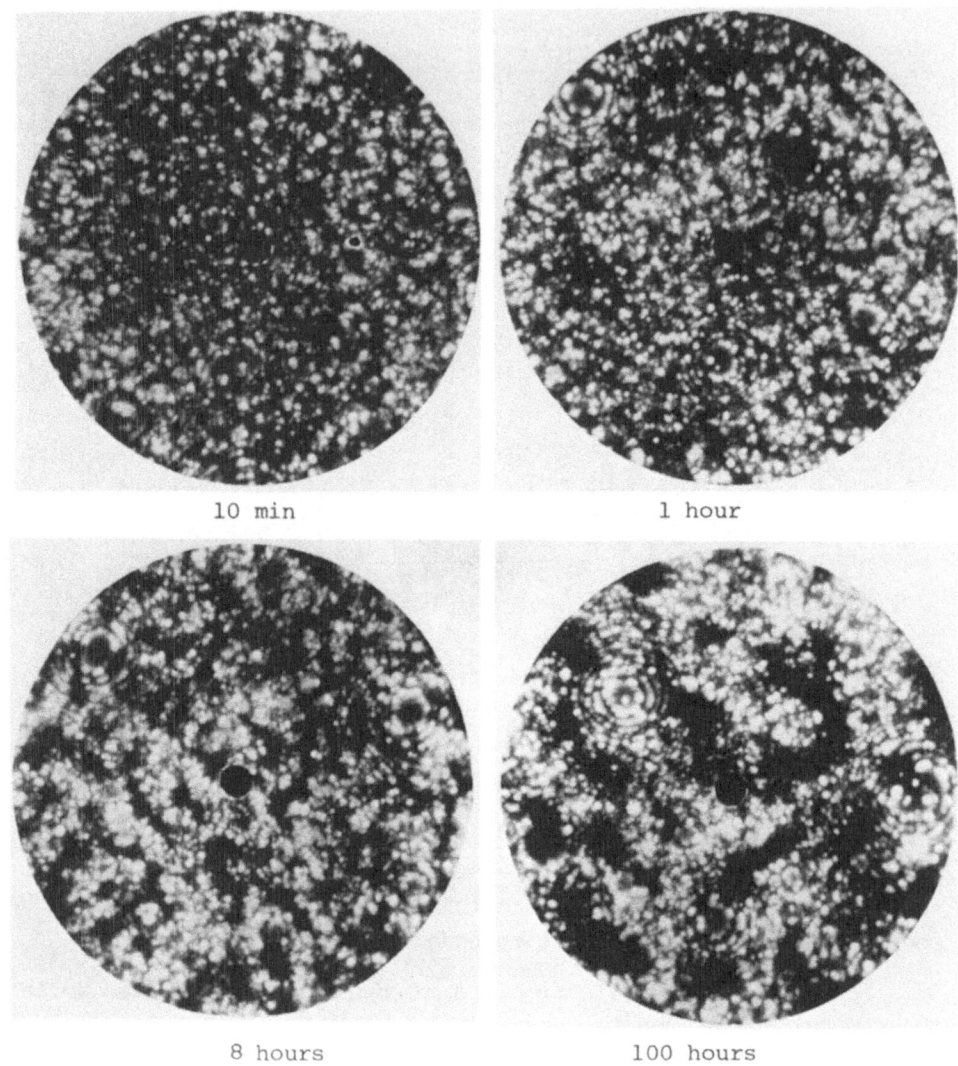

10 min   1 hour

8 hours   100 hours

Fig. 9. Fe-28.5% Cr-10.6% Co, aged at 560°C.

The precise time exponent or kinetics of the coarsening of these highly interconnected isotropic spinodal microstructures may intimately depend on the topological features of the microstructure.

4. CRYSTALLOGRAPHICALLY ALIGNED SPINODAL DECOMPOSITION IN Fe-Be ALLOYS

Coherency strains and elastic anisotropy can produce crystallographically aligned modulated microstructures. Decomposition of the supersaturated iron-beryllium solid solution is accompanied by significant coherency strains because of the large dissimilarity of the sizes of the iron and beryllium atoms. The preferential

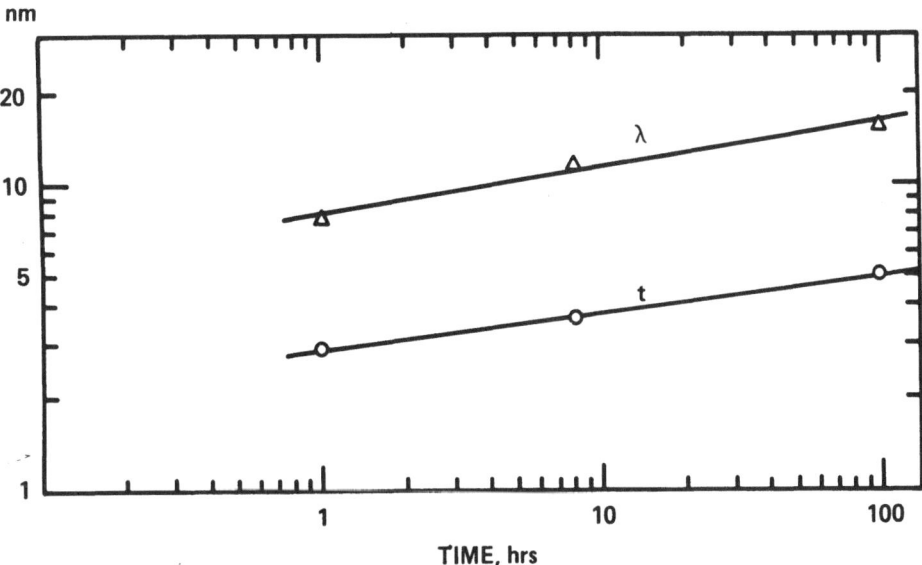

Fig. 10. Wavelength and thickness at Cr-rich regions in Fe-Cr-Co, aged at 560°C.

amplification of concentration waves along the elastically "soft" <100> directions results in an triaxially aligned microstructure.

A crystallographically aligned microstructure of spinodally decomposed iron-25 at.% beryllium alloy, aged 15 hours at 375°C, is shown in Figure 11. The three sets of orthogonal composition waves are revealed as the horizontal, vertical and circular dark bands in this [100] centered field-ion specimen because of the projection of the curved specimen surface. The microstructure consists of a macrolattice of block shaped particles aligned along the <100> matrix directions. Selected area analysis revealed that the darkly imaging continuous bands were the beryllium-enriched phase and the brightly imaging isolated regions were the iron-rich phase. A dark field transmission electron micrograph of this type of modulated microstructure is shown for comparison in Figure 12 for a specimen aged 2 hours at 350°C.

Some of the fine scale details of the morphology are best observed by comparing the field-ion micrograph with computer simulated images of the ideal microstructure, Figure 13, because of the geometrical effects of the projection. The field-ion micrograph revealed that the spacing between beryllium-enriched bands was not always identical although there were no abrupt changes in spacing along any pair of bands. The bands were not always exactly parallel to each other with some of the bands merging into each other. The size and shape of an individual iron-rich particle

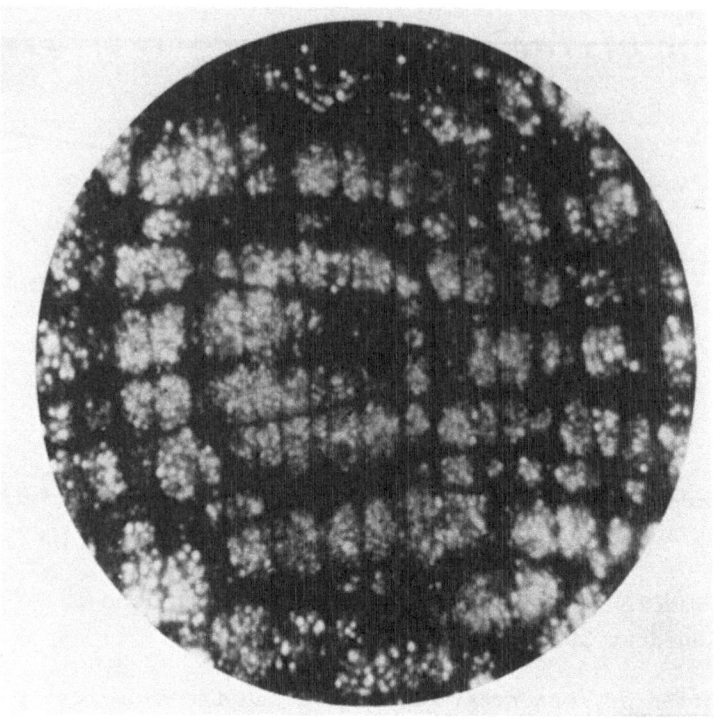

Fig. 11. Field-ion micrograph of Fe-25 at.% Be, aged 15 h, 375°C.

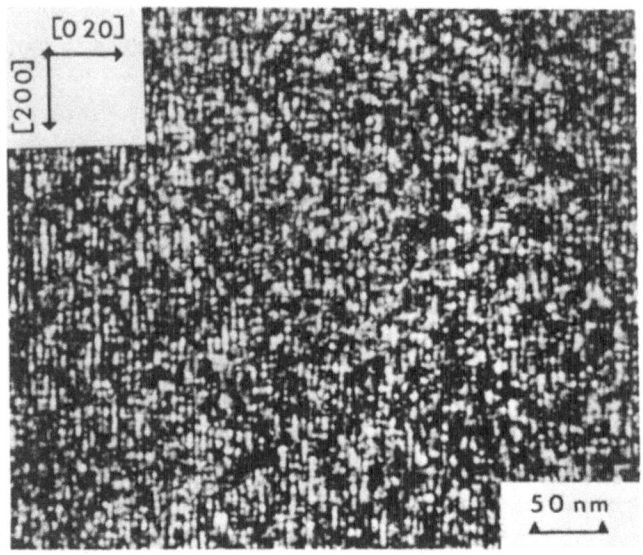

Fig. 12. TEM micrograph of Fe-25 at.% Be, aged 2 h, 350°C.

[100] Oriented Specimen

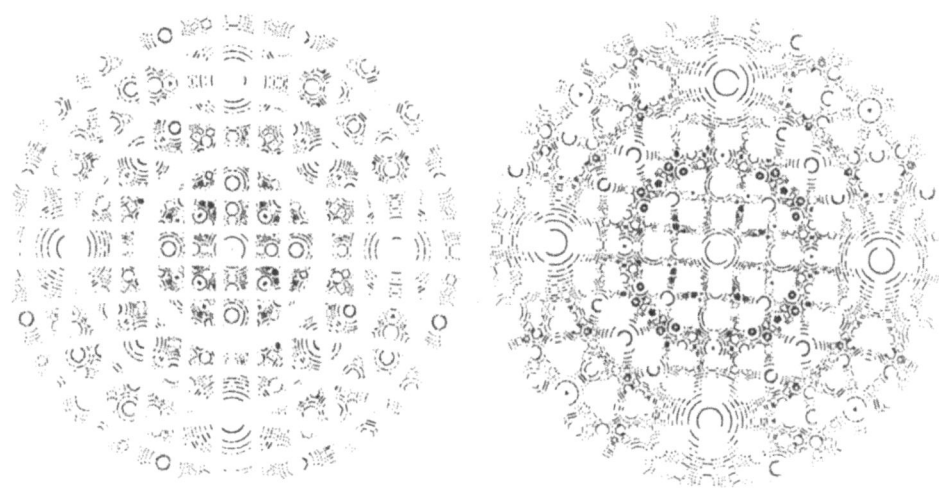

       Iron-rich Phase   Beryllium-enriched Phase
[111] Oriented Specimen

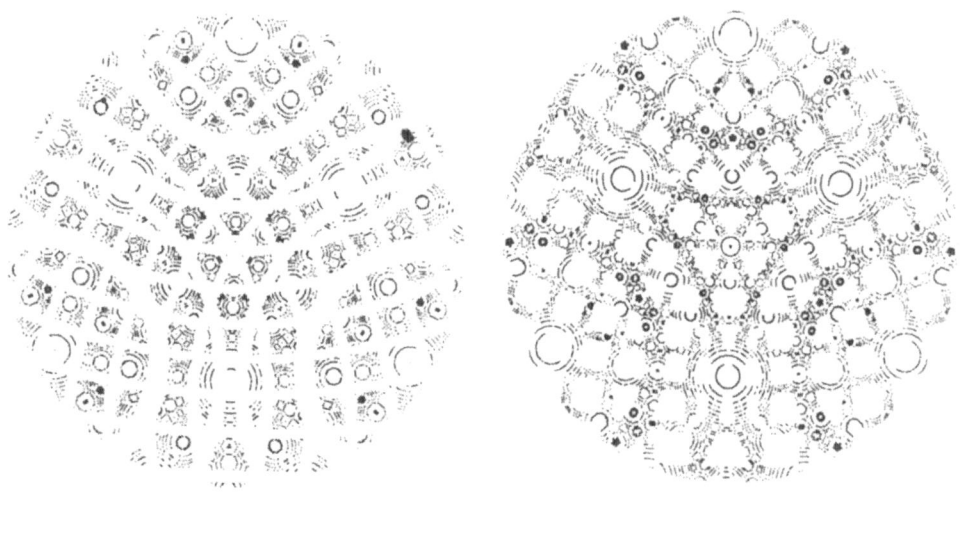

       Iron-rich Phase   Beryllium-enriched Phase

Fig. 13. Computer-simulated field-ion images. Each image shows
the atoms in either the iron-rich phase or the beryllium-enriched
phase. (Radius of specimen = 70 nm, shell thickness = 0.015 nm,
field of view = 120°.) Volume fraction of iron-rich phase = 51.2%.

322

is strongly influenced by its neighbors in the macrolattice.

Formation of the fine scaled modulated microstructure was observed at temperatures as low as 300°C although it was not discernible in a field-ion micrograph of a specimen aged for 96 hours at 265°C.

It was evident that coarsening occurred from the earliest stages that the decomposition products could be resolved. The coarsening proceeded without any major changes in the geometrical features of the morphology through a simple scaling of the modulated microstructure. This behavior persisted until the microstructure was consumed by another iron-beryllium intermetallic metastable precipitate.

6. SUMMARY

The atom probe field-ion microscope is a truely unique and powerful tool for studying the formation and nature of modulated microstructures resulting from spinodal decomposition in alloys. The combined microstructural and microchemical capabilities of the technique provides a means for characterizing the dynamics of phase separation. The technique can also reveal the true three-dimensional nature of the microstructure. The composition fluctuations that evolve during continuous phase separation can be directly monitored. The atom probe is an ideal instrument for the investigation of these ultra-fine scale microstructures.

7. ACKNOWLEDGMENTS

The work was supported by the National Science Foundation under a University/Industry cooperative grant, DMR-8022225. We thank Dr. M.G. Burke, U.S. Steel Research Laboratory and Mr. P.P. Camus, University of Pittsburgh for their assistance.

8. REFERENCES

1. Cahn, J.W., Trans. AIME 242, (1968), 166.
2. Wagner, R., "Field-Ion Microscopy," Vol.6, Crystals, Growth, Properties and Applications (Springer-Verlag, 1982).
3. Ralph, B., Hill, S.A., Southon, M.J., Thomas, M.P. and Waugh,A.R. Ultramicroscopy 8, (1982), 361.
4. Brenner, S.S. and Miller, M.K., Proc. 27th Int. Field Emission Symposium, Japan,(1980), 238.
5. Brenner, S.S. and Miller, M.K., Journal of Metals 35, (1983),54.
6. Watts, A.J. and Ralph, B., Surf. Sci. 70, (1978), 459.
7. Watts, A.J. and Ralph, B., Acta Met. 25, (1977), 1013.
8. Hill, S.A. and Ralph, B., Acta Met. 30, (1982), 2219.

9. Biehl, K.E. and Wagner, R., Proc. 27th Int. Field Emission Symposium, Japan, (1980), 267.

10. Biehl, K.E. and Wagner, R., Proc. Int. Conf. Solid-Solid Phase Transformations, Ed. Aaronson, H., et al. Pittsburgh (TMS-AIME, 1982), 185.

11. Brenner, S.S., Miller, M.K. and Soffa, W.A., Scripta Met. 16, (1982), 831.

12. Brenner, S.S., Miller, M.K. and Soffa, W.A., Proc. Int. Conf. Solid-Solid Phase Transformations, Ed. Aaronson, H., et al. Pittsburgh (TMS-AIME, 1982), 191.

13. Miller, M.K., Brenner, S.S., Camus, P.P., Piller, J. and Soffa, W.A., Proc. 29th Int. Field Emission Symposium, Goteborg, Ed. Andren, H-O and Norden, H. (Almqvist and Wiksell, Int., Stockholm, 1982), 489.

14. Soffa, W.A., Brenner, S.S., Camus, P.P. and Miller, M.K., Proc. 29th Int. Field Emission Symposium, Goteborg, Ed. Andren, H-O and Norden, H. (Almqvist and Wiksell, Int., Stockholm, 1982), 511.

15. Miller, M.K., Brenner, S.S., Burke, M.G. and Soffa, W.A., Proc. 30th Int. Field Emission Symposium, Philadelphia, (1983).

16. Chandra, D. and Schwartz, C.H., Met. Trans. 2, (1971), 511.

17. DeNys, T. and Gielen, P.M., Met. Trans. 2, (1971), 1423.

18. Cahn, J.W., J. Chem. Phys. 42, (1965), 93.

CHAPTER 4: MECHANICS OF
MODULATED STRUCTURES

THE ELASTIC THEORY OF THE DEFECT SOLID SOLUTION

J. W. Morris, Jr.[1], A. G. Khatchaturyan[2] and Sheree H. Wen[3]

[1]Department of Materials Science and Mineral Engineering,
University of California, Berkeley, California 94720,
[2]Institute of Crystallography,
USSR Academy of Sciences, Moscow, USSR,
[3]IBM Thomas J. Watson Research Center,
Yorktown Heights, New York 10598

ABSTRACT

This paper reviews the linear elastic theory of the defect solid solution. It is tutorial in nature, and is primarily intended to outline the development of the theory into the mathematical form that is most suitable for the solution of practical problems. The theory treats solid solutions containing distributions of solute defects that distort the solvent lattice and interact elastically with one another. It determines the total strain of the solvent lattice and the elastic contribution to the free energy of the solution in the strong harmonic approximation. The theory is specifically developed for a binary solution of point defects in the absence of external stress. It can be extended to treat multi-component solutions, stressed solutions, and solutions of finite defects or macroscopic inclusions; the equations governing these complex systems are also presented. The model is finally used to consider ordering and decomposition reactions in solutions whose components interact elastically.

INTRODUCTION

Most of the important phase transformations in solids involve the formation or reconfiguration of atoms, defects, or elements of new phase that introduce elastic strains into the parent lattice. These elastic strains influence the thermodynamics and kinetics of

328

phase transformations, and often have a striking effect on the product microstructure. Elastic interactions can be specifically important in creating or modifying the complex solid structures that are a central theme of this meeting.

Because many of the more important lattice distortions occur at the atomic level where the details of the interatomic interaction are important, the formulation of a complete elastic theory is no less difficult than the creation of an atomistic theory of the solid itself. It is, however, possible to construct simple models that contain important elements of the elastic effect and include many of its qualitative consequences. A number of theoreticians have participated in the development of such models; Cahn [1,2] has perhaps been the most influential in establishing their relevance to the behavior of real solutions. The simplest general model is based on a combination of two kinds of theory: the continuum elastic theory of macroscopic inclusions in elastic media, developed originally by Eshelby [3,4], and the theory of lattice vibrations in the harmonic approximation, as presented, for example, by Born and Huang [5]. The specific formulation that is most widely used was originally proposed by Khatchaturyan [6,7], and, in later but largely independent work, by Cook and de Fontaine [8].

The present paper is intended to provide a brief tutorial review of the elastic theory of the solid solution. For notational simplicity the detailed theoretical development is confined to the case of binary solutions of point defects. It is reasonably simple to generalize the model to include more complex systems such as multicomponent solutions, interstitial solutions, solutions of defects with finite size and shape, and distributions of macroscopic inclusions. These extensions are discussed briefly; more thorough treatments are available elsewhere [9-12].

THE FREE ENERGY OF THE DEFECT SOLID SOLUTION

The Problem

The fundamental problem that is considered here is illustrated in Figure 1. The figure shows a configuration of point defects on a simple square lattice. Since the defects do not 'fit' properly, each of them distorts the parent lattice in its neighborhood. The elastic distortions of the lattice cause an interaction between the defects that adds to the chemical interaction they would experience if the lattice were rigid. The problem is to compute the net lattice strain and the total increase in free energy due to the interaction of an arbitrary distribution of such defects.

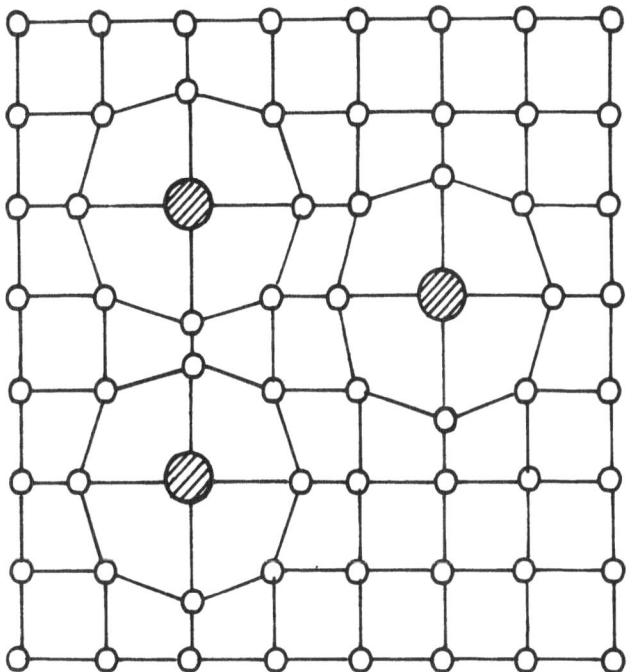

XBL 838-6144

Figure 1. A square lattice distorted by substitutional impurities.

Assumptions

The Reference Lattice. The development of the theory is simplest when the undistorted lattice of the solvent is used as the reference state. The solution is then made by adding solute defects. The positions (x) of the lattice points in the solution are related to their positions (r) in the reference state by the equations

$$x(r) = r + u(r) , \qquad (1)$$

where $u(r)$ is the vector displacement of the lattice point r. Any field over the lattice of the solution can then be written as a function of the reference position, r. In particular, the defect concentration is described by the function $c(r)$, which has the value one if there is a defect at the lattice site whose reference position is r, and is zero otherwise. The average of $c(r)$ over the lattice is $\bar{c}$. The concentration field may be expressed by the sum

$$c(r) = \bar{c} + \Delta c(r), \qquad (2)$$

where the concentration fluctuation at r, $\Delta c(r) = c(r) - \bar{c}$, is either $(1-\bar{c})$ or $(-\bar{c})$.

The Elastic Defect. The magnitude of the elastic defect is specified by giving the 'transformation strain', $s^{\bullet}$, that relates its effective size and shape to that of an equivalent element of the host lattice. A simple definition was suggested by Eshelby [3], and may be phrased as follows. Let a relaxed crystal of the solvent be transmuted into a relaxed crystal of the solute (as illustrated in Figure 2). The transmutation will usually change the overall crystal size and shape. The distortion can be expressed as a linear strain that defines the 'transformation strain', $s^{\bullet}$.

While its definition is straightforward, the evaluation of the transformation strain can be subtle. The relaxed (reference) state of the solute is not unambiguously defined, and may be a metastable structure of the pure solute or even an element of an ordered compound. Unless the reference shape of the solute differs from that of the solvent by a simple dilatation there is a further complication, since the transformation strain depends on the precise crystallographic orientation between the solute and solvent lattices. In that case there are several distinct variants of the solute that correspond to symmetrically equivalent choices of the orientation relation.

The preferred state of the solute is that which minimizes the total free energy (chemical plus elastic) of the solid solution. The preferred reference state can be identified by comparing the various ways of introducing the defects into the solution through the Eshelby cycle described below. It defines the transformation strain to within a choice of variant. The transformation strain can also be measured experimentally, provided that the types and variants of the solutes are known. As we shall see, the transformation

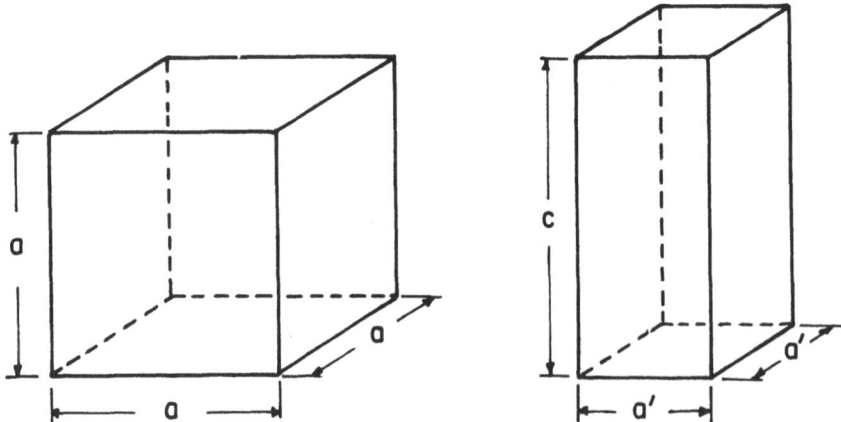

Figure 2. A tetragonal distortion of a cube.

strain is simply related to the average strain of the reference lattice. It can be found from the overall deformation of the reference crystal when the solution is made, or, in the case of a homogeneous solution, from the composition dependence of the lattice parameters.

<u>The Strong Harmonic Approximation</u>. The defect solid solution is derived from the pure solvent by introducing the defect distribution $c(r)$ and the lattice displacement field $u(r)$. The composition and displacement are functionally independent, since the instantaneous value of the displacement at a lattice site is determined by external forces and lattice vibrations as well as by the composition field. It follows that the free energy of solution is a functional of the functions $c(r)$ and $u(r)$. We shall assume that its values are adequately approximated by the harmonic series:

$$\Delta G[u(r),c(r)] = \sum_r G^c c(r) + (1/2)\sum_{r,r'} G^{cc}(r-r')c(r)c(r')$$
$$+ \sum_{r,r'} G_i^{uc}(r-r')c(r)u_i(r') \qquad (3)$$
$$+ (1/2)\sum_{r,r'} G_{ij}^{uu}(r-r')u_i(r)u_j(r')$$

where the sums are taken over the reference lattice. The coefficients are independent of composition and displacement. The absence of a term linear in the displacement, u, follows from the mechanical equilibrium of the reference lattice.

The second order tensor coefficient that appears in the last term of the harmonic series (G^{uu}) is the dynamical matrix of the reference lattice, and is usually given the symbol Λ. It is assumed to be independent of composition and sufficient to describe the response to lattice vibrations. These assumptions have three important consequences:

(1) In general, the instantaneous displacement, u, is the sum of a static displacement that shifts the equilibrium lattice positions and a dynamic displacement that is due to vibrations about the equilibrium positions. But since the dynamical matrix is constant the vibrational free energy of the solution is equal to that of the pure solvent. It follows that $u(r)$ need only represent the static displacement.

(2) The constant value of the dynamical matrix implies that the elastic constants of the solution are independent of composition and equal to those of the reference lattice.

(3) A matrix of macroscopic elastic constants can be derived from the dynamical matrix by probing the response of the lattice to homogeneous strains. The relevant equations are given below. But this procedure will not generate the full set of elastic constants.

In the important case of a primitive lattice with cubic symmetry the dynamical matrix yields only two independent elastic constants, while there are, in fact, three. This problem is inherent to the harmonic approximation (13), and arises from the failure of the model to include the many-body, anharmonic and thermal effects that influence the elastic rigidity of real crystals. The shortcomings of the model can be overcome in part by using the true elastic constants of the reference crystal in place of the computed constants wherever they appear.

Macroscopic Homogeneity. The static displacement, $u(r)$, is divisible into a uniform displacement that is due to the homogeneous strain, ε, of the reference lattice and a local displacement, $v(r)$, that varies from point to point:

$$u_i(r) = \varepsilon_{ij}r_j + v_i(r) .$$
(4)

The field $v(r)$ depends on precisely where the external boundary of the crystal is placed, with the result that the internal elastic field is sensitive to the overall shape of the solid. If, however, the solution is macroscopically homogeneous, in the sense that every part of it is like every other when viewed on a sufficiently coarse scale, then $v(r)$ is an oscillatory function whose mean is constant on any particular surface. The local displacement, $v(r)$, may therefore be assumed constant on the boundary, neglecting only terms that contribute to the surface energy and the strain in the surface layer.

Evaluation of the Free Energy: The Modified Eshelby Cycle

The final state of the solution can, hypothetically, be constructed by performing a series of elementary steps along the lines suggested by Eshelby (3). This procedure naturally yields an expression for the free energy that is the sum of terms that are individually well defined and permit a consistent separation between the 'chemical' and 'elastic' contributions.

In the modified Eshelby cycle an arbitrary distribution of solute defects is introduced into the reference lattice by performing the series of six steps that is diagrammed in Figure 3:

(1) Given a pure, stress-free solvent crystal of N lattice sites, identify an internal cluster of N^o atoms, where N^o (= $N\bar{c}$) is the number of solute defects in the solution (if the defects are interstitial, the cluster contains N^o interstitial sites of the proper type). Let this cluster be cut out of the parent lattice. The only change in the free energy of the assembly is a surface effect that will be neglected.

(2) Let solute atoms be substituted for each of the N^0 solvent atoms in the separated cluster (or placed in each of the N^0 interstitial sites) and let the transformed cluster relax into its stress-free reference state. The transmutation causes the free energy change

$$\Delta G_2 = N^0 \Delta\mu^c = N\bar{c}\Delta\mu^c , \tag{5}$$

where $\Delta\mu^c$ is the chemical energy difference per atom between the solvent and the solute in their reference states. The stress-free relaxation of the cluster deforms it with respect to the solvent lattice by the transformation strain, ε^*, but adds no energy since no mechanical work is done.

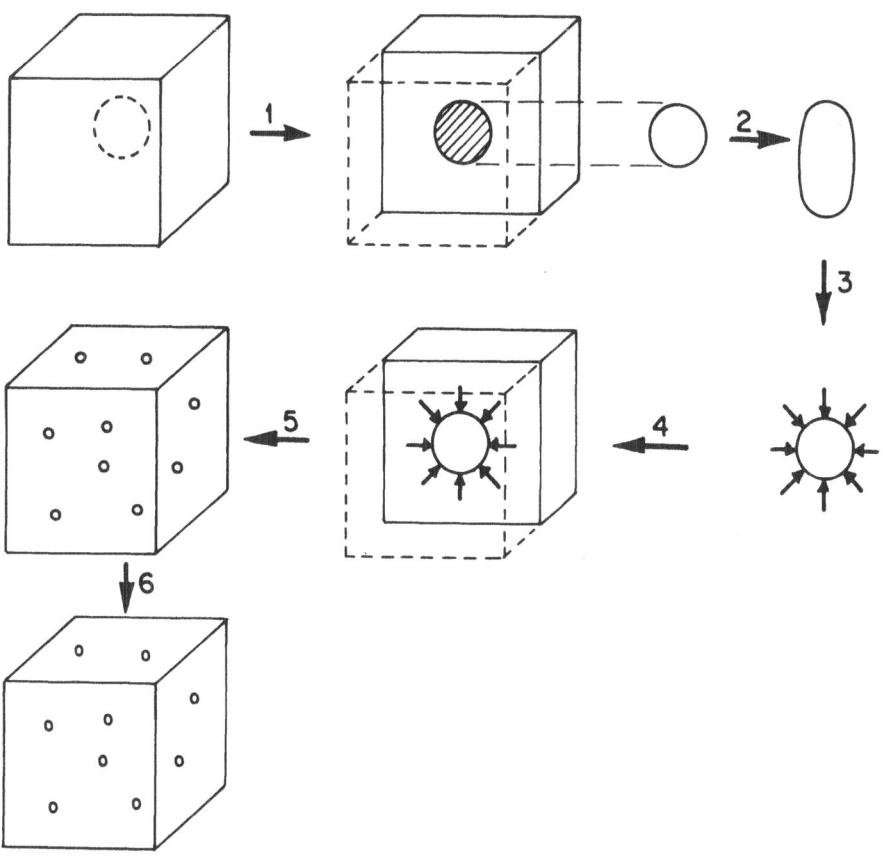

XBL838-6146

Figure 3. The modified Eshelby cycle.

(3) Let a set of surface tractions be applied to the cluster to restore it to the shape it had before the chemical change. The required strain is $\varepsilon = -\varepsilon^{0}$. Since the cluster is macroscopic its deformation is governed by the usual formulae of continuum elasticity. Neglecting the work done on the loading mechanism, which is recovered in a later step, the net free energy change is equal to the mechanical work done on the cluster:

$$\Delta G_3 = N v \bar{c} [(1/2)\lambda_{ijkl}\varepsilon^0_{ij}\varepsilon^0_{kl}]$$
$$= (V/2)\bar{c}\sigma^0_{ij}\varepsilon^0_{ij} \;, \tag{6}$$

where v is the volume per lattice site and V is the volume of the reference crystal, λ is the fourth order tensor of elastic constants, and σ^0 is the 'transformation stress':

$$\sigma^0_{ij} = \lambda_{ijkl}\varepsilon^0_{kl} \;. \tag{7}$$

At the end of this step the cluster is strained with respect to its own relaxed state, but is unstrained with respect to the reference lattice.

(4) Let the cluster be reintroduced into the solvent crystal. Since the cluster just fits into the hole from which it was drawn there is no free energy change.

(5) Let the cluster be broken up and dispersed through the host lattice to create the solid solution, while maintaining the size and shape of the atoms or elementary volumes of the solute. The associated free energy change defines the chemical part of the free energy of mixing. In the harmonic approximation the free energy of mixing is the sum of two terms: a free energy per defect, $\Delta\mu^s$, that results from changing the phase surrounding the solute (if the defect were an element of new phase with bulk properties this term would be the chemical contribution to its surface free energy), and a configurational energy that is due to the chemical interaction between defects. The total free energy change is

$$\Delta G_5 = N\bar{c}\Delta\mu^s + (1/2)\sum_{r,r'} V^{ch}(r-r')c(r)c(r') \;, \tag{8}$$

where $V^{ch}(r-r')$ is the binary chemical interaction potential.

(6) Lastly, let the distributed defects relax, introducing static displacements into the lattice. During the relaxation each defect acts as a center of force that tends to displace the lattice around it. The deformation is opposed by the elastic resistance of the lattice. The associated energy is

$$\Delta G_6 = - \sum_{r,r'} F_i(r-r')u_i(r)c(r')$$
$$+ (1/2)\sum_{r,r'} A_{ij}(r-r')u_i(r)u_j(r') \,, \tag{9}$$

where the symmetric second-order tensor, A, is the Born–von Karman tensor (dynamical matrix) of the host lattice, and the vector, F, is the 'Kanzaki force' that relates the lattice distortion at position r to the defect concentration at position r'. The free energy change, ΔG_6, represents the elastic relaxation of the solution, and is less than or equal to zero. Its net effect is to remove part of the elastic energy that was introduced when the solute was deformed to fit into the host lattice.

The sum of the free energies added in the six steps of the modified Eshelby cycle is the total free energy change on introducing the specific defect distribution $c(r)$ and static displacement $u(r)$ into a pure, stress-free solvent lattice. The free energy is divisible into chemical and elastic contributions, and can be written

$$\Delta G = \Delta G_{chem} + \Delta \emptyset_{el}, \tag{10}$$

where

$$\Delta G_{chem} = \Delta G_2 + \Delta G_5$$
$$= N\bar{c}\Delta\mu^o + (1/2)\sum_{r,r'} V^{ch}(r-r')c(r)c(r') \,, \tag{11}$$

with $\Delta\mu^o = \Delta\mu^c + \Delta\mu^s$, and

$$\Delta\emptyset_{el} = \Delta G_3 + \Delta G_6$$
$$= (V/2)\bar{c}\sigma_{ij}^o \varepsilon_{ij}^o - \sum_{r,r'} F_i(r-r')c(r)u_i(r') \tag{12}$$
$$+ (1/2)\sum_{r,r'} A_{ij}(r-r')u_i(r)u_j(r') \,.$$

Equation (10) has the form required by the strong harmonic approximation. The coefficients are independent of composition. Their symmetry is that of the reference crystal.

The Elastic Strain and Elastic Energy

The next step in the development of the theory is the simplification of the equation (12) that governs the elastic energy. Since the static displacement of the crystal is induced by the solute defects, its value is determined by the concentration field through the conditions of mechanical equilibrium. It follows that the static displacement can be eliminated from equation (12). To accomplish this, the macroscopic strain, ε, and the mean composition, $\bar{c}$, are separated from the internal displacement, $v(r)$, and the composition variation, $\Delta c(\mathbf{r})$. The macroscopic strain is found as a function of $\bar{c}$ from the condition that the solution obey the laws of linear elasticity when it is subjected to external stress. The internal displacement is evaluated as a function of $\Delta c(\mathbf{r})$ from the condition that the elastic energy be stable with respect to local displacements. The elastic energy can then be expressed as the sum of contributions from the mean concentration, $\bar{c}$, and the concentration deviation, $\Delta c(\mathbf{r})$. Finally, the mean field approximation is introduced to find the elastic contribution to the free energy of a solid solution of mobile defects.

To separate the macroscopic and microscopic strains equations (2) and (4) are substituted into equation (12). The result is

$$
\begin{aligned}
\Delta\emptyset_{el} = &\ (V/2)\bar{c}\sigma^0_{ij}\varepsilon^0{}_{ij} \\
&- N\bar{c}L^1_{ij}\varepsilon_{ij} + (N/2)L^2_{ijlm}\varepsilon_{ij}\varepsilon_{lm} \\
&- \sum_{\mathbf{r},\mathbf{r}'} F_i(\mathbf{r}-\mathbf{r}')v_i(\mathbf{r})\Delta c(\mathbf{r}') \\
&+ (1/2)\sum_{\mathbf{r},\mathbf{r}'} A_{ij}(\mathbf{r}-\mathbf{r}')v_i(\mathbf{r})v_j(\mathbf{r}') \ ,
\end{aligned}
\tag{13}
$$

where

$$
L^1_{ij} = (1/2)\sum_{\mathbf{r}} [F_i(\mathbf{r})r_j + F_j(\mathbf{r})r_i] \ ,
\tag{14}
$$

$$
\begin{aligned}
L^2_{ijlm} &= (1/N)\sum_{\mathbf{r},\mathbf{r}'} A_{ij}(\mathbf{r}-\mathbf{r}')r_l r_m \\
&= -(1/2)\sum_{\mathbf{r}} A_{ij}(\mathbf{r})r_l r_m \ .
\end{aligned}
\tag{15}
$$

The terms involving the macroscopic strain depend only on the mean composition, $\bar{c}$. Those containing the local displacement involve only the composition deviation, $\Delta c(\mathbf{r})$.

The Macroscopic Strain. The macroscopic strain is evaluated by requiring that the solution obey Hooke's Law,

$$
\sigma_{ij} = (1/V)[\partial(\Delta\emptyset_{el})/\partial\varepsilon_{ij}] = \lambda_{ijkl}\varepsilon^{el}_{kl} \ .
\tag{16}
$$

In this equation σ is the applied elastic stress, and ε^{el} is the elastic strain measured with respect to the relaxed state of the solution,

$$
\varepsilon^{el} = \varepsilon - \bar{\varepsilon} \ ,
\tag{17}
$$

where ε is the total strain with respect to the reference lattice and $\bar{\varepsilon}$ is the stress-free strain of the solution.

If equation (13) is differentiated with respect to ε_{ij} the result is

$$\sigma_{ij} = -(1/v)L^1_{ij}\bar{c} + (1/v)L^2_{ijlm}\varepsilon_{lm} \cdot \tag{18}$$

This relation is equivalent to equation (16) if we make the identification (recalling the discussion following equation (3))

$$(1/v)L^2_{ijlm} = \lambda_{ijlm} \cdot \tag{19}$$

and define the stress-free strain by the relation

$$L^2_{ijlm}\bar{\varepsilon}_{lm} = L^1_{ij}\bar{c} \cdot \tag{20}$$

Since $\bar{\varepsilon} = \varepsilon^\bullet$ when $\bar{c} = 1$, equation (20) has the solution

$$\bar{\varepsilon} = \bar{c}\varepsilon^\bullet \cdot \tag{21}$$

It follows that the defect solid solution obeys a tensor form of Vegard's Law in which the constant of proportionality is just the transformation strain, $\varepsilon^\bullet$. Hence the effective transformation strain is measured by the concentration dependence of the lattice parameters.

Equations (19-21) yield the result,

$$L^1_{ij} = v\sigma^0_{ij} \cdot \tag{22}$$

which completes the evaluation of the macroscopic terms in equation (13). The elastic energy of the solution in its stress-free state can now be written

$$\Delta\emptyset_{el} = (V/2)[\sigma^0_{ij}\varepsilon^0_{ij}]\bar{c}(1-\bar{c}) + \text{terms in } \mathbf{v}(\mathbf{r}) \cdot \tag{23}$$

The Internal Strain. The equilibrium values of the local displacement, $\mathbf{v}(\mathbf{r})$, must provide a minimum of the elastic energy. The condition

$$\delta\Delta\emptyset_{el}/\delta v_i(\mathbf{r}) = 0 \tag{24}$$

yields the linear equations

$$\sum_{\mathbf{r}'} A_{ij}(\mathbf{r}-\mathbf{r}')v_j(\mathbf{r}') = \sum_{\mathbf{r}'} F_i(\mathbf{r}-\mathbf{r}')\Delta c(\mathbf{r}') \cdot \tag{25}$$

Since, by the assumption of macroscopic homogeneity, $\mathbf{v}(\mathbf{r})$ is constant on the boundary (or on the boundary of a periodic subvolume), this system of equations can be solved by taking the Fourier transform of both sides, where the Fourier transform of a function, $f(\mathbf{r})$ is

$$f(\mathbf{k}) = \sum_{\mathbf{r}} f(\mathbf{r})\exp[-i\mathbf{k}\cdot\mathbf{r}] \tag{26}$$

Letting the (Green's) tensor, $\mathbf{G}(\mathbf{k})$, be the tensor inverse to the

338

Fourier transform of the dynamical matrix,

$$G(k) = A(k)^{-1} , \qquad (27)$$

the internal displacement is given by

$$v_i(k) = G_{ij}(k)F_j(k)\Delta c(k) . \qquad (28)$$

It follows that the Fourier components of the internal strain are

$$\varepsilon'_{ij}(k) = (1/2)[v_{i,j} + v_{j,i}]$$
$$= (1/2)[k_j G_{im}(k) + k_i G_{jm}(k)]F_m(k)\Delta c(k) . \qquad (29)$$

The total elastic strain energy of the stress-free solution can now be written

$$\Delta\emptyset_{el} = (V/2)[\sigma^0_{ij}\varepsilon^0_{ij}]\bar{c}(1-\bar{c})$$
$$- (1/2N)\sum_k [F_i(k)G_{ij}(k)F_j^*(k)] |\Delta c(k)|^2 , \qquad (30)$$

where $F^*(k)$ is the complex conjugate of $F(k)$ and the summation is taken over the N wave vectors in the first Brillouin zone of the crystal.

Separation of the Configuration-Dependent Term. Equation (30) shows that only part of the elastic energy is sensitive to the configuration of the solute defects; the remainder depends only on the mean defect concentration. But equation (30) does not completely separate the two kinds of elastic energy. Part of the configuration-independent contribution is embedded in the second term on the right-hand side. The identity

$$(1/N)\sum_k |\Delta c(k)|^2 = N\bar{c}(1-\bar{c}) \qquad (31)$$

has the consequence that the average value of the coefficient of $|\Delta c(k)|^2$ adds configuration-independent terms to the elastic energy. To separate out the average interaction we define the quantity

$$Q = \langle F_i(k)G_{ij}(k)F_j^*(k)\rangle = (1/N)\sum_k [F_i(k)G_{ij}(k)F_j^*(k)] , \qquad (32)$$

and the potentials

$$\emptyset^0 = [v\sigma^0_{ij}\varepsilon^0_{ij} - Q] , \qquad (33)$$

$$\emptyset(k) = [Q - F_i(k)G_{ij}(k)F_j^*(k)] , \qquad (34)$$

with the convention that $\emptyset(0) = 0$. The elastic energy can then be written

$$\Delta \emptyset_{e1} = (N/2)\emptyset^\circ \bar{c}(1-\bar{c}) + (1/2N)\sum_{k} \emptyset(k)|\Delta c(k)|^2 . \qquad (35)$$

The first term on the right hand side is the part of the elastic energy that is determined by the average defect concentration. The second term depends only on the configuration of the defects.

The elastic energy as a lattice sum. To express the elastic energy as a sum over the reference lattice in real space we take the reverse Fourier transform of the second term on the right in equation (35). The result is

$$\Delta \emptyset_{e1} = (N/2)\emptyset^\circ \bar{c}(1-\bar{c}) + (1/2)\sum_{r,r'} \emptyset(r-r')\Delta c(r)\Delta c(r') , \qquad (36)$$

where $\emptyset(r-r')$ is the reverse Fourier transform of $\emptyset(k)$:

$$\emptyset(r) = (1/N)\sum_{k}\emptyset(k)\exp[ik{\cdot}r] . \qquad (37)$$

This form of the equation preserves the separation between the configuration-dependent and the configuration-independent terms. The equation can also be written as the conventional lattice sum:

$$\Delta \emptyset_{e1} = (N/2)\sum_{r} \emptyset^\circ c(r)$$
$$+ (1/2)\sum_{r,r'} \emptyset^E(r-r')c(r)c(r') , \qquad (38)$$

where

$$\emptyset^E(r-r') = \emptyset(r-r') + \emptyset/N . \qquad (39)$$

The mean field approximation. Equation (35) or (36) gives the elastic contribution to the free energy of a solid solution that has a specific distribution of elastic defects. If the defects are mobile their configuration is not specified, and the elastic energy must be averaged over the appropriate ensemble to compute the free energy of the crystal. In the mean field approximation this average is made by replacing the discrete variables, $c(r)$, by their ensemble (or time) averages, $n(r)$, which may take any value between zero and one. The resulting equation for the elastic energy is

$$\Delta \emptyset_{e1} = (N/2)\emptyset^\circ \bar{c}(1-\bar{c}) + (1/2N)\sum_{k} \emptyset(k)|n(k)|^2 . \qquad (40)$$

In the particular case of a random solution, $n(k) = 0$ when $k \neq 0$, and only the configuration-independent term contributes to the energy.

The total free energy in the mean field approximation is the sum of the elastic and chemical free energies of the solution (eq. 11), plus a contribution from the configurational entropy of the ensemble. The result can be written

$$\Delta G = N\bar{c}V^0 + (1/2)\sum_{\mathbf{r},\mathbf{r}'} V(\mathbf{r}-\mathbf{r}')n(\mathbf{r})n(\mathbf{r}')$$

$$+ kT\sum_{\mathbf{r}}\{n(\mathbf{r})\ln[n(\mathbf{r})] + [1-n(\mathbf{r})]\ln[1-n(\mathbf{r})]\} \qquad (41)$$

where V^0 and $V(\mathbf{r}-\mathbf{r}')$ are found from the sum of equations (11) and (38).

Evaluation of the Coefficients Governing the Elastic Energy.

In order to calculate the elastic energy of a distribution of defects one needs solvable expressions for the coefficients appearing in equation (30) or (35). We have already discussed the determination of the transformation strain. The remaining coefficients are the Green's tensor, $G(\mathbf{k})$, and the Kanzaki force, F. As we shall show below, the Green's tensor can be found from the phonon dispersion relation for the reference crystal and the Kanzaki force can be approximated by considering interactions with first and second nearest neighbors. Alternatively, the whole coefficient in equation (30) can be evaluated in the long-wavelength limit, in which case it is given in terms of the transformation stress and the matrix elastic constants.

The Green's tensor. If the phonon dispersion relations for the reference crystal are known then the dynamical matrix, $A(\mathbf{k})$, can be diagonalized. Its principal axes are the three polarization vectors, $e^\alpha(\mathbf{k})$ ($\alpha = 1,2,3$), of the acoustic phonons of wave vector $\mathbf{k}$, and its diagonal elements are $[m\omega_\alpha(\mathbf{k})]^2$, where $\omega_\alpha(\mathbf{k})$ is the frequency of the αth acoustic mode of wave vector $\mathbf{k}$ and m is the mass of the host atom. It follows that the Green's tensor inverse to $A(\mathbf{k})$ can also be evaluated from the phonon dispersion relation. Its dyadic form is

$$G(\mathbf{k}) = \sum_{\mathbf{k}}[m\omega_\alpha(\mathbf{k})^2]^{-1}e^\alpha e^\alpha . \qquad (42)$$

The Kanzaki force. Equations (14) and (22) show that the Kanzaki force is related to the transformation stress by the equation

$$\sigma^0_{ij} = (1/2v)\sum_{\mathbf{r}}[F_i(\mathbf{r})r_j + F_j(\mathbf{r})r_i] . \qquad (43)$$

Since there are six independent components of $\sigma^\bullet$, the equation is solvable for six independent values of $F_i(\mathbf{r})$. It follows that the force $F(\mathbf{r})$ can be evaluated from equation (43) if it is assumed that the only important interactions are with the first and second nearest neighbors. Krivoglaz [14] and Khatchaturyan and Shatalov [15] have used this approach to write specific expressions for the Kanzaki force that governs the interactions of either octahedral or tetrahedral defects in BCC lattices. Others, particularly Kanzaki [16] and Cook [17], have described more general techniques for interpreting and evaluating the Kanzaki force.

The long wavelength approximation. If the defects are large on the scale of the crystal lattice, if the wavelength of the dominant composition fluctuations is long, or if one is prepared to accept approximations in the spirit of the Debye approximation in lattice dynamics, then the potential governing the elastic interaction can be evaluated in terms of the macroscopic elastic constants. In the long-wavelength limit ($|\mathbf{k}| \sim 0$),

$$F_j(\mathbf{k}) \sim -ik_m v\sigma^0_{jm} , \tag{44}$$

and

$$G(\mathbf{k}) \sim [v|\mathbf{k}|^2]^{-1}\Omega(\mathbf{e}) , \tag{45}$$

where $\mathbf{e}$ is a unit vector in the direction of $\mathbf{k}$ and $\Omega(\mathbf{e})$ is a second order tensor whose elements depend on the matrix elastic constants through the relation

$$\Omega^{-1}_{ij}(\mathbf{e}) = \lambda_{ik1j}e_k e_1 . \tag{46}$$

The potential $[F_i(\mathbf{k})G_{ij}(\mathbf{k})F^*_j(\mathbf{k})]$ that appears in equation (29) can then be replaced by the scalar

$$B'(\mathbf{e}) = e_i\sigma^0_{ij}\Omega_{jk}(\mathbf{e})\sigma^0_{km}e_m , \tag{47}$$

and the coefficient $\emptyset(\mathbf{k})$ becomes

$$\emptyset(\mathbf{k}) = \langle B(\mathbf{e})\rangle - B'(\mathbf{e}) . \tag{48}$$

Equations (35) and (46) have the consequence that in the long-wavelength limit the elastic energy of a composition wave, $\Delta C(\mathbf{k})$, is independent of its wavelength; the energy depends only on the direction of the wave vector, $\mathbf{k}$.

COMPLEX SYSTEMS

For simplicity, the derivation of the elastic contribution to the free energy has been restricted to the case of a single type of elastic point defect in a stress-free solvent lattice. This restriction is unnecessary. It is relatively simple to reformulate the equations for solutions that are externally stressed, solutions containing several defect types (or several variants of a single defect type), solutions of defects that have finite size and shape, and solutions of macroscopic inclusions. The relevant equations are given below. Their derivations are only briefly sketched; there are detailed presentations elsewhere (9-12).

Externally Stressed Solutions

The change in the Gibbs free energy on application of an external stress is the sum of the mechanical work done on the solution by the loading mechanism and the virtual work done on the loading mechanism by the solution. If the free energy is measured with respect to the relaxed state of the pure solvent crystal the total strain is $\varepsilon_{ij} = \varepsilon_{ij}^{el} + \bar{\varepsilon}_{ij}$ (eq. 17). The additional free energy is

$$\Delta G_\sigma = V\sigma_{ij}(\varepsilon_{ij}^{el} + \bar{\varepsilon}_{ij}) = V\sigma_{ij}\varepsilon_{ij} \ . \tag{49}$$

One-half of this energy adds to the internal internal energy of the solution, while one-half is stored in the potential energy of the loading mechanism. Note that the external stress does not interact with the internal strain, ε_{ij}' (eq. 29).

There are two situations in which the external stress has an important influence on the structure of the solution: when the defects are particles of new phase and when the defect has several variants that are mutually interconvertible. In the former case the external stress influences the conditions under which the new phase will appear, since it adds to or subtracts from the free energy of formation. In the latter case the external stress interacts differently with the distinct variants of the defect, breaking their degeneracy and promoting the formation of the more favored types.

Multicomponent Solutions (Multiple Defect Types)

Let there be m distinct defects in the solution, designated by the index α ($\alpha = 1,...,m$). The distinct defects may represent different solute species or distinguishable variants of a given solute. Each is characterized by its transformation strain, $\varepsilon^{\bullet\alpha}$, and is the source of a Kanzaki force, $\mathbf{F}^\alpha$. The free energy of the assembly can be found [10,11] by performing a modified Eshelby cycle in which m different clusters are removed from the reference lattice, transformed into reference clusters of each of the m defects, deformed, reinserted into the lattice, distributed, and permitted to relax. A treatment that parallels the one above shows that the homogeneous lattice strain is

$$\bar{\varepsilon}_{ij} = \sum_\alpha [\bar{c}_\alpha \varepsilon_{ij}^{\varrho\alpha}] \ , \tag{50}$$

while the internal strain is

$$\varepsilon_{ij}'(\mathbf{k}) = (1/2)[k_j G_{ik}(\mathbf{k}) + k_i G_{jk}(\mathbf{k})]\sum_\alpha [F_k^\alpha(\mathbf{k})\Delta c_\alpha(\mathbf{k})] \ . \tag{51}$$

The elastic energy is given by the equation

$$\Delta\emptyset_{e1} = (N/2)\sum_{\alpha\beta}[\emptyset^o_{\alpha\beta}\bar{c}_\alpha(\delta_{\alpha\beta}-\bar{c}_\beta)]$$
$$+ (1/2N)\sum_{\alpha\beta}[\emptyset^{\alpha\beta}(\mathbf{k})\Delta c_\alpha(\mathbf{k})\Delta c_\beta{}^*(\mathbf{k})] \ , \tag{52}$$

where $\delta_{\alpha\beta}$ is the Kronecker delta,

$$\emptyset^o_{\alpha\beta} = [v\lambda_{ijkl}\varepsilon^\alpha_{ij}\varepsilon^\beta_{kl} - Q_{\alpha\beta}] \ , \tag{53}$$

$$\emptyset^{\alpha\beta}(\mathbf{k}) = [Q_{\alpha\beta} - F^\alpha_i(\mathbf{k})G_{ij}(\mathbf{k})F^\beta_j{}^*(\mathbf{k})] \ , \tag{54}$$

$$Q_{\alpha\beta} = \langle F^\alpha_i(\mathbf{k})G_{ij}(\mathbf{k})F^\beta_j{}^*(\mathbf{k})\rangle \ , \tag{55}$$

and $\emptyset^{\alpha\beta}(\mathbf{k})$ is zero when $|\mathbf{k}| = 0$. The elastic energy of a multicomponent solution in the mean field approximation is found from equation (52) by replacing the compositions, $c_\alpha(\mathbf{r})$, by their ensemble averages, $n_\alpha(\mathbf{r})$.

Defects of Finite Size

The size and shape of an elastic defect influence the internal strain and the elastic energy. While the finite size of a solute atom can often be ignored, size and shape are almost always important when the defect is an inclusion, precipitate, or elementary volume of a new structure.

The size and shape of the defect is incorporated into the theory by redefining the concentration field and also, if necessary, redefining the reference lattice [18,19]. If the particle is centered at a lattice position denoted by $\mathbf{R}$, the region of space it occupies is given by the function $\xi(\mathbf{r}-\mathbf{R})$, which is equal to 1 when the vector $\mathbf{r}-\mathbf{R}$ is within the particle and is zero otherwise. The distribution of defect volume over the elements of space, $\mathbf{r}$, in the reference configuration is then specified by the function

$$\theta(\mathbf{r}) = \sum_{\mathbf{R}}[c(\mathbf{R})\xi(\mathbf{r}-\mathbf{R})] \ . \tag{56}$$

If the defect is larger than a lattice site it is often useful to let $\mathbf{R}$ designate the points of a superlattice whose cells correspond in size and shape to the elementary volume of the defect.

A solution of identical defects of finite size has the internal strain

$$\varepsilon'_{ij}(\mathbf{k}) = (1/2)[k_j G_{ik}(\mathbf{k}) + k_i G_{jk}(\mathbf{k})]F_k(\mathbf{k})\xi(\mathbf{k})\Delta c(\mathbf{k}) \ , \tag{57}$$

where the shape factor, $\xi(\mathbf{k})$, is the Fourier integral,

$$\xi(\mathbf{k}) = \int_V [\xi(\mathbf{r})\exp(-i\mathbf{k}\cdot\mathbf{r})] \ d^3r \ , \tag{58}$$

while the remaining terms are Fourier lattice sums (eq. 26). The elastic contribution to the free energy of the solution is given by equation (35), but with the new definitions

$$\phi(k) = [Q - F_i(k)G_{ij}(k)F_j^*(k)|\xi(k)|^2] \ , \tag{59}$$

$$Q = \langle F_i(k)G_{ij}(k)F_j^*(k)|\xi(k)|^2\rangle \ . \tag{60}$$

The formal expression for ϕ^0 remains the same. The generalization of this result to multicomponent solutions is straightforward [18,19] and can be inferred from equations (51)-(55).

Distributions of Macroscopic Inclusions

When the solutes are macroscopic inclusions the discrete reference crystal can be replaced by a continuum, and the elastic strain evaluated in the continuum limit. In the continuum limit the solute distribution is described by the field $\theta(r)$ (eq. 56), which is a function of the continuous vector variable, r. Whenever r falls on a lattice point, $c(r)=\theta(r)$. Hence $\bar{\theta}$, the volume fraction of the inclusions, is equal to $\bar{c}$. It follows via the long wavelength approximation (eqs. 42-45) that the internal strain is

$$\varepsilon'_{ij}(k) = [e_i\Omega_{jk}(e) + e_j\Omega_{ik}(e)]\sigma^0_{kl}\theta(k) \ , \tag{61}$$

where e is a unit vector in the direction of k. The elastic energy is

$$\Delta\phi_{el} = (V/2)\sigma^0_{ij}\varepsilon^0_{ij}\bar{\theta}(1-\bar{\theta})$$
$$- (1/16\pi^3)\int'[B'(e)|\theta(k)|^2]d^3k \ , \tag{62}$$

where the integral is primed to indicate that its principal value is meant, excluding the singularity at the origin. Alternatively,

$$\Delta\phi_{el} = (1/16\pi^3)\int'[B(e)|\theta(k)|^2]d^3k \ , \tag{63}$$

where

$$B(e) = \sigma^0_{ij}\varepsilon^0_{ij} - B'(e) \ . \tag{64}$$

Note that the kernal of the integral in either equation is the product of two terms that separate the elastic energy and the particle geometry. The function $B(e)$ is independent of the inclusion geometry and is also independent of the magnitude of k. The function $\theta(k)$ is a shape factor that is wholly determined by the size, shape and distribution of the inclusions.

APPLICATIONS

Overview

The linear elastic model has been widely used in the theory of phase transformations, microstructure, and material behavior. Many of the applications that are relevant to the topic of this meeting are discussed in recent reviews by Khatchaturyan (9,11) and de Fontaine (12), to which the reader is referred. There is, however, a simple organization to the common applications of the theory that is worth recognizing. Most of them address one of two generic problems with one of three theoretical techniques.

The first set of applications is to solids that contain quasi-macroscopic inclusions. The theory can be used to compute the strain and elastic energy introduced by isolated or distributed inclusions of one or many types. When the elastic moduli are uniform the generic problem has a straightforward solution that was provided by Khatchaturyan and Shatalov (10,11,20). Some progress has also been made in finding particular solutions that account for the difference between the moduli of the phases present (4,10,11,21,22). The computation of the elastic strain and energy has been used to explain the shapes, habit planes, internal structures, and apparent symmetries of precipitate phases (10,11,23-25), to interpret the spatially ordered configurations of new phase particles that sometimes arise during decomposition (19,26), and to clarify the influence of external stress on the nature and morphology of structural phase transformations (3,18,27,28).

The second set of applications concerns the behavior of solid solutions. The elastic theory has been used to supplement chemical models to explain and predict the structures of equilibrium phases and the conditions that lead to ordering or phase separation. The fundamental successes of the theory include interpreting the difference between the 'coherent' and 'incoherent' phase diagrams of multicomponent solids (2), explaining the preferred directions of spinodal waves during spinodal decomposition (1,10), and predicting the ordering observed in interstitial solid solutions (9,15,29-31).

The common theoretical approaches include the direct computation of the strain and energy of specific defect configurations, the analytic identification of the preferred mode of ordering or decomposition, and the computer simulation of solute reconfigurations and structural phase transformations. The direct computation of the elastic energy of a given defect configuration may proceed from either equation (35) or (36), depending on whether the defect distribution is more simply described in real or reciprocal space. In the former case the computation requires the evaluation of the two-body potential $\emptyset(r-r')$. Once this potential is known any discrete distribution of defects can be analyzed in a computer, and the agglomeration of defects into particles of new phase can be treated by defining elementary volumes of the new phase that fit together to fill space, as in equation (56). For that reason the real-space formulation is well suited for the simulation of ordering, decomposition and structural phase transformations in solids. The two-body potential can be used to identify the preferred motion of a diffusing species or the most favorable location for an elementary particle of a new phase, the indicated change in the defect configuration made in the computer, and the process iterated to model the path of a change of state. This approach has been used, for example, to simulate martensitic transformations (18,27) and solute reconfigurations (19,26) in model systems.

When the solute distribution has a simple representation in reciprocal space, as, for example, when it can be described by one or a few 'static concentration waves' with given wave vectors, k, then the energy is easily computed from equation (35). A simplified form of the equation forms the basis of Cahn's (1) original presentation of the theory of spinodal decomposition in solids. The same equation can be used to identify the preferred modes of ordering or decomposition in a solid solution whose interatomic potential includes a strong elastic contribution. This latter point has been particularly emphasized by Khatchaturyan (9,11), and is discussed below because of its importance to the subject of this meeting.

Ordering and Decomposition in Solid Solutions

There is a well-developed theory of ordering and decomposition reactions in solutions that obey both the harmonic and mean field approximations. The theory is based on the method of 'static concentration waves', and has been recently reviewed by Khatchaturyan [9,11] and by de Fontaine (12). We will sketch only the outline of the theory here to indicate some of the ways that the elastic interaction can promote behavior that falls within the subject of this meeting. It should be clear from the outset that the theory is in no sense peculiar to the elastic model of the solution. The method of static concentration waves is applicable to any solution whose elementary particles interact in pairs, irrespective of the physical source of the interaction.

The Stability of a Random Solid Solution.
The Gibbs free energy of a solid solution in the harmonic and mean field approximations is given by equation (41). This equation can be used to determine the equilibrium occupation probabilities, $n(r)$, by requiring that the set $n(r)$ provide a minimum of ΔG for a given value of the mean composition, $\bar{c}$. The technique of Lagrange multipliers yields the result

$$n(r) = [\exp\{(kT)^{-1}[\eta(r) - \mu]\} + 1]^{-1} , \tag{65}$$

where

$$\eta(\mathbf{r}) = \sum_{\mathbf{r}'} V(\mathbf{r}-\mathbf{r}')n(\mathbf{r}') \tag{66}$$

and μ is a constant chemical potential that can be found from the normalization condition

$$\sum_{\mathbf{r}} n(\mathbf{r}) = N\bar{c}. \tag{67}$$

Equation (65) is a set of N equations for the N values $n(\mathbf{r})$. Its right hand side depends on $n(\mathbf{r})$ through $\eta(\mathbf{r})$, so the set is difficult to solve directly. In the high temperature limit, however, $n(\mathbf{r})=\bar{c}$, and the solution is random. We may, therefore, gain some insight into the equilibrium states of the solution by beginning from the high temperature limit and identifying the solute redistributions that lead to instabilities in the random solution as the temperature is lowered. Instability results if a small perturbation in the defect distribution, $\delta n(\mathbf{r})$, leads to a decrease in the free energy.

The free energy change associated with the redistribution $\delta n(\mathbf{r})$ is, to second order,

$$\delta(\Delta G) = \sum_{\mathbf{r}}\delta n(\mathbf{r})\{\sum_{\mathbf{r}'}V(\mathbf{r}-\mathbf{r}')\delta n(\mathbf{r}') + kT[\bar{c}(1-\bar{c})]^{-1}\delta n(\mathbf{r})\}, \tag{68}$$

which is equal to zero when

$$\delta n(\mathbf{r}) = -(1/kT)[\bar{c}(1-\bar{c})]\sum_{\mathbf{r}'}V(\mathbf{r}-\mathbf{r}')\delta n(\mathbf{r}'). \tag{69}$$

The Fourier transform of equation (69) is

$$\delta n(\mathbf{k}) = -(1/kT)[\bar{c}(1-\bar{c})]V(\mathbf{k})\delta n(\mathbf{k}), \tag{70}$$

which has the solution

$$T_c(\mathbf{k}) = -(1/k)[\bar{c}(1-\bar{c})]V(\mathbf{k}). \tag{71}$$

The critical temperature, $T_c(\mathbf{k})$, is the temperature at which the random solution becomes unstable with respect to the appearance of a static composition wave of vector $\mathbf{k}$.

If $V(\mathbf{k})$ is defined to have a zero mean value (by the procedure leading to eq. (34), if necessary) then, barring the trivial case in which all $V(\mathbf{k})$ are zero, there are negative values of $V(\mathbf{k})$. These induce instabilities at positive temperatures, $T_c(\mathbf{k})$. The first instability encountered as the solution is cooled is that which corresponds to the minimum value of $V(\mathbf{k})$, $V(\mathbf{k}^*)$. It promotes the spontaneous appearance of a static composition wave with the wave vector $\mathbf{k}^*$.

This result suggests that a random solid solution will inevitably either order or a decompose as the temperature is lowered. If $k^\bullet$ is near the edge of the Brillouin zone then the composition fluctuation is short-range and the solution orders. If $k^\bullet$ is near the origin then the solution decomposes into two phases of different composition.

Constraints on the Instability Waves. As Khatchaturyan (9) has emphasized, there are constraints on the nature of the static composition waves that arise from instabilities in the random solution. The first constraint has its source in the degeneracy of the potential $V(k)$. This potential must have the same value at all points of reciprocal space that are connected by the symmetry operations in the symmetry group of the random solution. The solution is simultaneously unstable with respect to all wave vectors contained in the 'star' of $k^\bullet$, that is, all vectors that are related to $k^\bullet$ by symmetry operations and do not differ from it by a reciprocal lattice vector. (For example, the vector $-k^\bullet$ is a member of the star of $k^\bullet$ unless $2k^\bullet$ is a reciprocal lattice vector.) The instability wave is always a linear combination of waves in the star of $k^\bullet$. The development of the theory in terms of stars of the wave vectors is straightforward, though notationally complex, and is given in ref. [9].

The star of $k^\bullet$ also determines the order of the transformation through the Landau (32) symmetry rules. These may be rephrased (9) to state that a second-order transformation is only possible if no three vectors of the star of $k^\bullet$ can be combined to yield a reciprocal lattice vector. Otherwise the transformation is first-order, and equation (71) determines the instability temperature rather than the equilibrium temperature of the phase transformation.

It should be noted, however, that the symmetry of the potential $V(k)$ in a defect solid solution is the joint symmetry of the oriented reference states of the solvent and defect rather than the symmetry of the solvent lattice alone. The symmetry of the defect enters $V(k)$ through the Kanzaki force, $F(k)$, as can be seen from equation (43) or (44).

A second constraint on the instability wave concerns its ability to specify an ordered state of the solution. A single composition wave introduces a single order parameter (9), and hence necessarily divides the solution into two sublattices. If the wave that is derived from the star of $k^\bullet$ introduces more than two values of the occupation probability, $n(r)$, i.e., if it divides the solution into more than two sublattices, then it cannot completely determine the ordered structure and other, less favored waves must participate. The common ordered structures are discussed from this perspective in ref. (9).

Particular Properties of the Defect Solid Solution

Decomposition reactions. The interaction potential, $V(\mathbf{k})$, is the sum of chemical and elastic interaction terms:

$$V(\mathbf{k}) = V^{chem}(\mathbf{k}) + \emptyset(\mathbf{k}) , \tag{77}$$

where $V^{chem}(\mathbf{k})$ is the Fourier transform of the chemical interaction potential that is defined in equation (8) and $\emptyset(\mathbf{k})$ is the elastic potential. The chemical interaction dominates when the solution is substitutional and the species it contains do not differ greatly in size or shape, and also dominates when the species have a strong electronic interaction, as in the case of an ionic solution. But even when the chemical interaction is responsible for the equilibrium structure, the elastic potential may still influence the kinetics and morphology of the phase transformations that establish the equilibrium structure. Its importance is due to its long range, which has the consequence that the elastic energy is sensitive to the coherence of the lattice, and its anisotropy, which contrasts to the relative isotropy of typical chemical interactions. Both properties of the elastic potential affect the decomposition of solid solutions. The elastic energy determines the difference between the 'coherent' and 'incoherent' phase boundaries in the phase diagrams of solid solutions (2), and usually fixes the habit planes of the precipitates or spinodal waves that are the product of decomposition (1,10).

The physical source and significance of the difference between the coherent and incoherent phase boundaries may be easily seen. The configuration-independent part of the elastic energy (eq. 35),

$$\emptyset^1 = (N/2)\emptyset^0 \bar{c}(1-\bar{c}) , \tag{78}$$

is constant so long as the mean defect concentration does not change and the solution remains coherent, but decreases substantially if the solution decomposes into two disconnected lattices that are respectively very rich and very poor in the solute species. Hence this energy promotes decomposition. But decomposition into disconnected lattices requires the creation of a sharp interface, usually through the accumulation of interfacial dislocations, and is not energetically favored until the second-phase particles have grown to appreciable size. As a consequence most decomposition reactions initiate on a coherent lattice. The energy $\emptyset^1$ remains constant during the reaction, and decomposition cannot begin until the solution is cooled to a temperature below the coherent phase boundary.

Conversely, the energy $\emptyset^1$ inhibits the formation of the solution from disconnected phases. If a solid solution is decomposed and allowed to age until the product phases lose coherency, the solution will not reform until the two-phase mixture is heated to a

The two classes of instabilities: modulated structures. The minima of V(k) that lead to instabilities in the random solution may be usefully divided into two classes: special point minima and accidental minima. The special point minima are those that fall at the Lifshitz (33) special points of the Brillouin zone of the solution. These are points where two or more symmetry operations intersect, forcing an extremum

$$dV(k)/dk = 0 .$$
(72)

The accidental minima fall at other points in the Brillouin zone where equation (72) is satisfied because of the particular physical form of V(k).

The distinction between the two types of minima is particularly relevant to the theory of long-period superlattices and modulated structures. The Lifshitz special points are fixed by the symmetry of the disordered solution and are commensurate with its lattice. It follows that special point minima give rise to relatively simple, stable ordered structures. Accidental minima, on the other hand, may fall anywhere within the Brillouin zone, and may change their location with minor variations in temperature or composition. They need not be commensurate with the parent lattice. As Khatchaturyan has emphasized (9), the ordered phases that arise from accidental minima are likely to be long-period structures that are unstable with respect to the formation of antiphase boundaries and have very narrow stability ranges. Moreover, even a small perturbation in the position of an accidental minimum, caused, for example, by a change in temperature or composition, will usually induce a first-order phase transformation since a small change in n(k) causes a finite change in n(r).

Deviations from stoichiometry: secondary ordering. While the critical temperature, $T_c(k)$, depends on the concentration of the solution, the potential V(k) is independent of composition when the strong harmonic approximation applies. If it is further assumed, for simplicity, that V(k) is independent of temperature, then the hierarchy of the minima of V(k) is fixed by the nature of the solvent and defect. The instability wave that develops first on cooling is independent of the defect concentration. If this wave induces ordering rather than decomposition it will generally lead to a non-stoichiometric ordered phase. If the temperature is lowered further, the solution also becomes unstable with respect to static composition waves with other wave vectors. These may interact with the initial ordered structure to relieve the deviation from stoichiometry, either through secondary ordering or through decomposition into stoichiometric phases. Indeed, either secondary ordering or decomposition must happen eventually and must lead to a stoichiometric phase or phase mixture; the Nernst principle forbids a non-stoichiometric equilibrium phase as T approaches zero.

When the solution obeys both the harmonic and mean-field approximations the equations governing secondary ordering may be easily written down (9). If $n^0(r)$ specifies the composition of the non-stoichiometric phase that exists at high temperature, the perturbed composition is

$$n(r) = n^0(r) + \delta n(r) . \tag{73}$$

If this composition field is introduced into equation (41), it is easily shown that the variation in the free energy remains positive until

$$\delta n(r) = - (kT)^{-1}\{n^0(r)[1-n^0(r)]\}\sum_{r'}V(r-r')\delta n(r') , \tag{74}$$

but is negative at lower temperatures. The solution to the set of equations (74) that yields the highest critical temperature defines the preferred secondary ordering (if the high temperature phase is stoichiometric, the only solution is the trivial result $\delta n(r)=0$.)

The equations (74) are often difficult to solve. A much simpler set results if it is assumed that the secondary ordering involves only one sublattice of the high temperature phase, as will often be the case. Letting R designate the sites of this sublattice, the condition of instability is given by equation (74) with r replaced by R. But since $n(R)$ is constant the Fourier transform of the equation yields the simple result

$$\delta n(K) = - (kT)^{-1}[n^0(1-n^0)]V(K)\delta n(K) , \tag{75}$$

where the capital K indicates the Fourier transform

$$f(K) = \sum_R f(R)\exp[-iK{\cdot}R] . \tag{76}$$

and n^0 is the constant initial concentration on the sublattice. The preferred secondary ordering is determined by the value, K^0, that establishes the minimum of $V(K)$, and occurs at the critical temperature that solves equation (75).

If the secondary ordering predicted by equation (75) does not establish a stoichiometric phase, then the process must be iterated through successive ordering reactions until the solution either settles into a stoichiometric structure or decomposes. From this perspective it is not surprising that non-stoichiometric compounds often form complex structures. Pokrovskii and Khatchaturyan (34) have, for example, recently used the sequential ordering of static concentration waves to predict both the stable phases of the common oxides and the Magneli phases that occur in the non-stoichiometric oxides of titanium and vanadium.

temperature high enough for the entropy of mixing to overcome the
elastic energy increase that is required to reestablish a coherent
lattice, that is, until it is heated to a temperature above the
incoherent phase boundary.

The physical reason that elastic effects determine the shapes
and habits of the product phases lies in the anisotropy of the
elastic interaction. The static concentration waves that accomplish
decomposition have long wavelengths. Their chemical energy depends
on the wavenumber primarily because the imposition of a concentration wave introduces an incipient interface that has an effective
surface tension [1]. Since the surface tension is generally affected more by the sharpness of the interface than by its orientation,
the chemical term depends mainly on the magnitude, $|\mathbf{k}|$, of the wave
vector. The elastic interaction, on the other hand, loses its
dependence on the magnitude of $\mathbf{k}$ in the long wavelength limit (eq.
47), but usually retains a strong dependence on the direction, $\mathbf{e}$, of
$\mathbf{k}$ because of its anisotropy. Even if the chemical interactions
dominate the thermodynamics of the solution, elastic anisotropy may
still determine the orientation of the preferred concentration wave.
The experimental observations confirm this; the orientations of
spinodal waves and the habit planes of initial precipitates are well
predicted by the elastic theory [1,10,11,23-25].

Ordering reactions. There are at least two cases in which the
elastic interaction has an important influence on the equilibrium
structure: solutions of interstitials in metals, where the elastic
interaction between the defects is very strong, and ordering reactions involving a periodic structural distortion that is associated
with a strong minimum in the phonon spectrum of the parent phase.
While it is difficult to predict the phonon spectra, predictive
theories have been formulated for the interstitial solutions [9,11].

It is useful to distinguish the primary ordering of an interstitial solution, which determines the preferred interstitial site,
from the secondary ordering that fixes the precise arrangement of
the interstitial defects over the lattice of preferred sites. Primary ordering usually affects the variant of the defect. In the BCC
metals, for example, the distorted symmetry of the interstitial
sites has the consequence that the defect strain differs from one to
another. In particular, there are three distinct sublattices of
octahedral sites in BCC. They are distinguished by the direction of
the short axis, which lies along one of the three orthogonal axes of
the cube. The three distinct sites lead to three distinct variants
of the solute defect, that are interconvertible through a simple
diffusive jump of the interstitial atom from one sublattice to
another. The relative populations of interstitial defects on these
sublattices can be predicted through the use of equation (52) [9].
The theory has been used to explain the tendency of octahedral
interstitials such as carbon and nitrogen in BCC iron to populate a

single sublattice. The theory was also used to compute the relative sublattice populations and explain anomalies in the measured lattice constants of fresh martensites (30).

The secondary ordering of interstitial solutions usually leads to a specific arrangement of defects over a sublattice of interstitial sites. Note first that this type of interstitial ordering cannot be treated with the simple elastic theory of point defects in the long wavelength limit. In this limit (eq. 47) the interaction depends only on the direction, e, in reciprocal space, and cannot yield a static concentration wave with a finite wavelength. To predict an ordered phase from the elastic theory one must either evaluate the elastic potential from the Kanzaki force and the phonon spectrum of the reference lattice (eqs. 42 and 43) or one must let the defects have finite size and shape and obtain the interaction potential from equation (59). Interestingly, both approaches have been used with some success. Khatchaturyan and Shatalov (15) used the tensor form of Vegard's Law (eq. 21) to find the stress-free strain of a carbon defect in tantalum, and then employed the phonon spectrum of the Ta lattice to predict the ordered interstitial phases of carbon in Ta. Mori, Cheng and Mura (31) used the bulk elastic constants of BCC Fe together with the size and strain of a nitrogen atom in an octahedral interstitial site to predict the structure of the coherent nitride phase, $Fe_{16}N_2$. The success of the latter approach is surprising, since the wavelength of the preferred configuration is determined by the size and shape of the defect, but the simplicity and success of the model suggests that it deserves further exploration.

CONCLUSION

The linear elastic theory of the solid solution is based on strong assumptions, and cannot possibly contain the whole truth. Nonetheless it has been used with considerable success to explain and predict many of the important elastic effects in solution thermodynamics and phase transformations. The theory is reasonably simple to learn and apply, and can be a powerful tool in attacking the behavior of real solutions.

ACKNOWLEDGMENT

This work was supported by the Director, Office of Energy Research, Office of Basic Energy Sciences, Division of Materials Sciences, U.S. Department of Energy under Contract No. DE-AC03-76SF0098.

354

REFERENCES

1. Cahn, J. W. Acta Met. 9 (1961) 795: 10 (1962) 179.
2. Cahn, J. W. in The Mechanism of Phase Transformations in Crystalline Solids (Inst. of Metals, London, 1969), p. 1.
3. Eshelby, J. D. Proc. Roy. Soc. A241 (1957) 376.
4. Eshelby, J. D. Proc. Roy. Soc. A252 (1959) 561 : Prog. in Solid Mechanics 2 (1961) 89.
5. Born M. and K. Huang. Dynamical Theory of Crystal Lattices (Oxford University Press 1954).
6. Khatchaturyan, A. G. Sov. Phys.-Solid State 4 (1963) 2081.
7. Khatchaturyan, A.G.Fiz. Tverd. Tela. 9 (1967) 2595 : Sov. Phys.-Solid State 9 (1968) 2040.
8. Cook, H.E. and D. de Fontaine Acta Met. 17 (1969) 945: 18 (1970 189.
9. Khatchaturyan, A. G. Prog. Materials Science 22 (1978) 1.
10. Morris, Jr., J. W., A. G. Khatchaturyan and S. H. Wen, in Solid-Solid Phase Transformations (H.I. Aaronson, D.E. Laughlin, R.F. Sekerka and C.M. Wayman, eds., The Metallurgical Society-AIME, Warrendale, Pa., 1983) 101.
11. Khatchaturyan, A. G. The Theory of Structural Phase Transformations in Solids, John Wiley, New York 1983).
12. de Fontaine, D. Solid State Physics 34 (1979) 73.
13. Wallace, D. C. Thermodynamics of Crystals (John WileyNew York 1972).
14. Krivoglaz, M. A. The Theory of X-ray and Thermal Neutron Diffraction Scattering from Real Crystals (Plenum Press, New York 1969).
15. Khatchaturyan, A. G. and G. A. Shatalov. Acta Met. 23 (1975) 1089.
16. Kanzaki, H. J. Phys. Chem. Solids 2:24 (1957) 107.
17. Cook, H. E. Acta Met. 21 (1973) 1431.
18. Chen(Wen), S. H. PhD Thesis, Dept. Materials Science and Engineering, Univ. of California, Berkeley (1979) (LBL Tech. Rept. 9264, June, 1979)
19. Wen, S. H., A. G. Khatchaturyan and J. W. Morris, Jr. Met. Trans. 12A (1981) 581.
20. Khatchaturyan, A. G. and G. A. Shatalov. Sov. Phys.-Solid State 11 (1969) 118.
21. Lee, J. K. D. M. Barnett and H. I. Aaronson. Met. Trans. 8A (1977) 963.
22. Johnson W. C. and J. K. Lee. in Solid-Solid Phase Transformations (H. I. Aaronson, D. E. Laughlin, R. F. Sekerka and C. M. Wayman, eds., The Metallurgical Society-AIME, Warrendale, Pa. 1983) p. 151.
23. Wen, S. H. E. Kostlan, M. Hong, A. G. Khatchaturyan and J. W. Morris, Jr. Acta Met. 29 (1981) 1247.
24. Hong, M., D. E. Wedge and J. W. Morris, Jr. Acta Met. (in press)
25. Mayo, W. E. and T. Tsakalakos Met. Trans. 11A (1980) 1631.

26. Chen(Wen), S. H., A. G. Khatchaturyan and J. W. Morris, Jr. in Proc. Kona Conference on Modulated Structures, Kona, Hawaii, (Am. Inst. Physics, New York 1979).
27. Chen(Wen), S. H. A. G. Khatchaturyan and J. W. Morris, Jr. in Proc. Int. Conf. on Martensitic Transformations ICOMAT (Owen, W. ed., MIT Press (1979).
28. Shibata, M. and K. Ono. Acta Met. 23 (1975), 587 Trans. Japan Inst. Metals 17 (Suppl.) (1976) 35.
29. Blantner, M. E. and A. G. Khatchaturyan. Met. Trans.9A (1978) 753.
30. Kurdjumov, G. V. and A. G. Khatchaturyan. Acta Met. 23 (1975) 1077.
31. Mori, T., P. C. Cheng and T. Mura. Trans. Japan Inst. Metals 17 (Suppl.) (1976) 281.
32. Landau, L. D. Sov. Phys. 11:26 (1937) 545.
33. Lifshitz, E. M. Fiz. Zh. 7:61 (1942) 251.
34. Pokrovskii, B. I. and A. G. Khatchaturyan. Solid State Chem. (in press).

ON THE MECHANICS OF MODULATED STRUCTURES

E.C. Aifantis

Michigan Technological University, Houghton, MI 49931 USA
and University if Minnesota, Minneapolis, MN 55455 USA

PROLOGUE

The purpose of this lecture is to illustrate the appropriate-
ness and potential of the methods of continuum mechanics in modeling
modulated structures. Modulations are viewed, in general, as
occurrences which may involve one or more properties of a system and
extend from a submicroscopic to a macroscopic scale. They are also
viewed as capable of possessing wave lengths and amplitudes which
may vary from very small to very large values.

Within this broad definition of modulations one can easily
include phenomena ranging from liquid-vapor transitions and spinodal decomposition to slip banding, shear banding and necking. Even
though these phenomena clearly occur at different scales, nevertheless they are all characterized by the modulation of a proper
variable i.e. mass density, species concentration, dislocation density, or strain respectively. In other words, a particular microstructure is capable of localizing and producing spatial patterns
which, in effect, can control the overall properties of the structure.

To describe this large class of phenomena in a unifying manner
one can make use of the concept of "excited" and "normal" states:
excited are those associated with the modulations and normal those
accommodating such excitations to occur. For example, in spinodal
decomposition the excited states are associated with the diffusing
species while in slip banding with dislocations. Excited and normal states can interact by exchanging mass, momentum, energy and
entropy. These exchanges should be in accord with the laws of

mechanics and thermodynamics i.e. the conservation laws of mass, momentum, and energy, as well as the law of entropy growth.

While the attempt of specifying the general form of these laws within a modified framework of irreversible thermodynamics is not discussed here, the method is unambiguously described within a purely mechanical framework where only the conservation of mass and momentum are utilized for describing both excited and normal states. In this way, ambiguous thermodynamic definitions and procedures, concerned with chemical potentials, phase rules, minimization of free energies, etc., are avoided and localization of excited states and therefore nucleation and growth of new phases is interpreted via "loops" and "non-convex" equations of state or direct analogues of such behavior. Moreover, the long-range forces between states which are, in general, of non-local integral character are introduced here within a gradient approximation.

1. BASIC CONCEPTS & INTERDISCIPLINARY NOTIONS

1.1 Analogues from Chemistry and Biology

While the topic of modulations in the sense discussed in this Institute is a new one, in a broader sense modulations have occurred and analyzed previously in many biological and chemical systems. Dissipative and living structures have a tendency for pattern formation which is essentially considered responsible for the origin of life. The pattern formation in these systems have conveniently been studied through notions and methods of non-linear analysis. The concepts of non-convexity, multiplicity, bifurcation, and non-uniqueness are relevant in these studies. It was shown that in many chemical and biological cases these tools can serve usefully not only to a convenient interpretation of the qualitative behavior of the phenomena but also to an accurate quantitative modeling. However, the approach has not been applied to the field of materials science even though many of the relevant phenomena have a tendency to pattern formation.

The modest goal of this lecture is to show that modulated structures are indeed excellent candidates to be investigated by the methods of non-linear analysis. What one needs to identify is a characteristic quantity of the system α which varies with a parameter θ in a fashion shown in Figure 1a. When the parameter θ reaches a critical value θ_{CR} the graph is such that to one value of θ there correspond three values of α (multiplicity). At the same time and for $\theta > \theta_{CR}$ a characteristic property of the system p varies with α in a fashion shown by the graph of Figure 1b and again a multiplicity is involved. The quantity α can be identified with the density ρ in a problem of liquid-vapor transition, the concention c in a problem of spinodal decomposition, or the dislocation

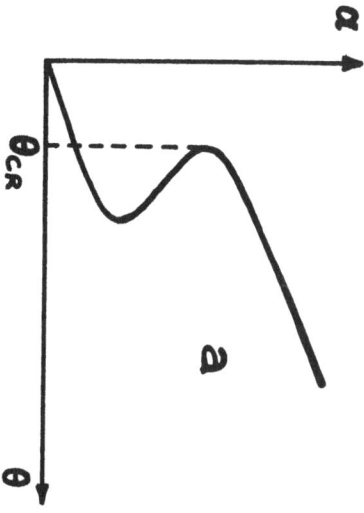

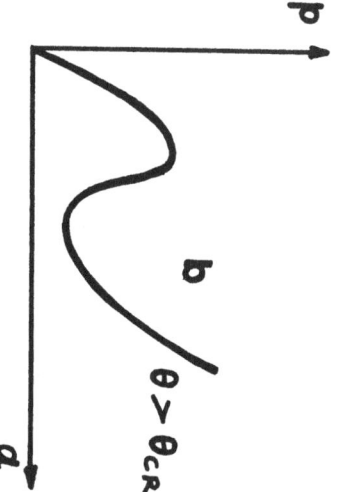

Figure 1

Loops and Multiplicity

density n in a problem of cell formation. Correspondingly, the property p can be an appropriate chemical potential μ or partial stress $\underset{\sim}{T}$ and the parameter θ is usually identified with temperature or applied stress.

1.2 Scale

Even though most of the modulated structures discussed in this Institute occur at the atomic scale, modulations are possible at larger scales such as the microscopic and macroscopic.

(a) Atomic: Modulations occurring at this scale are ordering of atoms, spinodal decomposition, and clustering.

(b) Microscopic: Among the modulations occurring at this scale most common are the localization of dislocations and the localization of microvoids. The formation of slip bands, dislocation cells, walls, and fabrics is a manifestation of such modulations.

(c) Macroscopic: Some of the most common modulations occurring at this scale can be identified with crazes, shear bands, necks, and plastic zones. Localization of deformation, damage, and microcracks are essentially the underlying mechanisms causing such modulations.

1.3 Proper Variable

The most commonly used variable to illustrate modulations in this Institute was a scalar quantity, i.e. a concentration of certain type of atoms. For example, spinodal decomposition and ordering can be adequately described by means of this variable. There are cases, however, where orientation is important and then one would need a vectorial or tensorial quantity. For example in the case of dislocation walls, a dislocation density tensor would be the appropriate quantity. This is reminiscent of the order parameter used by Landau in his minimization of free energy approach to phase transitions. In general, modulations can occur in terms of scalar, vector, or tensor quantities and this is usually determined by the nature of the problem. Such quantities are often called internal variables of the system.

To minimize the degree of arbitrariness involved in choosing the appropriate variables, it is suggested below a procedure for their unifying and rational identification. The material system is viewed as a superposition of two types of states: "excited" and "normal". Normal are the states which do not contribute directly to the nucleation and evolution of modulations but they simply support such excitations to come about. They have the same character as those that the material system can possess in the absence of the species (e.g. atoms, groups of atoms, molecules) that are responsible for the occurrence of such modulations. Excited states are those associated with the species which essentially make-up the modulations. The various states are allowed of course to interchange mass, momentum, energy, and entropy.

In writing formal statements of balance for these quantities one needs to introduce first the density ρ of excited states. In the case of spinodal decomposition the excited states are composed of the diffusing species while in the cases of dislocation and void localization the excited states are made up by the species (atoms) confined within the dislocation core and within the surface layer of the voids. It then turns out (1) that the following relations hold

Diffusion:	ρ = const. c ,
Dislocations:	ρ = const. n ,
Microvoids:	ρ = const. nr,
Microcracks:	ρ = const. nℓ,

where c is the concentration, n is the number of dislocations (microvoids or microcracks) per unit volume, r is the radius of microvoids, and ℓ is the average crack length. The factor const. is a molecular constant in the simplest cases but, in general, may be a function of stress, temperature, etc. Clearly, the quantities c, n, r, and ℓ which are commonly introduced on an empirical basis, were naturally introduced here on the basis of a systematic and rather unifying procedure.

1.4 Approach

Several methods can be employed in describing modulated structures and is not clear or always possible to interrelate them in a meaningful way.

(a) Discrete and Statistical Methods: Many of the results presented at this Institute were based on Monte-Carlo simulations, modifications of the ISING model and relevant assumptions on the form of Hamiltonians and appropriate intermolecular potentials.

(b) Variational Thermodynamic Models: These are essentially mean-field theories and are the oldest and perhaps the most popular method of analysis due to the influence of van der Waals and Landau and more recently of Cahn and Davis-Scriven. The assumption of existence of a free energy in unstable regions is a critical one and the usual procedure of minimizing this potential ($\delta\psi = 0$, $\delta^2\psi > 0$) is employed even though it has been noted recently that such criteria do not necessarily ensure a minimizer (for non-infinitely smooth minimizers). Moreover these methods are best suited for static problems and of course the problem of the second variation becomes very difficult in three dimensions even within the framework of the classical calculus of variations.

(c) Continuum Mechanics: Recently an attempt has been made to extend the use of continuum mechanics all the way down to the atomic scale. The central goal of this lecture is to illustrate the suitability of a "generalized" continuum mechanics framework to the modeling of modulated structures. General aspects are given in the next paragraph and particular applications are discussed in the following sections of the lecture.

362

1.5 Mechanics

There are two places where the structure of classical continuum mechanics should be modified in order to describe phenomena
associated with modulations.

First, the balance laws should allow the inclusion of "extra"
terms modeling possible exchanges of mass, momentum and energy between normal and excited states. Even in the case that such
detailed distinction is not necessary "extra" terms are usually
needed to account for the interaction between "surface" and "bulk"
points and properly include surface stress and surface energy
effects. For example, the statements of linear momentum and energy
balance read

$$\text{div}\underset{\sim}{T} + \overset{\wedge}{\underset{\sim}{f}} = \rho\dot{\underset{\sim}{v}}, \tag{1.1}$$

$$\rho\dot{\varepsilon} = \underset{\sim}{T}\cdot\text{grad}\underset{\sim}{v} - \text{div}\underset{\sim}{q} + \hat{e}, \tag{1.2}$$

with $(\rho, \underset{\sim}{T}, \underset{\sim}{v})$ denoting as usual density, stress, and velocity,
$(\varepsilon, \underset{\sim}{q})$ internal energy and heat flux and the fields $(\overset{\wedge}{\underset{\sim}{f}}, \hat{e})$ representing the "extra" terms of momentum and energy exchange. By
setting the extra terms equal to zero, (1.1) and (1.2) reduce to
the usual statements of balance of linear momentum and energy in
the absence of external body force and external heat supplies.

Second, the constitutive structure should be properly expanded
to account for higher-order symmetry phenomena associated with
modulations, as well as long-range forces and inhomogeneity
effects. Thus, higher-order symmetry considerations could be
incorporated by introducing a director field $\underset{\sim}{d}$ so that the stress $\underset{\sim}{T}$
will be of the functional form

$$\underset{\sim}{T} = \overset{\wedge}{\underset{\sim}{T}}(\rho, \underset{\sim}{d}, \ldots) , \tag{1.3}$$

with the vector $\underset{\sim}{d}$ (and in certain cases its evolution) denoting
internal symmetries of the structure. Non-local effects are introduced to describe with reasonable degree of accuracy the detailed
molecular interactions and long-range interatomic forces. Thus,
for a fluid-like structure B capable of supporting modulations, the
constitutive equation could be of the form

$$\underset{\sim}{T} = \int_B \underset{\sim}{G}[\rho(\underset{\sim}{x}), \rho(\underset{\sim}{y}), r(\underset{\sim}{x},\underset{\sim}{y})]\rho(\underset{\sim}{y}) \, d\upsilon(\underset{\sim}{y}) , \tag{1.4}$$

with $(\underset{\sim}{x},\underset{\sim}{y}) \in B$, $r(\underset{\sim}{x},\underset{\sim}{y}) = |\underset{\sim}{x} - \underset{\sim}{y}|$, $d\upsilon$ an elementary volume and G an
appropriate kernel function. In certain circumstances (1.4) can be
simplified to

$$\underset{\sim}{T} = \overset{o}{\underset{\sim}{T}}(\rho) + \int_{\underset{\sim}{x} \neq \underset{\sim}{y} \in B} G(r)\rho(\underset{\sim}{y}) d\upsilon(\underset{\sim}{y}). \tag{1.5}$$

with $\overset{o}{T}$ denoting the appropriate local component of stress in the case that modulations are absent. In passing, it is noted that the integral term in (1.5) can be replaced by a function of the gradients of ρ, thus substituting the long-range spatial dependence by a short-range spatial one. This is simply seen by adopting a Taylor's series expansion in the fashion

$$\rho(\underset{\sim}{y}) = \rho[\underset{\sim}{x} + (\underset{\sim}{y} - \underset{\sim}{x})] , \qquad (1.6)$$

about $\underset{\sim}{x}$ and retaining the first few terms in the series. It then turns out that

$$\underset{\sim}{T} = \underset{\sim}{\hat{T}}(\rho, \nabla\rho, \nabla^2\rho, \ldots) , \qquad (1.7)$$

which is the gradient-approximation of the non-local constitutive equation (1.4) and which will be utilized extensively in the following sections to illustrate modulation trends.

Finally, it is pointed out that in the case that internal variables are used to model modulated structures, their evolution should be determined in accordance with complete balance laws of the form

$$\dot{\alpha} + \operatorname{div}\underset{\sim}{j} = \hat{q} , \qquad (1.8)$$

where the second term of the left-hand side (which is traditionally neglected) models the flux of the internal variables within the elementary volume. It can be shown that $\underset{\sim}{j}$ is given by a differential form so that (1.8) is, in general, a partial differential equation to determine α.

In what follows, a properly modified framework of continuum mechanics will be adopted to model various modulation phenomena ranging from liquid-vapor transitions and spinodal decomposition to localization of dislocations and localization of deformation.

2. MOTIVATION FOR ANALYSIS

Since various classes of modulation phenomena that we consider here are diffusion-controlled, it is instructive to detail in this section how classical diffusion theories can be cast within the formalism outlined at the later part of previous section without invoking usual thermodynamic arguments in terms of chemical potentials.

2.1 Thermodynamic Basis

A thermodynamic basis for diffusion processes is provided by the flux postulate

$$\underset{\sim}{j} = -M\nabla\mu \; , \tag{2.1}$$

where M is the mobility and μ the chemical potential.

Various classes of diffusion processes are then obtained by introducing appropriate assumptions for the chemical potential μ. Thus, by assuming that

$$\mu = \text{const.}\rho \; , \tag{2.2}$$

we have Fick's first law of diffusion

$$\underset{\sim}{j} = -D\nabla\rho \; , \tag{2.3}$$

with D being a constant diffusion coefficient.

If instead of (2.2) we assume the relation

$$\mu = \text{const.}\rho - (\text{const.})* \; \Delta\rho \; , \tag{2.2}*$$

we have Cahn's modification of Fick's law

$$\underset{\sim}{j} = -D\nabla\rho + E\nabla(\Delta\rho) \; , \tag{2.4}$$

with D and E taken as constants.

On inserting (2.3) into the equation expressing conservation of mass

$$\dot{\rho} + \text{div}\underset{\sim}{j} = 0 \; , \tag{2.5}$$

we obtain the classical diffusion equation

$$\dot{\rho} = D\Delta\rho \; , \tag{2.6}$$

which requires the diffusivity D to be positive (down-hill diffusion). On the other hand, substitution of (2.4) into (2.5) yields Cahn's diffusion equation

$$\dot{\rho} = D\Delta\rho - E\Delta^2\rho \; , \tag{2.7}$$

which allows D to be positive or negative (up-hill diffusion) since $E\Delta^2\rho$ is the leading term of this equation and E must be positive for stable behavior.

It is well known, however, that often the chemical potential is not an unambigiously defined quantity and even though statements like (2.6) or (2.7) are utilized, a fundamental difficulty still remains with postulating the existence of a potential μ in (2.1). The purpose of the next paragraph is to show that (2.6) and (2.7) can be derived within a mechanical theory which overcomes the question of existence of chemical potentials and an underlying thermodynamic structure.

2.2 Mechanical Basis

A mechanical basis for diffusion is provided through the use of momentum balance for the diffusing species. Since diffusion is a type of motion it should conform, in principle, with the law of conservation of momentum properly modified to take into account the momentum exchange between interdiffusing species. In expressing this statement more precisely we have

$$\text{div}\underset{\sim}{T} + \hat{\underset{\sim}{f}} = j_t , \tag{2.8}$$

where $\underset{\sim}{T}$ is the stress tensor supported by the diffusing species (a surface measure of the interatomic forces among the diffusing species alone or the stress that the diffusing species exert upon themselves); $\hat{\underset{\sim}{f}}$ is the resistance force or diffusive drag (a volume measure of the interatomic forces between the diffusing species and the matrix); and j is as usual the flux of diffusing species with the subscript t denoting time differentiation. Thus, (2.8) instead of (2.1) is the starting point of a mechanical theory for diffusion.

The next step is to introduce constitutive equation for $\underset{\sim}{T}$ and $\hat{\underset{\sim}{f}}$. It is the nature of these equations which define various classes of diffusion behavior. For simple diffusion processes where long-range interactions can be neglected, it is reasonable to assume on the basis of simple mechanical analogies of either continuum or statistical nature that the appropriate constitutve assumptions will be of the form

$$\{\underset{\sim}{T}, \ \hat{\underset{\sim}{f}}\} \xrightarrow{\text{fcts}} \{\rho, \ j\} , \tag{2.9}$$

and, in particular,

$$\underset{\sim}{T} = -\pi\rho\underset{\sim}{1} \ , \ \hat{\underset{\sim}{f}} = -\frac{1}{M}j , \tag{2.10}$$

where $\underset{\sim}{1}$ denotes the unit matrix and π and M are constants. It then turns out (2),(3), that substitution of (2.10) into (2.8) yields with $D \equiv \pi M$

$$j + Mj_t = -D\nabla\rho , \tag{2.11}$$

which is a generalization of the classical statement (2.3) to account for inertia effects. On inserting (2.11) into (2.5) we obtain

$$\dot{\rho} + M\ddot{\rho} = D\Delta\rho , \tag{2.12}$$

which also generalizes (2.6). In usual diffusion situations inertia effects are negligible and then (2.11) and (2.12) are reduced to the classical statements (2.3) and (2.6).

In the case that long-range forces cannot be neglected and surface tension effects are dominant, the form (2.9) is replaced by

$$\{\underset{\sim}{T}, \hat{\underset{\sim}{f}}\} \xrightarrow{\text{fcts}} \{\rho, j, \nabla\rho, \mathbf{\nabla}^2\rho\} , \tag{2.13}$$

where $(\nabla^2\rho)_{ij} = \rho_{,ij}$ is a second-order tensor denoting the second gradient of ρ. A simple model for $\underset{\sim}{T}$ and $\hat{\underset{\sim}{f}}$ consistent with (2.13) is

$$\underset{\sim}{T} = (-\pi\rho + \pi^*\Delta\rho)\underset{\sim}{1} , \quad \hat{\underset{\sim}{f}} = -\frac{1}{M}\underset{\sim}{j} , \tag{2.14}$$

with π^* denoting a new constant measuring surface tension effects. Upon substitution of (2.14) into (2.8) we obtain with $E = \pi^* M$

$$j + Mj_t = -D\nabla\rho + E\nabla(\Delta\rho) , \tag{2.15}$$

which is a generalization of Cahn's statement (2.4) to account for inertia effects within a purely mechanical framework. On inserting (2.15) into (2.5) we obtain

$$\dot\rho + M\ddot\rho = D\Delta\rho - E\Delta^2\rho , \tag{2.16}$$

which also generalizes (2.7). Additional diffusion classes can be derived by elaborating on different versions of the constitutive forms (2.9) and (2.13). A rather detailed set of such classes of diffusion behavior is derived in (2),(3) where terms like $D^*\Delta\dot\rho$ and $D(0)\Delta\rho + \int_0^\infty \dot D(s)\rho(t-s)ds$ can appear in the diffusion equations. While the role of such terms could be significant for certain classes of diffusion-controlled phase transformations, we do not explore this question further in the present lecture. We rather confine attention to the effect of gradients as illustrated in (2.13) and assess their importance in various linear and non-linear phenomena of phase changes and modulations.

3. THE SIMPLEST MODULATION

In this section we utilize a simple mechanical framework to derive a non-linear partial differential equation whose solutions give rise to elementary modulations in a single component system. The relevant variable in terms of which these modulations are expressed is the density ρ of this fluid-like system. For simplicity, we consider equilibrium states of a single-component substance, so that

$$\text{div}\underset{\sim}{T} = 0. \tag{3.1}$$

When modulations of the density ρ are possible the expression for the stress T should reflect the effect of surface tension which, in some sense, provide an approximate measure of the long-range interatomic forces. It is essentially this surface tension which under certain conditions leads to the formation of fluid microstructures or modulated structures for this one-component system. In conformity with the discussion of previous section we account for long-range forces by adopting a gradient-approximation representation into the constitutive equation for $\underset{\sim}{T}$. We thus have

$$\underset{\sim}{T} = -p(\rho)\underset{\sim}{1} + \underset{\sim}{T}^E , \qquad (3.2)$$

where the first part of the right-hand side is the appropriate stress in the absence of modulations and $\underset{\sim}{T}^E$ is the surface tension dependent extra stress component. This is, in general, of the form

$$\underset{\sim}{T}^E = \underset{\sim}{T}^E(\rho, \nabla\rho, \nabla^2\rho) , \qquad (3.3)$$

which by assuming to be a polynomial of second degree and order reduces to

$$\underset{\sim}{T}^E = [\alpha(\rho)\Delta\rho + \beta(\rho)|\nabla\rho|^2]\underset{\sim}{1} + \gamma(\rho)\nabla^2\rho + \delta(\rho)\nabla\rho \otimes \nabla\rho , \qquad (3.4)$$

where the various coefficients are functions of ρ and the symbol $\otimes$ denotes dyadic so that $(\nabla\rho\otimes\nabla\rho)_{ij} \equiv \rho_{,i}\,\rho_{,j}$. Thus, the system of equations which is to be solved for generating elementary modulated structures consists of

$$\underset{\sim}{T} = (-p + \alpha\Delta\rho + \beta|\nabla\rho|^2)\underset{\sim}{1} + \gamma\nabla^2\rho + \delta\nabla\rho \otimes \nabla\rho ,$$

$$\text{div}\underset{\sim}{T} = 0. \qquad (3.5)$$

Upon substitution of $(3.5)_1$ into $(3.5)_2$ we obtain the following overdetermined system of non-linear differential equations for the density

$$\nabla(-p + a\square\rho + \tilde{b}|\nabla\rho|^2) = (d\square\rho)\nabla\rho , \qquad (3.6)$$

where

$$\square \equiv \Delta + \frac{1}{2}\frac{d'}{d}|\nabla|^2 , \quad \tilde{b} \equiv b +\frac{1}{2}(d - a\frac{d'}{d}) ,$$
$$a \equiv \alpha + \gamma, \, b \equiv \beta + \delta , \, d \equiv \gamma' - \delta. \qquad (3.7)$$

As is pointed out in (4) one can establish the following

Theorem: If $\tilde{b} \neq 0$, then the solutions of (3.6) are of either, of the forms

$$\rho = \rho(x) ,$$

$$\rho = \rho(r) ; r = \sqrt{x^2 + y^2} , \qquad (3.8)$$

$$\rho = \rho(R) ; R = \sqrt{x^2 + y^2 + z^2} ,$$

i.e., they are either parallel planes, or concentric cylinders, or concentric spheres.

Remark: The theorem does not hold when $\tilde{b} \equiv 0$. In that case more general shapes for three-dimensional modulations are possible. Moreover, for radially symmetric cases the condition $\tilde{b} \equiv 0$ allows the derivation of results for three-dimensional modulations similar to those established for one-dimensional ones as discussed below.

For planar modulations the density $\rho = \rho(x,y,z)$ and the stress $\underset{\sim}{T} = \underset{\sim}{T}(x,y,z)$ are functions of one coordinate only, say x. It follows that the only nonvanishing components of stress are

$$T_{xx} = -p + a\rho_{xx} + b\rho_x^2 , \quad T_{yy} = T_{zz} = -p + \alpha\rho_{xx} + \beta\rho_x^2, \qquad (3.9)$$

with the coefficients a and b defined in (3.7). Accordingly, the only non-trivially satisfied component of the equilibrium equations (3.1) i.e.,

$$\frac{\partial T_{xx}}{\partial x} = 0 , \qquad (3.10)$$

gives

$$a\rho_{xx} + b\rho_x^2 = p - \bar{p} , \qquad (3.11)$$

where $\bar{p}$ is a constant.

Equation (3.11) governs the structure of one-dimensional modulations and possesses three types of bounded solutions for infinite domains in addition to the trivial constant solution. The analysis of (3.11) is detailed in (5) but we provide here a summary of results as they relate to the form of all possible modulations.

(a) Monotone Modulations: These correspond to the following conditions at infinity ($\rho_1 < \rho_2$)

$$\rho \to \rho_1 \text{ as } x \to -\infty , \ \rho \to \rho_2 \text{ as } x \to +\infty , \qquad (3.12)$$

or

$$\rho \to \rho_2 \text{ as } x \to -\infty , \ \rho \to \rho_1 \text{ as } x \to +\infty , \qquad (3.13)$$

and are depicted in Figure 2a. Given a set of numbers $(\bar{p},\rho_1,\rho_2)$, the modulations are uniquely determined (up to choice of x_o) by the expressions

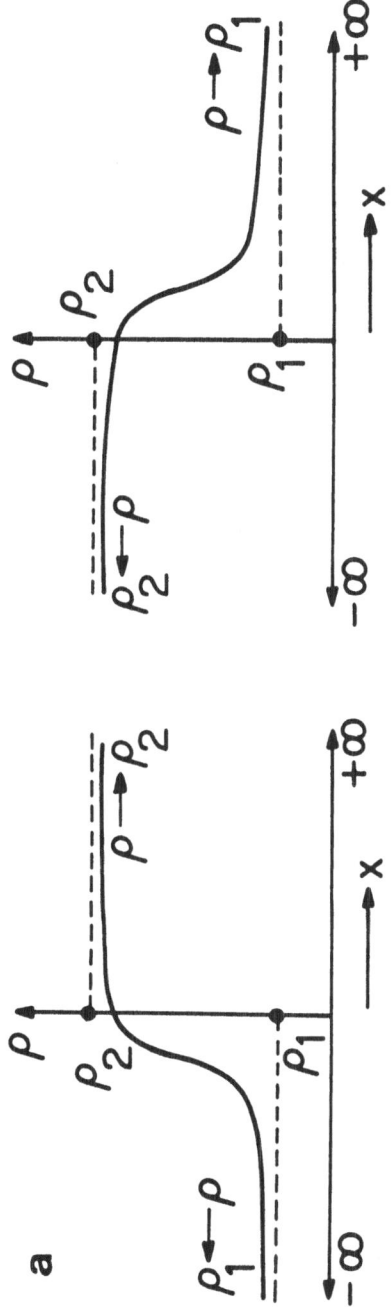

Figure 2a

One-dimensional Modulations

$$x = x_0 + \int_{\rho(x_0)}^{\rho(x)} \frac{d\rho}{\sqrt{2F(\rho)/G(\rho)}} \quad , \tag{3.14}$$

for (3.12), or

$$x = x_0 - \int_{\rho(x_0)}^{\rho(x)} \frac{d\rho}{\sqrt{2F(\rho)/G(\rho)}} \quad , \tag{3.15}$$

for (3.13), where $F(\rho)$ and $G(\rho)$ are given by

$$F(\rho) \equiv \int_{\rho_1}^{\rho} (p - \bar{p})E(\rho)d\rho \; , \; G(\rho) \equiv aE(\rho) \; , \tag{3.16}$$

and the weighting function $E(\rho)$ is defined by

$$E(\rho) \equiv \frac{1}{a} \exp\left[2\int \frac{b}{a} \, d\rho\right]. \tag{3.17}$$

The characteristic numbers $(\bar{p}, \rho_1, \rho_2)$ are determined uniquely from the area condition

$$\int_{\rho_1}^{\rho_2} (p - \bar{p})E(\rho)d\rho = 0 \; , \tag{3.18}$$

and the pressure conditions

$$p(\rho_1) = p(\rho_2) = \bar{p}. \tag{3.19}$$

(b) Symmetric Modulations: These correspond to the following conditions at infinity $(\rho_1 < \rho_2)$

$$\rho \to \rho_1 \text{ as } x \to \underline{+}\infty \; , \; \rho_{max} = \rho_2 \; , \tag{3.20}$$

or

$$\rho \to \rho_2 \text{ as } x \to \underline{+}\infty \; , \; \rho_{min} = \rho_1 \; , \tag{3.21}$$

and are depicted in Figure 2b. Given a set of numbers $(\bar{p}, \rho_1, \rho_2)$, symmetric modulations are uniquely determined (up to choice of x_0) by (3.14) in the case of (3.20); or by (3.15) in the case of (3.21). These representations hold up to the point $\bar{x}(-\infty < x \leq \bar{x})$ where $\rho(x)$ attains its maximum $(\rho(\bar{x}) = \rho_2)$ in the case of (3.20), or minimum $(\rho(\bar{x}) = \rho_1)$ in the case of (3.21); in both cases the graph of $\rho(x)$ is symmetric about $\bar{x}$. The characteristic numbers $(\bar{p}, \rho_1, \rho_2)$ form a one-parameter family, determined by the area condition (3.18) and the pressure conditions

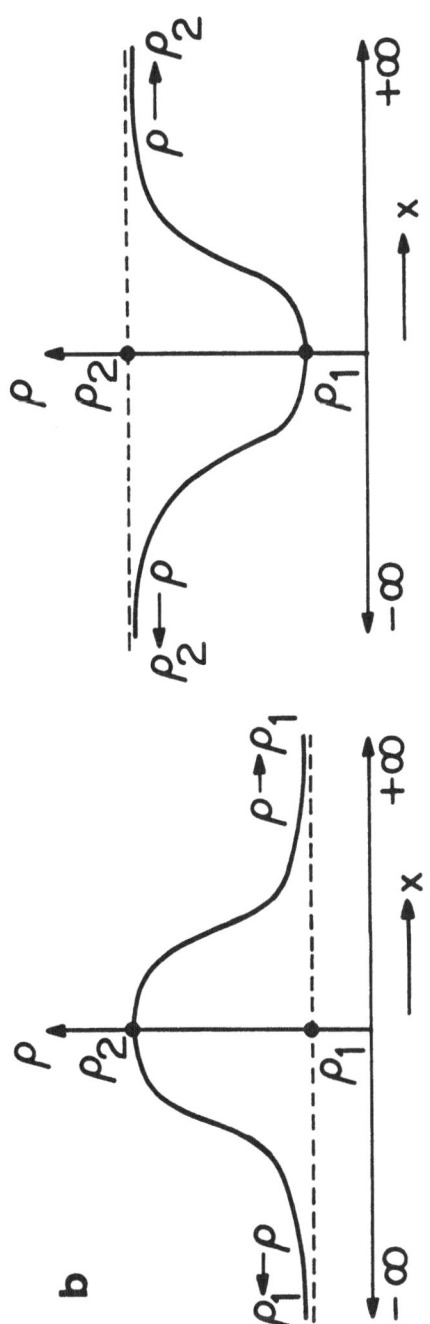

Figure 2b

One-dimensional Modulations

$$p(\rho_1) = \overline{p} \ , \ p(\rho_2) < \overline{p} \ , \tag{3.22}$$

in the case of (3.20), or

$$p(\rho_1) > \overline{p} \ , \ p(\rho_2) \cong \overline{p} \ , \tag{3.23}$$

in the case of (3.21).

(c) <u>Periodic Modulations</u>: These correspond to no specific conditions at infinity; they are depicted in Figure 2c. Given a set of numbers $(\overline{p}, \rho_1, \rho_2)$, the modulations are uniquely determined (up to choice of x_0) by (3.14). This representation holds for any interval $\overline{x} \leq x \leq \overline{\overline{x}}$, where $\overline{x}$ and $\overline{\overline{x}}$ are consecutive minima and maxima of ρ, with $\rho(\overline{x}) = \rho_1$ and $\rho(\overline{\overline{x}}) = \rho_2$; the complete graph is symmetric about both $\overline{x}$ and $\overline{\overline{x}}$. The characteristic numbers $(\overline{p}, \rho_1, \rho_2)$ form a two-parameter family, determined by the area condition (3.18) and the pressure conditions

$$p(\rho_1) > \overline{p} \ , \ p(\rho_2) < \overline{p}. \tag{3.24}$$

From the above discussion, it follows that equation (3.11) gives rise to three types of modulated structures. The first two types do not possess a finite wave length but the third type has a definite wave length 2λ given by the formula

$$\lambda = \int_{\rho_1}^{\rho_2} \frac{d\rho}{\sqrt{2F(\rho)/G(\rho)}} \ . \tag{3.25}$$

The area condition (3.18) should be satisfied by all three types of modulations. In (4) it is discussed how (3.18) relocates Maxwell's line and how it can be used to test the compatibility of statistical models for the stress coefficients $\{\alpha, \ \beta, \ \gamma, \ \delta\}$ with van der Waals thermodynamic theory.

The results listed in this section are also relevant to other types of modulation phenomena, as those induced by spinodal decomposition and discussed in the next section. Instead of the variable ρ (the density), modulations are induced there by another variable c (the composition). A non-linear theory of spinodal decomposition can be constructed which for stationary states in the absence of strain effects is quite similar with the one discussed here. A linear coupled theory of spinodal decomposition is also outlined.

4. SPINODAL DECOMPOSITION: NON-LINEAR UNCOUPLED THEORY

In conformity with the analysis of Section 2, the equation of

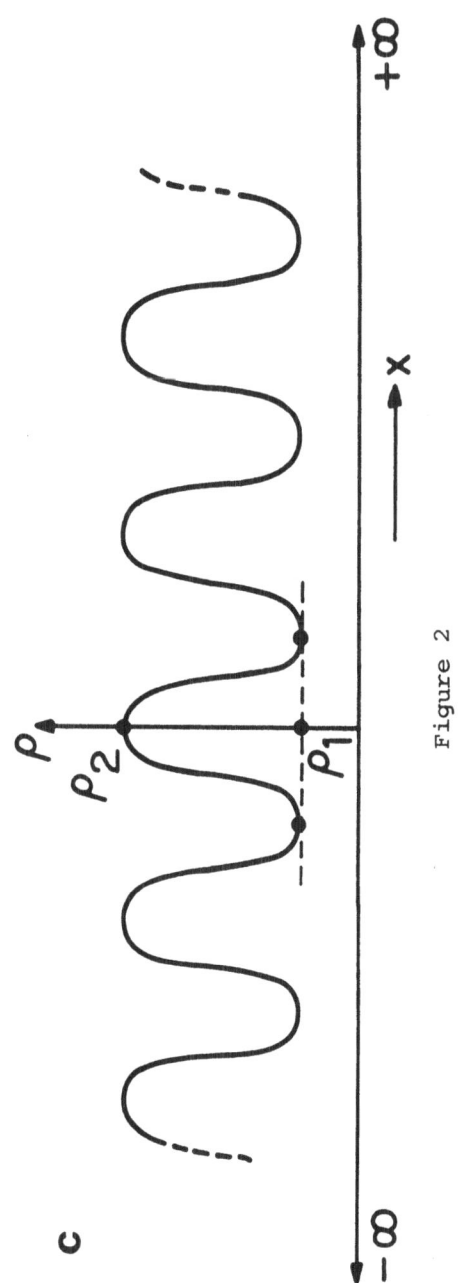

Figure 2

One-dimensional Modulations

momentum for the diffusing species reads

$$\text{div}\underset{\sim}{T} + \overset{\wedge}{\underset{\sim}{f}} = \underset{\sim}{j}_t . \tag{4.1}$$

Also motivated by the discussion of the two earlier sections, we assume constitutive equations for $\underset{\sim}{T}$ and $\overset{\wedge}{\underset{\sim}{f}}$ of the form

$$\left. \begin{aligned} \underset{\sim}{T} &= (-\pi + \alpha\Delta c + \beta |\nabla c|^2)\underset{\sim}{1} + \gamma\nabla^2 c + \delta\nabla c \otimes \nabla c, \\ \overset{\wedge}{\underset{\sim}{f}} &= -\frac{1}{M}\underset{\sim}{j}. \end{aligned} \right\} \tag{4.2}$$

Upon substitution of (4.2) into (4.1) we obtain the following non-linear expression for the flux $\underset{\sim}{j}$

$$\underset{\sim}{j} + M\underset{\sim}{j}_t = M\{\nabla(-\pi + a\Box c + \tilde{b}|\nabla c|^2) - (d\Box c)\nabla c\} , \tag{4.3}$$

where all the symbols have been defined in (3.7).

For equilibrium states the first two terms of (4.3) drop and the results of theorem (3.8) hold for spinodal modulations.

Similarly, it can be shown that for one-dimensional situations, equilibrium and stationary states are equivalent, and one can reach (3.11) with p replaced by π and ρ by c. It thus follows that all the results of previous section are still valid and the three types of one-dimensional modulations are also appropriate for spinodal decomposition problems.

Another point of interest which seems to merit more careful analysis is the case where the coefficient d is zero. It turns out that this condition is critical in distinguishing spinodal decomposition from ordering and also allows for more complex profiles than those depicted in Figure 2. Some results for this situation were obtained in [6] and a typical profile for the case that d has three zeros at $\bar{c}_1$; $\bar{c}_2$, $\bar{c}_3$ i.e. $d(\bar{c}_1) = d(\bar{c}_2) = d(\bar{c}_3) = 0$ is given in Figure 3. As it might be expected, there is non-uniqueness associated with this case. Even more complex bounded modulations can be obtained in the case that the graph of $\pi(c)$ is characterized by several loops as indicated in Figure 4 but this problem will be discussed in detail elsewhere.

Next, we specialize the general non-linear theory given by (4.3) to derive special classes of spinodal decomposition which are described by differential equations not previously noted in the literature. Thus, by taking

$$\tilde{b} = 0 , \quad a = \text{const.}, \quad d = \text{const.} \tag{4.3}$$

we can derive

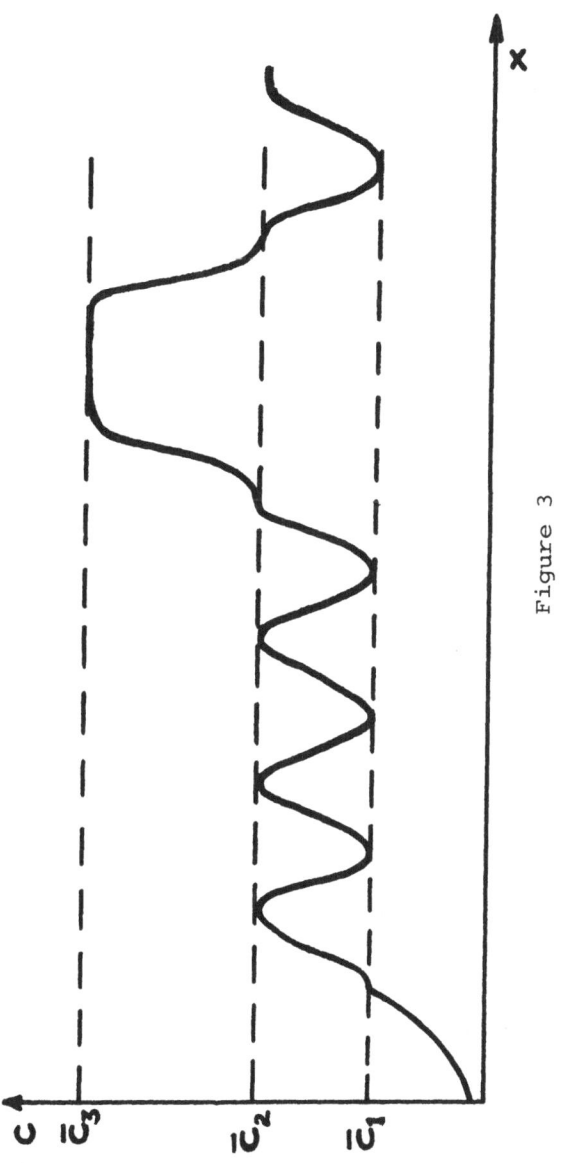

Figure 3

More Complex Modulations

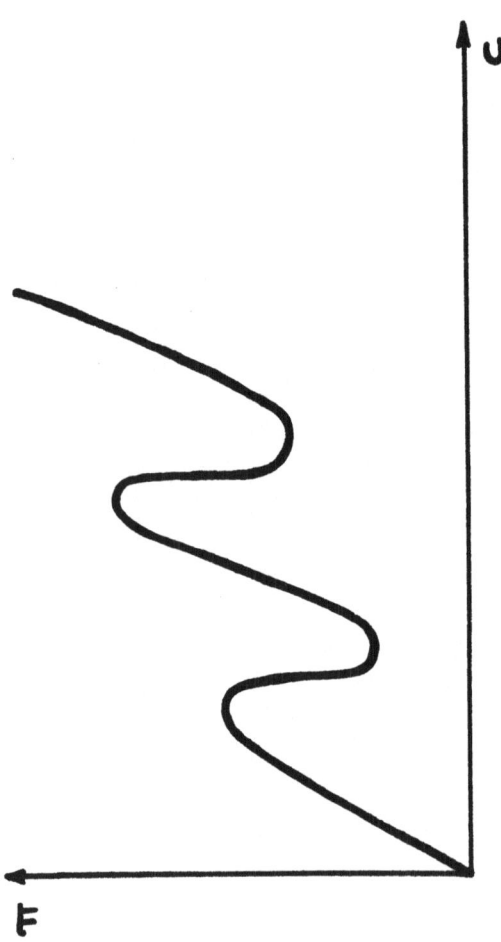

Figure 4

Multiple Loops

$$\underset{\sim}{j} + M\underset{\sim}{j}_t = -(D + D*\Delta c)\nabla c + E\nabla(\Delta c) , \qquad (4.4)$$

with

$$D \equiv M\pi' , \quad D* \equiv Md , \quad E \equiv Ma , \qquad (4.5)$$

Linearization of (4.4) yields

$$\underset{\sim}{j} + M\underset{\sim}{j}_t = -D\nabla c + E\nabla(\Delta c) , \qquad (4.6)$$

which is identical in form with (2.15).

Finally, we note that if attention is confined to one dimension and inertia effects are neglected, (4.3) is reduced to

$$j = -M(\pi - ac_{xx} - bc_x^2)_x , \qquad (4.7)$$

which together with the equation of mass balance

$$\overset{\bullet}{c} + j_x = 0 , \qquad (4.8)$$

gives

$$\overset{\bullet}{c} = [M(\pi - ac_{xx} - bc_x^2)_x]_x. \qquad (4.9)$$

This is a fourth-order non-linear differential equation whose solutions model spinodal modulations under transient conditions. It is noted that (4.9) has not been utilized as yet, even though it appears to be the simplest version of a fully non-linear theory for one dimensional spinodal decomposition.

5.   SPINODAL DECOMPOSITION: LINEAR COUPLED THEORY

In the previous section, effects due to the strain of the solid were neglected. These effects can best be illustrated in the case of small elastic deformations. If we neglect inertia ($\underset{\sim}{j}_t = 0$) and effects of external body forces ($\underset{\sim}{b} = 0$), the equations of dynamic equilibrium for the system are

$$div\underset{\sim}{S} = \rho_R\underset{\sim}{\overset{\bullet\bullet}{u}} , \qquad (5.1)$$

where $\underset{\sim}{S}$ is the total stress of the deformed spinodally decomposed solid, ρ_R is a reference density, and $\underset{\sim}{u}$ the displacement. The corresponding equation for the diffusing species is as before

$$div\underset{\sim}{T} + \overset{\wedge}{\underset{\sim}{f}} = 0. \qquad (5.2)$$

The appropriate constitutive equations are of the form

$$\{\underset{\sim}{S}, \underset{\sim}{T}, \hat{\underset{\sim}{f}}\} \xrightarrow{\text{fcts}} \{c, \underset{\sim}{j}, \underset{\sim}{E}, \nabla c, \nabla^2 c\}, \tag{5.3}$$

where now the total stress $\underset{\sim}{S}$ appears among the constitutive functions and the strain $\underset{\sim}{E}$ appears among the constitutive variables. The simplest possible physical considerations can lead to the following special forms for (5.3)

$$\left.\begin{array}{l}
\underset{\sim}{S} = (\lambda e - \gamma c + \lambda^* \Delta c)\underset{\sim}{1} + 2G\underset{\sim}{E} + 2G^* \nabla^2 c, \\[2mm]
\underset{\sim}{T} = (-\pi c + \pi^* \Delta c + \pi^{**} e)\underset{\sim}{1}, \\[2mm]
\hat{\underset{\sim}{f}} = -\frac{1}{M}\underset{\sim}{j}.
\end{array}\right\} \tag{5.4}$$

We note that in the case where c vanishes identically (5.4) reduces to the classical statement of linear elasticity. The constant γ measures the expansion of the lattice as a result of the concentration c and λ^* and G^* the expansion of the lattice as a result of the gradients of the concentration c. Equation (5.4) states that the stress supported by the diffusing species is hydrostatic with π^{**} measuring the effect of strain E (note that $e \equiv \text{tr}\underset{\sim}{E}$ is the trace of the strain tensor).

Upon substitution of (5.4) into (5.2) and (5.1) we obtain the following system of coupled equations

$$\left.\begin{array}{l}
\dot{c} = D\Delta c - E\Delta^2 c - L\Delta(\text{div}\underset{\sim}{u}), \\[2mm]
G\Delta\underset{\sim}{u} + (\lambda + G)\nabla(\text{div}\underset{\sim}{u}) = \gamma\nabla c - \delta\nabla(\Delta c) + \rho_R\ddot{\underset{\sim}{u}}.
\end{array}\right\} \tag{5.5}$$

In certain cases where relative flux effects are significant a term $c_R \text{div}\dot{\underset{\sim}{u}}$ with c_R a reference concentration needs to be added to the left-hand side of $(5.5)_1$. The last term in $(5.5)_1$ and the last two terms in $(5.5)_2$ were not accounted for in Cahn's original theory of spinodal decomposition in strained isotropic solids. Equations (5.5) were analyzed in some detail in [7] and [8] and the effect of the new terms was evaluated in certain cases. For example, it was shown how a modulation in concentration produces a corresponding modulation in displacement which can be persistent in time. Similarly it was shown how fluctuations in an externally applied body force can lead to persistent modulations of concentration and displacement, thus producing an explicit and analytically traced mechanism of a mechanically induced phase transformation. Moreover, it was shown that the coupling of strain as described by (5.5) imposes extra conditions for stability, in addition to the one originally discovered by Cahn [9]. Finally the problem of wave propagation through spinodally decomposing and linearly deforming elastic media was addressed.

6. PERSISTENT SLIP BANDS

This is another type of modulation that occurs in somewhat larger scale and involves the localization of dislocations in forming patterns which, in effect, control the mechanical behavior. The appropriate variable which leads to modulated structures in this case is the dislocation density n. To simplify the discussion we confine attention to one dimension by considering a monocrystal with a single slip system oriented such that the slip plane is parallel to the applied shear stress. It has been observed that a periodic distribution of dislocations or persistent slip bands is established under conditions of both monotonic or cyclic loading as illustrated in Figure 5. A rather general framework for describing the formation of such structural patterns and their relation to macroscopic quantities was given in (10). Here we only attempt to illustrate how the theoretical arguments presented in earlier sections can be utilized to predict stationary periodic one-dimensional dislocation structures as those depicted in Figure 5.

The equilibrium equation for the configuration of Figure 5 is simply

$$\tau_x = 0 \rightarrow \tau = \tau_0 , \tag{6.1}$$

where τ is the total stress and τ_0 the applied shear. The momentum equation for the dislocated state reads

$$T_x + \hat{f} = 0 , \tag{6.2}$$

and the appropriate constitutive equations for T and $\hat{f}$ are

$$T = \pi n , \quad \hat{f} = -\alpha j + \beta \tau. \tag{6.3}$$

Upon substitution of (6.3) into (6.2) we obtain

$$j = Dn_x + D*\tau , \tag{6.4}$$

with the constants D and D* expressed simply in terms of those in (6.3).

The equation of mass balance for the dislocated state reads

$$\dot{n} + j_x = \hat{q} , \tag{6.5}$$

where the source term $\hat{q}$ models the production or annihilation of dislocations. An appropriate constitutive representation for $\hat{q}$ is

$$\hat{q} = q_0 n + q_1 n^2 + q_2 n^3 , \tag{6.6}$$

where the coefficients q_0 , q_1 , q_2 are functions of the stress τ.

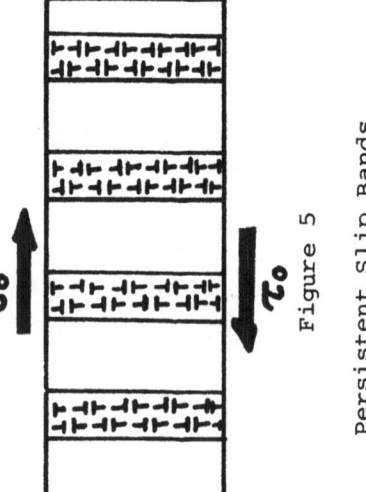

Figure 5

Persistent Slip Bands

On combining (6.4), (6.5) and (6.6) for steady-states and on noting (6.1) we arrive at the equation

$$n_{xx} = g(n) ,$$ (6.7)

with

$$g(n) = an - bn^2 + cn^3 ,$$ (6.8)

and the coefficients (a,b,c) being functions of τ_0.

The graph g(n) is depicted in Figure 6a and suggests that when the state of stress τ_0 is such that

$$\Delta \equiv b^2 - 4ac < 0 ,$$ (6.9)

the dislocation density remains uniform and modulations do not develop. On the contrary when the level of stress is such that $\Delta > 0$ and the roots of g(n) = 0

$$n_1 = (b - \sqrt{\Delta})/2c , \quad n_2 = (b + \sqrt{\Delta})/2c ,$$ (6.10)

define a loop shown in Figure 6a, periodic solutions are obtained as shown in Figure 6b. It turns out that the maximum and minimum values n_1^* and n_2^* of the periodic structure should satisfy the conditions

$$\int_{n_1^*}^{n_2^*} g(\rho)d\rho = 0 , \quad g(n_1^*) > 0 > g(n_2^*) ,$$ (6.11)

and that the portion of the modulation between n_1^* and n_2^* is represented by

$$x = x_0 + \int_{n(x_0)}^{n(x)} dn/\sqrt{2G(n)} , \quad G(n) = \int_0^n g(n)dn.$$ (6.12)

7.   SHEAR BANDS AND NECKING

This is another type of modulation occurring at a macroscopic scale and involves localization of deformation. Under certain circumstances it can be shown that a constitutive equation of the form

$$\underset{\sim}{S} = \hat{\underset{\sim}{S}}(\underset{\sim}{E} , \alpha) ,$$ (7.1)

with $\underset{\sim}{S}$ denoting stress, $\underset{\sim}{E}$ strain and α an internal variable (possibly dislocation density or void density) combined with a complete balance law for α as suggested by (1.8), i.e

382

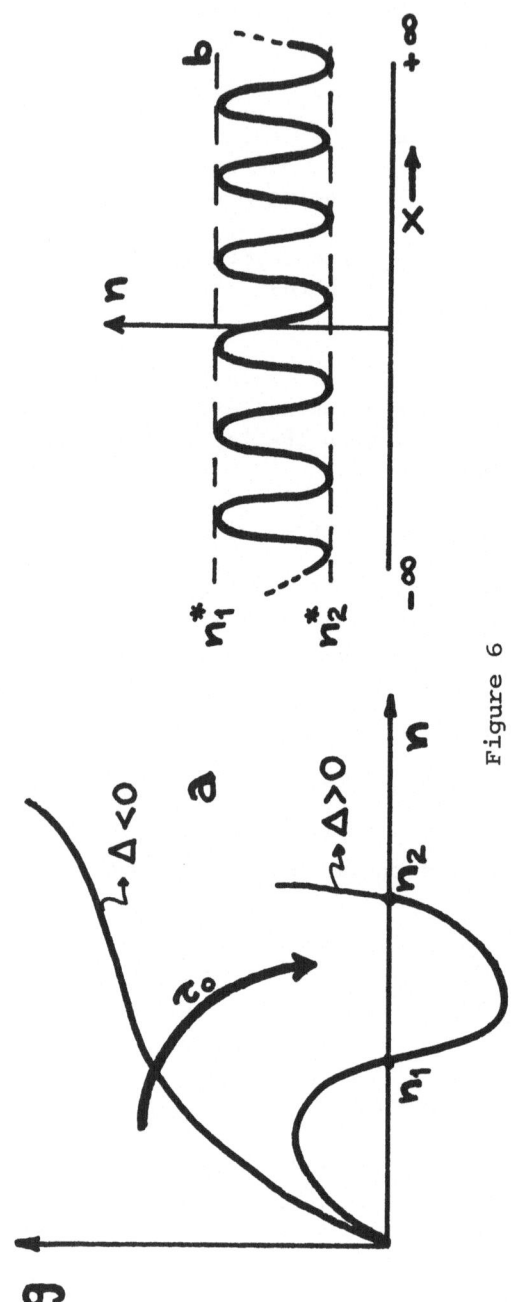

Figure 6

Periodic Dislocation Structures

$$\dot{\alpha} + D_x \alpha = \hat{q} \; , \tag{7.2}$$

where D_x denotes an appropriate partial differential equation, can lead to an elimination of α from (7.1) which now is replaced by

$$\underset{\sim}{S} = \hat{\underset{\sim}{S}}(\underset{\sim}{E}, \; \nabla\underset{\sim}{E}, \; \nabla^2\underset{\sim}{E}) \; , \tag{7.3}$$

i.e., a gradient representation of the stress-strain relation.

With suitable simplification (7.3) can lead to the following relation

$$\tau = \hat{\tau}(\gamma) + \hat{\alpha}(\gamma) \; \gamma_{yy} + \hat{\beta}(\gamma) \; \gamma_y^2 \; , \tag{7.4}$$

in the case of the simple shear of Figure 7a and for Figure 7b to

$$\sigma = \hat{\sigma}(\varepsilon) + \hat{\alpha}(\varepsilon)\varepsilon_{yy} + \hat{\beta}(\varepsilon)\varepsilon_y^2 \; , \tag{7.5}$$

where γ denotes shear strain, ε tensile strain and graphs of $\hat{\tau}(\gamma)$ and $\hat{\sigma}(\varepsilon)$ are again of the usual loop-type form. The equilibrium equations are simply

$$\tau = \tau_0 \; , \tag{7.6}$$

in the case of (7.4) or

$$\sigma = \sigma_0 \; , \tag{7.7}$$

in the case of (7.5).

By combining properly (7.4) through (7.7) it turns out that the form of modulations depicted in Figure 2 are also possible for strain modulations. In the case of Figure 7a the relevant physical phenomenon is termed localization of shear, while in the case of Figure 7b it is commonly known as necking.

In closing, we note that coupled modulations are also possible but the system of equations describing them can be quite complex. For example, in one dimension it is possible to arrive at the system

$$\left.\begin{array}{l} a_1(\varepsilon)\,\varepsilon_{xx} + b_1(\varepsilon)\varepsilon_x^2 = \sigma(\varepsilon,\alpha) - \bar{\sigma} \; , \\[2mm] a_2(\alpha)\,\alpha_{xx} + b_2(\alpha)\alpha_x^2 = \pi(\alpha,\varepsilon) - \bar{\pi} \; , \end{array}\right\} \tag{7.8}$$

where, for example, ε models the strain of a one-dimensional bar and α an additional variable for the system such as composition of second species, density of dislocations, or a birefrigence parameter. Obviously the analysis of (7.8) in far more complex than the single-component equations.

384

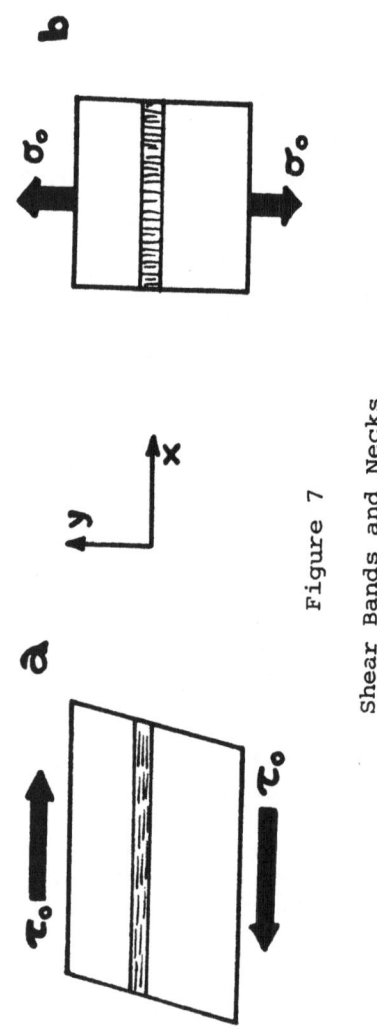

Figure 7

Shear Bands and Necks

ACKNOWLEDGMENT

Support of the National Science Foundation, the Corrosion Center and the Mathematics Institute (IMA) of the University of Minnesota is gratefully acknowledged. The support of the MM program of MTU has also been substantially helpful in the final stages of preparation.

REFERENCES

1. Aifantis, E.C., Microscopic Processes and Macroscopic Response, Mechanics of Engineering Materials, Eds. C.S. Desai and R.H. Gallagher, (John Wiley & Sons, 1984), pp. 1-22.

2. Aifantis, E.C., On the Problem of Diffusion in Solids, Acta Mechanica 37 (1980) 265-296.

3, Aifantis, E.C., Elementary Physicochemical Degradation Processes Mechanics of Structured Media, Ed., A.P.S. Selvadurai, pp. 301-317, Elsevier, Amsterdam-Oxford-New York (1981).

4. Aifantis, E.C. and J.B. Serrin, The Mechanical Theory of Fluid Interfaces and Maxwell's Rule, J. of Colloid and Interface Science, 96, 517-529 (1983).

5. Aifantis, E.C. and J.B. Serrin, Equilibrium Solutions in the Mechanical Theory of Fluid Microstructures, J. of Colloid and Interface Science, 96, 530-547 (1983).

6. Miklavcic, M. and E.C. Aifantis (forthcoming).

7., Unger, D.J. and E.C. Aifantis, Coupled Stress-Concentration Spinodals, in: Environm. Degradation of Engng. Matl's. Eds. M.R. Louthan, R.P. McNitt, R.D. Sisson, pp. 585-591, VPI (1981).

8. Unger, D.J. and E.C. Aifantis, Wave Propagation in Spinodally Decomposing Media, Proc. Workshop Media with Microstructures and Wave Propagation, Eds. E.C. Aifantis and L. Davison, Houghton, Jan. 1983 (to appear in Int. J. Eng. Sci.).

9. Cahn, J.W., On Spinodal Decomposition, Acta Metallurgica 9 (1961) 795-801.

10. Aifantis, E.C., Dislocation Kinetics and the Formation of Deformation Bands, Defects Fracture and Fatigue, Eds. G.C. Sih & J.W. Provan, (The Hague, Martinus Nijhoff Publishers 1983) pp. 75-84.

THE EFFECT OF STRAIN ON THE ELASTIC CONSTANTS OF COPPER

T. Tsakalakos and A. F. Jankowski

Rutgers University
Department of Mechanics and Materials Science
New Brunswick, NJ 08903

I INTRODUCTION

Several years ago a technique was developed to produce thin films containing one-dimensional composition modulation (16). These foils were suited for studies of the effects of structure on their mechanical properties because of their well-defined and controllable structure. It was found that their elastic moduli were substantially higher than the homogeneous ones (9). It became apparent that the enhanced elastic modulus effect must be due to some very fundamental changes introduced by the one-dimensional composition modulation. The concept of an elastic strain corresponding with the state of stress produced in these layered structures is considered as an explanation for the increased moduli experimentally observed.

Calculations on the elastic constants of metals have been widely performed in several investigations by means of pseudo-potential energy formulation (1,2,3). This method was adopted here to calculate elastic constants of copper as a function of lattice deformation. Special consideration was given to the symmetry change of the cubic crystal structure of Cu to that of a tetragonal symmetry for a biaxial mode of deformation.

2 TOTAL CRYSTAL ENERGY

The total crystal energy E for noble metals in a pseudo-potential approach is given as follows:

$$E = E_{es} + E_{fe} + E_{bs} + E_r. \tag{1}$$

E_{es} represents the electrostatic Coulomb energy of positive point charges in the uniform negative charge background and is often called the Madelung energy. E_{fe} represents the free electron energy which depends only on the crystal volume. E_{bs} represents the deviation of the energy of conductive electrons from that of a free electron gas and is called the band structure energy. E_r is the Born-Mayer type of potential for nearest neighbor ions and is called the ion-core repulsive energy.

The electrostatic energy E_{es} can be expressed as (1):

$$E_{es} = \frac{1}{2} \sum_{q \neq 0} \frac{4\pi Z^2 e^2}{q^2 \Omega_0} (\delta_{q,k} - \frac{1}{N}) \tag{2}$$

The free electron energy is given by (1,2,3):

$$E_{fe} = Z(\frac{2.21}{r_s^2} - \frac{.916}{r_s} - .115 + .031 \ln r_s) + \frac{bZ^2}{\Omega_0} \tag{3}$$

The ion-core repulsive energy E_r is given by (1,3):

$$E_r = \frac{\alpha}{a} \sum_{R=R_0} e^{-\beta R} \tag{4}$$

The band structure energy E_{bs} is expressed in terms of the pseudo-potential as (2):

$$E_{bs} = \sum_{q \neq 0} F(q) \tag{5}$$

The Nomenclature includes a detailed explanation of the above equations and pertinent terms.

3 ELASTIC CONSTANTS

A body accumulates potential energy while undergoing deformation. The expression for the potential energy of deformation per unit volume (or the elastic potential) for a tetragonal system is:

$$\overline{V} = \frac{1}{2} C_{11}(\varepsilon_{11}^2 + \varepsilon_{22}^2) + \frac{1}{2} C_{33}\varepsilon_{33}^2 + \frac{1}{2} C_{66}\varepsilon_{12}^2$$

$$+ \frac{1}{2} C_{44}(\varepsilon_{23}^2 + \varepsilon_{13}^2) + C_{12}\varepsilon_{11}\varepsilon_{22} + C_{13}\varepsilon_{33}(\varepsilon_{11} + \varepsilon_{22}) \tag{6}$$

The matrix of elastic constants corresponding to a tetragonal system is:

$$
C_{ij} = \begin{pmatrix}
C_{11} & C_{12} & C_{13} & 0 & 0 & 0 \\
 & C_{11} & C_{13} & 0 & 0 & 0 \\
 & & C_{33} & 0 & 0 & 0 \\
 & & & C_{44} & 0 & 0 \\
 & & & & C_{44} & 0 \\
 & & & & & C_{66}
\end{pmatrix}
\tag{7}
$$

The elastic constants may be found from the elastic potential by considering various forms of crystal lattice deformation. Formally,

$$
C_{ijk\ell} = \frac{\partial^2 \overline{V}}{\partial n_{ij} n_{k\ell}}
\tag{8}
$$

where n_{ij} are the components of the Lagrangian strain tensor:

$$
n_{ij} = \frac{1}{2} (\alpha_{ik}\alpha_{jk} - \delta_{ij})
\tag{9}
$$

in which

$$
\alpha_{ij} = \frac{\partial r_i}{\partial \ell_j}
\tag{10}
$$

where ℓ and r are the initial and final coordinate vectors. In tetragonal crystals the elastic constants can be determined as follows: C_{11} and C_{33} are derived by considering the second-order differential of the elastic potential with respect to a uniaxial volume expansion; C_{66} and C_{44} are derived by considering shear in one plane with volume conservation; C' and C'' are formulated considering compression in one direction with volume conserved. The transformation from the position vector R_{0i} to R_i corresponds to the crystal lattice deformation described above. The expressions for the elastic constants are as follows, where ε denotes the strain corresponding to the mode of deformation considered:

$$C_{11} = \left. \frac{\partial^2 \overline{V}}{\partial \varepsilon^2} \right|_{\varepsilon=0} \qquad (11)$$

where

$$R_{ix} = (1+\varepsilon)R_{oix} \qquad (11.1)$$

$$R_{iy} = R_{oiy} \qquad (11.2)$$

$$R_{iz} = R_{oiz} \qquad (11.3)$$

$$C_{33} = \left. \frac{\partial^2 \overline{V}}{\partial \varepsilon^2} \right|_{\varepsilon=0} \qquad (12)$$

where

$$R_{ix} = R_{oix} \qquad (12.1)$$

$$R_{iy} = R_{oiy} \qquad (12.2)$$

$$R_{iz} = (1+\varepsilon)R_{oiz} \qquad (12.3)$$

$$C_{66} = \left. \frac{\partial^2 \overline{V}}{\partial \varepsilon^2} \right|_{\varepsilon=0} \qquad (13)$$

where

$$R_{ix} = R_{oix} + \varepsilon R_{oiy} \qquad (13.1)$$

$$R_{iy} = R_{oiy} \qquad (13.2)$$

$$R_{iz} = R_{oiz} \qquad (13.3)$$

$$C_{44} = \left. \frac{\partial^2 \overline{V}}{\partial \varepsilon^2} \right|_{\varepsilon=0} \qquad (14)$$

where

$$R_{ix} = R_{oix} + \varepsilon R_{oiz} \qquad (14.1)$$

$$R_{iy} = R_{oiy} \tag{14.2}$$

$$R_{iz} = R_{oiz} \tag{14.3}$$

$$C' = \frac{1}{4} \frac{\partial^2 \overline{V}}{\partial \varepsilon^2} \Bigg|_{\varepsilon=0} \tag{15}$$

where

$$R_{ix} = (1+\varepsilon)R_{oix} \tag{15.1}$$

$$R_{iy} = R_{oiy} / (1+\varepsilon) \tag{15.2}$$

$$R_{iz} = R_{oiz} \tag{15.3}$$

$$C'' = \frac{1}{4} \frac{\partial^2 \overline{V}}{\partial \varepsilon^2} \Bigg|_{\varepsilon=0} \tag{16}$$

where

$$R_{ix} = R_{oix} \tag{16.1}$$

$$R_{iy} = (1+\varepsilon)R_{oiy} \tag{16.2}$$

$$R_{iz} = R_{oiz} / (1+\varepsilon) \tag{16.3}$$

and

$$C' = \frac{1}{2} (C_{11} - C_{12}) \tag{17}$$

$$C'' = \frac{1}{4} (C_{11} + C_{33} - 2C_{13}) \tag{18}$$

4 THE EFFECT OF APPLIED ELASTIC STRAIN

In order to study the effect of strain on the elastic constants, we will adopt the following scheme. We will strain the
lattice in such a mode as defined by the stresses produced by the
concept of coherency in layered structures as shown in Fig. 1. In
general, under homogeneous strain, each initial lattice site $\underset{\sim}{r}$
undergoes transformation into r' by a homogeneous distortion matrix $\underset{\sim}{A}$:

$$\underline{r}' = \underline{A} \, \underline{r} \qquad (19)$$

Since the total crystal energy involves summation over the reciprocal lattice vectors it is essential to establish the corresponding reciprocal lattice transformation. If q and q' are reciprocal lattice vectors before and after the deformation it can be easily shown that:

$$\underline{q}' = \underline{B} \, \underline{q} \qquad (20)$$

where

$$\underline{B} = (\underline{A}^{-1})^{t} \qquad (21)$$

That is, the reciprocal lattice transformation corresponding to the crystal lattice rearrangement described by the matrix A is its transposed inverse. The distortion matrix corresponding to the biaxial state of stress is

$$\underline{A} = \begin{pmatrix} 1+e_1 & 0 & 0 \\ & 1+e_2 & 0 \\ & & 1+e_3 \end{pmatrix} \qquad (22)$$

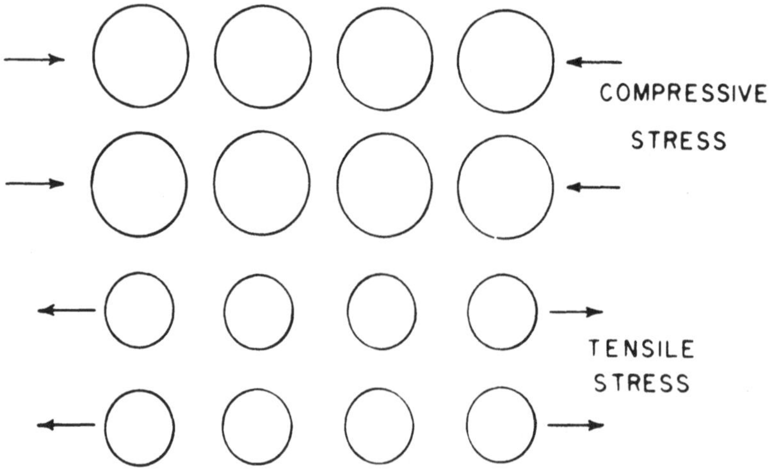

Fig. 1. Coherency between atomic layers in the [001] direction for a composition modulated structure.

where the principal strains are

$$\epsilon_1 = \epsilon \tag{22.1}$$

$$\epsilon_2 = \epsilon \tag{22.2}$$

$$\epsilon_3 = -2 \frac{C_{13}}{C_{33}} \epsilon \tag{22.3}$$

with ϵ defined as the applied elastic strain.

The biaxial state of stress in layered structures transforms a cubic crystal system into a tetragonal system. Thus, the six elastic constants in Eq. (7) are considered. The corresponding biaxial modulus for a tetragonal system is

$$Y[100] = C_{11} + C_{12} - \frac{2C_{13}^2}{C_{33}} \tag{23}$$

In order to calculate the elastic constants, considering the effect of applied elastic strain ϵ, the potential energy of deformation is equated with the total crystal energy:

$$\Omega_o \bar{V} = E \tag{24}$$

Accordingly, a second-order differential of the total crystal energy with respect to ϵ is taken, as given in Eqs. (11) to (16). Calculations to follow will reveal the effect of varying ϵ. When the strain is compressive (-) or tensile (+), the elastic constants reflect the difference in ϵ from zero.

An example of an elastic constant with expansion in terms of the total crystal energy is

$$
\begin{aligned}
C' = \frac{1}{4} \sum_{R'=R_o} & \left\{ \frac{\alpha\beta}{2\Omega_o} e^{-\beta R} \left[(\beta R+1) \left(\frac{R_x^2-R_y^2}{R}\right)^2 - R_x^2-3R_y^2 \right] \frac{1}{R} \right\} \\
& - \frac{1}{4\Omega_o} \sum_{q'\neq 0} \left\{ \left[(K_y^2+3K_x^2)\frac{1}{K} - (K_y^2-K_x^2)\frac{1}{K^3} \right] \frac{\partial F(q)}{\partial q} + \left(\frac{K_y^2-K_x^2}{K}\right)^2 \frac{\partial^2 F(q)}{\partial q^2} \right\} \\
& - \frac{1}{4\Omega_o} \sum_{q'\neq 0} \left\{ -4\pi Z^2 \left[((K_y^2+3K_x^2)\frac{1}{K} - (K_y^2-K_x^2)^2\frac{1}{K^3})\frac{-2}{\Omega_o q^3} + \left(\frac{K_y^2-K_x^2}{K}\right)^2 \frac{6}{\Omega_o q^4} \right] \right\}
\end{aligned}
\tag{25}
$$

A second example is

$$C_{66} = \sum_{R^i = R_0} \left\{ \frac{\alpha\beta}{2\Omega_0} e^{-\beta R} R_y^2 \left[\beta R + 1 \right) \left(\frac{R_x}{R} \right)^2 - 1 \right] \frac{1}{R} \right\}$$

$$- \frac{1}{\Omega_0} \sum_{q^i \neq 0} \left\{ \frac{K_x^2}{K} \left(1 - \left(\frac{K_y}{K} \right)^2 \right) \cdot \frac{\partial F(q)}{\partial q} + \left(\frac{-K_y K_x}{K} \right)^2 \frac{\partial^2 F(q)}{\partial q^2} \right\}$$

$$+ \frac{1}{\Omega_0} \sum_{q^i \neq 0} \left\{ -4\pi Z^2 \left[\frac{K_x^2}{K} \left(1 - \left(\frac{K_y}{K} \right)^2 \right) \frac{-2}{\Omega_0 q^3} + \left(\frac{-K_y K_x}{K} \right)^2 \frac{6}{\Omega_0 q^4} \right] \right\} \quad (26)$$

5 COMPUTATIONAL RESULTS

The noble metal copper is considered. Experimental values for the elastic constants of copper are well known, as shown in Table 1. These values correspond to those which are produced in the current investigation for e=0, the case of zero applied elastic strain. The current theoretical values for the elastic constants at zero applied elastic strain fall within 5% of the known experimental values at room temperature (8).

C_{IJ}	COPPER	NICKEL	GOLD
C_{11}	.168	.251	.192
C_{12}	.121	.150	.163
C_{44}	.075	.124	.420
AT ROOM TEMPERATURE			

Table 1. Experimental values for the elastic constants (TPa) of metals (8).

The applied elastic strain will vary from 3.0% compressive to 3.5% tensile. The results are dramatic for C_{11}, C_{12} and C_{66} as seen in Table 2 and Figure 2. An increase in magnitude occurs for compressive strain and a decrease in magnitude for tensile strain. For example, for 3% compressive strain: C_{12} increases 45%; C_{66} increases over 60%; and C_{11} increases approximately 20%. For 3% tensile strain: C_{11} decreases 10%; C_{12} decreases over 40%; and C_{66} decreases almost 35%. In contrast, however, the values of C_{33}, C_{13} and C_{44} remain relatively unchanged as seen in Figure 3 and Table 3. The effect of applied elastic strain is compounded for the biaxial modulus Y[100] as evidenced singularly in C_{11} and C_{12}. The magnitude of Y[100] doubles for just 3% compressive strain as shown in Figure 4.

ELASTIC CONSTANTS AND BIAXIAL MODULUS
OF COPPER VARY WITH ELASTIC STRAIN
UNITS: TPa

%STRAIN	C_{11}	C_{12}	C_{66}	Y[100]
-3.0	.200	.168	.127	.237
-2.5	.189	.153	.118	.217
-2.0	.183	.142	.109	.200
-1.5	.178	.135	.101	.178
-1.0	.174	.127	.092	.164
-0.5	.170	.120	.086	.146
0.0	.168	.116	.079	.124
0.5	.161	.103	.073	.111
1.0	.158	.095	.068	.092
1.5	.152	.084	.063	.084
2.0	.149	.079	.061	.077
2.5	.152	.077	.055	.064
3.0	.151	.070	.052	.053
3.5	.154	.069	.047	.041

Table 2. The elastic constants C_{11}, C_{12}, C_{66} (TPa) and the biaxial modulus Y[100] (TPa) for copper varying with the applied elastic strain (%).

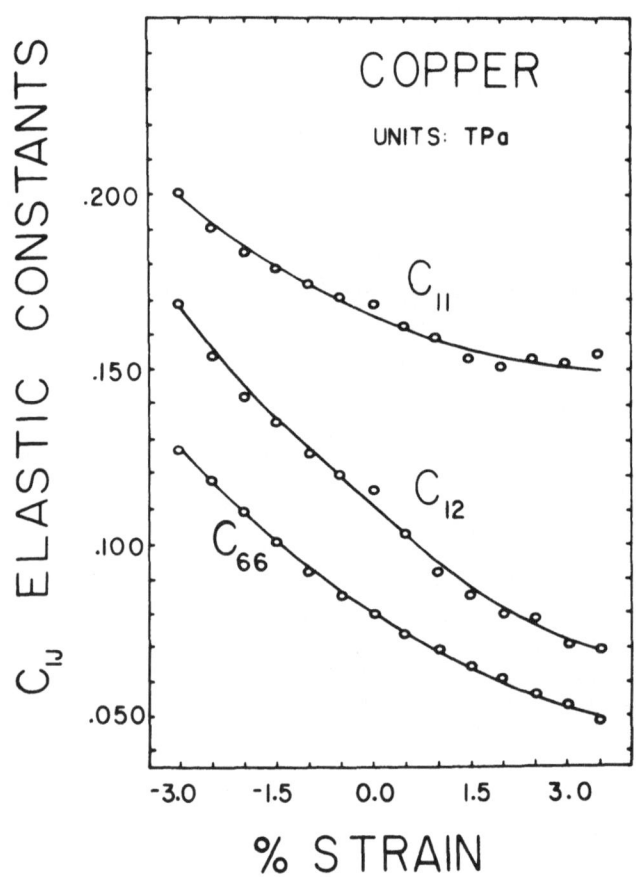

Fig. 2. C_{11}, C_{12} and C_{66} (TPa) of copper versus the applied elastic strain ϵ (%).

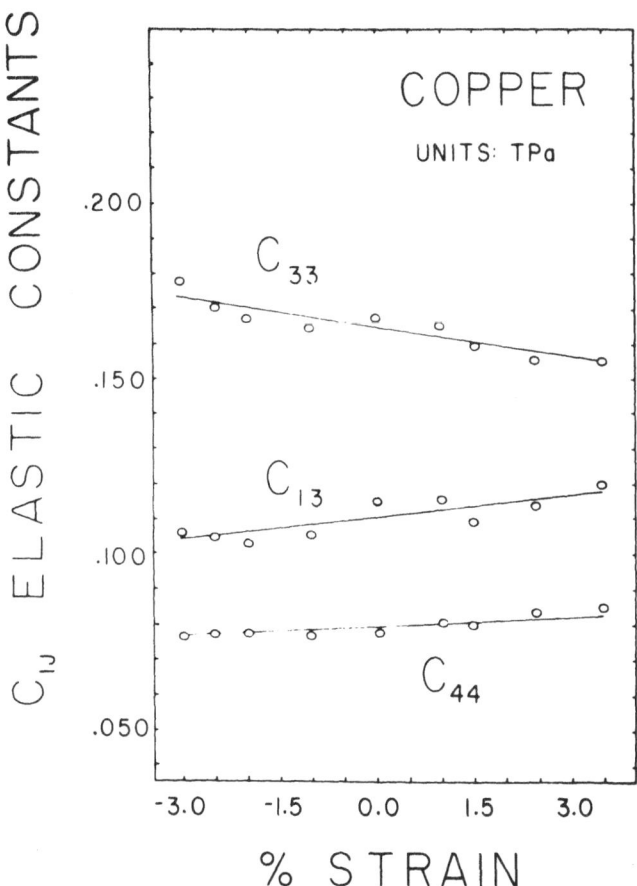

Fig. 3.   C₃₃, C₁₃ and C₄₄ (TPa) of copper versus the applied
          elastic strain ε (%).

ELASTIC CONSTANTS (TPa) OF COPPER

VARY WITH ELASTIC STRAIN

% STRAIN	C_{33}	C_{13}	C_{44}
-3.0	.177	.107	.077
-2.5	.170	.104	.078
-2.0	.167	.103	.078
-1.0	.165	.106	.078
0.0	.168	.116	.079
1.0	.166	.115	.081
1.5	.160	.110	.081
2.5	.155	.113	.084
3.5	.155	.119	.085

Table 3. The elastic constants C_{33}, C_{13} and C_{44} (TPa) of copper varying with the applied elastic strain (%).

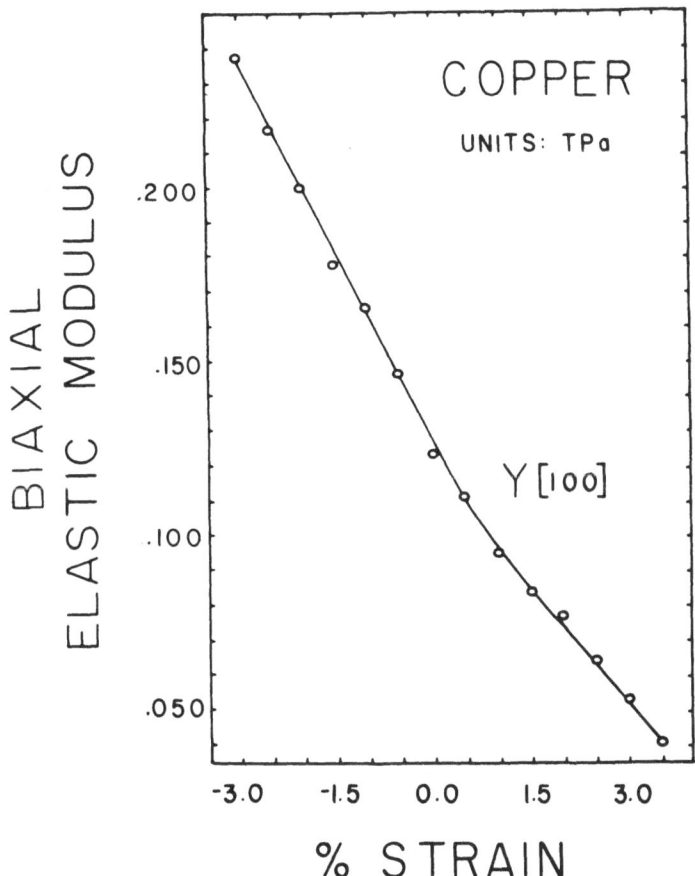

Fig. 4. The biaxial modulus Y[100] (TPa) of copper versus the
applied elastic strain ε (%).

6 THIRD ORDER ELASTIC CONSTANT APPROXIMATION

The theory of elasticity provides the means to approximate
the effect of elastic strain on the second order elastic constants.
A model in which the second-order elastic constants of the stressed
material are expressed as a function of the strains, and the elastic
constants at zero stress, has been developed [20,21]. This approach
will utilize values of the third-order elastic constants, experimentally obtained, in the approximation. A comparison with the
model formulated using the pseudopotential energy approach will be
made.

The second-order elastic compliances can be expressed,
utilizing the Lagrangian strain tensor, to the first order as:

$$C_{ijk\ell} = (\frac{\rho'}{\rho_0})\alpha_{im}\alpha_{jn}\alpha_{kp}\alpha_{\ell q}[C^0_{mnpq}+C^0_{mnpars}\eta_{rs}+\dots] \qquad (27)$$

where ρ_0 and ρ' are the specific masses of the material in the unstressed and stressed configurations, while C_{mnpq}^0 and C_{mnpars}^0 are the elastic constants of the second and third order in the unstressed state. For the biaxial state of stress (21):

$$\alpha_{11} = \alpha_{22} = 1+e; \; \alpha_{33} = 1+e_3; \; \alpha_{ij} = 0 \text{ for } i \neq j,$$

$$\eta_{11} = \eta_{22} = \frac{1}{2}[(1+e)^2-1]; \; \eta_{33} = \frac{1}{2}[(1-e_3)^2-1];$$

$$\eta_{ij} = 0 \text{ for } i \neq j,$$

$$(\frac{\rho'}{\rho_0}) = \frac{1}{(1+e)^2(1-e_3)}, \text{ and}$$

$$C_{ijk\ell} = (\frac{\rho'}{\rho_0})\alpha_{ij}\alpha_{jj}\alpha_{kk}\alpha_{\ell\ell}[C^0_{ijk\ell}+C^0_{ijk\ell mm}\eta_{mm}] \qquad (28)$$

Upon substitution we find:

$$C_{11} = \frac{(1+e)^2}{1+e_3}\{C^0_{11} + \frac{1}{2}(C^0_{111}+C^0_{112})[(1+e)^2-1] +$$

$$\frac{1}{2}C^0_{113}[(1-e_3)^2-1]\} \qquad (29)$$

$$C_{12} = \frac{(1+e)^2}{1+e_3}\{C^0_{12}+ \frac{1}{2}(C^0_{121}+C^0_{122})[(1+e)^2-1] +$$

$$\frac{1}{2}C^0_{123}[(1-e_3)^2-1]\} \qquad (30)$$

$$C_{44} = (1-e_3)\{C^0_{44} + \frac{1}{2}(C^0_{441}-C^0_{442})[(1+e)^2-1] +$$

$$\frac{1}{2}C^0_{443}[(1-e_3)^2-1]\} \qquad (31)$$

The third order elastic constants for Cu have been measured at room teperature (2): $C^0_{111} = -1.702$, $C^0_{112} = -.965$, $C^0_{123} = -.010$, $C^0_{144} = .034$, $C^0_{166} = -.832$, and $C^0_{456} = .012$ (units in TPa). Tetragonal symmetry relations are expressed in Voigt notation as follows: $11\longrightarrow1$, $22\longrightarrow2$, $11\longleftrightarrow22$, $33\longrightarrow3$, $23\longleftrightarrow32\longrightarrow4$, $13\longleftrightarrow31\longrightarrow5$, $44\longleftrightarrow55$, $12\longleftrightarrow21\longrightarrow6$. To equate the unknown third-order elastic constants, consider the cubic symmetry conditions, i.e., $C^0_{113} = C^0_{112}$ and $C^0_{443} = C^0_{442}$, for the unstressed state. In addition

$C_{112}^0 = C_{121}^0 = C_{122}^0$ and $C_{144}^0 = C_{441}^0 = C_{442}^0 = C_{443}^0$. The result for the calculations on Cu are (units in TPa) in the following table:

% Strain	C_{11}	C_{12}	C_{44}	Y[100]
-2.0	.180	.144	.081	.094
-1.0	.174	.130	.080	.110
-0.5	.172	.124	.080	.117
-0.4	.171	.122	.079	.119
-0.3	.170	.120	.079	.120
-0.2	.169	.119	.079	.120
-0.1	.168	.117	.079	.122
0.0	.168	.116	.079	.124
0.1	.168	.115	.079	.126
0.2	.167	.113	.079	.127
0.3	.166	.112	.079	.127
0.4	.165	.110	.079	.128
0.5	.163	.108	.078	.128
1.0	.160	.100	.078	.135
2.0	.152	.083	.077	.144

The biaxial modulus has been calculated in the previous table considering a cubic structure formulation. That is:

$$Y[100] = C_{11} + C_{12} - 2\frac{C_{12}^2}{C_{11}}$$

In comparison, the following results have been obtained utilizing the pseudopotential energy approach (units in TPa) for a tetragonal crystal structure.

% Strain	C_{11}	C_{33}	C_{12}	C_{13}	C_{44}	C_{66}	Y[100]
-0.4	.165	.162	.112	.105	.078	.084	.143
-0.3	.162	.160	.108	.102	.079	.084	.139
-0.2	.162	.160	.107	.103	.079	.081	.136
-0.1	.166	.162	.113	.104	.079	.080	.133
0.0	.168	.168	.116	.116	.079	.079	.124
0.1	.158	.163	.101	.105	.079	.079	.122
0.2	.155	.159	.096	.102	.080	.078	.120
0.3	.156	.163	.097	.105	.079	.076	.117
0.4	.156	.160	.097	.105	.080	.075	.115

A significant discrepancy exists upon comparison of the results for the biaxial modulus Y[100]. The pseudopotential energy approach predicts a 15% increase for 0.4% compressive strain whereas the approximation utilizing third-order elastic constants predicts a 4% decrease. So far, the comparative analysis has been made considering only those elastic constants associated with cubic crystal symmetry. However, in a stressed state, the structure is tetragonal. This introduces the following elastic constants:

$$C_{33} = \frac{(1+e)^2}{1-e_3}\{C_{33}^0 + \frac{1}{2}(C_{331}^0 + C_{332}^0)[(1+e)^2-1]+\frac{1}{2}C_{333}^0[(1-e_3)^2-1]\}$$

$$(32)$$

where, $C_{331}^0 = C_{333311}^0 = C_{222211}^0 = C_{111122}^0 = C_{112}^0$

$C_{332}^0 = C_{333322}^0 = C_{222233}^0 = C_{111133}^0 = C_{112}^0$

$C_{333}^0 = C_{333333}^0 = C_{222222}^0 = C_{111111}^0 = C_{111}^0$

$$C_{13} = \frac{(1+e)^2}{1-e_3}\{C_{13}^0 + \frac{1}{2}(C_{131}^0+C_{132}^0)[(1+e)^2-1] + \frac{1}{2}C_{133}^0[(1-e_3)^2-1]\}$$

$$(33)$$

where, $C_{131}^0 = C_{113311}^0 = C_{112211}^0 = C_{112}^0$

$C_{132}^0 = C_{113322}^0 = C_{112233}^0 = C_{123}^0$

$C_{133}^0 = C_{113333}^0 = C_{112222}^0 = C_{221111}^0 = C_{112}^0$

$$C_{66} = (1-e_3)\{C_{66}^0+\frac{1}{2}(C_{661}^0+C_{662}^0)[(1+e)^2-1] + \frac{1}{2}C_{663}^0[(1-e_3)^2-1]\}$$

$$(34)$$

where, $C_{661}^0 = C_{662}^0 = C_{663}^0 = C_{166}^0$.

The biaxial modulus should be computed using equation (23) for tetragonal symmetry. The results for the calculations on Cu are (units in TPa) in the following table:

% Strain	C_{33}	C_{13}	C_{66}	Y[100]
-2.0	.146	.100	.090	.187
-1.0	.157	.108	.085	.155
-0.5	.163	.112	.081	.142
-0.4	.164	.113	.081	.137
-0.3	.165	.113	.080	.135
-0.2	.166	.114	.080	.131
-0.1	.167	.115	.079	.127
0.0	.168	.116	.079	.124
0.1	.169	.117	.079	.121
0.2	.170	.118	.078	.116
0.3	.171	.119	.078	.112
0.4	.173	.119	.078	.111
0.5	.174	.120	.077	.105
1.0	.179	.124	.075	.088
2.0	.190	.133	.071	.049

Comparison of the calculated biaxial modulus Y[100] between the approximation utilizing the third order elastic constants, considering a tetragonal structure in the stressed state, and the pseudopotential energy show agreement for elastic strains between 1% tensile and 1% compressive. The necessity to consider the stressed state as tetragonal is clear. For example, for 0.5% compressive strain, the biaxial modulus increases 18% as compared to 15% for the pseudopotential and C_{ijk} approaches, respectively. For 1% compressive strain, both approaches show a 3.6% increase for C_{11}.

7 DISCUSSION

The concept of coherency in layered structure materials has widely utilized to explain a number of interesting and sometimes unusual phenomena, such as large discontinuities in the interdiffusivities of binary alloys. It is evident, however, that the same concept might have intriguing implications in the fundamental properties of metallic materials when the layer thickness approaches atomic spacings. In a conventional deformation process of a metallic material it is not possible to achieve large elastic strains (usually less than 1%) because plastic deformation sets in. In a modulated structure material which has layer thickness three to five atomic spacings, it is energetically favorable to develop large elastic strains, compressive or tensile, by a biaxial mode of deformation, rather than dislocations in the interfaces. This idea implies that the crystal lattice has undergone a large deformation which results in displacements of atoms up to 20% from their equilibrium positions. It is thus obvious that all layers developing compressive stresses (which actually constitute almost half of the material) will have an increase of the elastic constants.

The results of this study indicate a drastic increase in the elastic constants C_{11}, C_{12} and C_{66} as well as the biaxial modulus Y[100] of a single Cu layer under the biaxial coherency stress. These results are consistent with the experimental works of Tsakalakos and Hilliard (14) in the Cu-Ni system. For example, a composition modulated Cu-Ni foil of 66% Cu has a reported Y[100] value of 0.23 TPa. This value is over 50% greater than that of a bulk Cu-Ni alloy of the same Cu composition (Y[100]~.14 TPa). Similar results have been reported for Au-Ni, Cu-Pd and Ag-Pd composition modulated foils (9,15).

The observed increase in the elastic moduli of composition modulated structures is a consequence of elastic strains present in the atomic layers of the constituent metals. The "stiffening of the interatomic potentials" is due to atomic displacements.

The portion of the total crystal energy which has the greatest effect on the elastic constants is the ion-core repulsive energy term. This observation has similarly been made by Thomas (2), as well as Hiki and Granato (17).

It is therefore proposed that the supermodulus effect of composition modulated films is a result of the large coherency strains. The evidence of this study demonstrates that one might be able to increase the modulus by displacing atoms from their equilibrium positions which they would occupy in a perfect, infinite modulated structure, thereby "stiffening" the interatomic potential. On the other hand, when the spacial periodicity becomes large enough, the lattice loses its coherency as soon as dislocation grids at the interface between A-rich and B-rich layers (for an A/B modulated structure) become more favorable energetically. The loss of the coherency strains will have, as a result, the substantial decrease in the modulus to its expected value. Two basic reasons might cause the decrease of the modulus for ultrathin layers (one or two lattice parameters). First, the layers in such a small thickness are not well defined since some diffusion takes place during deposition. Thus, the modulation cannot achieve a high amplitude for small wavelengths. On the other hand, the effective layer thickness for coherency strains across each interface is about three to five atomic distances. Thus, the maximum value of the supermodulus effect occurs in a 0.8-1.2 nm layer thickness. This can explain why the modulus in composition modulated films has a maximum at about 1.6-2.5 nm wavelength.

In addition to the concept of coherency strains affecting the metallic bond and consequently the elastic constants, several other models have been proposed to explain the origin of the supermodulus effect. One of them, namely, the screening singularities in the electron energy of a crystal due to a Fermi Surface and Brillouin Zone interaction has recently received some support by several researchers (9,18,19). The present study, however, demonstrates clearly that the effect of large compressive elastic strains in a crystal influence substantially the metallic bonds and therefore the elastic constants of metals. It is a factor that cannot be neglected in explaining the unusual phenomena observed in modulated structures.

NOMENCLATURE

Z   charge valency = 1

Ω_0 $\frac{4}{3}\pi Z r_s^3 a_B^3 = \frac{1}{4}a^3 =$ atomic volume of primitive cell

Ω $N\Omega_0 \alpha R^3 \alpha K^{-3}$

N   multiplicity

r_s interelectronic distance

a_B Bohr radii

a   lattice parameter = 1.912 au (Cu)

K   reciprocal lattice vector = $(K_x^2 + K_y^2 + K_z^2)^{1/2}$

R   lattice vector = $(R_x^2 + R_y^2 + R_z^2)^2$

R_0 nearest neighbor distance

$\frac{bZ^2}{\Omega_0}$ first order perturbational energy in terms of the pseudo-potential and the parameter b is usually determined by the steady-state condition for the total energy E with the lattice constant used: $\frac{dE}{d\Omega_0} = 0$

α repulsive energy parameter = $.3415 \times 10^6$ ryd (1)

β repulsive range parameter = 13.84 au^{-1}

K_f Fermi wave number = $(\frac{9}{4}\pi)^{1/3} \cdot \frac{1}{r_s}$

q   K

$F(q)$ — energy wave number characteristic in terms of a local model pseudopotential and a linear screening function corrected for exchange and correlation among the conduction electrons

$$F(q) \quad \omega_0(q)^2 \frac{\chi(q)}{\epsilon(q)} D(q)$$

$\omega_0(q)$ — bare-ion pseudopotential form factor using Harrison's modified point-ion form (4)

$$\omega_0(q) \quad \frac{1}{\Omega_0}\left[\frac{-8\pi}{q^2} + \beta*/(1+q^2 r_c^2)^2\right]$$

$\epsilon(q)$ — dielectric screening function using Hubbard-Sham form (5)

$$\epsilon(q) \quad 1 - \frac{16\pi}{\Omega_0 q^2}(1-f(q))\chi(q)$$

$$\chi(q) \quad \frac{-3}{8K_f^2}\left(1 + \frac{1-\eta^2}{2\eta} n\left|\frac{\eta+1}{\eta-1}\right|\right)$$

$$f(q) \quad \frac{1}{2}\eta^2/\left[\eta^2+\frac{1}{4}+(2K_f\pi)^{-1}\right]$$

$$D(q) \quad e^{-.0256\eta^4}$$

$\eta \quad K/2K_f$

$\beta* \quad 59.10 \text{ Ryd } a_0^3$

$r_c \quad .516\, a_0 \text{ (Cu)}$

Notes: A normalizing multiplicity factor is applied to those summations taken over reciprocal space, as in Eqs. (26) and (27).

Differentials are numerically solved using a 3 point first order differential procedure and a 5 point second order differential procedure.

ACKNOWLEDGEMENTS

I wish to thank the support and encouragement shown to me by the National Science Foundation (NATO Advanced Study Institute Travel Award). and Grant DMR-78-26503.

REFERENCES

1. Soma, T. J. Phys. F: Metal Phys. 4 (1974) 2157.
2. Thomas, Jr., J. F. Phys. Rev. B 7 (1973) 2385.
3. Soma, T. Phys. Stat. Sol. (A) 29 (1975) 443.
4. Harrison, W. A. Pseudopotentials in the Theory of Metals (Benjamin, N. Y., 1966).
5. Sham, L. J. Proc. Roy. Soc. (London) A283 (1965) 33.
6. Lekhnitskii, S. G. Anisotropic Plates (Gordon and Breach Sci. Publ., N. Y., 1968).
7. Nye, J. F. Physical Properties of Crystals (Oxford, 1969).
8. Kittel, C. Introduction to Solid State Physics (Wiley, 3rd Ed., 1968).
9. Yang, W. M. C., T. Tsakalakos and J. E. Hilliard. J. Appl. Phys. 48 (1977) 876.
10. Ashcroft, N. W. and D. C. Langreth. Phys. Rev. 155 (1966) 682.
11. Ashcroft, N. W. and D. C. Langreth. Phys. Rev. 159 (1967) 500.
12. Tsakalakos, T. Thin Solid Films 75 (1981) 293.
13. Tsakalakos, T. Interdiffusion and Enhanced Elastic Modulus Effect in Composition Modulated Copper-Nickel Thin Foils, Ph.D. Thesis, Northwestern Univ., Illinois (1966).
14. Tsakalakos, T. and J. E. Hilliard. J. Appl. Phys. 54 (1983) 734.
15. Hénein, G. E. and J. E. Hilliard. J. Appl. Phys. 54 (1983) 728.
16. Cook, H. E. Ph.D. Thesis, Northwestern Univ., Illinois (1966).
17. Hiki, Y. and A. V. Granato. Phys. Rev. 144 (1966) 411.
18. Dunaeu, N. M. and M. S. Zakharora. JETP Lett. 20 (1974) 336.
19. Tsakalakos, T. Thin Solid Films 86 (1981) 79.
20. Wallace, D.C., Phys. Rev., 162 (1967) 776.
21. Henein, G.E., Ph.D. Thesis, Northwestern University, Illinois (1979).

CHAPTER 5: SPINODAL STRUCTURES

MECHANICAL BEHAVIOR OF SPINODAL ALLOYS

L. H. Schwartz

Department of Materials Science and Engineering and
Materials Research Center, Northwestern University,
Evanston, IL 60201, U.S.A.

ABSTRACT

The mechanical properties of spinodal alloys are reviewed
and related to structural features of the alloys. Following a
brief discussion of techniques for and results of structural
characterization, the empirical observations of yield strength
are compared with theory. In closing, the response of these
alloys to cyclic deformation is compared with that of homogeneous
and precipitation hardened alloys.

1 INTRODUCTION

Phase separation via spinodal decomposition (SD) has been
the subject of intense study since Hillert (1) first clarified
the origin of x-ray side bands accompanying the decomposition of
Cu-Ni-Fe alloys. These results were followed closely by the
linearized early stage theory of Cahn (2). The interest in these
SD alloys stems not only from the subtleties of the phase
transformation itself, but from the substantial improvements in
mechanical and magnetic properties which may result. Extended
reviews of these topics have appeared elsewhere and the reader is
referred to Soffa and Laughlin (3) for a summary of the
experimental characterizations of SD alloys and to Ditchek and
Schwartz (4) and Wagner (5) for reviews of the resultant property
changes. This paper will concentrate on the response of SD
alloys to monotonic and cyclic stress and a discussion, where
appropriate, of the relation of measured properties to
microstructure. While detailed discussion of the phase

decomposition is beyond the primary intent of this presentation, some of the most salient features will be touched on in the next section. The following two sections will focus on mechanical behavior of SD alloys under monotonic and cyclic loading.

2 MICROSTRUCTURE OF SPINODAL ALLOYS

While SD alloys derive their microstructure from a common origin, phase separation below the coherent spinodal, they differ markedly in actual structure as a result of the influence of several factors (2): kinetics - many alloys decompose so rapidly into two distinct phases during the quench that only late stage coarsening effects are observed; elastic anisotropy - while decomposition of AlZn (6) and FeCr (7) alloys is essentially isotropic at early stages of decomposition, others such as CuNiFe (8), CuNiSn (9) and CuNiCr (10) decompose more rapidly along [100] directions producing highly anisotropic microstructures; compositional assymetry - alloys close to the center of the two phase field produce interconnected microstructures, while those far from the center exhibit microstructures indistinguishable from those produced by nucleation and growth. Many details of these microstructural features are of interest in discussions of the phase transformation itself, but an understanding of the mechanical properties requires characterization of at least the extent of decomposition (amplitude of composition fluctuations), separation between regions of maximum and minimum solute concentration (wavelength), and spatial arrangement of such regions (morphology). Several experimental techniques are available for such studies: transmission electron microscopy for wavelength and morphology (3,4), x-ray (or neutron) diffraction for amplitude and wavelength (4), and field ion atom probe microscopy (5) for all three parameters in many alloy systems.

The generalized composition profile for an alloy of average composition C_o may be written in terms of its Fourier components as:

$$C(\tilde{r}, t) - C_o = A(\tilde{\beta}, t)\exp(i\tilde{\beta}\cdot\tilde{r})d\tilde{\beta}, \tag{1}$$

where the magnitude of the wave vector $\beta = 2\pi/\lambda$. Theories of spinodal decomposition (11,12,13) are more conveniently expressed in terms of the auto correlation function which is directly proportional to the intensity scattered in a diffraction experiment

$$I(\tilde{\beta}, t) = A(\tilde{\beta}, t)^2 \tag{2}$$

An example of such a scattering pattern is shown in Fig. 1 for an Fe-32Cr alloy aged for various times at $500^\circ C$ and studied using

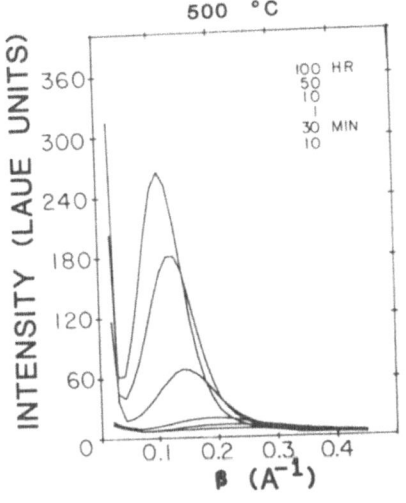

Fig. 1 Small angle scattering intensity, $I(\beta,t)$ vs β for Fe-32Cr quenched from 860°C and aged for 500°C (14).

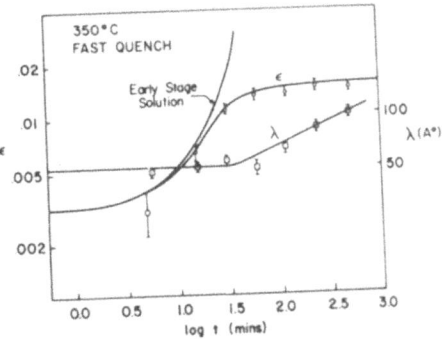

Fig. 2 Strain amplitude and wavelength obtained from analysis of high angle side bands for Cu-10w/oNi-6w/oSn aged at 350°C (17).

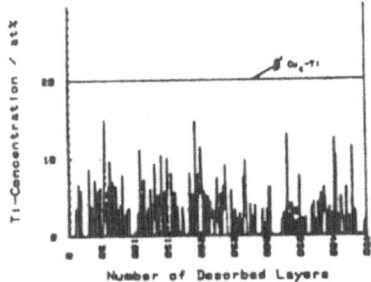

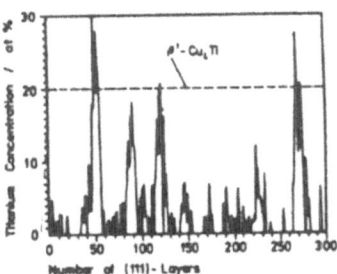

Fig. 3 Concentrations profiles of Cu-2.7a/oTi aged for 10 min (left) and 500 min (right) at 350°C (22).

small angle neutron scattering (14). The general features of
increasing intensity and shift toward smaller angles with
increasing aging times are in qualitative accord with nonlinear
theories of SD (11,12,13). One specific feature is worthy of
note here. In these theories, the position of the intensity
maximum, β_m is found to vary in time as

$$\beta_m = \beta_0 t^{-\phi} \tag{3}$$

where $\phi < 0.2$ for early stages while the amplitude of the
composition fluctuation is increasing, and $\phi = 1/3$ in the late or
coarsening stages when two phase separation is essentially
complete and particle size coarsening dominates. Results
confirming elements of these predictions have now been obtained
by several workers including a recent study of Al-22Zn (14) and
Fe-32Cr (14,16). These observations emphasize the point made by
both Ditchek and Schwartz (4) and Wagner (5) that since for most
alloys studied $\phi \sim 1/3$, most of the available data for strength
in SD alloys is for samples in the late coarsening stages in
which the amplitude has neared its maximum and wavelength is
changing.

Small angle scattering allows direct comparison with
theories of SD formulated in reciprocal space, but mechanical
behavior is best understood in terms of real space structure.
Simplifying models of the structure have been used by many
workers as a basis for calculating the strength from structural
parameters and for extracting these approximate structural
parameters from either the small angle scattering or the
corresponding satellites which appear in the vicinity of every
Bragg peak and are referred to as "side-bands" or "satellites".
The simplest of these models approximates the structure as three
perpendicular cosinusoidal waves with wavenumber β_m and
composition profile:

$$C(t) - C_0 = A(t)(\cos\beta_m x + \cos\beta_m y + \cos\beta_m z). \tag{4}$$

The time development of β_m is obtained empirically and $|A(t)|^2$ is
proportional to the area of the small angle scattering peak. The
situation is more complicated when data is sought from high angle
side bands, since these include scattering effects from
variations in interatomic distance as well as composition.
Separation of these effects requires further resort to models.
The most exacting of these model dependent analyses of sidebands
was carried out by Ditchek and Schwartz (17) in their study of
the Cu-Ni-Sn system. The results of that study for a Cu-10w/oNi-
6w/oSn alloy aged at 350°C are shown in Fig. 2. Instead of
amplitude, A, the extent of decomposition is characterized by the
quantity $\varepsilon = A\eta$ where $\eta = (d1na/dc)/a$ is the elastic distortion
parameter and a is the alloy lattice parameter. In this study

the structural model evolved from cosinusoidal wave-like for the earliest aging times to square wave-like for later times, guided by a one-dimensional nonlinear theory of SD due to Tsakalakos (18). Solid lines in Fig. 2 are best theoretical fits to the data.

No matter how subtle the models may be, model dependent analyses of side-bands are no substitute for the direct atomic compositional analysis offered by the field ion microscopy and atom probe techniques (FIM and AP). This is particularly true in highly assymetric alloys in which the assumption of a sinusoidal composition profile is totally inappropriate. Wagner (5) has shown, for example, that in the Cu-5a/oTi system discrete particles of composition near Cu_4Ti have been revealed by TEM (19) and FIM (20) at aging times and temperatures for which the x-ray sideband analysis of Miyazaki et al (21) suggests maximum Ti compositions of only 9.5 at.pct.

Using the atom probe technique, Biehl and Wagner (20) have determined the Ti concentration profiles along a [111] direction in a Cu-2.7a/oTi alloy. The development of these profiles is revealed in Fig. 3 for 10 min and 500 min aging at 350°C (22). Note the rapid development of regions of small extent with compositions in the range of Cu_4Ti. Fourier analysis of the concentration profiles showed that in the early decomposition stages the composition modulations may be described by quasi-sinusoidal fluctuations having a mean wavelength λ_m but with variation $\lambda_m \pm \lambda_m/10$. Assuming a Gaussian distribution of wavelengths Biehl and Wagner synthesized the arrangement of the centers of Ti enriched regions in a (100) plane as shown in Fig. 4. Although quite irregular, the modulated structure is easily recognizable. Thus even in this alloy far from the center of the miscibility gap, one may not be too far wrong in using Eq. 4 to approximate the composition profile, but great caution must be observed in interpreting scattering data using this model to extract $A(t)$.

3 MONOTONIC DEFORMATION

Phase separation, whether via nucleation and growth or spinodal decomposition, may lead to increased yield strength in alloys. For example, when Fe-32Cr is quenched from 860°C, the observed room temperature yield strength in polycrystalline samples is 210 MPa (for grain size of 65 μm). Subsequent aging below the spinodal results in the dramatic increases in yield stress depicted in Fig. 5 (23). This section of this paper reviews the search for understanding of the origin of this effect.

416

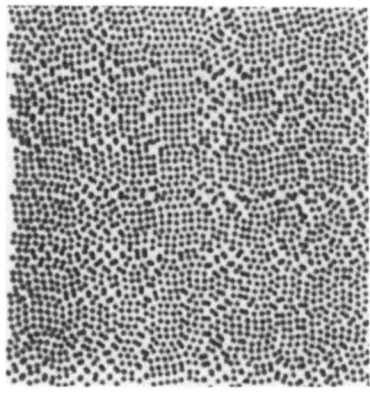

Fig. 4 Computer simulation of the arrangement of the centers of Ti enriched regions in a (100) plane for a narrow distribution of wavelengths in Cu-2.7Ti aged at 350°C (20).

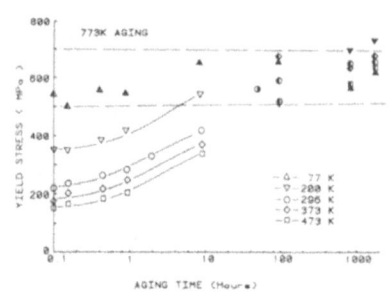

Fig. 5 Increase in yield stress of Fe-32Cr which results from aging for the indicated times at 773 K. Measurements carried out between 77 K and 473 K. Open symbols refer to samples in which no twinning was evident while in all other samples twinning was observed (23).

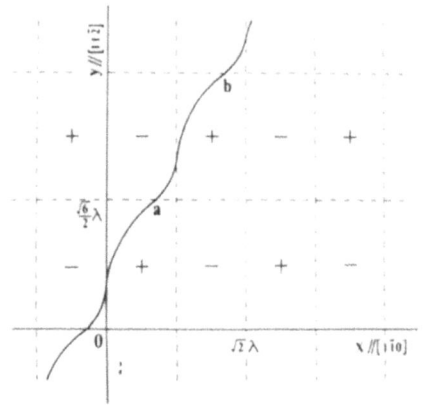

Fig. 6 Schematic representation of a mixed dislocation on a (111) plane in SD fcc alloy. The positions of maximum (+), minimum (-) and zero force on the dislocation are indicated (29).

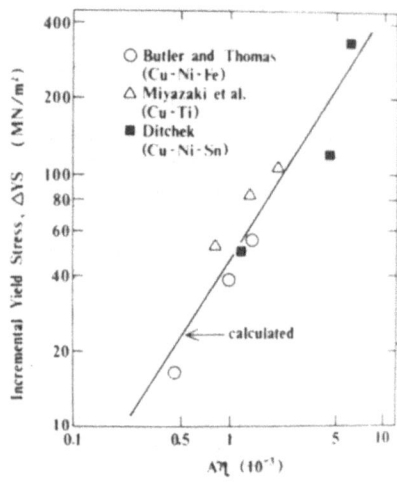

Fig. 7 Experimental data of incremental yield stress for three alloy systems compared to the theoretical expression of Eq. 6 (29).

Shortly after the development of the linearized theory of SD, Cahn (24) examined the effect of the internal coherency stress on dislocation mobility. The coherency stress, a consequence of the separation of atoms of different sizes, was evaluated for a composition profile of the form given in Eq. 4. Cahn solved the force balance equation for a screw or edge dislocation under the action of the coherency stress and an applied stress, and predicted that the incremental yield stress (increase above that of the homogenous alloy) should be:

$$\Delta YS = BA^2\eta^2\lambda \tag{5}$$

where B is a constant. It should be emphasized that this calculation, based on a composition model which is most nearly realistic in the earliest stages of decomposition, can only be expected to apply there, if at all. The effects of subsequent coarsening of the microstructure on hardening have been reviewed by Wagner (5), and this present paper will concentrate on the early stage effects.

In an early test of Cahn's prediction, Butler and Thomas (8) studied yielding in Cu-Ni-Fe alloys, using TEM to determine the average λ from sidebands and magnetic Curie point measurements to estimate the amplitude of the decomposition. Their results disagreed in every respect with Cahn's prediction: the observed incremental yield stress increased linearly with A, was independent of (or weakly dependent on) λ, and was at least an order of magnitude larger than predicted. This general disagreement with theory was confirmed in subsequent studies of Cu-Ni-Cr (25), Cu-Ti (21), Ni-Ti (26) and Cu-Ni-Sn (27) and led to the search for alternative theories.

The first of these alternatives actually applied only to the very late stage of SD in Cu-Ni-Fe. In this case coarsening has proceded and a realistic model may be based on large well defined regions with differing lattice constants joined at a coherent interface. Dahlgren (28) developed a model of the coherency stress for such a system and calculated the resultant yield stress in good agreement with experiments in magnitude, independence of λ and first order dependence on A.

The clear disaccord between theory and experiment for the strength of early stage SD alloys and success of a coherency stress model in explaining the late stage strength of Cu-Ni-Fe, led Kato, Mori and Schwartz (KMS) to reexamine the early stage theory (29). Using the same model proposed by Cahn, KMS solved the force balance equation describing stability of a dislocation of mixed character in the presence of coherency and applied stresses. The calculated critical resolved shear stress for such mixed dislocations was found to be

$$\tau_{fcc} = A\eta Y/\sqrt{6} \tag{6}$$

for (111) slip in fcc alloys, where Y is a combination of elastic constants. The origin of the independence of Eq. 6 on λ and the first order dependence on A may be seen by examination of Fig. 6. A pure edge (lying along $[11\bar{2}]$) or screw (lying along $[1\bar{1}0]$) experiences forces of alternating sign along its line resulting from the internal stress minima and maxima and the averaging of these effects leads to a low net force. A mixed dislocation, however, experiences stress extrema of one sign along its length leading to a net force proportional to the maximum coherency stress (proportional to A) and independent of the separation of stress maxima. KMS suggested that while micro-yielding may be associated with the movement of screw or edge dislocations, pinning would lead to dislocations of mixed character. The macroscopic yield stress is then attributed to the movement of these mixed dislocations. This view is supported by the position taken by Kocks (30) who emphasized in this context that "even those dislocations that are hardest to move must do so to make long range slip possible."

Quantitative comparison of experiment with theory is difficult since so few systems have been studied, and of these, one expects Eq. 6 to apply only to the very earliest stages. KMS selected the results for Cu-Ni-Fe (8), Cu-Ti (21) and Cu-Ni-Sn (27) and found good accord as displayed in Fig. 7. In this comparison, KMS multiplied the composition amplitude quoted by Miyazaki et al (21) by a factor of three to account for the fact that it is A/3 which is measured in the x-ray side-band experiment. Since this cannot correct for the errors inherent in analysis of side-bands for assymmetric alloys, one must recognize that the extent of the good agreement for Cu-Ti may be fortuitous. Nevertheless, these results suggest that the origin of the hardening of copper-based fcc spinodal alloys is indeed the internal coherency stress.

The KMS theory has been recast in terms of dislocation energy rather than force (31) and this formalism used to discuss the effects of thermal activation on dislocation motion in a periodic force field (32). These calculations confirm the empirical observations that hardening by spinodal decomposition is temperature independent (Cu-Ni-Sn Fig. 8 (33) and Fe-Cr (ref. 34 and Fig. 5)) and that the observed activation volumes for dislocation motion (33,34) are not related to the periodic internal stress field, but to other solid solution or chemical short range effects.

Kato (35) has recently extended the calculation of coherency stress hardening to bcc spinodal alloys. For the commonly observed slip systems, he estimates that the critical resolved

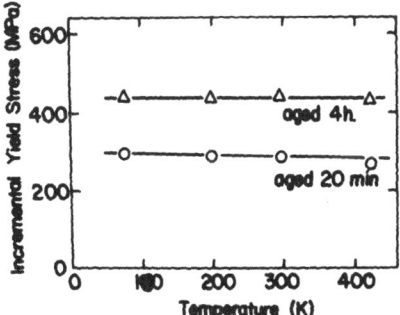

Fig. 8 The temperature
independence of the
incremental yield stress in
Cu-10w/oNi-6w/oSn aged at
350°C for 20 min and 4 hr
(33).

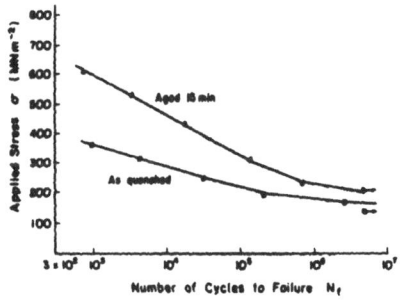

Fig. 9 Applied stress vs
number of cycles to failure
for as-quenched and aged Cu-
10w/oNi-6w/oSn. The arrows on
the data points indicate that
the specimen had not failed at
the termination of the test
(39).

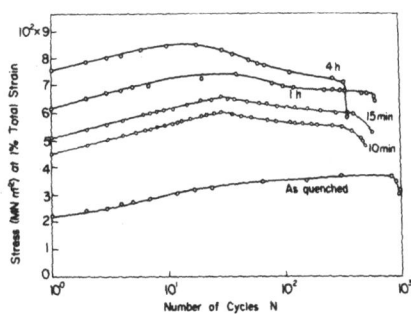

Fig. 10 Stress at 1.0% total
strain limit as a function of the
number of cycles for Cu-10w/oNi-
6w/oSn aged for 15 min at 350°C
(40).

shear stress

$$\tau_{bcc} = gA\eta Y \tag{7}$$

where $0.4 \leq g \leq 0.5$ depending on slip system. In this same
paper, he points out that hardening due to periodic variations in
elastic modulus may be quite important in the Fe-Cr system and
estimates this contribution as

$$\tau'_{bcc} = 0.65 \ \Delta G b/\lambda \tag{8}$$

where ΔG is the difference in shear modulus between the two
phases and b is the magnitude of the Burgers vector. An
expression of the form of Eq. 8 had been obtained earlier by
Ghista and Nix (36), but no spinodal alloy has exhibited the
decrease in strength in the coarsening regime which is implied by
the λ^{-1} dependence of this formula. Using the formulae of Eqs. 7
and 8 and the assumption that the composition difference between
Cr-rich and Cr-poor regions is 70 at.pct., Kato estimated a
tensile yield stress of 360 MPa in reasonable agreement with the
results of Lagneborg (34) and of Park and Schwartz (23). However
a detailed test comparing theory and experiment in this system
awaits completion of ongoing mechanical properties studies by
Park and Schwartz (23), neutron small angle scattering studies by
LaSalle and Schwartz (14) and field ion microscopic studies by
Brenner, Miller and Soffa (7).

4   CYCLIC DEFORMATION

While considerable attention had been paid to the response
of spinodal alloys to monotonic deformation, prior to 1980 only
one paper had appeared describing results of experiments in which
SD alloys were subjected to cyclic stress. It had been
speculated by Ham et al (37) that the homogeneous decomposition
of the spinodal alloys might reduce the likelihood of localized
soft regions which were believed to be precursors to the
development of deformation leading to ultimate failure.
Consequently, Ham et al. investigated the fatigue behavior of SD
Cu-Ni-Fe in high cycle stress-controlled tests. They reported
that after appropriate heat treatment the alloy has a fatigue
ratio of 0.5 at 10^6 cycles and exhibits an apparent fatigue
limit. Since most copper based precipitation hardened alloys
have fatigue ratios (stress to fail at 10^6 cycles $\div$ ultimate
tensile strength) in the range of 0.3-0.4, these results were
considered evidence that SD might be a mechanism by which both
high strength and fatigue properties could be achieved. It must
be noted that the heat treatments used by Ham et al. brought the
Cu-Ni-Fe well into the coarsening regime, possibly to the point
where loss of lattice coherency occurs and a dislocation network

develops (38). It may be speculated that it was that dislocation network rather than the spinodal structure which stabilized this alloy to cyclic deformation, for subsequent studies of early stage SD alloys show quite different results.

A fatigue study of early stage SD Cu-10w/oNi-6w/oSn was recently carried out by Quin and Schwartz (39,40). In contrast with the results of Ham et al, Quin and Schwartz found no substantial improvement in fatigue resistance. In fact, the fatigue ratio dropped from 0.36 for as-quenched to 0.33 for samples aged 15 min. Displayed in Fig. 9 are the stress to failure tests (S-N) for these samples (39), showing only modest improvement in the apparent fatigue limit while monotonic yield strength is improved by a factor of 2.5 for this same heat treatment. Results similar to these were noted by Sinning and Haasen (41) and Sinning (42) in their study of Cu-4Ti single crystals. In fact their S-N curves for as-quenched and 10 min age at 390°C look remarkably similar to those of Fig. 9 while for 1000 min aging at 390°C the fatigue limit is actually equal to that of the as-quenched alloy. Sinning and Haasen (41) did not carry out transmission electron microscopy studies of their samples, but did see regular surface markings which they attributed to the development of persistent slip bands. In their study, Quin and Schwartz also saw regular surface slip markings, but TEM studies showed only extensive deformation on single planes (dislocations piled up with average spacing of 55 ± 5 nm). This highly localized deformation appeared to be the precursor for crack formation.

Low cycle fatigue studies on Cu-Ni-Sn (40) do lead to the development of the TEM microstructure known as persistent slip bands (PSB) (43), and provide a clue about the nature of cyclic deformation in SD alloys. The stress at 1.0% total strain is plotted in Fig. 10 for Cu-10w/oNi-6w/oSn aged for various times at 350°C. Following a brief increase in stress similar to that observed for the as-quenched samples, all SD heat treatments produced cycle softening. When examined, x-ray sideband intensities decreased in this softening regime, giving strong evidence for a mechanism of localized demodulation of the spinodal microstructure by dislocation motion in the PSB. This result was confirmed by TEM examination which showed the lack of the microstructural features of SD in the PSB: The details of this demodulation effect are not understood. Quin and Schwartz (39) noted that the motion of the first dislocation in a composition modulated structure has the effect of decreasing the modulation amplitude on the slip plane. This would decrease the stress necessary for the motion of the next dislocation since yield stress is proportional to amplitude. This effect would lead to extensive deformation on a single plane, as seen in the high cycle (low stress) experiments; however, the development of

422

the PSB and demodulation of the spinodal microstructure may require cross slip which occurs at higher strains.

ACKNOWLEDGEMENTS

The author wishes to acknowledge his students, former and present, who have also been his teachers in many of the studies described in this review: Drs. Brian Ditchek, Mary Quin, Masaharu Kato for their earlier contributions and Messrs. Jerry LaSalle and Kyung-Ho Park for kindly allowing presentation of their unpublished results. This research was sponsored by the National Science Foundation under grant #DMR82-11089 and earlier grants. Much of the research at Northwestern was carried out in the facilities of the Materials Research Center, supported by the NSF-MRL under grant #DMR79-23573.

REFERENCES

1. Hillert, M. D.Sc. Thesis, Mass. Inst. Technol 1956.
2. Cahn, J. W. Acta Metall. 9 (1961) 795; 10 (1962) 179.
3. Soffa, W. A. and D. E. Laughlin. Proceedings of an International Conference on Solid-Solid Phase Transformations (Warrendale, PA, Am. Inst. Min. Met. and Pet. Eng., 1982, p.159).
4. Ditchek, B. and L. H. Schwartz. Ann. Rev. Mater. Sci. (1979) 219.
5. Wagner, R. Czech. J. Phys. B31 (1981) 198.
6. Rundman, K. B. and J. E. Hilliard. Acta Metall. 15 (1967) 1025.
7. Brenner, S., D. Miller and W. A. Soffa. Proceedings of an International Conference on Solid-Solid Phase Transformations (Warrendale, PA, Am. Inst. Min. Met. and Pet. Eng., 1982, p. 191).
8. Butler, E. P. and G. Thomas. Acta Metall. 18 (1970) 347.
9. Schwartz, L. H., S. Mahajan, and J. T. Plewes. Acta Metall. 22 (1974) 601.
10. Chou, A., A. Datta, G. H. Meier, and W. A. Soffa. J. Mat. Sci. 13 (1978) 541.
11. Langer, J. S., M. Bar-On, and H. D. Miller. Phys. Rev. A 11 (1975) 1417.
12. Binder, K. and D. Stauffer. Adv. Phys. 25 (1976) 343.
13. Marro, J., A. B. Bortz, M. Kalos, and J. L. Lebowitz. Phys. Rev. B 12 (1975) 2000.
14. LaSalle, J. and L. H. Schwartz. Unpublished research.
15. Farouhi, A. R. Ph.D. Thesis, University of California-Berkeley 1982.
16. Katano, S. and M. Iizimi. Unpublished research.
17. Ditchek, B. and L. H. Schwartz. Acta Metall. 28 (1980) 807.

18. Tsakalakos, T. Ph.D. Thesis, Northwestern University 1977.

19. Laughlin, D. E. and J. W. Cahn. Acta Metall. 23 (1975) 329.

20. Biehl, K.-E. and R. Wagner. Proc. 27th Int. Field Emission Symposium (Tokyo, 1980, p. 267).

21. Miyazaki, T., E. Yajima, and H. Suga. Trans. JIM 12 (1971) 119.

22. Biehl, K.-E. and R. Wagner. Proceedings of an International Conference on Solid-Solid Phase Transformations (Warrendale, PA, Am. Inst. Min. Met. and Pet. Eng., 1982, p. 185).

23. Park, K. -H. and L. H. Schwartz. Unpublished research.

24. Cahn, J. W. Acta Metall. 11 (1963) 1275.

25. Mihalisin, J. R., E. L. Huston and F. A. Badia. Second International Conference on the Strength of Metals, vol. 2 (1970, p. 679).

26. Dawance, M. M., D. H. Ben Israel and M. E. Fine. Acta Metall. 12 (1964) 705.

27. Ditchek, B. Ph.D. Thesis. Northwestern University 1978.

28. Dahlgren, S. D. Metall. Trans. 8A (1977) 347.

29. Kato, M., T. Mori and L. H. Schwartz. Acta Metall. 28 (1980) 285.

30. Kocks, U. F. Mater. Sci. Eng. 27 (1977) 291.

31. Kato M., T. Mori and L. H. Schwartz. Mater. Sci. Eng. 51 (1981) 25.

32. Kato, M., T. C. Lee and S. L. Chan. Mater. Sci. Eng. 54 (1982) 145.

33. Kato, M. and L. H. Schwartz. Mater. Sci. Eng. 41 (1979) 137.

34. Lagneborg, R. Acta Metall. 15 (1967) 1737.

35. Kato, M. Acta Metall. 29 (1981) 79.

36. Ghista, D. N. and W. D. Nix. Mater. Sci. Eng. 3 (1969) 293.

37. Ham, R. K., J. S. Kirkaldy and J. T. Plewes. Acta Metall. 15 (1967) 861.

38. Livak, R. J. and G. Thomas. Acta Metall. 19 (1971) 497.

39. Quin, M. P. and L. H. Schwartz, Mater. Sci. Eng. 54 (1982) 121.

40. Quin, M. P. and L. H. Schwartz, Mater. Sci. Eng. 46 (1980) 249.

41. Sinning, H.-R. and P. Haasen. Scripta Metall. 15 (1981) 85.

42. Sinning, H.-R. Mater. Sci. Eng. 55 (1982) 247.

43. Kuhlmann-Wisldorf, D. and C. Laird. Mater. Sci. Eng. 27 (1977) 137.

A NUMERICAL STUDY OF KINETICS IN SPINODAL DECOMPOSITION

T. Tsakalakos and M.P. Dugan

Department of Mechanics and Materials Science, Rutgers University, College of Engineering, Piscataway, NJ 08854 and RCA Solid State Technology Center, Route 202, Somerville, NJ 08876

SUMMARY

A numerical scheme is provided for the time evolution of a periodic composition wave leading to a stationary configuration. Two specific cases of growth are considered in which the initial conditions defined as a single cosinusoidal wave and as a super-position of eighty cosinusoidal waves of different wavelengths. A comparison of the numerical solution with existing analytical solution is shown, resulting in a simple semi-empirical growth law for studying the kinetics of spinodal decomposition in alloys.

1. INTRODUCTION

J. W. Cahn(1,2) developed a generalized diffusion equation that explained the growth of the composition waves and their crystallographic orientation. The solution of the diffusion equation by Cahn predicted an exponential growth rate and is valid only for the very early stages, i.e., for infinitesimal fluctuations. This solution fails to describe the time evolution of the microstructure at long times. In a number of subsequent(3,4) investigations of the later stages of spinodal decomposition, some qualitative observations have been provided as to how a composition wave of large amplitude will grow. Most of the investigations state that at later stages the system spends most of its time in configurations which are called stationary states. These states are periodic waves of certain shape and amplitude depending on the wavelength and are solutions of the time-independent non-linear diffusion equation. Recently, Tsakalakos(5) has shown that the analytical expressions of the stationary states in a one-dimensional system are Jacobian elliptic functions.

Despite the success of the linear theory in describing the early stages of spinodal decomposition and the analytical solutions for the stationary states there is no current data that actually link the infinitesimal periodic wave with its corresponding stationary state. Nor has there been any quantitative growth rate derived at intermediate and later stages of the decomposition. Pertubation techniques(5) have shown there is a "slow down" effect from the exponential growth rate of the linear regime when the system approaches its stationary state. It is the purpose of this paper to provide an exact numerical solution of the non-linear diffusion equation in order to obtain time evolution of a spinodal structure. The generalized diffusion equation in one-dimension as derived by Cahn(1,2) is given by

$$\frac{\partial u}{\partial t} = \frac{\partial}{\partial x} \left(\tilde{D} \frac{\partial}{\partial x} u \right) - 2\tilde{K} \frac{\partial^4 u}{\partial x^4} \tag{1}$$

where $u(x,t) = c(r,t) - c_0$, in which $c(x,t)$ is the composition at the distance x and at time t and c_0 is the average composition of the binary alloy, $\tilde{D}$ is the interdiffusion coefficient which in general for large amplitudes, depends on the composition, and the last term in the diffusion equation is called "gradient" energy term and was introduced by Cahn and Hilliard(6) to account for the increase of the free energy due to the gradient of composition. In order to maintain non-linear terms up to the fourth order in this differential equation a quadratic dependence of $\tilde{D}$ on the composition variation is required:

$$\tilde{D} = D_0 + D_1 u + D_2 u^2 \tag{2}$$

It can be readily shown that by using the scaling quantities

$$\tau = 4\tilde{\kappa}/D_0^2 \text{ and } \ell = 2\sqrt{\tilde{\kappa}/|D_0|}, \quad \nu = D_2/|D_0|$$

with $D_0 < 0$, $D_1 = 0$, $D_2 > 0$ we obtain the reduced equation:

$$\frac{\partial u}{\partial t} = \frac{\partial}{\partial x} \left[(-1+\nu u^2) \frac{\partial u}{\partial x} \right] - \frac{1}{2} \frac{\partial^4 u}{\partial x^4} \tag{3}$$

Since the solution is periodic with period λ, the initial and boundary conditions are:

$$u(x,o) = Q \cos 2\pi x/\lambda \tag{4a}$$

and

$$\frac{\partial u}{\partial x}(o,t) = \frac{\partial u}{\partial x}(\lambda,x) = 0, \quad t>0 \tag{4b}$$

2. CENTRAL DIFFERENCE SCHEME

In order to insure stability and convergence of the solution, the method of finite differences employing a five point formula based on Taylor series expansion about X_0 was selected. The partial derivative with respect to time is expressed as the forward difference:

$$\frac{\partial u(x,t)}{\partial t} = \frac{1}{\Delta t}\left[u(x, t+\Delta t) - u(x,t)\right] \tag{5}$$

Thus, if we discretize in space, using the five point central differences for the derivatives, a system of ordinary differential equations is obtained, which after time discretization by forward differences gives:

$$U^{n+1} = AU^n \qquad n = 0, 1, 2, \ldots$$

where, $U = \left[U_1, U_2, \ldots U_n\right]^T$

$$A = \begin{pmatrix}
F_0 & F_1 & F_2 & 0 & 0 & 0 & 0 & & & & 0 & F_2 & F_1 & 0 \\
F_1 & F_0 & F_1 & F_2 & 0 & 0 & 0 & & & & 0 & 0 & F_2 & 0 \\
F_2 & F_1 & F_0 & F_1 & F_2 & 0 & 0 & & & & 0 & 0 & 0 & 0 \\
0 & F_2 & F_1 & F_0 & F_1 & F_2 & 0 & & & & & & & \\
& & & & & & & \cdot & \cdot & \cdot & & & & \\
& & & & & & & \cdot & \cdot & \cdot & & & & \\
& & & & & & & 0 & F_2 & F_1 & F_0 & F_1 & F_2 & 0 \\
0 & 0 & 0 & 0 & & & & 0 & 0 & F_2 & F_1 & F_0 & F_1 & F_2 \\
0 & F_2 & 0 & 0 & & & & 0 & 0 & 0 & F_2 & F_1 & F_0 & F_1 \\
0 & F_i & F_2 & 0 & & & & 0 & 0 & 0 & 0 & F_2 & F_1 & F_0
\end{pmatrix} \tag{6}$$

where F_i, $i = 0, 1, 2$ are not constants but non-linear functions of U_i^j's given by:

$$F_0 = 1 - r(\frac{5}{2}\tilde{D} + \frac{3}{\Delta x^2} - 2\nu U'^2) \tag{7a}$$

$$F_1 = r(\frac{4}{3}\tilde{D} + \frac{2}{\Delta x^2}) \tag{7b}$$

$$F_2 = -r(\frac{\tilde{D}}{12} + \frac{1}{2\Delta x^2}) \tag{7c}$$

where $r = \frac{\Delta t}{\Delta x^2}$, $\tilde{D} = -1 + \nu(U_i^j)^2$, $\tilde{D}^1 = 2\nu U_i^j$

and $U'^2 = (\frac{\partial U}{\partial x})^2 = \frac{1}{144\Delta x^2} (U_{i-2} - 8U_{i-1} + 8U_{i+1} - U_{i+2})^2$

It should be emphasized that the matrix elements of A are not constants but functions of the composition variations U_i^j . Such a pseudo-linear representation of the non-linear diffusion equation has a significant advantage for it assists in developing the criteria for stability and convergence. Although the coefficients of this linear representation are not constant as indicated in Eqs. 7a, b, c fruitful information can be obtained as a direct result of the physical behavior of the non-linear solution throughout its time evolution. By direct inspection of the non-linear diffusion Equation (3), the following important conclusions about its solution can be stated:

i) Any periodic solution cannot have amplitude larger than the $\Delta C_e = C_e - C_0$ where C_e is the equilibrium concentration. Since u is normalized to ΔC_e we therefore have the condition $|U| \leq 1$. This also implies that:

$$-1 \leq \tilde{D} \leq -1 + \nu$$

ii) Analytical solutions of the time-independent equation has shown (4,5,6) that the maximum slope of any periodic configuration cannot exceed the slope of the equilibrium interface which can be described by a hyperbolic tangent.

Based upon these two physical conditions, we can now conclude that the F_i's are bounded quantities for a given choice of the spatial and time intervals of the descretization.

3. Results and Discussion

In order to test the stability of the numerical scheme the following non-linear case was chosen for investigation. A sinusoidal composition fluctuation was assumed as the initial condition at $t = 0$ for the profile of the composition wave, having the following characteristic values:

$$t = 0 \qquad U = 0.1 \cos 2\pi x/\lambda; \; \lambda = 16.142664$$

A spatial discretization of 80 intervals was assigned and the following values of the numerical parameters were chosen to satisfy the stability criterion which dictates that the errors due to the central difference approximation and the truncation which occurs at each step must decay with each subsequent iteration rather than accumulate [7]:

$$\Delta t = 0.00025, \; \Delta x = 0.2017833 \text{ which give a value for } r = 0.00614$$

For our particular case the wavelength $\lambda = 16.142664$ corresponds to $K^2 = 0.995$ for which the amplitude of the stationary states can be calculated [5] to be:

$$Q^*_{th} = 0.9987$$

The numerical solution provided an amplitude:

$$Q_{cal} = 0.9984$$

which is in remarkable agreement with the theoretical value.

The profile of the wave is shown in Figure (1). The initial profile at $t = 0$ is shown in curve A and the numerical solution is plotted at $t = 60$ (240,000 iterations) together with the analytical solution (curve B). It is seen that the numerical and analytical solution agree with an accuracy better than 0.03 percent.

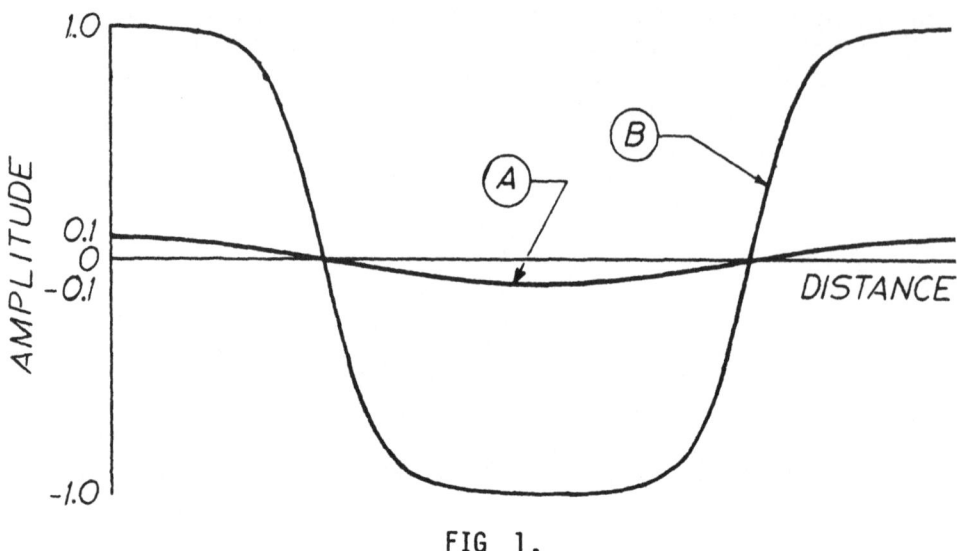

FIG 1.

Profile of a single wavelength. The initial profile "A" at t=0
and the numerical and analytical square-like wave "B" at t=60.

Perhaps the most important point in these numerical computations is the fact that the numerical scheme provides the actual kinetics from the early exponential growth to the later stages where a gradual retardation occurs and the wave reaches its stationary state. Although the analytical solutions exist for both the very early stages and the stationary state (infinite time) there has been no solution that links the two extreme cases together. Nor has there been any kinetics study of the intermediate configurations.

Tsakalakos [5] has derived an approximate analytical formula for time evolution of the spinodal structure. This formula was based on perturbation techniques and is given by:

$$Q(t) = Q_1^* \tanh \left[(Q_0/Q_1^*)e^{R(k)t} \right] \qquad (8)$$

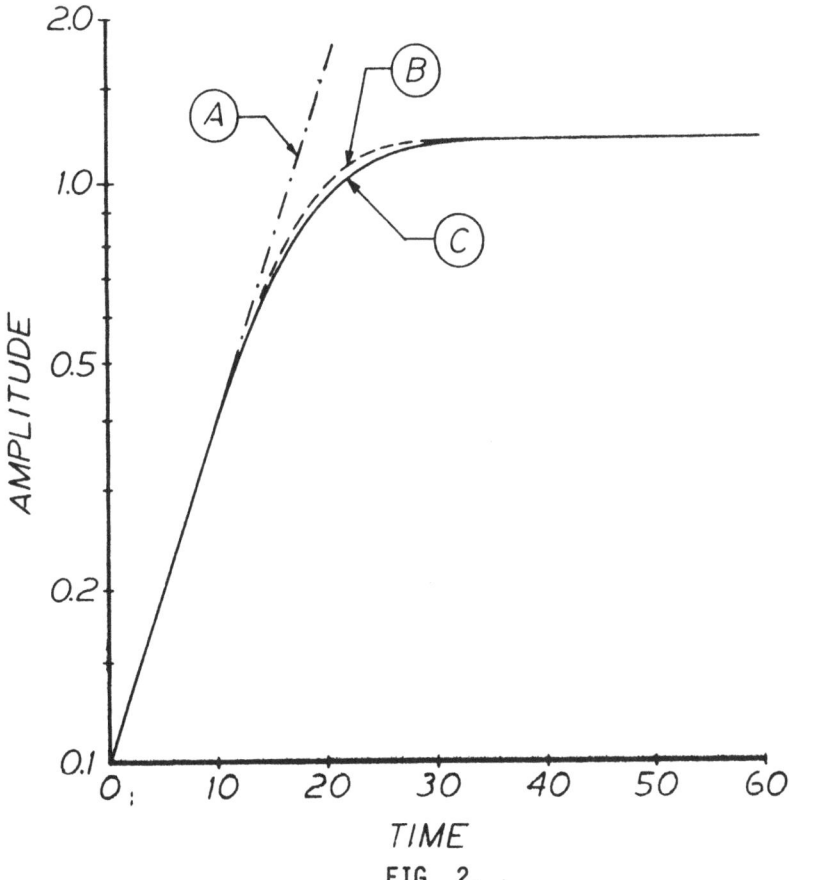

FIG 2.

The increase in amplitude of the composition wave vs. time. Curve "A" represents Cahn's linear solution, "B" is Tsakalakos' approximate solution and "C" is the solution by our numerical model.

432

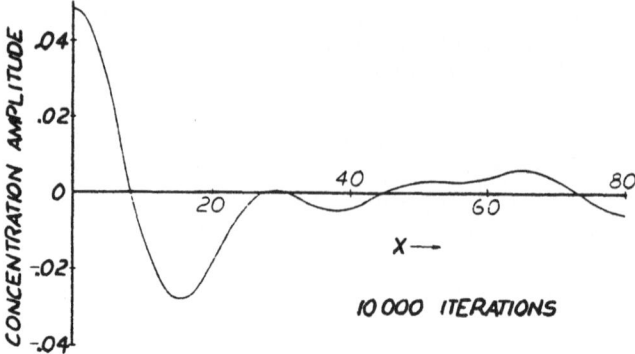

FIG 3a.

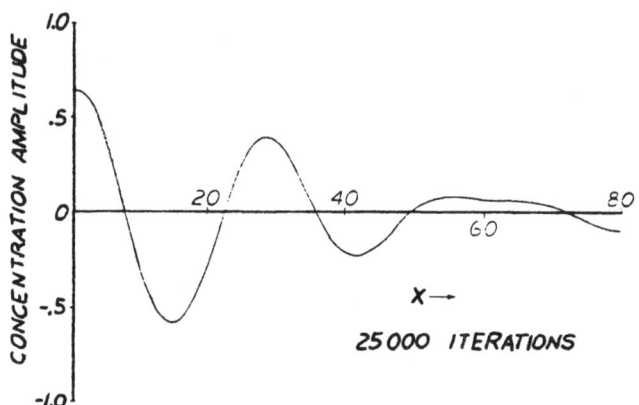

FIG 3b.

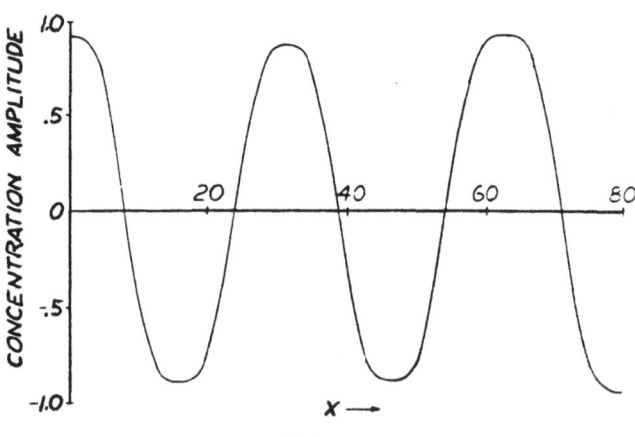

FIG 3c.

Refer to text for discussion.

This approximate solution is plotted in Figure (2) as curve B together with the numerical solution (curve C) and the linear solution (curve A). It can be seen that the agreement is remarkably good and indicates why Equation (1) provides the basis for kinetics studies of spinodal alloys. Such studies have been made by Ditchek and Swartz(8) for the Cu-9Ni-6Sn spinodal alloys. It should be finally noted that non-linear behavior at intermediate stages (t~10) is due to the fact that the wave is trying to assume its square-like configuration by faster growth of the higher harmonics. This is indeed observed by the numerical Fourier transform performed during the numerical solution. After this intermediate stage, the wave grows towards its stationary state.

A simulation of concentration wave growth in a spinodal alloy was performed by defining the initial concentration profile as the summation of 80 cosinusoidal waves. The amplitude of each component was set equal to 0.005 on a normalized scale. This concentration profile was discretized into 160 intervals and operated upon by the difference equation obtained from equation (3).

Figures (3a, b, and c) illustrate three stages in the evolution of the spinodal structure. The concentration is plotted against distance for three values of elapsed time (iterations of the program). In Figure (3a) an early stage is illustrated which shows that the initial distribution has not been altered significantly after 10,000 iterations. Figure (3b) shows a time corresponding to an intermediate stage where the spinodal wavelength is becoming established.

The stationary state is illustrated in Figure (3c). The wave has grown in amplitude and assumed a square-like profile. A careful examination of this reveals that the profile is not exactly periodic but that there is a slight deviation from the Jacobian Elliptic Function. The computer simulation was continued to a total of 100,000 iterations without a change in the amplitude and wavelength of Figure (3c).

In addition to the profile analysis in real space, a Fourier Transform of the profile was performed at several time intervals. The increase in the amplitude of three Fourier components is shown in Figure (4). The growth obeys the predictions of the nonlinear theory of spinodal decomposition in that it is initially exponential and begins to level off as time progresses. The wave number ($K = 2.8$) associated with the maximum amplitude has the greatest Fourier amplitude. The wave number ($K = 1.45$) associated with a larger wavelength has a smaller stationary state amplitude and takes a longer time to reach it. This corresponds to the establishment of the square-like profile of the Jacobian Elliptic function which requires the inclusion of additional Fourier components.

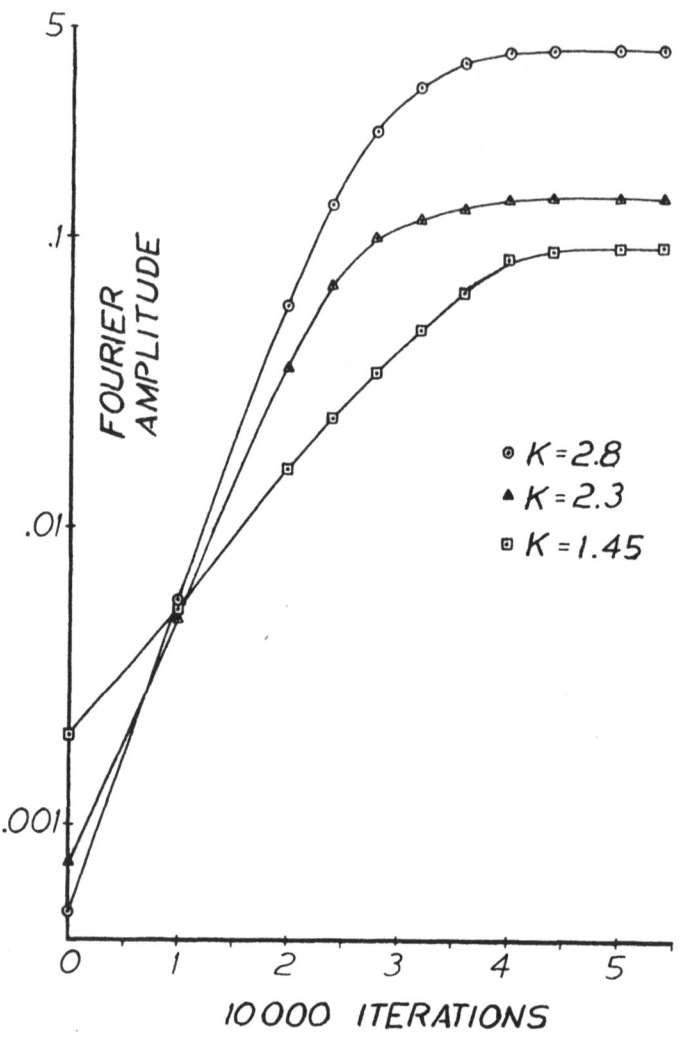

FIG 4.

Growth of three Fourier components show exponential growth at early stages which levels off as time progresses.

Figure (5) depicts the increase in intensity vs wave number as a function of time. The slow down effect is clearly evident as well as the shift to longer wavelengths of the real intensity as the stationary state is reached. Another feature to note is the growth of small peaks at smaller wave numbers. These are the result of computational errors and the finite size of the interval being considered in our model.

We have, therefore, demonstrated that the five point central difference numerical scheme can be successfully employed to solve

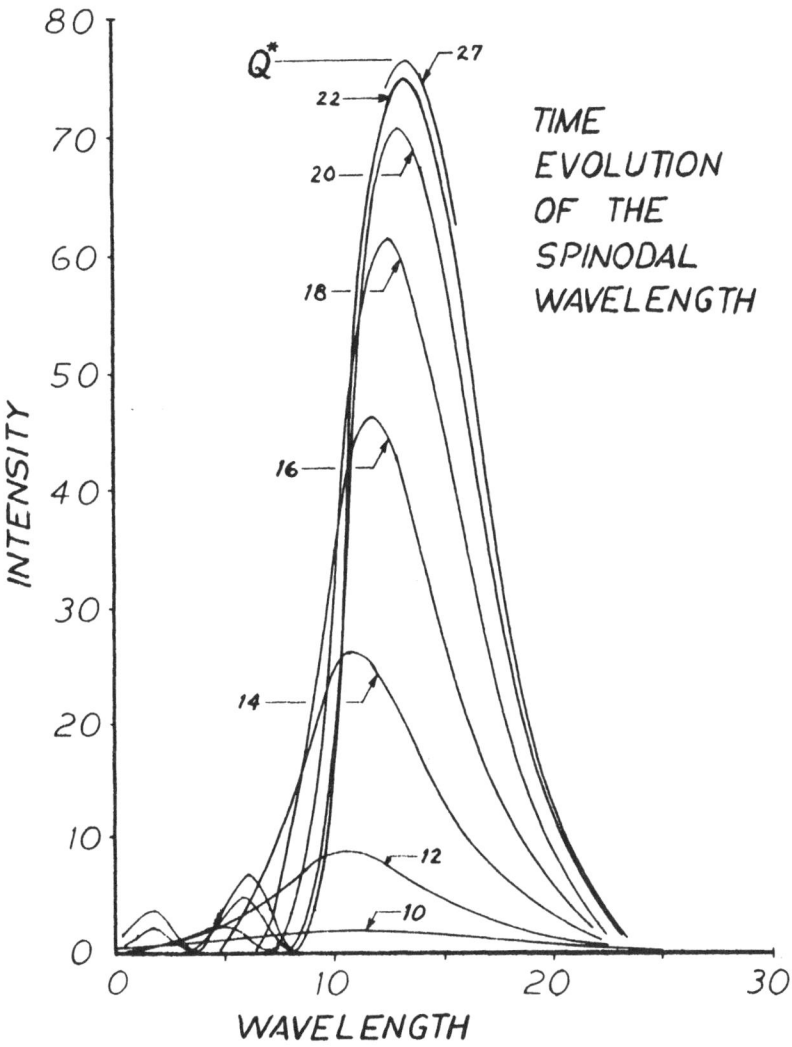

FIG 5.

Time evolution of the spinodal wavelength showing increase in intensity vs. wavelength. The slow-down effect is clearly evident as well as a shift to longer wavelengths as the stationary state (Q*) is reached.

the non-linear diffusion equation particularly in the case of uphill diffusion (negative diffusion coefficient). On the practical side, this scheme provides an excellent method of studying and predicting the properties of spinodal alloys as their superior properties depend on the microstructure which in turn is developed through the non-linear diffusion process. A rigorous approach to the solution of this problem will require the numerical solution of the three dimensional non-linear diffusion equation. In addition, a modification of initial conditions is necessary to allow for a more realistic distribution of infinitesimal composition fluctuations.

Such an attempt is currently in progress. Nevertheless, our present investigation has set forward the basis for such future development by clearly demonstrating the stability of the numerical scheme and, in general, the potential it has for solving more complex non-linear problems.

ACKNOWLEDGEMENT

This work was supported by a National Science Foundation Grant DMR-78-26503.

REFERENCES

1. J.W. Cahn, "On Spinodal Decomposition," Acta Metall. 9, 795-801 (1961).

2. J.W. Cahn, "Spinodal Decomposition in Cubic Crystals," Acta Metall. 10, 179 (1962).

3. J.W. Cahn, "The Later Stages of Spinodal Decomposition and the Beginnings of Particle Coarsening," Acta Metall. 14, 1685-1692 (1966).

4. J.S. Langer, "Theory of Spinodal Decomposition in Alloys," Annals of Physics 65, 53-86 (1971).

5. T. Tsakalakos, "Interdiffusion and Enhanced Elastic Modules Effect in Composition Modulated Copper-Nickel Thin Foils," Ph.D. Thesis, Northwestern University, Evanston, Illinois (1977).

6. J.W. Cahn and J.E. Hilliard, "Free Energy of a Nonuniform System. I. Interfacial Free Energy," J. Chem. Phys. 28, 25-267 (1958).

7. T. Tsakalakos and M.P. Dugan, "Nonlinear Diffusion in Spinodal Decomposition," to be published.

8. B. Ditcheck and L.H. Schwartz, "Applications of Spinodal Alloys," Ann. Rev. Mater. Sci. 9, 219-253 (1979).

CHAPTER 6: COMPOSITION MODULATED FILMS

MANUFACTURE OF TERNARY THIN FILM LAYERED FOILS BY CONTROLLED EVAPORATION

C.N. Manikopoulos[*] and T. Tsakalakos[+]

Department of Electrical Engineering* and Department of Mechanics and Materials Science, College of Engineering, Rutgers University, Piscataway, New Jersey 08854

ABSTRACT

An evaporator system and associated electronic bias and control circuits have been built to produce bulk foils made of a large number of periodic ternary very thin film structures. This is accomplished by chopping the evaporant stream of three evaporation sources each set to a constant evaporation rate. The electronic bias and control circuits made of inexpensive IC's, transistors, and general purpose laboratory instruments are presented. It is found that evaporation rates, typically about 3 Å/sec with long term stability of about 5% can be achieved. We have manufactured several bulk foils consisting of about 10,000 periodic ternary thin film (10-50 Å) layered structures. Characteristic properties of samples consisting of CuNiCr are reported.

1. INTRODUCTION

The interaction between thin film layers is very critical in the operational properties of many solid state devices. The interface conductivity, the interface diffusion, the magnetic interlayer coupling, tunneling and other phenomena may play a decisive role on the operating properties and the long term stability and reliability of a device. A very useful method to study these effects, presented in this work, is to build up a bulk structure of numerous such very thin layers by controlled vapor deposition (1,2,3). A typical sample produced by our apparatus would be made up of 10,000 or more layers, each 10-50 Å thick. The sample may be either ternary or binary with a uniform third component. It is characterized by a specified layer thickness and specified ratios of the three components in the deposited layers. It is known that such layered structures give characteristic X-ray patterns with satellite reflections around the Bragg peak (4,5,6). Diffusion in the samples can be studied by sequential analysis of the X-ray patterns. Moreover, interface conductivity can be investigated by four point resistivity measurements. The presence of a very large number of interfaces enhances the magnitude of these phenomena and thus greatly facilitates their investigation.

We present here a simple and very inexpensive feedback controlled three-source evaporator system especially designed and built to manufacture multilayered samples. It is made up of low cost IC's and transistors and readily available general purpose laboratory instruments.

Several CuNiCr samples have been prepared. Some of their characteristic properties have been analyzed and the results presented for illustration purposes.

2. THE VACUUM HARDWARE

The vacuum components utilized in the sample presentations were housed in an 18" glass vacuum bell jar evacuated by a 6" diffusion pump, a mechanical pump, and a Meisner cold trap. Typically a vacuum in the range of $2\text{-}4\times10^{-7}$ Torr is achieved.

Much care must be taken to design the geometry of the components in the vacuum system so that every layer in the manufactured foil from all three sources be of uniform thickness. A sketch of the evaporator system hardware is shown in Fig. 1. The substrate holder of dimensions 3.75" x 4.25" carrying two 3" x 1.5" samples symmetrically is mounted at a distance of 12" from the material sources. A quartz crystal thickness monitor is placed as nar the substrate as possible. The substrate holder is mounted on a thick copper block the temperature of which is controlled by a temperature controller.

Immediately above the source crucible in Fig. 1 a control flow rate pressure gauge is shown enveloping the stream of evaporant material as it leaves the source. Three such gauges are mounted each directly above a source; the geometry of their location is shown in Fig. 2(a) and a detailed schematic of the flow rate gauge in Fig. 2(b). A rotating sectored wheel allows sources A, B or C to be alternatively shielded or free; the shielding time ratios are specified by the geometrical design of the sections carved on the chopping wheel. The grid and filament of this pressure gauge are taken from ordinary ionization gauges. The collector is made of 0.010" Molybdenum wire housed for support in thin ceramic tubing. A fourth gauge shielded from the direct stream of evaporant material was used to monitor the local background pressure.

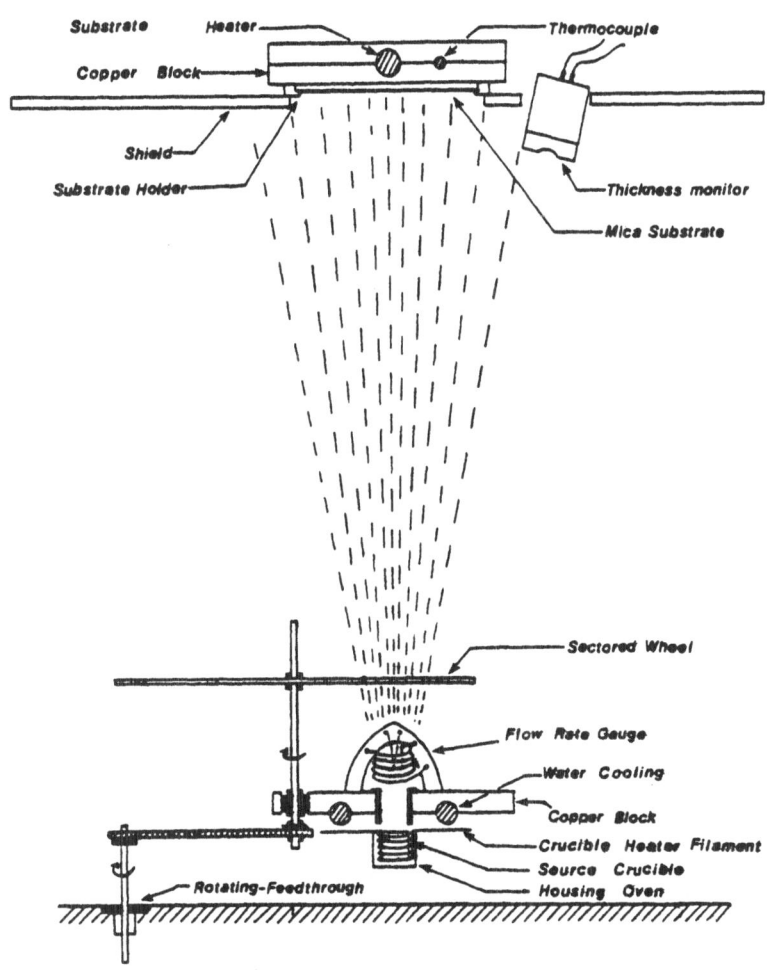

FIGURE 1 - SCHEMATIC OF THE EVAPORATOR SYSTEM SHOWING ALL
IMPORTANT VACUUM HARDWARE

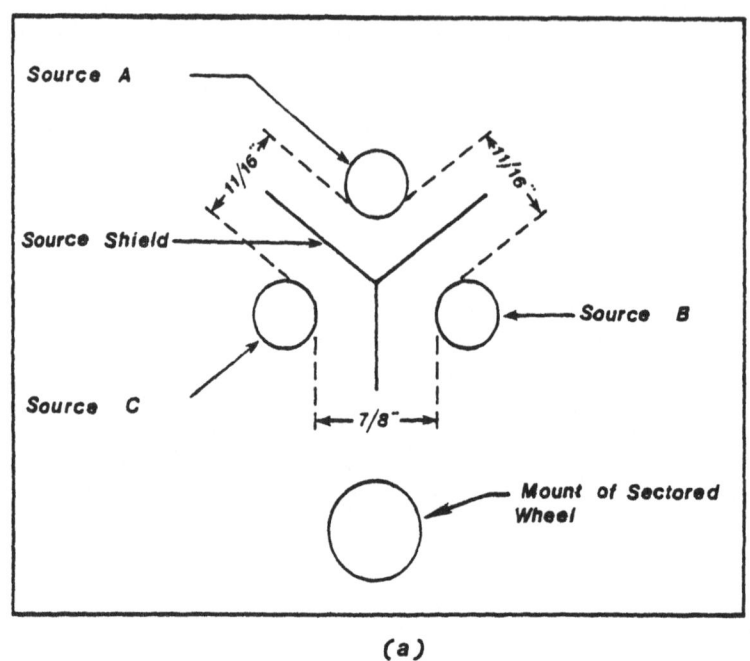

(a)

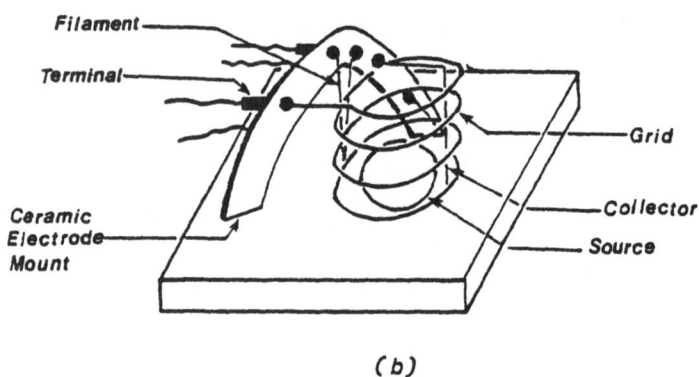

(b)

FIGURE 2 - (a) SCHEMATIC OF THE GEOMETRY OF THE THREE MATERIAL
EVAPORATION SOURCE,
(b) DETAILED SCHEMATIC OF A FLOW RATE GAUGE
MONITORING THE PRESSURE OF THE EVAPORANT STREAM

The evaporation sources are carefully constructed and placed to reduce heat losses in order to reach high temperatures and to hold as much material as possible since in practice only a fraction of it ends up as a deposited layer. The source crucible is housed within a small oven; the interspace is filled with alumina slurry and contains the tungsten heating filament. The heating power required for normal operation is in the range of 200 to 500 watts for each crucible. About +100 V DC bias is applied to each heating filament to prevent any thermionically emitted electrons from reaching the flow rate gauges.

3. THE ELECTRONIC CONTROL CIRCUITRY

The ion current flowing to the collector of a property biased ionization gauge is directly proportional to the local vapor pressure at the gauge. After detection and amplification this current provides the measure of the local evaporation rate. This is compared to a set voltage representing the desired evaporation rate. The difference of the two can be used as a correction voltage which controls the output of a proportional power controller feeding the crucible filament. Three such electronic control chains were built each powering one of the evaporation source filaments.

A block diagram of the evaporation rate control circuitry is shown in Fig. 3. The collector current I_c of the evaporation rate ionization gauge generates a voltage drop $V_1 = I_c R$ at the detection resistor R which is amplified to a value V_e by the amplifier and then compared to the output of the background reference ionization gauge V_b. The difference of the two $V_{ne} = V_e - V_b$ is obtained at the output of the first comparator and represents the net evaporation rate of material at the source considered. This is compared at the second comparator to a set voltage V_s externally adjustable which represents the desired value of the material evaporation rate. The difference $V_r = V_s - V_{ne}$ at the output of the second comparator acts as a correction voltage at the input of the power controller which heats the crucible filament. The output of the controller is zero when $V_r = 0$ and adjusted to be maximum when $V_r = V_0 = 10$ volts.

In Fig. 4 we show a detailed schematic of the evaporation rate control instrumentation. It is made of inexpensive IC's in simple arrangements. It consists of two sections. Shown in Fig. 4(a) is the bias circuit for the ionization gauge which generates an ion current I_c at the anode proportional to the local pressure. It is important that the emission current I_e to the grid be kept constant. To achieve this the grid bias circuit includes feedback control which changes the heating current of the filament and thus the thermionic emission current density resulting in a constant emission current. The output signal of this section is $V_i = I_c R$ proportional to the local pressure. The resistance R can be set

444

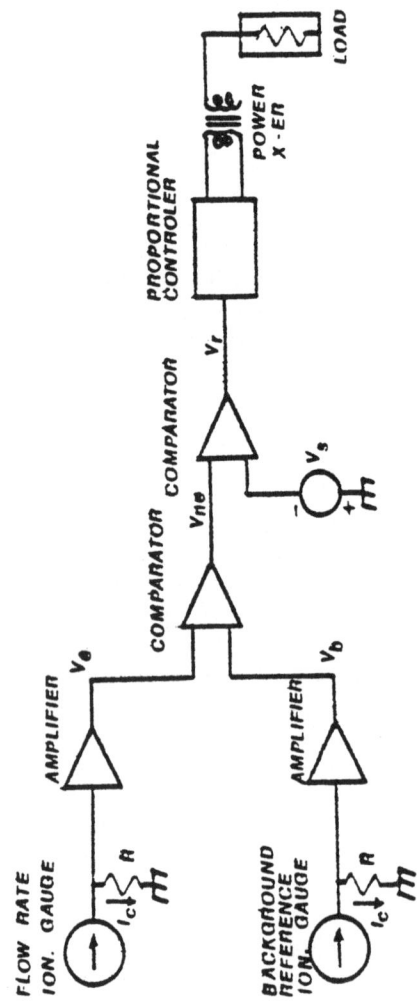

FIGURE 3 – BLOCK DIAGRAM OF THE EVAPORATION RATE CONTROL CIRCUITRY

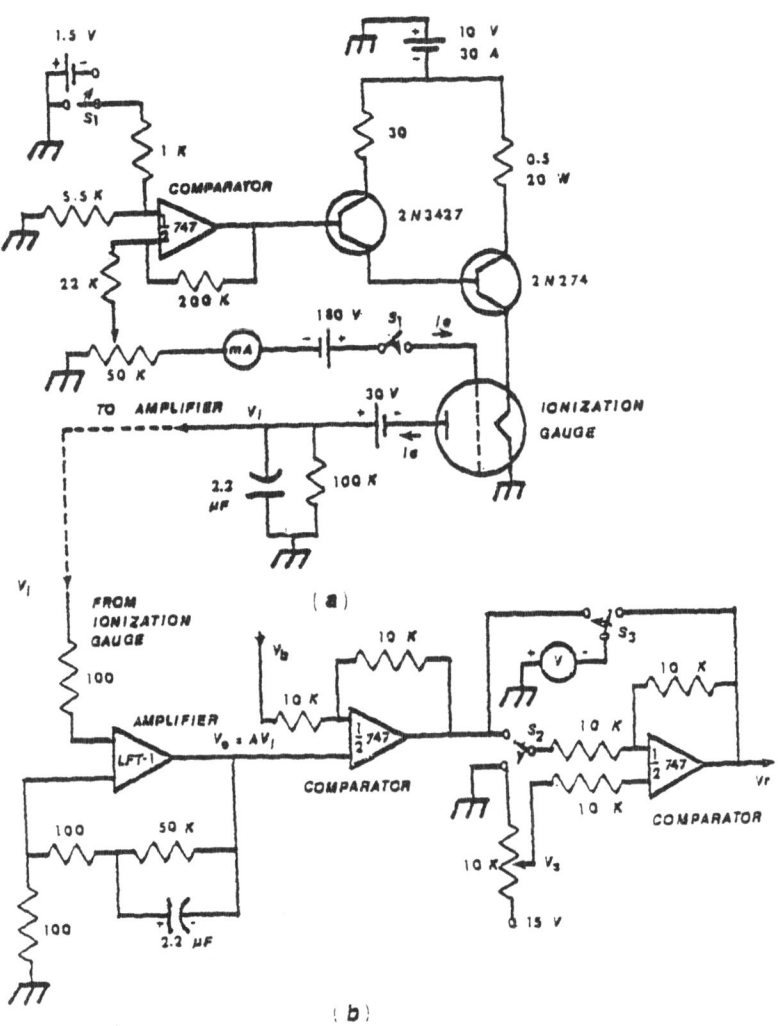

FIGURE 4 - (a) SCHEMATIC OF THE BIAS CIRCUIT OF THE IONIZATION CIRCUIT.
THE GRID BIAS CIRCUIT INCLUDES FEEDBACK CONTROL OF THE
FILAMENT HEATING CURRENT DESIGNED TO KEEP THE EMISSION
CURRENT I_e CONSTANT

(b) SCHEMATIC OF THE CIRCUIT GENERATING THE CORRECTION
SIGNAL VOLTAGE WHICH DRIVES THE POWER CONTROLLER

to either 100 K or 1 M depending on the sensitivity required.

The section shown in Fig. 4(b) provides the correction signal voltage driving the proportional controller of the power delivered to the crucible heater. The output signal V_i from the ionization gauge is amplified by a DC amplifier to $V_e = A V_i$ where the amplification factor A can be as high as 500. Any high quality low drift operational amplifier can be used. The signal from the background reference gauge V_b is subtracted from the signal V_e which represents the local evaporant pressure; the difference V_{ne} which appears at the output of the first comparator represents the net material evaporation rate. This voltage is compared to a set voltage V_s provided by an externally adjusted potentiometer at a second comparator; the voltage V_s is a measure of the desired evaporation rate. The output $V_r = V_s - V_{ne}$ represents the error signal driving the proportional power controller. The controller is a Robicon model 401 of maximum power output $W_m = 3.3$ kVA; it is adjusted to deliver maximum power at an error voltage V_c usually set at 10 volts. Two switches S_2 and S_3 strategically connected allow for all important voltages to be monitored at will.

When the equilibrium temperature of the evaporation source crucible is reached upon heating the filament the total heat loss rate $W_L(T)$ of the source should be equal to the total heat input rate from the controller $W_H(T)$. Since at temperatures above the melting point of the material there is a one to one correspondence between the temperature T and pressure P of the evaporant material we may write W_L and W_H as functions of pressure, i.e., $W_L = W_L(P)$ and $W_H = W_H(P)$. The evaporant pressure P_e will be obtained at the condition $W_L(P) = W_H(P)$. It is clear that $W_L(P)$ is a mono-tonically increasing nonlinear function of the pressure which is determined by the dimensions and geometry of the crucible. It turns out that $W_H(P)$ is a straight line defined by the two points where maximum and minimum power is delivered, i.e., $W_L(P_1) = W_m$ at $P_1 = V_s + V_b - V_c / V_b^0$, and $W_L(P_2) = 0$ at $P_2 = V_s + V_b / V_b^0$.

Here $V_b^0 = A I_c^0 R_L$, where I_c^0 is the initial ion current and V_b^0 is the output reading in volts of the amplifier; both are a measure of the background pressure P_0 just before evaporation. The slope of this line is

$$\frac{dw}{dP} = - \frac{W_m}{P_0} \cdot \frac{V_b^0}{V_c}$$

which depends on the value of V_c and V_b^0; the latter implies a dependence of the slope on the amplification factor A. Thus a shift of the straight line $W_H(P)$ which would result in a different operating evaporant pressure P_e can be affected either by altering V_s which shifts P_1 and P_2 and thus the whole line $W_L(H)$ in parallel or by changing V_c or A which would change the slope of the line.

We have found it more convenient to move to a different operating pressure P_e by adjusting the value of the set voltage V_s keeping all other operating parameters fixed.

4. EXPERIMENTAL RESULTS

Following the operating procedure devised for controlled evaporation we start by setting this amplification factor so that all gauges read a convenient voltage; typically we choose $V_b^0 = 0.1V$ at a pressure value of $P_0 = 5 \times 10^{-6}$ Torr. This setting usually provides a net evaporation voltage V_{ne} around 5 volts for practical deposition rates which is a good working range for the input of our power controller. The relation between the gauge reading after amplification, V_e, as a function of local pressure for different values of V_b^0 is shown in Fig. 5(a). The dependence is linear, as expected, over a wide range in pressures. The deposition rate is controlled by the set voltage V_s. A plot of actual measurements for a particular set of conditions is shown in Fig. 5(b). This is a calibration curve for deposition of copper for a setting of $V_b^0 = 0.1V$. If a particular set voltage and thus deposition rate has been selected it is found to remain stable for long periods of time to within a very narrow range. A plot of instantaneous deposition rates taken every 4 seconds for 5 minutes is shown in Fig. 6. The variation over time is seen to be less than 5%. It is found that the stability over longer periods of time, i.e., 30-60 minutes is similarly characterized.

The primary objective for building the evaporator system was to produce foils containing small thickness ternary film layers. Depending upon the geometry of the rotating wheel and its position with respect to the crucibles, a number of modulated structures could be produced. Several possible such structures are shown in Table 1. It is apparent from this table that many intriguing possibilities exist for fundamental studies not only of pure metallic layer structures (such as Type No. 1 and No. 5) but also of various compounds with given stoichiometric composition within each layer. Of particular interest is the special case of modulated films, No. 6, which represents a superposition of a periodic structure B/C/B/C... and a constant deposition of A throughout the thickness of the foil.

In order to test our production technique, several foils of the ternary CuNiCr alloy system were prepared as described above. The composition Cu-33 a/o Ni-6 a/o Cr was chosen because it exhibits spinodal decomposition (8) at temperatures as low as 400°C, thus providing a unique means of studying the stability of the solid solution and of its concentration fluctuations. The modulated structure (Cr)Cu(Cr)Ni/(Cr)Cu/(Cr)Ni/... (case No. 6 of Table 1) at the above ternary concentration was selected and produced. This system can be easily manufactured and shows a simple X-ray

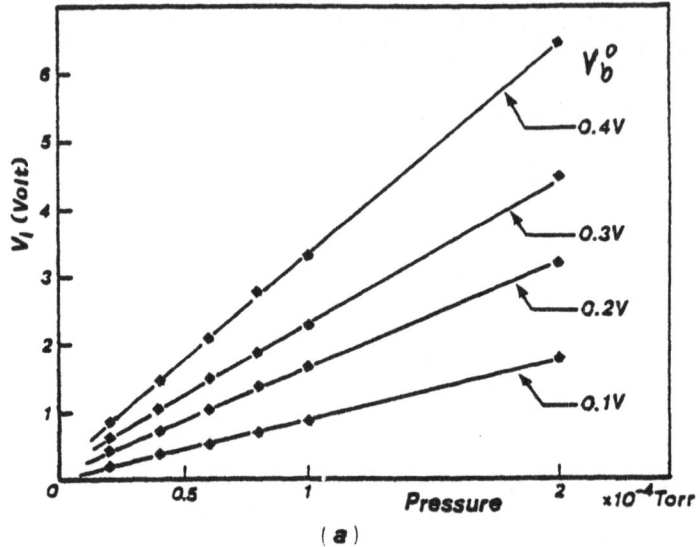

(a)

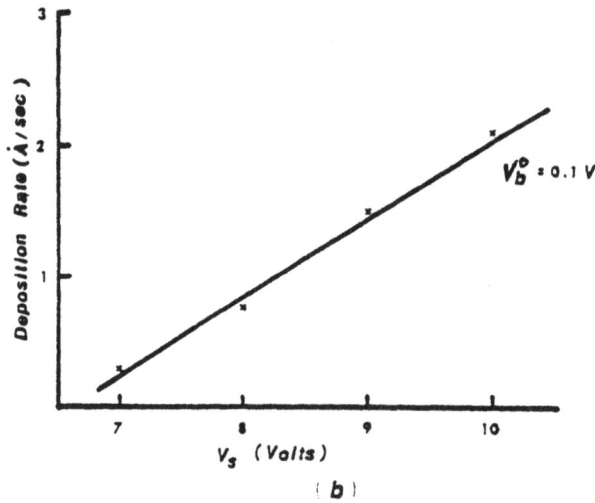

(b)

FIGURE 5 - (a) THE READING OF THE IONIZATION GAUGE IN VOLTS
AFTER AMPLIFICATION IS PLOTTED VERSUS THE
LOCAL PRESSURE FOR DIFFERENT INITIALIZATION
VALUES V_b^o.

(b) A CALIBRATION CURVE FOR COPPER SHOWING THE
DEPOSITION RATE MEASURED VERSUS SET VOLTAGE
FOR THE INITIAL VALUE OF $V_b^o = 0.1$ VOLTS

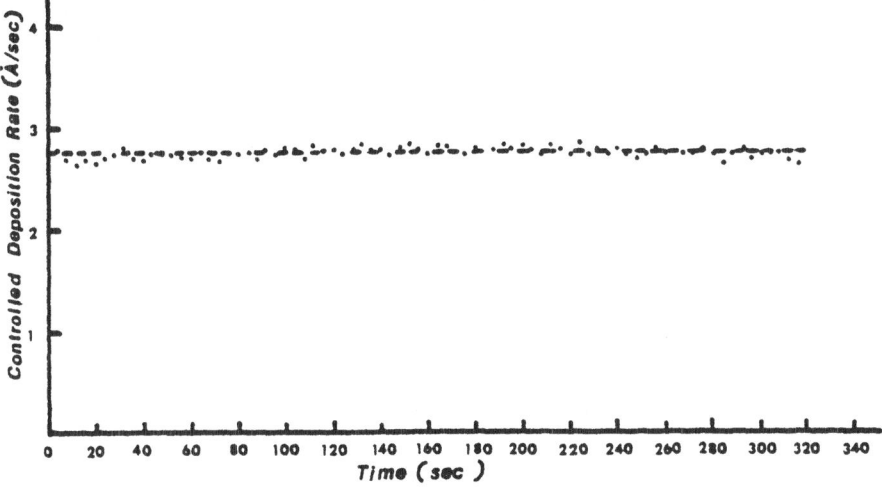

FIGURE 6 - THE MEASURED INSTANTANEOUS DEPOSITION RATE
IS PLOTTED VERSUS TIME OVER A PERIOD OF
5 MINUTES

Table 1

POSSIBLE MODULATED STRUCTURES

No.	TYPE	REMARKS
1	A/B/A/B...	Binary
2	A/(AB)/A/(AB)/...	Binary, Pure A/Given Composition of A-B/...
3	A/(BC)/(AB)/(BC)/...	Ternary, Pure A/Given Composition of B-C/...
4	(AB)/(BC)/(AB)/(BC)/...	Ternary, Given Composition of A-B/Given Composition of B-C/...
5	A/B/C/A/B/C/...	Ternary, A/B/C/...
6	(A)B/(A)C/(A)B/(A)C/...	Ternary Continuous A + B/C/B/C...

diffraction pattern.

Prior to any test it was necessary to obtain the maximum possible information regarding the films and their composition modulation. The existence of a layer structure of small thicknesses results in a modulation of both the atomic scattering factor and the lattice parameter. This in turn yields sharp Bragg peaks flanked by a pair of satellites (9) situated at a distance $s = 1/d \pm 1/\lambda$ where $s = 2\sin\theta/\lambda_{X-rays}$, d is the lattice spacing and λ is the wavelength of the modulated structure. Thus the wavelength of the composition wave can be measured from the location of the satellite. For a sinusoidal modulation the intensity of a satellite is proportional to the square of the amplitude which thus can be estimated accordingly. An X-ray diffraction pattern is shown in Fig. 7. In this figure the location of the 111 Bragg peak is $2\theta_B = 43.91$, while the low and high angle satellites are located at $2\theta_- = 40.56$ and $2\theta_+ = 47.30$ correspondingly. The wavelength can be calculated from the above formula to be $\lambda = 2.83$ nm. It should be remarked that there was no evidence of a 200 Bragg peak present in all diffraction patterns. This indicates that the films exhibited a strong 111 texture, as expected, due to the high temperature (350°C) of the substrate during the deposition. The X-ray intensity measurements were made with GE XRD-5 diffractometer. CuK_α monochromatic radiation was used from a singly bent graphite monochrometer. The horizontal divergence of the incident and diffracted beam were 1.5 and 0.3°. The diffracted intensity was measured by a scintillation counter and solid state electronics.

SUMMARY

An evaporator was constructed to produce ternary layered structures of small layer thickness (10-50 Å). The composition modulated foils were prepared by coevaporating the three components through a rotating pinwheel shutter onto mica substrate for <111> texture. The electronic bias and control circuits for control evaporation rates were presented. They are made of inexpensive IC's and transitory and general purpose easily available laboratory instruments. These circuits have performed very satisfactorily. Typical evaporation rates of 1-10 Å/s were accomplished.

X-ray diffraction patterns of modulated Cu-33 a/o Ni-6 a/o Cr foils showed the existence of satellites about the Bragg peak which provides strong evidence for the presence of periodic modulated structures. A number of modulated structures can be produced by a simple metals or compounds of given composition.

ACKNOWLEDGEMENTS

This work was supported in part by the National Science Foundation Grants No. DMR-78-26503 and No. ENG-79-08490 to which

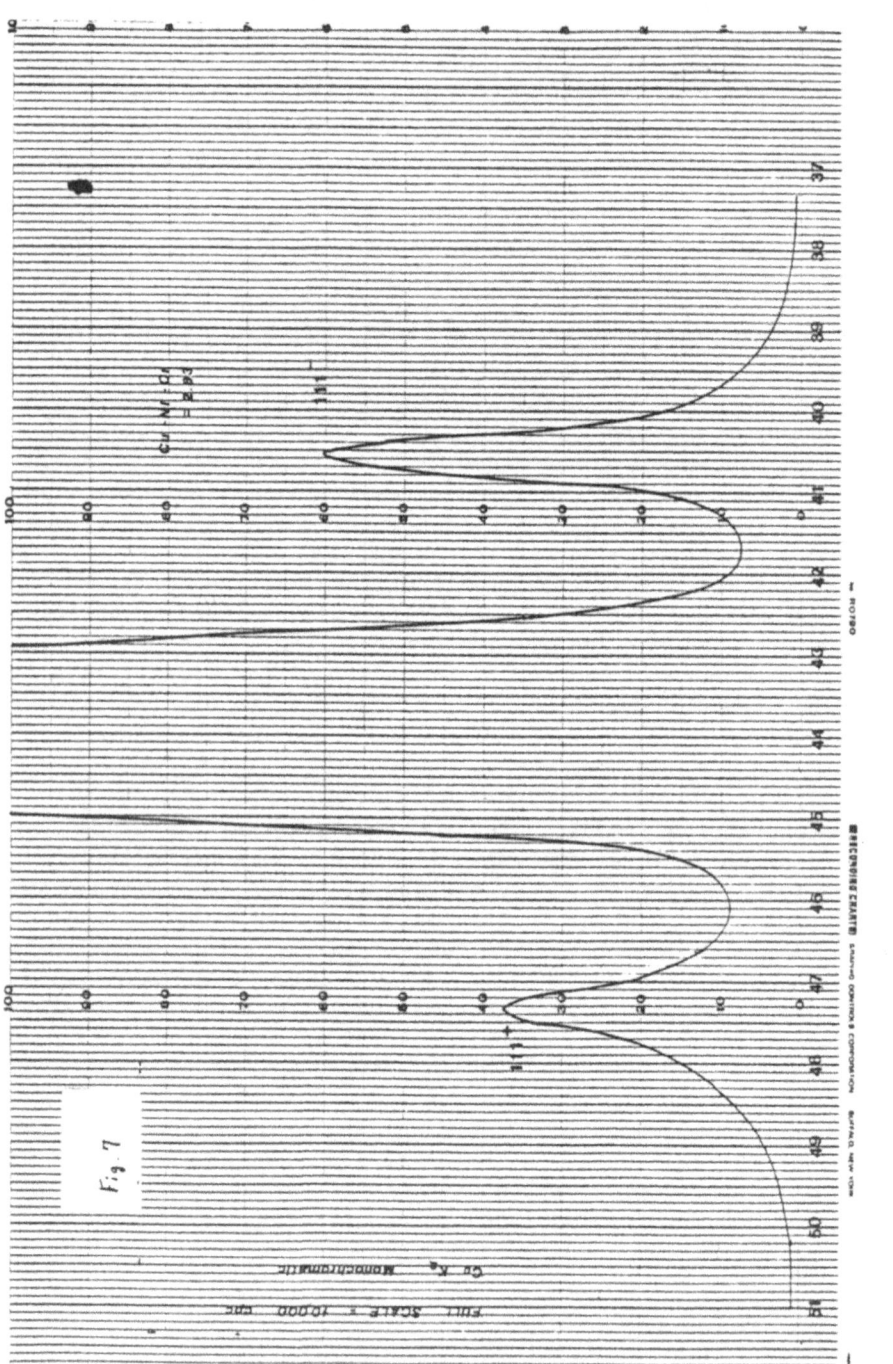

FIGURE 7 — THE X-RAY DIFFRACTION PATTERN MEASURED FOR A Cu-33 a/o Ni-6 a/o Cr MODULATED
TERNARY FOIL MANUFACTURED BY OUR APPARATUS IS SHOWN. THE MAIN 111 BRAGG PEAK

the authors are greatly indebted.

REFERENCES

(1) Giedd, G.R. and M.H. Perkins. Rev. Sci. Instrum. 31 (1960), 773.
(2) Cook, H.E. and J.E. Hilliard. J. Appl. Phys. 40 (1969) 2191.
(3) Murakami, M., D. deFontaine, P. Shiplery and W. Rodgers. J. Vac. Sci. Technol. 10, 3 (1973) 445.
(4) Du Mont and J.P. Yontz. J. Appl. Phys. 11 (1940) 357.
(5) Dinklage, J. and R. Frerichs. J. Appl. Phys. 34 (1963) 2633.
(6) Philofsky, E.M. and J.E. Hilliard. J. Appl. Phys. 40 (1969) 2191.
(7) Erb, D.M. Ph.D. Thesis, Northwestern University (1969).
(8) Wu, C.K., R. Sinclair, G. Thomas. Met. Trans. 9A (1977) 381-387.
(9) Guinier, A. X-Ray Diffraction in Crystals, Imperfect Crystals and Amorphous Bodies (W.H. Freeman and Co., San Francisco (1963)).

THEORETICAL APPROACHES TO UNDERSTANDING THE PROPERTIES OF MODULATED
STRUCTURES - A REVIEW

Philip C. Clapp

University of Connecticut, Storrs, Connecticut 06268

ABSTRACT

Experimental studies on compositionally modulated alloys in
systems like Cu-Ni, Cu-Pd, Ag-Pd and Au-Ni have shown remarkable
increases (factors of 2 to 4) in the bi-axial elastic modulus when
the modulation wavelength is close to a critical wavelength (on the
order of 20Å) depending on the system. This has been sometimes
called the "supermodulus effect". In contrast, compositionally
modulated Cu-Au alloys appear to have no enhancement of elastic
modulus at any wavelength.

These results present a very great challenge to our theoretical understanding of material properties because elastic moduli
have heretofore been found to be almost totally insensitive to
microstructural changes. It is generally believed that only if the
nature of the forces in the solid change in some significant way
can the elastic moduli alter appreciably. There are substantial
anomalies in the magnetic properties of some of the compositionally
modulated alloys which also strongly suggest fundamental altera-
tions in the electron distribution within the solid.

The several theories which have been offered to date to
explain the "supermodulus effect", ranging from Fermi surface-
Brillouin zone interactions to coherency strain phenomena, will be
reviewed and critically examined. The general outlines of future
theoretical developments in this field will also be suggested.

I. NEED FOR A THEORY

The "supermodulus effect" which has been discovered in a number of composition modulated alloys produced by vapor deposition is
a very startling phenomena and a great challenge to our theoretical
understanding of materials. So far the effect has been found in
four systems Cu-Ni, Cu-Pd, Ag-Pd, and Au-Ni. It has been searched
for but not found in Cu-Au. The "supermodulus effect" refers to
the remarkable increase in the biaxial elastic modulus by a factor
of between 2 and 4 when the wavelength of the modulation is in the
neighborhood of 2 nm. Usually the effect has been found when the
foils were deposited as (111) epitaxial layers, and has appeared as
an increase in Y(111). It is worth noting that when Cu-Ni foils
were produced with (100) epitaxial layers no supermodulus effect
was found.

The supermodulus effect is very dependent on the wavelength of
composition modulation, showing a sharp peaking at a particular
wavelength (depending on the alloy system involved) but usually
somewhere in the vicinity of 2 nm. The effect shows a marked dependence on the amplitude of the composition modulation and appears to
be proportional to the square of the amplitude in all of the systems in which the effect has been precisely measured. One further
aspect of the effect which is important to note is that the stress-
strain curves of the foils show a markedly non-linear behavior for
all values of the strain. The slope of the stress- strain curve is
a monotonic decreasing function of the strain. It may be added
that magnetic anomalies have been found in some, but not all, of
the systems that exhibit the supermodulus effect. Although the
magnetic anomalies may give a clue as to possible mechanisms we
will not deal with magnetic aspects of the problem in this paper.

Given the fact that a wide variety of experience has shown
that any sort of mechanical deformation of metallic alloys can
change the elastic moduli by at most 1 or 2%, and that order-
disorder transformations rarely change the elastic moduli by more
than 10 to 20 percent, the supermodulus effect is truly remarkable
and would seem to indicate that a very fundamental change has
occurred in the materials which exhibit this effect. What this
fundamental change might be is a very worthy subject of theoretical
investigation.

II. THEORETICAL APPROACHES TO EXPLAINING THE SUPERMODULUS EFFECT

The possibility that the supermodulus effect can be explained
by the changes in the elastic constants due to coherency strains in
the foils can be quickly examined and discarded. The third order
elastic constants (which give the rate of change of the second

order elastic constants with strain at small values of strain) are
about the same order of magnitude as the second order elastic constants themselves in the alloys of concern here. Since the coherency strains are at most a few percent, the change in the second
order elastic constants predicted in this way can only be a few
percent. This does not even begin to approach the 100-300 percent
change in the elastic constants observed in the supermodulus
effect.

In 1970 before the supermodulus effect was discovered by
Hilliard and coworkers (1), Koehler (2) wrote a very provocative
theoretical paper which proposed a method for producing a very
strong solid. He proposed that by using alternate layers of
materials with high and low elastic constants that resolved shear-
ing stresses of the order of 0.01 of the lower shear constant would
be required in order to drive dislocations through the combination,
and that the layers should be so thin that a Frank-Read source
could not operate inside one layer. He further specified that the
specimens should be prepared by epitaxial crystal growth consisting
of alternate layers of the two crystals and that their lattice
parameters should be nearly equal. He claimed that such a combination would resist deformation to a much greater degree than conventional materials and would not be susceptible to brittle fracture.
He went on to specify some particular combinations which he felt
were especially promising. Remarkably enough his leading candidate
was copper-nickel. He made the general recommendation that the
layers of each material be no greater than 100 atomic layers thick.

Despite the remarkable similarity between Koehler's model and
the preparation of materials which exhibit the supermodulus effect,
there was no suggestion in Koehler's work that a supermodulus
effect would exist. However, his model was directed at building
materials in which the mobility and generation of dislocations
would be practically nil, and this raises the question of whether
the greatly restricted dislocation motion that these materials
would have could lead in some way to the supermodulus effect.

This possibility can be quickly discounted because the motion
of dislocations plays very little part in the magnitude of the
elastic constants of any material. This can be seen from the fact
that the elastic constants, as measured by ultrasonic sound velocity
measurements (and which do not involve the motion of dislocations),
agree for virtually all materials with the measurements done by
conventional stress strain experiments (which do involve significant motion of dislocations). It is interesting to ask the reverse
question of whether the compositionally modulated materials that
have been produced recently provide a good test for Koehler's model
of a very strong solid. The Cu-Ni system has been very extensively
studied but it is very difficult to get meaningful fracture strength
data with the very thin foils that have been produced so far because

of edge tearing. Nevertheless there are some preliminary indications that these materials may indeed be much stronger (in the
sense of ultimate tensile strength) than the same material as a
homogeneous alloy. This should be a very interesting question for
future experimental research.

Most of the theoretical models that have made a serious attempt
to explain the supermodulus effect have been based on the idea that
the Brillouin zone introduced by the periodic composition modulation will be in contact with a flat part of the Fermi surface of
the alloy over the range of modulation wavelength that produces a
supermodulus effect. Hilliard's group at Northwestern have developed this approach in considerable detail starting with the early
work of Purdes (3) and Tsakalakos (4). The basic mechanism is an
old one and was proposed by Jones to explain the electron per atom
numbers that Hume-Rothery found governed the progression of phases
in copper base alloys as the composition varied. This basic mechanism of an energy stabilization arising from a Brillouin zone-Fermi
surface contact was also used by Sato and Toth (5) to explain in
quantitative detail the appearance of long period superlattices in
a variety of ordered alloys. However the use of this mechanism to
explain large changes in elastic moduli is a new and interesting
idea.

A simple picture of how the mechanism might work can be given
as follows. Assuming that the modulation wavelength does produce a
Brillouin zone that touches a flat portion of the Fermi surface,
this will cause a lowering of the energies of the conduction electrons that are in the states just at the Fermi surface and a corresponding increase in the energy of the conduction electron states
that are empty just above the Fermi surface in reciprocal space.
If the alloy is now strained in some fashion, the Brillouin zone of
the alloy (which is determined by its crystal structure parameters)
will change, and if it can be assumed that the Fermi surface configuration does not change significantly at the same time, the
Brillouin zone will be driven to detach from the Fermi surface.
This will cause a loss of the energy advantage that the conduction
electrons had previously experienced and will translate into an
increase in the energy of the lattice as whole. One of the critical
tests for this type of model is whether the energy change that
would occur as a result of the deformation is of the correct order
of magnitude to explain a supermodulus effect quantitatively. A
second critical test is whether the composition modulation wavelengths that produce a peak in the supermodulus effect do indeed
correspond with the Brillouin zone making contact with a flat
portion of the Fermi surface. We shall now present an overview of
the work that has been done in this direction.

In Purdes' Ph.D thesis work he made a detailed study of the
structure of composition modulated Cu-Ni and Au-Ag thin films

including precise determinations of the lattice parameter in the direction of the composition modulations. He showed that for the Cu-Ni, Cu-Pd, and Au-Ni systems (which all show a supermodulus effect) that there was a strong correlation between the wavelength dependence of the modulus and the magnitude of the variation in planar spacing along the modulation direction. He also showed that for the Au-Ag system (which exhibited no modulus enhancement) that the variation in planar spacing from the average was independent of the modulation wavelength. Furthermore he showed that both the average planar spacing and its variation was much larger in the systems which showed the supermodulus effect than one could expect just on the basis of the composition variation along the modulation direction that one would calculate from the rate of lattice parameter change with composition. These results, he argued, were in qualitative agreement with the effects that would be obtained from a Brillouin zone-Fermi surface contact model.

Tsakalakos and Hilliard (6) added significantly to the credibility of this model by their studies of Cu-Ni foils. They showed that the peak in the supermodulus effect occurred at 1.7 nm for the composition modulation wavelength of (111) planes. They compared this result with positron annihilation measurements of the Fermi surface in a homogeneous alloy of the same composition and observed that a composition modulation of about 2.0 nm would result in a Brillouin zone contact in reasonable agreement with the observed optimal wavelength. They also showed that specimens having a very poor (111) texture did not show any increase of modulus in the as-deposited state. This would be consistent with the model because the Brillouin zone produced by layerings other than (111) planes would not be in a critical contact situation. Tsakalakos made semi-quantitative calculations of the expected magnitude of the effect on the elastic constants but there was sufficient uncertainty in several of the input parameters to lead to a rather large uncertainty in the numbers resulting from the calculation. This work did point the way, however, to the necessity for a much more detailed energy band calculation.

One approach to a more precise energy calculation has been taken by Gonis and Flevaris (7,8) who have modified the coherent potential approximation (CPA) to be applicable to compositionally modulated alloys. They find that their modification of the CPA can produce increased structure in the density of states with the appearance of energy gaps for certain directions and strengths of modulation. However they have not as yet reported a full scale calculation for a real system under varying amounts of strain to determine the magnitude of the change in elastic moduli that might be expected. Jarlborg and Freeman (9,10) have used more of a brute force energy band calculation approach in which they carried out self consistent semi-relativistic spin polarized linear muffin tin orbital energy band calculations. These were performed for a 50-50 composition modulated structure consisting of three atomic layers each of Cu and Ni modulated along the (111) direction. Their initial calculation was done without strain and was primarily motivated by the question of whether complex magnetization distributions existed in these modulated alloys. They were able to determine the magnetization distribution in considerable detail and concluded that the spatial distribution of spin magnetic moments should be much smaller than those deduced from ferromagnetic resonance experiments. Their predictions were later confirmed by direct magnetization and neutron diffraction measurements.

Freeman et. al (11) have also studied the effect of coherency strains on the density of states at the Fermi surface in their calculations. It is a troubling fact that the density of states is quite sensitive to small assumed changes in the local strains, especially if the strain produces a change in local symmetry. Since any calculation of the elastic moduli via this approach will require computing the second derivative of the density of states with respect to strain this situation may present a formidable difficulty. An additional problem, at least in the case of Cu-Ni modulated films, is that neutron diffraction studies by Felcher et. al. (14) have shown that there is quite strong evidence of chemical clustering within the (111) layers. This means that the simplified picture of each layer being either pure nickel or pure copper as envisaged in the band structure calculations may be a considerable distortion of the real situation.

Henein (17) used a rigid band model and the band structure of Cu to ascribe the supermodulus effect in the Cu-Ni system to a singularity in the electronic response when the new superlattice Brillouin zone boundary became tangential to the alloy Fermi surface. Pickett (12) carried this idea a step further by using better Fermi surface calculations of average composition alloys in the Cu-Ni and Ag-Pd systems. He showed that the modulation wavelengths that led to the supermodulus effect in these two systems were about right to achieve contact with the Fermi surface in an extended zone scheme for those two cases. He went on to make a reasonable guess for the shape of the Fermi surface in a Cu-Au alloy of 50-50 composition from which he showed that no (111) modulations of any wavelength could produce contact with the Fermi surface. If Pickett's guess at the Fermi surface in Cu-Au is topologically correct this would provide a simple way of understanding why Cu-Au does not show a supermodulus effect with (111) modulations whereas the other two systems do. However there is still no quantitative calculation of the expected magnitude of the elastic modulus change in this work and thus the question of whether this critical contact phenomenon can explain the very large changes in elastic moduli remains unanswered in this approach.

Wu (13) has taken the approach of trying to calculate the change in elastic moduli by determining the change in the screened ion-ion interaction energy caused by a Brillouin zone-Fermi surface contact. He notes that the structure dependent part of the lattice energy can be separated into a term which depends on the average lattice added to a term which depends on the degree of short range order or compositional variation. He argues that only the compositional variation term need be calculated and this term can be related to the x-ray diffraction intensity that the compositional modulation produces. By making some assumptions about the shape of this diffraction peak, the shape of the Fermi surface, and the dielectric constant for a nearly free electron gas he estimates the magnitude of the change in the energy arising from the composition modulations. Although he gets large changes in this contribution to the elastic constants his results are extremely sensitive to the assumed width of the diffraction peak and also to the assumed topology of the Fermi surface. There is also the question as to how safe it is to ignore the contribution from the average lattice which is expected to provide a much greater contribution to the elastic constants than the order dependent term, especially when one bears in mind Purdes' result that the lattice parameters varied quite significantly in conjunction with the supermodulus effect in alloys that showed that phenomenon. Nevertheless Wu's approach is an interesting one and if the input parameters can be given with sufficient precision this technique could provide some informative results.

III. FUTURE DIRECTIONS

Although the Fermi surface-Brillouin zone models have achieved some successes, especially in the area of critical wavelength prediction and modulation amplitude dependence of the supermodulus effect, a large question remains as to whether these models can explain the enormous increases observed in the elastic moduli. Another concern which can be expressed about this type of model is that in the past this model has been used to explain the existence of a stabilization energy for the structure that achieved contact of the Brillouin zone with the Fermi surface. If this contact is essential for an understanding of the supermodulus effect why are the modulated structures not stable with respect to the homogeneous alloy? This is somewhat surprising in view of the fact that all of the foils showing the supermodulus effect can be converted to homogeneous composition foils at relatively low temperature anneals. In any case this type of model will only be convincing once an elastic moduli calculation is successfully performed.

In a sense of irresponsible speculation, several possibilities can be suggested that have not yet been considered to this author's knowledge. The first of these is that there may be a substantial change in bonding character occurring in these alloys as a result of the compositional modulation. To be more specific, it may be that the forbidden energy gap produced by the introduction of a new Brillouin zone causes a promotion of electrons from the the d-band in these materials into s and p bands. The nearly empty s and p bands are known to lie very close in energy to the nearly filled d bands in these alloys. In fact, in the calculations of Freeman et. al. (11) there is already an indication that there is a transfer of electrons into s and p states to some degree. Since in any band the lowest energy states are bonding states whereas the highest energy states are anti-bonding states the electron transfer envisaged would be from antibonding d states into bonding s and p states. It is possible that the best way to test this question is by using a density functional type of calculation which has been so successfully employed to predict the bulk modulus, cohesive energy, and equilibrium lattice spacing for many of the pure metals (15). The density functional approach has also shown the considerable importance of kinetic energy increase associated with magnetic ordering which can produce very large changes in bulk moduli in the 3-d transition series (16). This latter effect should be raised as a second serious possiblility for explaining the supermodulus effect because all of the alloys in which it has been found have at least one element which is magnetic or strongly paramagnetic. Again the density functional approach might be the best way of testing this possibility.

I believe that the reader can judge from what has been said above that there are some very exciting theoretical and experimental investigations waiting to be done. There are many possibilities for other alloy systems to test for the presence of the supermodulus effect. It may well be that much more spectacular results lie waiting to be discovered. Hopefully a theory that can quantitatively explain the existing experimental results can be successfully employed to predict the most interesting alloy systems to investigate.

REFERENCES

1.   W.M.C. Yang, T. Tsakalakos and J.E. Hilliard, J. Appl. Phys. 48, 876-879 (1977).
2.   J.S. Koehler, Phys. Rev. B 2, 547-551 (1970).
3.   A.J. Purdes, Ph.D. Thesis, Northwestern Univ., Evanston, Ill. (1976).
4.   T. Tsakalakos, Ph.D. Thesis, Northwestern Univ., Evanston, Ill. (1977).
5.   H. Sato and R.S. Toth, Phys. Rev. 6, 1833 (1961).
6.   T. Tsakalakos and J.E. Hilliard, J. Appl. Phys. 54, 734-737 (1983).

7. A. Gonis and N.K. Flevaris, Sol. State Commun. $\underline{37}$, 595-7 (1981).
8. A. Gonis and N.K. Flevaris, Phys. Rev. B $\underline{25}$, 7544-57 (1982).
9. T. Jarlborg and A.J. Freeman, Phys. Rev. Letts. $\underline{45}$, 653-6 (1980).
10. T. Jarlborg and A.J. Freeman, J. Appl. Phys. $\underline{53}$, 8041-5 (1982).
11. A.J. Freeman, J. Xu and T. Jarlborg, J. Mag. & Mag. Mat. $\underline{31-34}$, 909-14 (1983).
12. W.E. Pickett, J. Phys. F $\underline{12}$, 2195-204 (1982).
13. T-B Wu, J. Appl. Phys. $\underline{53}$, 5265-8 (1982).
14. G. P. Felcher et al., J. Mag. & Mag. Mat. $\underline{21}$, L198-202 (1980).
15. V.L. Moruzzi, A.R. Williams and J.F. Janak, Phys. Rev. B $\underline{15}$, 2854-57 (1977).
16. J.F. Janak and A.R. Williams, Phys. Rev. B $\underline{14}$, 4199-4204 (1976).
17. G. E. Henein, Ph.D. Thesis, Northwestern Univ., Evanston, Ill. (1979).

MECHANICAL AND THERMOELECTRIC BEHAVIOR OF COMPOSITION MODULATED FOILS

D. Baral*#, J.B. Ketterson[+] and J.E. Hilliard*

*Department of Materials Science and Engineering, Materials Research Center, Northwestern University, Evanston, IL 60201, #Presently at Memorex Corporation, Santa Clara, Calif., [+]Department of Physics and Astronomy, Northwestern University, Evanston, IL

ABSTRACT

From measurements of independent elastic moduli it has been shown that the large, modulation wavelength dependent enhancement of the biaxial modulis in composition modulated foils is due to the changes in the primary elastic constants. These results provide an explanation for some of the apparently conflicting data in the literature. The plastic and thermoelectric properties of these materials were also found to be anomalous and wavelength dependent.

1. INTRODUCTION

The discovery of a large increase in the biaxial modulus of composition modulated Au-Ni and Cu-Pd foils by Yang, Tsakalakos and Hilliard (1) has prompted additional measurements (2-8). Attempts have been made (9-13) to understand this unusual phenomena using different theoretical approaches. Unfortunately, none of these can explain all the experimental observations.

The various measured elastic moduli are different combinations (14) of the primary elastic constants (c_{11}, c_{12} and c_{44}) and the composition and lattice spacing modulation modify them. Experiments which determine independent combinations of elastic moduli have been performed (15) on composition modulated Cu-Ni foils (that had been removed from the substrate) as a function of modulation parameters (wavelength, composition and amplitude). In this paper we will review the results of measuring the elastic and plastic properties of these films along with the independent thermoelastic measurements.

2. HISTORICAL REVIEW

It is well known that a strength enhancement can be obtained (16,17) in composite materials. Macroscopically laminated materials have shown (18-21) an increase in both their yield and breaking stress. However, only small changes in the elastic moduli were observed. The enhancement of the biaxial modulus in compositionally modulated structures at certain modulation wavelengths prompted the measurement of other moduli and some of the results appear to be in conflict with those given in reference (1).

For example, in a collaborative effort with Berry and Pritchet, Cu-Ni films were prepared (2) at Northwestern University by Purdes in 1974. The films were deposited on fused silica reeds. In a study of the flexural vibration, it was found that when the modulation was eliminated (by in-situ annealing) there was no detectable change in the resonance frequency. (However, there was a small change in the internal friction). The apparent contradiction of these results with the observed enhancement in the biaxial modulus is, in retrospect, probably due to one or more of the following reasons:

a) The films deposited on the silica reeds were not characterized by X-ray diffraction because of the fragility of the reeds. It has since been shown (14,18) that the texture and grain size are important in determining the elastic moduli. If the films deposited did not have a strong [111] texture, then no significant change in the modulus upon annealing would be expected.

b) The films measured had modulation wavelengths of 9 and 35A. Subsequent biaxial modulus measurements (22,23) revealed that there is a maximum in the modulus with decreasing wavelength and that at 9 and 35A the modulus was approximately the same as that of bulk, homogeneous alloys.

c) A further complication arises from the fact (14) that the biaxial and flexural moduli involve different combinations of the primary elastic constants and are therefore not directly comparable.

Measurements of the flexural modulus by the vibrating reed method and Young's modulus by the PUCOT technique were performed (3,24) on a composition modulated Cu-Ni foil (wavelength ~16A) by Testardi and his collaborators to check the enhancement of the modulus. They reported that the moduli had values larger than those for annealed foils. They also reported similar behavior of the shear (torsion) modulus.

The shear moduli of Cu-Nb and Mo-Ni systems were reported by Kueny et al. (7) and Khan et al. (8) using a Brillouin scattering technique. It was found that the velocity of propagation of Rayleigh (surface) waves (that was related to the shear modulus) went through a minimum at a critical wavelength. Using a molecular dynamical calculation, an attempt to explain such an observation in Mo-Ni has recently been made (13) after noting that the observed wavelength dependence of moduli cannot be understood (25) using classical theories of elasticity. The model used in that work (13) uses several assumptions and does not appear to be powerful enough to treat compositionally modulated films.

The biaxial moduli of Cu-Pd films were remeasured (6) by Itozaki. Using modified equipment and a different data analysis procedure, he concluded that there is no enhancement of the moduli at short wavelengths. However, his measured values of the moduli for pure copper and palladium films with a [111] texture are not in agreement (by as much as 50%) with those calculated using the reported single crystal elastic constants of bulk Cu and Pd. In addition his method of determining the biaxial modulus from the experiental parameters contained assumptions that were probably not valid.

Although the measured films were known (26) to be textured and multi-grained, no attempt was made until 1981 (14) to relate the measured moduli to the crystallographic (primary) elastic constants. It is important to note that as a result of composition and lattice spacing modulations, the nature of the forces between atoms change and this in turn changes the primary elastic constants. Is is thus reasonable that measured moduli involving different combinations of the primary elastic constants will have a different wavelength dependence and, as already noted, this is a possible explanation of the conflicting results in the literature.

In an attempt to resolve the issue a detailed study (15) was made of different elastic moduli of compositionally modulated Cu-Ni foils having approximately the same average composition and amplitude of modulation, but differing in the modulation wavelength. On the basis of the X-ray diffraction the films were considered to be fcc with a [111] texture having a composition modulation along the [111] direction. It was also assumed that the grains were randomly oriented in the plane of the foil. The consistency in the experimental results suggest that these approximations are realistic. The plastic and thermoelectric properties of these films were also determined.

3. THEORY

The effective strain tensor for a textured film with randomly oriented grains has recently been calculated (14). For a [111] texture, the various moduli can be written as

Biaxial modulus, $Y_b = 6AC/(A+4C)$
Flexural modulus, $Y_v = Y_b/2 + (4C+B)/6$
Young's modulus, $Y_y = (Y_b/3Y_v)(4C+B)$
Torsion modulus, $Y_t = (4C+B)/6$
Shear modulus, $Y_s = (B+C)/3$

where,

$A = (c_{11} + 2c_{12})/3$
$B = (c_{11} + c_{12})$ and
$C = c_{44}$

In general these moduli are independent of each other. However, because of the transversely isotropic symmetry present in the new coordinate system, the moduli, Y_b, Y_v, Y_t and Y_y are related by:

$$Y_b = Y_v + (Y_v^2 - Y_y Y_v)^{1/2} \text{ and } Y_y Y_v = 2Y_b Y_t$$

Thus, from the measurement of the three moduli Y_v, Y_y and Y_t, one cannot determine c_{11}, c_{12} and c_{44} uniquely. One can, however, use these relationships to check the internal consistency of different measurements.

4. EXPERIMENTAL RESULTS

The techniques of film preparation and moduli measurement are described elsewhere (15,27). In this section we summarize the results.

Figure 1 shows the modulation wavelength dependence of the flexural, Young's and torsional moduli of as-deposited, compositionally modulated Cu-Ni films removed from the substrate. The foils were approximately 2 um in thickness and had an average composition of 50 at.% copper and a modulation amplitude in the range 0.4 ± 0.1. (The amplitude of modulation was estimated using a sinusoidal approximation and a computer program developed by Tsakalakos (4).) The flexural modulus (Fig. 1a) had a behavior similar to that previously observed (4,22) in the biaxial modulus. There was a maximum in the flexural modulus at a wavelength of ~20A.

The Young's modulus, which is a different combination of the primary elastic constants, showed a different behavior (Fig. 1b). It had two maxima instead of one. Using the relationship between different moduli, one can write the Young's modulus as the inverse of the flexural modulus with a prefactor that is a slow function of the wavelength. Thus changes in the primary elastic constants (produced by the modulation) that cause a maximum in the flexural or biaxial modulus, produces a minimum in the Young's modulus in the same wavelength regime. Since the prefactor is a slowly varying

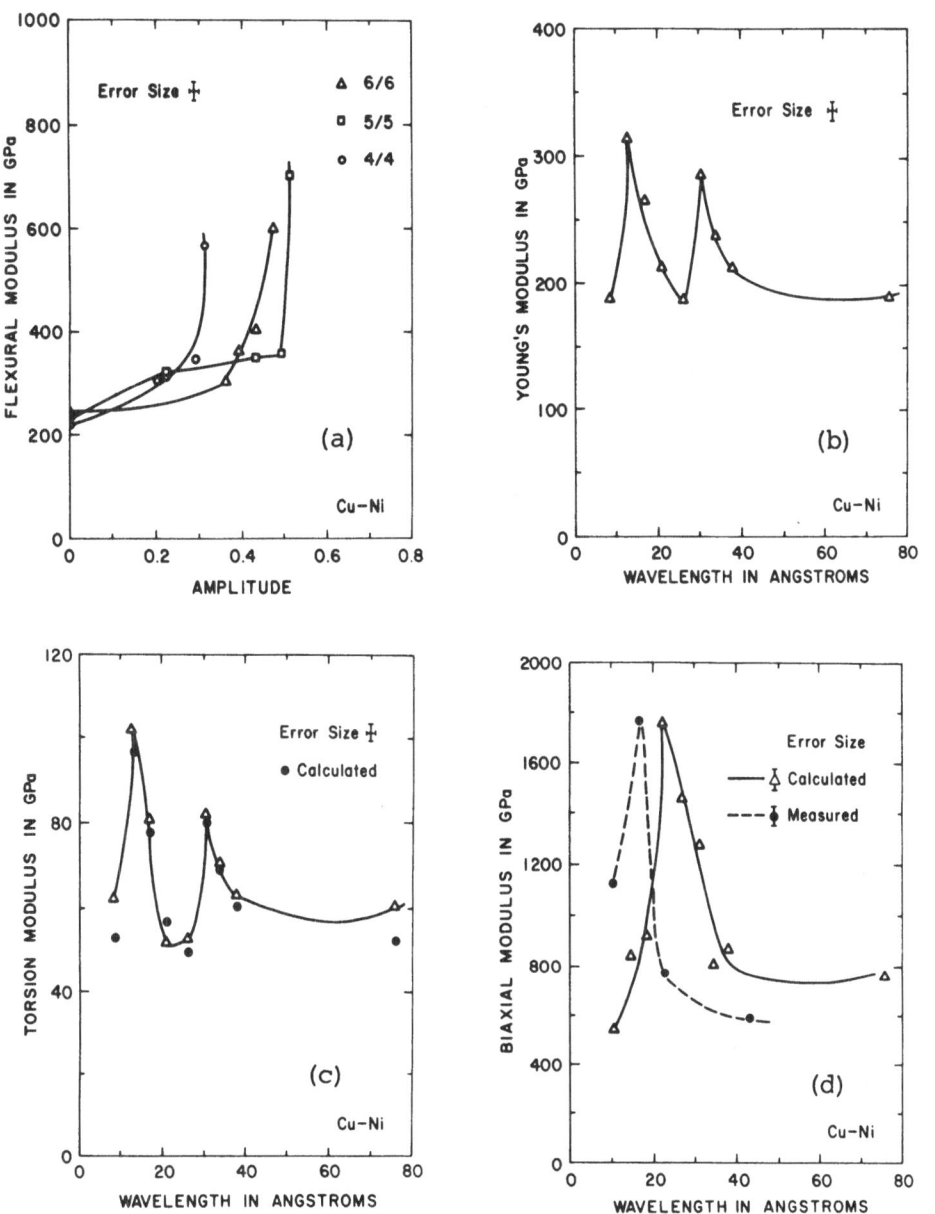

Fig. 1. Modulation wavelength dependence of (a) flexural modulus,
(b) Young's modulus, (c) torsion modulus and (d) biaxial modulus for as-
deposited Cu-Ni foils with average composition of 50 at.% Cu and
modulation amplitude of ~0.4±0.1.

function of wavelength in the region of interest, two maxima were observed in Young's modulus. A similar behavior is seen in the torsion modulus. From the relatioship between the flexural, Young's and torsion moduli one can calculate the torsion moduli from the other two. The results are shown in Fig. 1c. The good agreement between the measured and calculated results indicate that the approximations made in deriving the relationships are valid.

Figure 1d shows a comparison between the experimental data (filled circle) of the measured biaxial modulus for an amplitude normalized to 0.5 (4,22) and the values calculated using the experimental data on the flexural and Young's moduli on as deposited foils with amplitude in the range 0.4 ± 0.1. There is a striking similarlity between the two curves although they are shifted by ~3A. Lack of experimental data points in the vicinity of the critical wavelength might account for this shift. Independent measurements thus confirmed the enhancement of the biaxial modulus that was previously observed.

The breaking stress and microhardness (both relate to the plastic behavior) of composition modulated Cu-Ni foils can be seen in Fig. 2. It will be seen that the wavelength at which a maximum in biaxial and flexural moduli were observed, gives the maximum strength to foil (Fig. 2a). It is interesting to note that the magnitude of this strength compares with the ultimate tensile strength of high strength steels. (Cu-Ni bulk alloys have an UTS of ~345 MPa (50k PSI), homogeneous foils have a breaking strength 657 MPa (95k PSI) whereas for a modulated foil of wavelength ~20A it is 1036 MPa (150k PSI)).

Another measure of the plastic deformation of composition modulated Cu-Ni foils is the Vickers hardness number (determined from the diameter of a diamond indentation having a wedge angle of 150°). It will be seen that the plastic deformation is a minimum at the wavelength at which a maximum in the flexural modulus and breaking stress were observed. We also notice that as one goes to a longer wavelength the hardness increases again. This might be due to the interfacial dislocations that are formed as the structure becomes incoherent at long (> 40A) wavelengths.

Predictions have been made (28) with respect to the plastic behavior of composition modulated foils using classical dislocation theory. However, the present experimental data, where a modulation wavelength dependence has been observed, cannot be explained by such a continuum model.

To explain the observed changes in the mechanical properties, Purdes (26) conjectured that the superlattice Brillouin zone contacted the Fermi surface at a critical wavelength which, via a

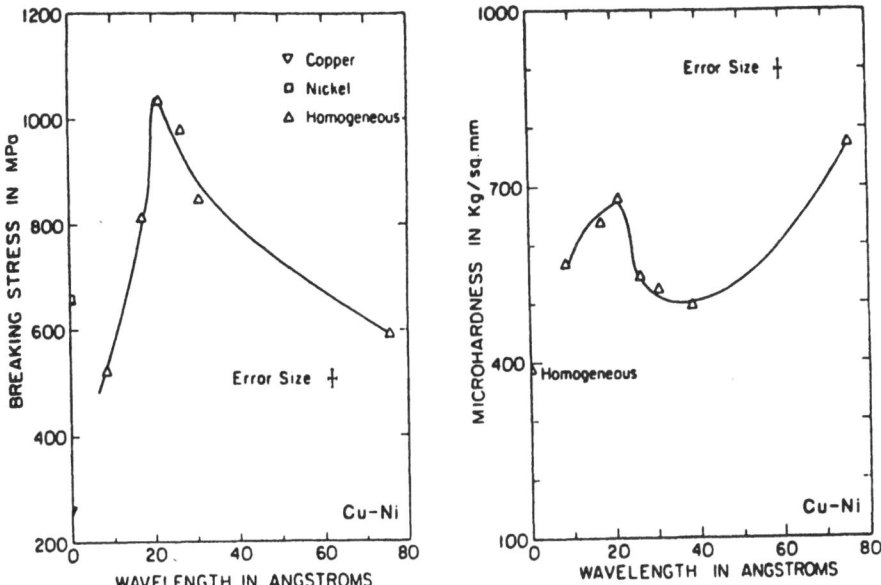

FIG. 2. Plastic behavior of compositionally modulated Cu-Ni foils
with average composition of 50 at.% Cu and modulation
amplitude ~0.4 ± 0.1. Modulation wavelength dependence
of (a) breaking stress and (b) microhardness.

singular change in the electronic structure, results in the
observed sudden change in modulus. This idea was later used by
Henein (5), and more quantitatively by Pickett (9) and Wu (10)
where the effect of the Brillouin-zone - Fermi surface interaction
on the screening was studied. The other explanation (3) involves
a change in mechanical behavior caused by coherency strain effects.
This theory, however, cannot explain the observed reduction of
biaxial or flexural modulus in short wavelength films.

In order to derive information which is more directly linked
to electronic structure, thermo-emf measurements were performed
(29). Since the measured emf is related (30) to the electron
density of states near the Fermi surface, amplitude and composition
normalized thermo-emf measurements were performed to obtain the
effects of modulation wavelength on the thermo-emf. Figure 3
shows the dependence for Cu-Ni and Ag-Pd foils. It was found that
in both cases, the foil which had shown a maximum in biaxial
modulus showed a maximum in thermo-emf. It is interesting to note
(29) that when the foils are homogenized (by annealing) the thermo-
emf increases. A much lower emf was found in short wavelength
films. This suggests that these foils are electronically different
from a homogeneous foil.

472

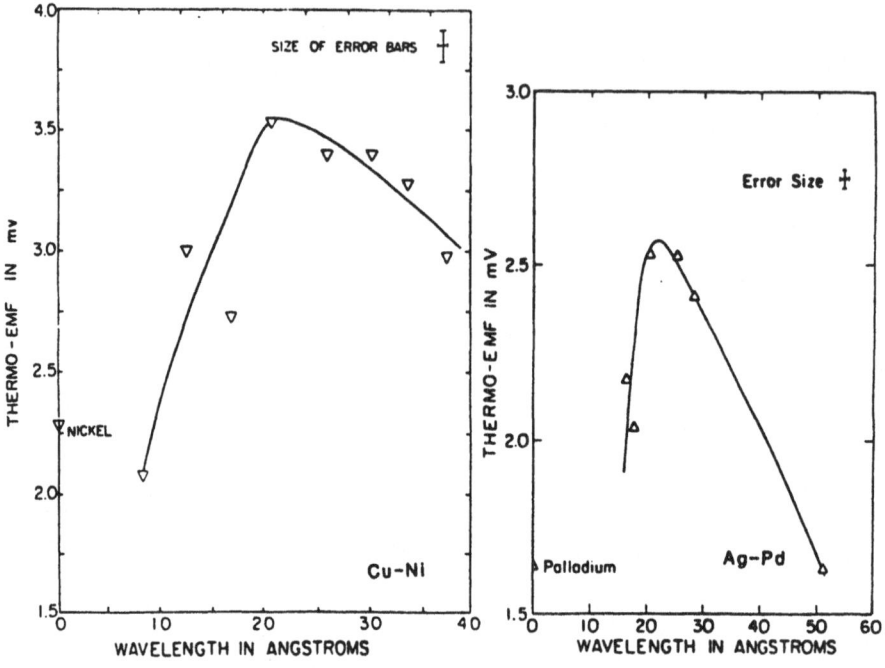

FIG. 3. Thermoelectric behavior of composition modulated Cu-Ni
and Ag-Pd foils (measured with respect to copper for a
100°C temperature differential).

SUMMARY

The results summarized in this review confirm the enhancement
of the biaxial modulus that has been previously reported for
compositionally modulated structures at wavelengths of ~20A. An
anomaly in the thermo-emf is also observed at approximately the
same wavelength. These results in conjunction with earlier ones,
emphasize that any theory attempting to explain these effects must
account for a nonmonotonic dependence of the properties as a
function of modulation wavelength.

ACKNOWLEDGEMENT

We would like to thank N.K. Flevaris for his help in film
preparation and K. Miyano for introducing us to the measurement of
the flexural modulus.

This work was supported by National Science Foundation through
the Materials Research Center of Northwestern University.

REFERENCES

(1) Yang, W.M.C., T. Tsakalakos and J.E. Hilliard. J. Appl. Phys. 48 (1977) 876.
(2) Berry, B.S. and W.C. Pritchet. Thin Solid Films 33 (1976) 19.
(3) Testardi, L.R., R.H. Willens, J.T. Krause, D.B. McWhan and S. Nakahara. J. Appl. Phys. 52 (1981) 510.
(4) Tsakalakos, T. Ph.D. Thesis, Northwestern University (1977).
(5) Henein, G., Ph.D. Thesis, Northwestern University (1979).
(6) Itozaki, H., Ph.D. Thesis, Northwestern University (1982).
(7) Kueny, A., M. Grimsditch, K.Miyano, I. Banerjee, C.M. Falco and I.K. Schuller. Phys. Rev. Lett. 48 (1982) 166.
(8) Khan, M.R., C. Chun, G. Felcher, M. Grimsditch, A. Kueny, C. Falco and I.K. Schuller. Phys. Rev. B (to be published).
(9) Pickett, W.E. J. Phys. F12 (1982) 2195.
(10) Wu, T.B. J. Appl.Phys. 53 (1982) 5265.
(11) Yamauchi, H. Phys. Rev. B 27 (1983) 905.
(12) Clapp, P. NATO Advanced Study Instutute on Modulated Structure Materials, 15-25 June, Crete Greece (1983).
(13) Schuller, I.K. and A. Rahman. Phys. Rev. Lett. 50 (1983) 1377.
(14) Baral, D., J.E. Hilliard, J.B. Ketterson and K. Miyano. J. Appl. Phys. 53 (1982) 3552.
(15) Baral, D., Ph.D. Thesis, Northwestern University (1983).
(16) Vinson, J.R. and T.Y. Chou. Composite Materials and Their Use In Structures (John Wiley & Sons, New York (1975)).
(17) Kelly, A. and G.J. Davies, Met. Rev. 10 (1965) 1.
(18) Lehoczky, S.L. J. Appl. Phys. 49 (1978) 5479.
(19) Palatnik, L.S. and A.I. Il'yurskiy. Fiz. Met. Metall. 15 (1963) 620.
(20) Frommeyer, G. and G. Wasserman. Acta. Met. 23 (1975) 1353.
(21) Bevk, J., J.P. Haribson and J.L. Bell. J. Appl. Phys. 49 (1978) 6031.
(22) Tsakalakos, T. and J.E. Hilliard. J. Appl. Phys. 54 (1983) 734.
(23) Henein, G. and J.E. Hilliard. J. Appl. Phys. 54 (1983) 728.
(24) Willens, R.H. NATO Advanced Study Institute on Modulated Structure Materials, 15-25 June, Crete, Greece (1983).
(25) Kueny, A. and M. Grimsditch. Phys. Rev. 26 (1982) 4699.
(26) Purdes, A., Ph.D. Thesis, Northwestern University (1976).
(27) Baral, D., J.B. Ketterson and J.E. Hilliard, to be published.
(28) Koehler, J.S. Phys. Rev. B 2 (1970) 547.
(29) Baral, D. and J.E. Hilliard. Appl. Phys. Lett. 41 (1982) 156.
(30) Bernard, R.D. Thermoelectricity in Metals and Alloys (Halsted, New York (1972)).

EFFECTS OF SHORT WAVELENGTH COMPOSITION MODULATIONS ON INTERDIFFUSION IN SILVER-PALLADIUM THIN FOILS

G. E. Hénein and J. E. Hilliard

Department of Materials Science and Engineering
Materials Research Center, Northwestern University
Evanston, Illinois 60201

ABSTRACT

The effects of short wavelength composition modulations upon interdiffusion were evidenced in the Ag-Pd system by an x-ray diffraction technique. The general behavior of the interdiffusity versus the wavelength of the concentration wave was found in agreement with the nonlinear diffusion theory of Tsakalakos. Superimposed on this general behavior, two anomalies were found which can be attributed respectively to, 1) a loss of coherency between the Ag-rich and Pd-rich regions, and 2) a screening singularity or an increase in the shear modulus (or both) such as was found in the Au-Ni system. The wavelength dependence of the interdiffusion coefficient yielded the first four gradient-energy for this system.

INTRODUCTION

The treatment of interdiffusion in multilayered binary thin foils by Cook, de Fontaine and Hilliard(1) was an extension of the continuum models developed by Hillert(2) and Cahn(3,4) to describe chemical diffusion over short penetration distances. This treatment, which takes into account the discrete nature of the crystal lattice, has been applied to describe interdiffusion in various binary systems: Au-Ag(5), Cu-Pd(6), AuNi(8) and Cu-Ni(9,10). The present work is an attempt to apply this model to the silver-palladium system.

Cook, de Fontaine and Hilliard(1) considered a binary cubic crystal containing a composition modulation defined by:

$$c(p,t) = Q(t)\cos[2\pi x(p)/\lambda], \tag{1}$$

where c is the concentration of component 2, Q is the amplitude of the modulation and λ its wavelength, t is time and p denotes the p'th atomic plane. If the diffusion equation is linearized, the solution to Eq. (1) is:

$$Q(t) = Q(0)\exp[-\tilde{D}_B B^2(\lambda)t], \tag{2}$$

in which:

$$\tilde{D}_B[hk\ell] = \tilde{D}[1+(2\eta^2 Y[hk\ell]/f'') + (2K/f'')B^2(\lambda)], \tag{3}$$

in which η = dℓna/dc (where a is the lattice parameter), f'' the second derivative of the Helmholtz free energy, K the gradient-energy coefficient, Y[hkℓ] the biaxial elastic modulus in the (hkℓ) plane and B^2 the dispersion relation which, for modulations along $\langle 100 \rangle$ or $\langle 111 \rangle$, can be written in terms of the interplane spacing d as follows:

$$B^2(\lambda) = (2/d^2)[1-\cos(2\pi d/\lambda)]. \tag{4}$$

The term in Y[hkℓ] in Eq. (3) is to allow for the effect of coherency strains and that in K for the effect of concentration gradients. For an arbitrary periodic modulation, Eq. (2) can be used to determine the time dependence of the amplitudes of the Fourier components of the modulation.

The analysis of the diffracted x-ray intensity from a material containing a modulation in lattice spacing and in scattering factor (11,12) yields sharp Bragg peaks flanked by satellites of increasing order occurring in pairs about each Bragg peak. A pair of satellites of order m arises from the m'th Fourier component of the modulation and their intensities are proportional to the square of the amplitude. The first order satellites are the most intense, and therefore were the ones used for estimating λ. Denoting their intensity by I(t), it follows from Eq. (2) that:

$$I(t)/I(0) = \exp(-2B^2\tilde{D}_B[hk\ell]t). \tag{5}$$

In the framework of the linear diffusion model therefore, a plot of $\ell n[I(t)/I(0)]$ versus isothermal annealing time should be linear. From the slope it is then possible to determine the diffusion coefficient $\tilde{D}_B[hk\ell]$. This model also predicts that the "effective" interdiffusion coefficient $\tilde{D}_B$ is linear with B^2.

EXPERIMENTAL PROCEDURE

Preparation of Specimens

Ag-Pd composition modulated thin foils were produced by vacuum evaporation of the two components. In order to achieve a high degree of (111) texture a 40 nm-thick layer of silver was deposited on the heated (250°C) Muscovite mica substrate immediately prior to the deposition of the modulated film. The latter deposition was

conducted at a substrate temperature of 200°C to avoid interdiffusion between the Ag and Pd layers.

The diffusion anneals were performed inside a quartz tube furnace evacuated to 0.1 mPa. All the anneals were performed at 375°C and correction was allowed for the heating-up and cooling-down transients(13).

The x-ray diffraction was used to obtain the average composition and the modulation wavelength. The goniometer arrangement was similar to that devised by Schwartz, Morrison and Cohen(14). The integrated peak intensities were corrected for polarization, absorption(15) and temperature(15). The atomic scattering factors for Ag and Pd were calculated from a 9-parameter fitting formula (16). The average composition was determined from the lattice parameter data of Coles(17).

EXPERIMENTAL RESULTS AND DISCUSSION

Satellite Intensity Decay Curves

Of the 92 samples prepared, 13 were selected for their high degree of (111) texture and uniformity (0.50 ± 0.03 at. fract.) in average composition from one specimen to another. Their wavelengths ranged from 0.94 to 5.08 nm. Uniformity in composition was important because high temperature data(18,19) for the diffusion of Ag and Pd in Ag-Pd alloys indicate a fairly strong composition dependence.

The plots of $\ln I(t)/I(0)$ versus t are shown in Figs. 1, 2 and 3. It is apparent from all the curves that the linear diffusion equation is not obeyed at short times, the decay of the satellite intensities being much faster than predicted. This enhanced diffusivity (which has been observed in all the systems previously studied) is primarily due to recrystallization and grain growth during the early stages of annealing as indicated by an approximately 10 pct. increase in the intensity of the 111 Bragg peak. A second factor contributing to this transient is the non-linearity of the diffusion equation when the composition amplitude is not small enough to justify the assumption that $\tilde{D}$, f" and K are independent of composition. A theoretical analysis(20) has shown that the amplitude $Q(t)$ of the first Fourier component of the composition modulation decays as a sum of negative time exponentials. At short times all the exponential terms contribute to the diffusion which is therefore more rapid than predicted by the linear model. At longer times all the terms but the first become

478

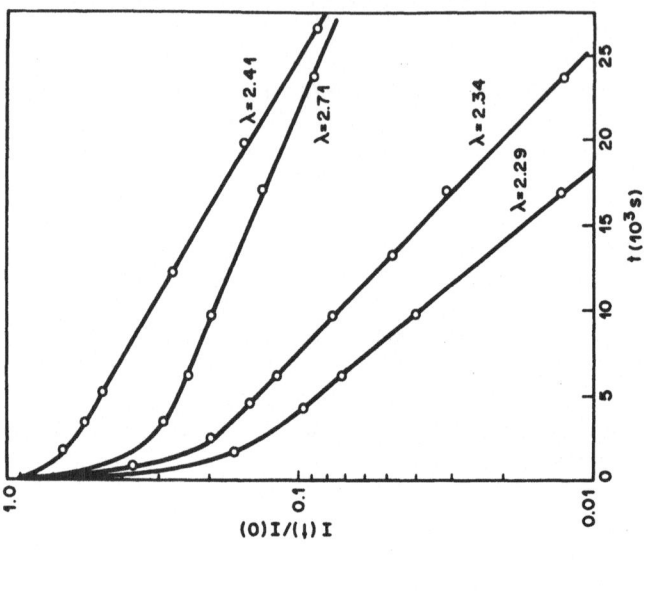

Fig. 1 — Time decay of satellite intensities. Wavelengths 0.94, 1.15, 1.21, 1.52 and 1.94 nm.

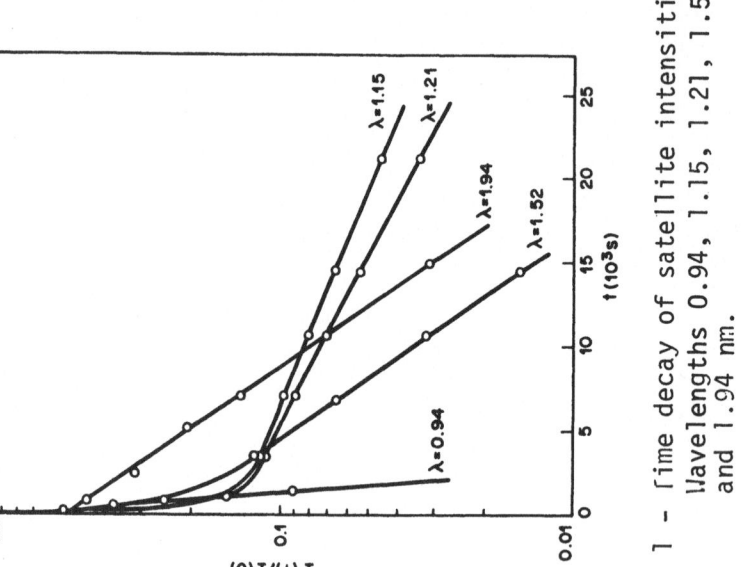

Fig. 2 — Time decay of satellite intensities. Wavelengths 2.29, 2.34, 2.41 and 2.71 nm.

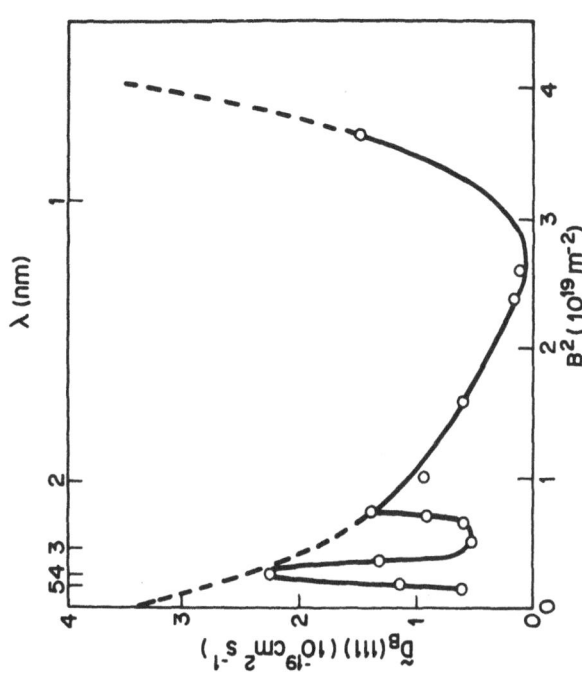

Fig. 4 – Observed diffusivities $\tilde{D}_B[111]$ versus B^2 at 375°C. The dashed line represents the fit to Eqs. 6 and 7 for $c_o = 0.50$ at fract.

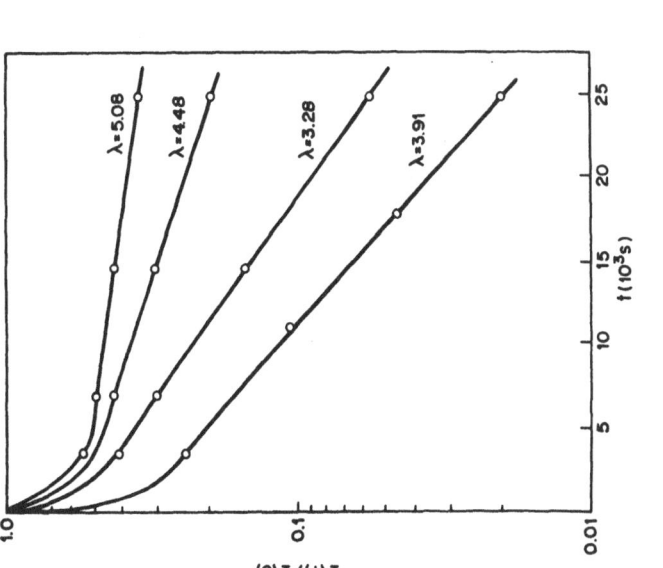

Fig. 3 – Time decay of satellite intensities. Wavelength 3.28, 3.91, 4.48 and 5.08 nm.

Table I. Observed diffusivities $\tilde{D}_B[111]$ at 375°C.

c_0 at.fract. Ag	λ nm	B^2 $10^{19}\,m^{-2}$	$\tilde{D}_B[111]$ $10^{-19}\,cm^2/s$	$\tilde{D}_B[111](c_0 = .50)$ $10^{-19}\,cm^2/s$
0.53	0.94	3.64	1.48	1.615
0.50	1.15	2.61	0.10	0.10
0.49	1.21	2.39	0.142	0.159
0.52	1.52	1.58	0.615	0.608
0.49	1.94	1.0	0.94	1.05
0.52	2.29	0.73	1.40	1.385
0.50	2.34	0.70	0.91	0.91
0.49	2.41	0.66	0.59	0.66
0.47	2.71	0.52	0.53	0.579
0.51	3.28	0.36	1.31	1.261
0.53	3.91	0.256	2.25	2.456
0.48	4.48	0.195	1.14	1.128
0.51	5.08	0.152	0.61	0.587

Key: c_0 = average composition (atom fraction)

λ = wavelength

B^2 = dispersion relation defined by Eq. (4)

$\tilde{D}_B[111]$ = observed diffusivity

$\tilde{D}_B[111](c_0 = .050)$ = diffusivity extrapolated to $c_0=0.5$ using Eqs. (6) and (7).

negligible and it is the first exponential term which governs the behavior of Q(t) and yields the linear portions of the $\ln I$ versus t plots. [The leading term is the same as that predicted by the linear model in Eq. (2).]

Data for the 13 samples are given in Table I and the coefficient $\tilde{D}_B$ is plotted versus B^2 in Fig. 4. Three features characterize this curve: (1) A polynominal behavior (dashed line), (2) A sharp drop between 2.3 and 2.7 nm and (3) A second drop between 3.9 and 5.1 nm.

General Behavior of $\tilde{D}_B$ Versus B^2 and Gradient Energy Coefficients

It is apparent from Fig. 4 that a single gradient energy coefficient cannot describe the behavior of $\tilde{D}_B$ versus B^2 because otherwise the relationship would have been linear. The gradient energy coefficients were formally introduced by Cahn(2) as the coefficients in the expansion of the free-energy in terms of the partial derivatives of the latter with respect to the composition gradients. The number of such coefficients is theoretically infinite. Physically(21) each gradient energy coefficient represents the interaction of an atom with the successive atomic shells surrounding it. If long-range interactions are present in the alloy Tsakalakos and Hilliard(9) have shown that it is necessary to include several gradient energy coefficient K_i in the free-energy expansion and the expression for $\tilde{D}_B$ becomes:(20)

$$\tilde{D}_B = (D/f'')(f''+2\eta^2 Y + 2 \sum_{i=1} K_i B^{2i}) \tag{6}$$

and leads to a polynominal behavior $\tilde{D}_B$ versus B^2.

In order to estimate these coefficients from the experimental data, we used a method(20) in which the above expression (Eq. 6) is used for $\tilde{D}_B$ and the composition dependence of the quantity $\tilde{D}(1 + 2\eta^2 Y/f'')$ is expressed as:

$$\tilde{D}(1+2\eta^2 Y/f'') = D_0 + D_1 q + D_2 q^2 \tag{7}$$

where $q = c - c_0$ in which c is the atomic fraction of palladium and c_0 the average composition (0.5 in this case). The parameter η was estimated from the lattice-parameter data of Coles(17) and f'' was determined using a method described by Rundman and Hilliard(22) with the thermodynamic data of Myles.(23)

The biaxial elastic modulus Y for the case of [111] texture is given by:(2,24)

$$Y[111] = 6C_{44}(C_{11}+2C_{12})/(C_{11}+2C_{12}+4C_{44}) \tag{8}$$

and was estimated using an interpolation between the elastic constants of Ag and Pd.(25)   The results of the latter procedure gave good agreement with experimental determinations of Y[111] at long wavelengths by the authors.(26)   The following values were used for an alloy of 0.50 average composition:

$$\eta = 4.9 \times 10^{-2}$$

$$f'' = 6.31 \times 10^{9} \text{ J/m}^3$$

$$Y = 0.23 \text{ TPa.}$$

A non-linear least squares fit was then performed on the data using a regression routine. The following "singular" points (cf. Table I and Fig. 4): λ = 2.34, 2.41, 2.71, 3.28, 4.48 and 5.08 nm, were omitted from the regression fitting.   The following values were obtained:

$$D_0 = 3.457 \times 10^{-19} \text{ cm}^2\text{s}^{-1}$$

$$D_1 = -24.96 \times 10^{-19} \text{ cm}^2\text{s}^{-1}$$

$$D_2 = -1154 \times 10^{-19} \text{ cm}^2\text{s}^{-1}$$

$$K_1 = -5.0 \times 10^{-19} \text{ Jnm}^{-1}$$

$$K_2 = 3.6 \times 10^{-20} \text{ Jnm}$$

$$K_3 = -1.3 \times 10^{-21} \text{ Jnm}^3$$

$$K_4 = 1.9 \times 10^{-23} \text{ Jnm}^5$$

It has been shown(9) that the gradient energy coefficients can be calculated from the Fourier components of the interatomic potentials.   However, to our knowledge, no data are available from diffuse x-ray or neutron scattering in the Ag-Pd alloy in order to determine these components and it was therefore impossible to check the values which we derived for these coefficients.

A plot of the experimental $\tilde{D}_B$'s together with the fitted expression for $\tilde{D}_B$ at c = 0.5 at. fract. is shown in Fig. 4. It has been shown(27) that when only nearest-neighbor interactions are taken into account, the first gradient energy coefficient K_1 is proportional to the heat of mixing of the two components at c_0 = 0.5 at. fract. The heat of mixing of Ag and Pd is negative(28,29) and, since K_1 was found to be negative, we can infer that long-range interactions do not predominate in this system.   The same correspondence was found in Cu-Pd,(5) Au-Ag,(4) Al-Zn(22) and Al-Ag(30) systems but not in Cu-Ni.(20)   In the latter system long-range interactions were proposed as being responsible for a negative K_1 despite the positive heat of mixing.

Loss of Coherency in the Ag-Pd System

The second feature of the experimental $\tilde{D}_B$ versus B^2 curve is the decrease in $\tilde{D}_B$ in the wavelength range of 3.9-5.1 nm. We believe that this occurs because of a gradual loss of coherency between the (111) silver-rich and palladium-rich layers. This would eliminate the term $2\eta^2 Y$ in Eq. (4) resulting in a decrease in the value of $\tilde{D}_B$.

Cahn(3) has shown that the coherency strain energy per unit volume is given by:

$$E_c = \eta^2 Y Q^2 \tag{9}$$

for a square composition wave, and

$$E_c = \eta^2 Y Q^2/2 \tag{10}$$

for a sinusoidal composition wave, where Q is the amplitude of the wave. There is, however, the possibility for the lattice to lose its coherency along the <111> direction as soon as the presence of dislocation grids at the interfaces between Ag-rich and Pd-righ layers becomes more favorable energetically. Van der Merwe(31) has given the following expression for the energy per unit area of a grid of parallel dislocations regularly spaced so as to accommodate exactly the misfit between two adjacent layers:

$$\tau = (\mu c/4\pi^2)\{1 + \beta - (1 + \beta^2)^{1/2} - \beta \ell n[2\beta(1 + \beta^2)^{1/2} - 2\beta^2]\} \tag{11}$$

If a and b represent the natural spacings perpendicular to the dislocation lines in Pd and Ag respectively, c is given by:

$$(P+1)a = Pb = (P+1/2)c$$

where P is a positive integer, μ is the average shear modulus and β is given by:

$$\beta = \pi(1-\nu)/P$$

where ν is the Poisson's ratio. Ditchek(32) showed that Eq. (11) predicts the correct wavelengths for loss of coherency in several spinodal alloys and found that this expression was preferable to de Fontaine's criterion.(33)

In a unit volume of a modulated foil consisting of a cube containing n wavelengths λ, the energy associated with such grids of dislocations at each interface is:

$$2n\tau = 2\tau/\lambda. \tag{12}$$

Total loss of coherency should therefore be expected when:

$$E_c \geq 2\tau/\lambda,$$

that is when the composition modulation wavelength satisfies:

$$\lambda \geq 2\tau/E_c. \tag{13}$$

The interfaces between the Ag-rich and the Pd-rich regions in our films were parallel to the (111) planes and a cross-grid of dislocations is necessary to relieve all the coherency stresses: one grid parallel to the <110> direction and a second grid, perpendicular to the first, parallel to the <112> direction. Van der Merwe(31) showed that in such a case the total interfacial energy was just the sum of the energies associated with each grid. Taking the lattice parameters(34) of Ag and Pd as respectively 0.408 and 0.389 nm and the shear modulus(25) of a 50/50 alloy as $\mu = 0.059$ TPa we obtain:

$$\tau[1\bar{1}0] = 0.114 \text{ J/m}^2$$

$$\text{and } \tau[11\bar{2}] = 0.066 \text{ J/m}^2 \tag{14}$$

The total interfacial energy per unit volume is:

$$\tau = (2/\lambda)(\tau[1\bar{1}0] + [11\bar{2}])$$

$$\tau = 0359/\lambda \text{ J/m}^3.$$

Using the following values for a 50/50 alloy,

$$Y = 0.230 \text{ TPa}$$

$$\text{and } \eta = 4.9 \times 10^{-2}$$

Eqs. (9) and (10) yield, for the coherency strain energy per unit volume of a foil containing a modulation of maximum amplitude ($Q = 0.5$):

$$E_c = 1.38 \times 10^8 \text{ J/m}^3 \text{ for a square modulation, and}$$
$$E_c = 6.90 \times 10^7 \text{ J/m}^3 \text{ for a sinusoidal modulation.}$$

Using these last values and Eq. (14), Eq. (13) predicts that for a square composition wave of maximum amplitude, coherency will be completely lost for wavelengths longer than 2.6 nm. For a sinusoidal wave of maximum amplitude, loss of coherency will occur at wavelengths greater than 5.2 nm. The composition waves present in our samples were neither square (some diffusion always occurs during the deposition of the foils) nor purely sinusoidal (Fourier components of order higher than the first were occasionally seen in the diffraction patterns). We thus expect loss of coherency at ~ 4.0 nm. The average distance between adjacent dislocations (given by the product Pc) is 5.8 nm for the <110> dislocations and 3.3 nm for the <112> dislocations. In view of these short separation distances, a more accurate estimation of the interfacial energy should include the interaction between dislocations. This would increase the wavelength at which coherency is lost.

The ratio $R = I^-/I^+$ of the intensity of the low-angle to that of high angle first order satellites can be used(5,11) to determine experimentally the state of coherency of modulated materials. Our values of R indicated that coherency was maintained for wavelengths no longer than 3.28 nm. For longer wavelengths, the tails of the (111) Bragg peaks impinged on the satellite peaks and rendered the estimation of the ratio too uncertain for estimating the state of coherency. We conclude, therefore, that, without ruling out any other possible effect, the drop in $\tilde{D}_B$ observed between 4.0 and 5.0 nm can be accounted for by a loss of coherency in the films.

Anomalous Behavior of $\tilde{D}_B$

A complementary investigation(26) has shown that, in the same range of wavelengths 2.3-2.7 nm where $\tilde{D}_B$ exhibits a sharp drop (Fig. 4), the elastic modulus Y [111] shows a marked enhancement (of at least 100 pct.). The modulus Y enters the expression for $\tilde{D}_B$(Eq. 3); we show, however, in the Appendix that this enhancement in Y had no direct influence on our measured values of $\tilde{D}_B$. It is therefore believed that a change in the gradient energy coefficients is responsible for this anomaly in $\tilde{D}_B$. A screening singularity has been predicted by Dunaev and Zakharova(35) for the interdiffusion coefficient $\tilde{D}_B$ in the Au-Ni system and verified experimentally by Yang and Hilliard.(36) Such a screening anomaly would occur when the Fermi surface of the alloy comes into contact with the additional Brillouin zone planes created by the composition modulation. This would affect the interatomic potentials and therefore the gradient energy coefficients because they are directly related to them.(19) Although the latter hypothesis is supported theoretically(35) and experimentally(36) and also by the calculation developed in the Appendix, another explanation is the following: In Zener's model of diffusion(37,38) the shear modulus μ enters directly into the expression for the atomic diffusion coefficient. We do not know at the present time whether the enhancement(26) of Y was due primarily to an anomaly in the shear modulus but in the latter instance, some anomaly in the bulk interdiffusion $\tilde{D}$ would be expected: if μ increases, that is if the lattice becomes more "rigid," one would expect a decrease in the diffusivity.

SUMMARY AND CONCLUSIONS

Measurements of the effective interdiffusion coefficient $\tilde{D}_B$ [111] have shown the following:

(1) The results could not be accounted for by a single gradient-energy coefficient. This indicates that atomic interactions beyond first nearest-neighbors are significant in the Ag-Pd system.

(2) There was a decrease in $\tilde{D}_B$ for $\lambda \geqslant 5$ nm which we believe is due to a loss in coherency.

(3) There was an abrupt drop in $\tilde{D}_B$ for λ's in the range 2.0-2.7 nm. This drop occurred in the same range where an enhancement of the elastic modulus has been observed in another investigation. We have considered two possible explanations for the anomalous variation in $\tilde{D}_B$: (a) A screening singularity which influences the gradient energy coefficients, or (b) a decrease in $\tilde{D}_B$ because of an increase in the shear modulus.

The set of experiments described in the present work exemplifies the power of the technique used, which allows the simultaneous collection of several fundamental pieces of information concerning a particular alloy system. It offers a way to gain access to the interatomic potentials, although indirectly and incompletely without complementary studies. The extent to which long-range interactions are present in the alloy can be obtained by the number and relative weights of the gradient energy coefficients. Critical phenomena occurring in the $\tilde{D}_B$ versus B^2 curve. It allows the determination of the state of coherency across epitaxial inter-faces.

On the practical side, a minimum in the interdiffusion coefficient was obtained at short wavelengths ($\lambda = 1.1 - 1.2$ nm, $\tilde{D}_B$ $\sim 0.1 \times 10^{-19}$ cm^2/s). Extrapolation(18,19) to 375°C of measurements of interdiffusivities in Ag-Pd couples(39) yields $\tilde{D}$ $\sim 1.5 \times 10^{-19}$ cm^2/s. The minimum in $\tilde{D}_B$ thus represents a factor 15 decrease in interdiffusivity. The existence of such minima could therefore allow composition-modulated films to become an alternative to "diffusion barrier" layers widely used in the semiconductor technology.

ACKNOWLEDGEMENTS

The National Science Foundation is gratefully acknowledged for its Grant DMR 77-15893 sponsoring this work.

APPENDIX

A recent study(26) has revealed a substantial increase of Y[111] in Ag-Pd foils having composition modulations with wavelengths in the range of 2.0-3.0 nm. The purpose of the following discussion is to demonstrate that the increase in the modulus cannot account for the anomalies observed in the dependence of $\tilde{D}_B$ on B^2.

For wavelengths between 2.0 and 3.0 nm the observed modulus could be fitted to:

$$Y[\lambda,Q] = Y_0 + \phi[\lambda]Q^2, \tag{A1}$$

in which Y_0 is the modulus of a homogeneous foil having the same composition as the average for the modulated foil and ϕ depends only on the wavelength for a given average composition.

Substituting Eq. (A1) in Eq. (3) we obtain (omitting the direction arguments on the D's):

$$\tilde{D}_B/\tilde{D}_B^- = 1 + (2\eta^2\phi[\lambda]\tilde{D}/\tilde{D}_B f'')Q^2, \tag{A2}$$

in which $\tilde{D}_B^-$ is the diffusivity that would be observed for $\phi[\lambda] = 0$. The second term on the right-hand side of Eq. (A2) is therefore the fractional increase in $\tilde{D}_B$ caused by the enhancement in the elastic modulus. The maximum value of ϕ observed(26) for Ag-Pd was 3.3 TPa/(at. fract.)^2 and occurred at $\lambda = 2.3$ nm (which corresponds to $B^2 = 6.6 \times 10^{18} m^{-2}$). Also, $2\eta^2/f'' = 0.76$ (TPa)$^{-1}$, $\tilde{D} \approx 2.3 \times 10^{-19} cm^2/s$ (based on an extrapolation of the polynomial fitting) and $\tilde{D}_B \approx 0.6 \times 10^{-19} cm^2/s$. Substituting these values in Eq. (A2) we obtain:

$$\tilde{D}_B/\tilde{D}_B^- \approx 1 + 10Q^2; \quad [Q \text{ in at. fract.}]. \tag{A3}$$

The maximum amplitude in the as-deposited foils was ~ 0.4 at. fract. which yields an approximately 1.6 fold increase in $\tilde{D}_B$. However, as already stated, the estimation of $\tilde{D}_B$ was based on the slope of the $\ln(I/I_0)$ versus t plot after the initial non-linear transient had decayed out. Typically, this occurred at an amplitude of 0.15 at. fract. which would produce a 25 pct. increase. This is appreciably smaller than the discontinuity observed in the plot of $\tilde{D}_B$ versus B^2. However, what is perhaps more convincing than the foregoing calculation (which, of necessity, is only approximate) is that any significant contribution from the enhanced modulus effect will produce a non-linearity in the $\ln(I/I_0)$ versus t plot at sufficiently large amplitudes. And, since the values of $\tilde{D}_B$ were determined from the linear region of the plot, this automatically excludes the Q^2 term in Eq. (A1).

REFERENCES

1. H. E. Cook, D. de Fontaine and J. E. Hilliard, Acta Met., 17, 765 (1969).
2. M. Hillert, Acta Met., 9, 525 (1961).
3. J. W. Cahn, Acta Met., 9, 795 (1961).
4. J. W. Cahn, Acta Met., 10, 179 (1962).
5. H. E. Cook and J. E. Hilliard, J.A.P., 40, 2191 (1969).
6. E. M. Philofsky and J. E. Hilliard, J.A.P., 40, 2198 (1969).
7. W. M. Paulson and J. E. Hilliard, J.A.P., 48, 2117 (1977).
8. W. M. C. Yang, PhD Thesis, Northwestern University, 1971.
9. T. Tsakalakos and J. E. Hilliard, submitted for publication, J. Appl. Phys., companion paper.
10. T. Tsakalakos and J. E. Hilliard, J. de Physique, C7, 38, 404 (1977).
11. D. de Fontaine, Local Atomic Arrangements Studied by X-Ray Diffraction, J. B. Cohen and J. E. Hilliard, Eds. (Gordon and Breach, New York, 1967) Chap. 2.
12. A. Guinier, X-Ray Diffraction in Crystals, Imperfect Crystals and Amorphous Bodies, P. Lorrain and D. Sainte Marie Lorrain, Transl. (W. H. Freeman and Co., San Francisco, 1963), pp. 279-282.
13. P. G. Shewmon, Diffusion in Solids, McGraw Hill Book Co., 1963, p.32.
14. L. H. Schwartz, L. A. Morrison and J. B. Cohen, Advan. X-Ray Anal., 7, 281 (1964).
15. International Tables for X-Ray Crystallography, The Kynoch Press (1965).
16. D. T. Cromer and J. T. Waber, Acta Cryst., 18, 104 (1965).
17. B. R. Coles, J. Inst. Met., 84, 346 (1956).
18. N. H. Nachtrieb, J. Petit and J. Wehrenberg, J. Chem. Phys., 26, 106 (1957).
19. R. L. Rowland and N. H. Nachtrieb, J. Phys. Chem. 67, 2817 (1963).

20. T. Tsakalakos, Ph.D. Thesis, Northwestern University 1977.
21. H. Yamauchi, M.S. Thesis, Northwestern University, 1972.
22. K. B. Rundman and J. E. Hilliard, Acta Met., 15, 1025 (1967).
23. K. M. Myles, Acta Met., 13, 109 (1965).
24. W. M. C. Yang, T. Tsakalakos and J. E. Hilliard, J.A.P., 48, 876 (1977).
25. J. A. Rayne, Phys. Rev., 118, 1545 (1960).
26. G. E. Hénein and J. E. Hilliard, J. Appl. Phys., 54, 728 (1983).
27. J. W. Cahn and J. E. Hilliard, J. Chem. Phys., 28, 258 (1958).
28. J. N. Pratt, Trans. Faraday Soc., 56, 975 (1960).
29. J. P. Chan and R. Hultgren, J. Chem. Thermod., 1, 45 (1969).
30. D. Erb, Ph.D. Thesis, Northwestern University, 1969.
31. J. van der Merwe, Proc. Phys. Soc. London, A63, 616 (1950).
32. B. Ditchek, Scripta Met., 11, 207 (1977).
33. D. de Fontaine, Acta Met., 17, 477 (1969).
34. W. P. Pearson, A Handbook of Lattice Spacings and Structures of Metals and Alloys, Pergamon Press Ltd., London, 1958.
35. N. M. Dunaev and M. I. Zakharova, JETP Lett., 20, 336 (1974).
36. W. M. C. Yang, Ph.D. Thesis, Northwestern University, 1970.
37. C. Zener, J.A.P., 22, 372 (1951).
38. A. D. Le Claire, Progress in Metals Physics, 4, 265 (1953).
39. J. M. Guglielmacci and M. Gillet, Surf. Sci. 105, 386 (1981).

COMPOSITIONALLY MODULATED METALLIC GLASSES

A.L. Greer

Division of Applied Sciences, Harvard University, Cambridge,
MA 02138, U.S.A.

1   INTRODUCTION

Metallic glasses (amorphous alloys) are of considerable scientific and technological interest (reviewed in, e.g., (1)). They were first produced by rapid quenching of the melt (2), but other techniques can be used, including deposition by evaporation or sputtering. Alloys which have been produced in glassy form fall into two main categories: noble or transition metals with 15-25 atomic % of metalloids; and combinations of early and late transition metals, which typically form glasses over wider composition ranges.

Diffusion measurements in metallic glasses have to be made below the glass transition temperature, T_g, in order to avoid rapid crystallization, and the diffusivities are consequently very low ($< 10^{-20}m^2s^{-1}$). Such diffusivities have been estimated by diffusing a species in from the specimen surface and measuring the composition profile using sputter sectioning or Rutherford back-scattering (3). While these techniques are useful, much lower diffusivities can be determined by X-ray measurements of the inter-diffusion in compositionally modulated thin films, a technique first used by DuMond and Youtz (4), and developed by Cook and Hilliard (5). As-produced metallic glasses can lower their free energy not only by crystallizing but also by relaxing their glassy structure. It is important to take this into account in interpreting diffusion measurements. The structural relaxation is accompanied by a slight densification ($\lesssim 0.4\%$) and by a marked reduction in atomic mobility. The modulated film technique can be used to monitor the diffusivity continuously. The composition-profiling techniques, on the other hand, can measure the diffusivity only after long anneals, during which substantial relaxation also occurs.

The ability to make time-resolved measurements of very low
diffusivities makes the use of compositionally modulated thin films
especially valuable in studying metallic glasses. The analysis of
interdiffusion measurements is outlined in Section 2, and in Section
3 the experimental results to date are reviewed. Possibilities for
further work on metallic glasses are described in Section 4.

2   INTERDIFFUSION IN MODULATED AMORPHOUS ALLOYS

In an amorphous material a composition modulation gives rise to
satellites about the (000) X-ray reflection. The intensity of the
first-order satellite is proportional to the square of the amplitude
of the first Fourier component of the modulation. In a linear
analysis the Fourier components are considered independent, and the
effective interdiffusivity, $\tilde{D}_\lambda$, is given by the decay rate of the
first-order satellite intensity, $I(t)$,

$$\tilde{D}_\lambda = - \frac{\lambda^2}{8\pi^2} \frac{d}{dt} \ln [I(t)/I_o] \tag{1}$$

where I_o is the initial intensity and λ is the modulation wavelength (4). $\tilde{D}_\lambda$ is dependent on the modulation wavelength. For
crystalline materials, an analysis based on discrete atomic planes
is necessary (6). For amorphous materials, however, a continuum
analysis, first developed by Cahn and Hilliard (7), is more appropriate. In their analysis, for a binary solid solution, $\tilde{D}_\lambda$ is
related to the macroscopic interdiffusion coefficient, $\tilde{D}$, by

$$\tilde{D}_\lambda = \tilde{D} \left[1 + \frac{8\pi^2 \kappa}{f''_o \lambda^2} \right] , \tag{2}$$

where f''_o is the second derivative, with respect to composition, of
the Helmholtz free energy per unit volume of the homogeneous phase,
and κ is the gradient energy coefficient. Assuming only nearest
neighbor interactions in a regular solution, κ is given by

$$\kappa = \frac{2\Delta H r_o^2}{3V} , \tag{3}$$

where ΔH is the heat of mixing per mole at the equiatomic composition, r_o is the interatomic distance, and V is the molar
volume. With the same assumptions f''_o is found to be:

$$f''_o = \frac{4}{V} [RT - 2\Delta H] \tag{4}$$

where R is the gas constant. Given adequate thermodynamic data, f_o'' would be known, and it would be possible to test eq. (3) directly. For novel, glass-forming compositions such data are often not available, and would be difficult to obtain at the temperatures of interest because of the metastability of metallic glasses.

3 SURVEY OF RESULTS

In this section the work on compositionally modulated metallic glasses will be reviewed. This work has been carried out at Harvard University. Compositionally modulated films were produced by DC sputtering from two targets of different compositions. Experimental details are given elsewhere (8,9).

3.1 Preliminary Measurements

The first compositionally modulated metallic glasses were made by Rosenblum et $al.$ (8,9). The films, of composition $(Pd_{80}Au_7Si_{13})_{70}/Fe_{30}$, showed only a first-order X-ray satellite, indicating a roughly sinusoidal composition modulation. On annealing, the satellite intensities decayed, and using eq. (1), $\tilde{D}_\lambda$ was derived. It was found that $\tilde{D}_\lambda$ decreased with annealing time, rapidly at first, and then tended to stabilize. This behavior was attributed to structural relaxation. Because of problems with crystallization, new films, of composition $(Pd_{85}Si_{15})_{61}/(Fe_{85}B_{15})_{39}$, were produced, in which each type of layer was of glass-forming composition. Interdiffusion measurements on these films confirmed the earlier results.

Greer et $al.$ (10,11) made measurements on films of composition $(Pd_{85}Si_{15})_{50}/(Fe_{85}B_{15})_{50}$, and the remainder of the results quoted in this paper will be from their work. Films with a total thickness of ~ 5 μm and modulation wavelengths of 2.2 Å to 74.6 Å were produced. Only a weak first-order satellite was observed for the smallest wavelength film, while for the $\lambda = 74.6$ Å film peaks out to fourth order were detected. The metals dominate the X-ray scattering, and the measured $\tilde{D}_\lambda$ is the Pd-Fe interdiffusivity. Although the composition of the films is complex, the results are analyzed according to the treatment for a binary system, as outlined in Section 2.

This simplification is open to question since, if the diffusivities of the metalloids were quite different from those of the metals, there could be measurable effects on the Pd-Fe interdiffusivity. For example, in a ternary system, $\tilde{D}_\lambda$ was found to decrease with time because of such effects (12). In metallic glasses, however, the metalloid and metal diffusivities are similar (13), and the treatment of the $(Pd_{85}Si_{15})_{50}/(Fe_{85}B_{15})_{50}$ films, in which the metalloids are minor components, as a binary system is a good approximation.

494

3.2 Gradient Energy Effects

Two films, with $\lambda = 20.6$ Å and 39.4 Å, were annealed at 250°C. The behavior of the first order satellite intensities is shown in Fig. 1. For each film there is a rapid initial drop in intensity, the origins of which are considered in the next section. After ~ 15 hours, however, $\ln(I(t)/I_O)$ falls linearly with time, indicating a constant interdiffusivity and a negligible rate of structural relaxation in the films. Since the annealing temperature was the same in each case, the degree of relaxation at a given time is the same in each film. From the final gradients of the plots, $\tilde{D}_\lambda$ was calculated using eq. (1) and was found to be: 7.09×10^{-26} m^{2}s^{-1} for the $\lambda = 20.6$ Å film; and 1.37×10^{-24} m^{2}s^{-1} for the $\lambda = 39.4$ Å film. The difference in these values is a manifestation of gradient energy effects. Analyzing these effects according to eq. (2), it is found that $\tilde{D} = 1.86 \times 10^{-24}$ m^{2}s^{-1} and that $K/f_O'' = -5.18 \times 10^{-20}$ m^2. Since $\tilde{D}$ is positive, f_O'' is positive, and K must be negative. Negative values of the gradient energy coefficient are expected for ordering systems. Although no directly relevant thermodynamic data are available, ordering might be expected from the behavior of Pd-Fe solid solutions.

Using eqs. (3) and (4), the heat of mixing, ΔH, at 250°C is found to be 2.45 kJmol^{-1}. This positive value is inconsistent with the other observations on this system. It is clear that the regular solution model with nearest neighbor interactions only is not capable of explaining the diffusion behavior. In a more general form the model can include longer range interactions, characterized by a r.m.s. interaction distance, δ. Defining δ as in (7), it must be at least 4.6 Å for the value of ΔH inferred from the model to be negative, as it should. This large value of δ may be a consequence of the metal-metalloid bonding in the system.

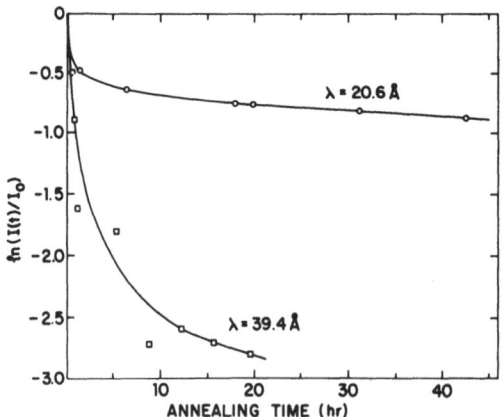

Fig. 1. Decay of first-order satellite intensities from films with $\lambda = 20.6$ Å and $\lambda = 39.4$ Å on annealing at 250°C.

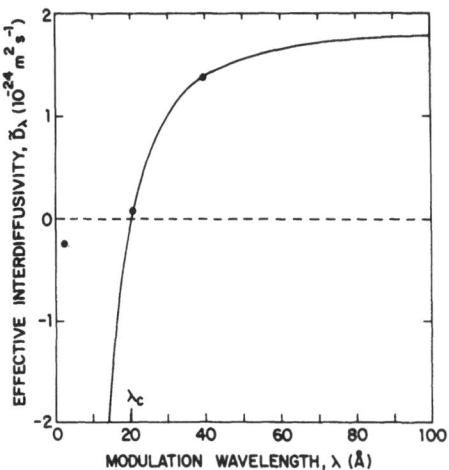

Fig. 2. $\tilde{D}_\lambda$ vs λ calculated according to eq. (2) and fitted to the measurements at $\lambda = 20.6$ Å and $\lambda = 39.4$ Å.

Figure 2 is a plot of $\tilde{D}_\lambda$ vs λ, calculated from eq. (2) and fitted to the measured interdiffusivities. The negative values of $\tilde{D}_\lambda$ indicate phase-separation at low wavelength, i.e. ordering. Experimental evidence for this will be considered later. The critical wavelength, λ_c, for which $\tilde{D}_\lambda$ is zero, is 20.22 Å, a value which is unusually large compared to typical crystalline systems (5). In earlier work (10), it was assumed that for $\lambda > 20$ Å, gradient energy effects could be ignored. Clearly this is not the case in this system.

3.3 Structural Relaxation

The initial non-linear decay in satellite intensity, shown in Fig. 1, could be due to a number of causes. $\tilde{D}$ is composition-dependent, and with a large initial modulation amplitude this could give rise to non-linear effects. The calculations of DeFontaine (14), however, suggest that such effects should be small. There is small angle X-ray scattering from the films which decays in about 2 hours at 250°C. This may arise from changes in the films such as elimination of voids or reduction of argon content, which could have an effect on the very early stages of interdiffusion. The continuing decrease in $\tilde{D}_\lambda$ after ~ 2 hours is unlikely to be due to either of the above causes and is attributed to structural relaxation. A film with $\lambda = 32.2$ Å was annealed at 300°C in order to study the relaxation kinetics. The behavior of the first- and second-order satellite intensities is shown in Fig. 3. The complex behavior of the second-order satellite is believed to arise from the composition dependence of $\tilde{D}$ (14,15), and the form of the curve is in qualitative agreement with the results of Tsakalakos (15) on crystalline Cu-Ni films. The first-order satellite was used to determine $\tilde{D}_\lambda$, and $\tilde{D}$ was

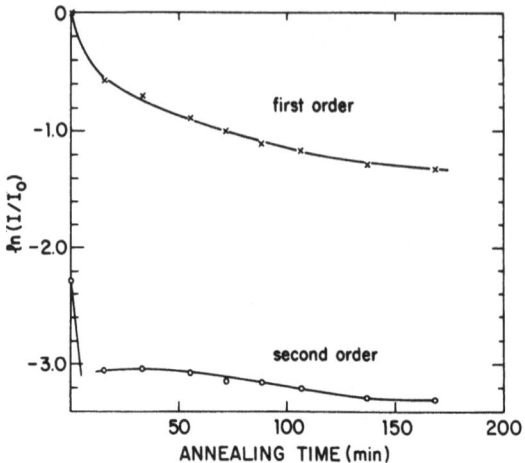

Fig. 3. Decay of satellite intensities on annealing of a λ = 32.2 Å film at 300°C.

calculated using the value of κ/f_0'' derived above. It was found that $1/\tilde{D}$ rose linearly with time (Fig. 4). The low value of $\tilde{D}$ after 120 minutes is believed to be due to the onset of crystallization.

It is evident that structural relaxation has a large effect on $\tilde{D}$. In order to obtain a physically meaningful activation energy for

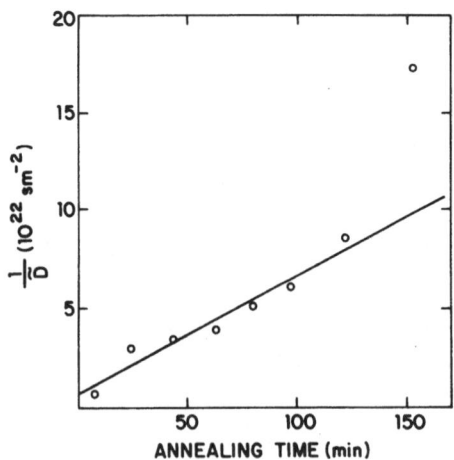

Fig. 4. Relaxation kinetics of $1/\tilde{D}$ in λ = 32.2 Å film annealed at 300°C.

$\tilde{D}$ it is essential that the experiment be carried out under isoconfigurational conditions. This can be done conveniently and accurately using a single modulated film. As can be seen from Fig. 1, after the $\lambda = 20.6$ Å sample had been held at 250°C for ~ 40 hours the changes in $\tilde{D}_\lambda$ were undetectable (although the sample was not in configurational equilibrium). The temperature was then lowered to 230°C, $\tilde{D}_\lambda$ was measured at that temperature, and on returning to 250°C it was verified that $\tilde{D}_\lambda$ at 250°C was unchanged; i.e. that no further relaxation had taken place. This procedure was repeated for excursions to 220°C, 210°C and 270°C. As shown in Fig. 5, the temperature dependence of $\tilde{D}$ is Arrhenius-type, with an isoconfigurational activation energy of 195 ± 15 kJ mol^{-1}. The value of the preexponential factor, 5.6×10^{-5} m^{2}s^{-1}, is reasonable.

In understanding atomic transport in metallic glasses it is important to study the relationship between the diffusivity, D, and the viscosity, η. For liquid metals, experimental results are in fair agreement ($\pm 50\%$) with the Stokes-Einstein relation:

$$D = kT/6\pi\eta r \quad , \tag{5}$$

where k is Boltzmann's constant, T is the temperature and r is an average ionic radius. In metal-metalloid glasses, as in the present case, it is expected that the metal atom diffusivity would be related to η. The viscosity data of Taub and Spaepen (16) on Pd$_{82}$Si$_{18}$ glass were chosen for comparison. The viscosity was measured over the range 151°C to 264°C and exhibited an activation energy of 192×17 kJ mol^{-1}, very close to the value for interdiffusion (Fig. 5). At all temperatures structural relaxation caused η to rise linearly with annealing time. Fig. 4 shows similar behavior for $1/\tilde{D}$. By comparing the rates of increase, $\tilde{D}$ and η can be related, without concern about the degree of relaxation.

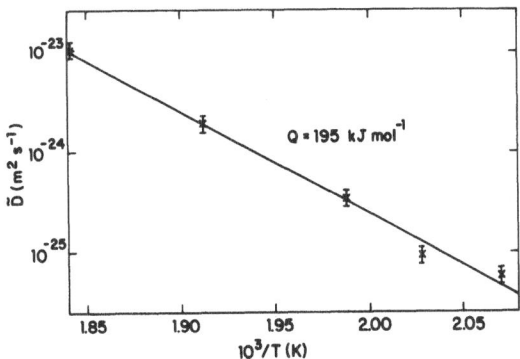

Fig. 5. The temperature dependence of $\tilde{D}$ in a $\lambda = 20.6$ Å film, indicating the isoconfigurational activation energy.

At 300°C, $d\eta/dt = 1.04 \times 10^{10}$ Nm^{-2} (extrapolated) and
$d(1/\tilde{D})/dt = 1.01 \times 10^{19}$ m^{-2}. From these values it is calculated that
$\tilde{D}$ is about 160 times the value that would be estimated from η
using eq. (5). Thus, although $\tilde{D}$ and η are closely related,
showing the same activation energy and relaxation behavior, there
are diffusional jumps that do not lead to viscous flow. The
Stokes-Einstein relation is not valid in these glasses.

3.4 Interdiffusion at Very Short Wavelength

An interesting feature of amorphous materials is that it is
possible for the wavelength of a composition modulation to be less
than the interatomic spacing. In a crystal this would be possible
only if the layers were high-index atomic planes. The Pd-Fe interatomic distance is estimated to be ~ 2.65 Å, and a Pd-Si/Fe-B
modulated film was produced with $\lambda = 2.275$ Å. The behavior of the
first-order satellite intensity on annealing at 250°C is shown in
Fig. 6. The initial rise in intensity is expected in ordering
systems at short wavelength. An approximate value of $\tilde{D}$ is plotted
in Fig. 2; the value does not lie close to the curve, but eq. (2) is
expected to break down at very short wavelength. The subsequent
drop in intensity is attributed to ordering within the layers. This
removes the one-dimensional order and homogenizes the film.

3.5 Discussion

The values of interdiffusivity determined above are in good
agreement with metal atom diffusivities in metallic glasses determined by composition profiling. In previous work (10), the values
from modulated films appeared somewhat low because gradient energy

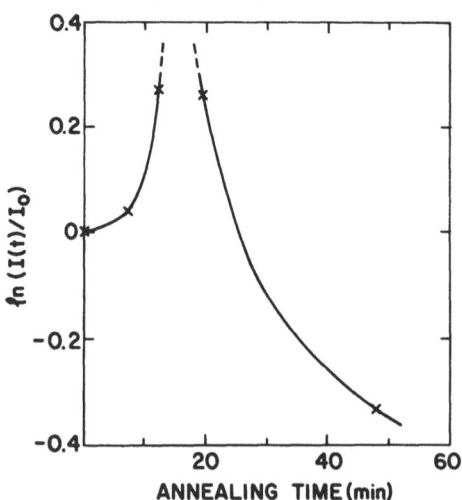

Fig. 6. The behavior of the first-order satellite intensity from a
$\lambda = 2.275$ Å film annealed at 250°C.

effects were not taken into account. The values of $\tilde{D}$ determined in metallic glass films are greater than the values expected for single crystal films at the same temperature, but smaller than the values expected for fine-grained films. In the work described above, the lowest measured $\tilde{D}_\lambda$ was 1.31×10^{-27} m^2s^{-1}. If this $\tilde{D}_\lambda$ were to be measured by composition profiling, an anneal of over 200 years would be required for the necessary penetration distance of 30 Å.

4   PROSPECTS FOR FURTHER WORK

Not only are modulated films useful in studying metallic glasses, but the use of glasses in the films is useful in studying the effects of composition modulation. Some current research directions are indicated below.

4.1   Studies of Metallic Glasses

Interdiffusion studies on a wider range of metallic glass systems would provide valuable information on the stability and thermodynamics of non-crystalline phases. The use of films containing only two elements (possible in binary systems with a wide glass-forming range) would simplify the interpretation of results. Some metallic glasses phase-separate, and it would be interesting to study these systems in the same manner as that used by Tsakalakos for crystalline Cu-Ni films (17).

4.2   Studies of Composition Modulation Effects

The major advantages of using amorphous alloys, instead of crystalline ones, to study these effects are:

(i)   The modulation wavelength can be adjusted continuously, whereas in crystalline materials effects due to interference with the crystal periodicity can occur.

(ii)   Coherency strains due to lattice matching are avoided. Internal strains are likely to be lower then in crystalline films, and this simplifies the interpretation of certain effects, such as the dependence of $\tilde{D}_\lambda$ and elastic modulus on modulation wavelength.

(iii)   In amorphous alloys the composition can be adjusted continuously, often over a wide range. This is not possible in crystalline materials because of the formation of stoichiometric compounds.

ACKNOWLEDGEMENTS

This work was supported by the Office of Naval Research under contract N00014-77-C-0002 and by the National Science Foundation under contract DMR80-20247.

REFERENCES

1. Cahn, R.W. Contemporary Physics 21 (1980) 43.
2. Duwez, P., R.H. Willens, and W. Klement. Nature 187 (1960) 869.
3. Kijek, M., M. Ahmadzadeh, B. Cantor, and R.W. Cahn. Scripta Metallurgica 14 (1980) 1337.
4. DuMond, J. and J.P. Youtz. Journal of Applied Physics 11 (1940) 357.
5. Cook, H.E. and J.E. Hilliard. Journal of Applied Physics 40 (1969) 2191.
6. Cook, H.E., D. DeFontaine, and J.E. Hilliard. Acta Metallurgica 17 (1969) 765.
7. Cahn, J.W. and J.E. Hilliard. Journal of Chemical Physics 28 (1958) 258.
8. Rosenblum, M.P., F. Spaepen and D. Turnbull. Applied Physics Letters 37 (1980) 184.
9. Rosenblum, M.P. Ph.D. Thesis, Harvard University (1979).
10. Greer, A.L., C.-J. Lin, and F. Spaepen. Proc. 4th Int. Conf. on Rapidly Quenched Metals (eds. T. Masumoto and K. Susuki). Japan Institute of Metals (1982).
11. Cammarata, R.C. and A.L. Greer. Proc. 5th Int. Conf. on Liquid and Amorphous Metals, Los Angeles, August 1983, to be published in Journal of Non-Crystalline Solids.
12. Murakami, M., D. DeFontaine, J.M. Sanchez, and J. Fodor. Thin Solid Films 25 (1975) 465.
13. Greer, A.L. Proc. 5th Int. Conf. on Liquid and Amorphous Metals, Los Angeles, August 1983, to be published in Journal of Non-Crystalline Solids.
14. DeFontaine, D. Ph.D. Thesis, Northwestern University (1967).
15. Tsakalakos, T. Ph.D. Thesis, Northwestern University (1977).
16. Taub, A.I. and F. Spaepen. Acta Metallurgica 28 (1980) 1781.
17. Tsakalakos, T. Scripta Metallurgica 15 (1981) 235.

ELECTRICAL PROPERTIES OF MULTILAYERED Cr/SiO_2 THIN FILMS

D. Niarchos, B.J. Papatheofanis, G. Monfroy,
Physics Department, Illinois Institute of Technology, Chicago, IL
60616

M. Tanielian, J. Willhite,
Gould Research Center, Rolling Meadows, IL 60008

INTRODUCTION

Novel systems like thin metallic films (1,2), thin wires (3), granular metallic films (4) and mixed metal/insulator systems (5) exhibit behavior in their electrical properties, which can be explained by localization (6), electron interaction (7) and or percolation theories (8). In our present study we are presenting electrical resistivity measurements on a new series of multilayered metal/insulator thin films. According to the ratio of metal/insulator the behavior is metallic or insulator-like, the critical concentraction being approximately 50%. Another interesting feature is that in the metallic like samples a metal-insulator transition occurs around 80 ± 10 K which is concentration dependent. Annealing affects the conduction from a 2-D like to a 3-D like.

PREPARATION AND CHARACTERIZATION

The films were RF sputtered in an Ar atmosphere from a 99.95% Cr target and 99.9% SiO_2 target in an alternating fashion and deposited on glazed alumina substrates kept at room temperature. The films were made up of six layers of Cr and six layers of SiO_x ($1.8 \leq x \leq 2.0$) as seen in fig. 1. The thickness of one Cr and one SiO_2 layer was kept fixed to 100Å. The metal insulator ratio was changed by changing the Cr or SiO_2 thickness from 90 Å to 20 Å. The overall film thickness was 600 Å.

The samples were characterized using x-ray, transmission electron and Auger spectroscopy. X-ray diffraction failed to give any diffraction peaks, indicative of amorphous or fine grain state.

502

Cr/SiO$_2$ FILMS

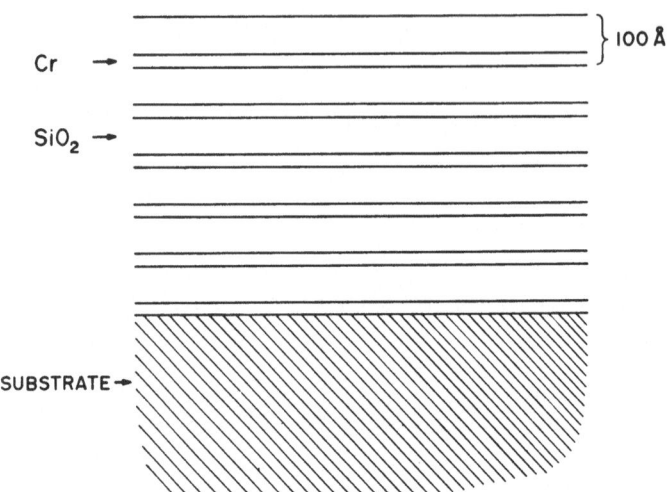

Fig. 1: Idealized film structure. Thickness per pair of Cr/SiO$_2$ layer 100 Å, total film thickness 600 Å.

This is supported by transmission electron studies, where the average grain size is 50-70 Å for a 60 Å Cr/40 Å SiO$_2$ film. Auger data analysis gave not only the composition of the samples, but also revealed an undulation in the Cr, Si and O signals as shown in fig. 2.

RESULTS AND DISCUSSION

A conventional low frequency a-c four probe technique was used for the resistivity measurements. Occasionally a dc four probe technique was used and compared to the data from the a-c technique. No significant difference was found. The contacts were evaporated Al pads 1μm thick, and Au or Cu wires were bounded to the pads. A conventional L.He cryostat was used for the resistivity measurements and a 2T superconducting magnet was used for the magnetoresistivity measurements.

Fig. 3 shows the sheet resistance ($\Omega/\square$) at R.T. vs the volume percentage of Cr concentration. A rapid change from metallic to an insulating behavior occurs at approximately 50% concentration. This concentration is assuming percolation, the same found for homogeneous metal/insulator mixtures or granular metallic systems (9,10).

The fractional resistivity for some samples is shown in fig. 4. For films with Cr concentration higher than 50% a minimum occurs at

503

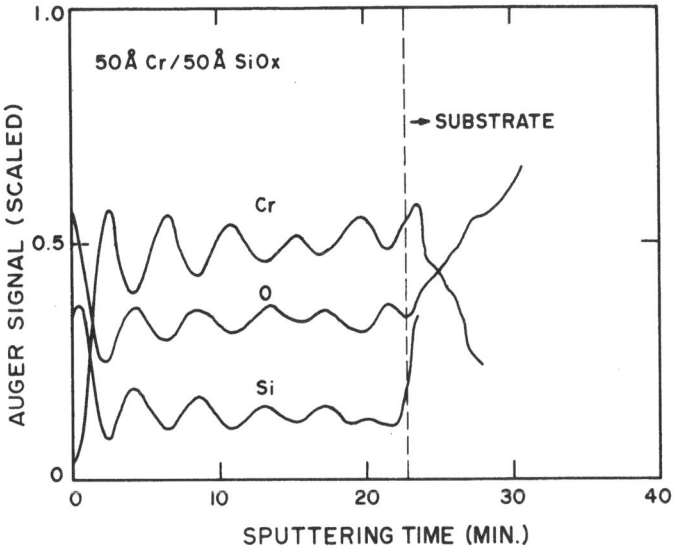

Fig. 2: Auger spectroscopy data for a 50 Å G/50 Å SiO₂ film. An undulation is apparent indicative of the layered structure.

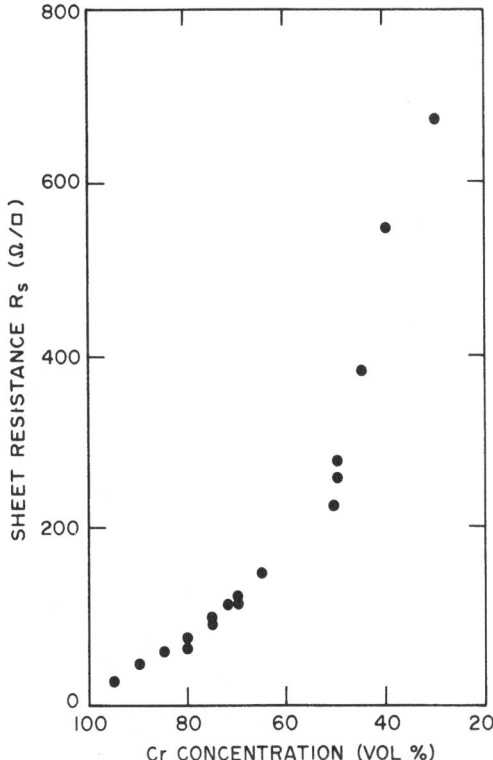

Fig. 3: The sheet resistance ($\Omega/\square$) as a function of Cr (vol %) concentration.

some temperature Tm which is concentration dependent. The minima are shifted to higher temperatures as the Cr concentration is reduced. Above Tm the behavior is metallic-like with the fractional resistivity increasing as the Cr concentration increases. Below Tm the fractional resistivity does not follow any power law alone, but the behavior can approximately be described by a log T and $T^{-\frac{1}{2}}$ behavior. This is consistent with the nature of the system which is made up of conducting thin layers of Cr and insulating SiO_2 layers, with some mixture at the interface.

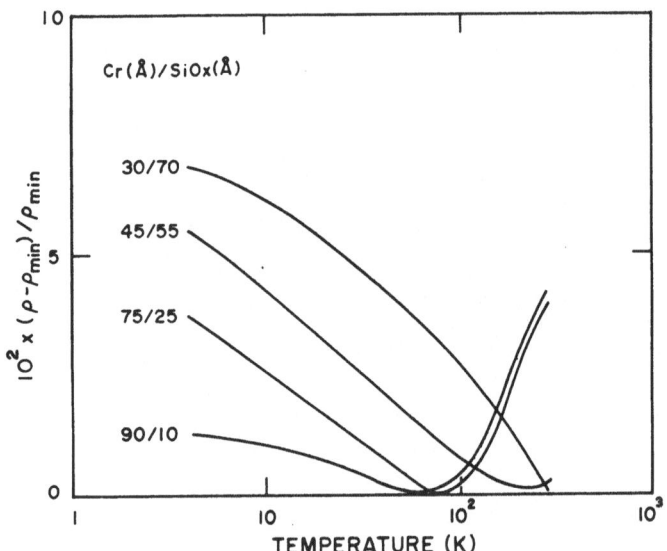

Fig. 4: The fractional resistivity for representative samples vs log T. ρ min is the lowest value for resistivity.

When annealing, the samples become less disordered resulting in behavior which is more bulk metallic as shown in fig. 5. The fractional resistivity below Tm decreases and Tm is shifted to lower values as the annealing temperature increases. This argument, of destroying the layering with annealing, is supported by Auger data on an annealed sample, where the undulation was less pronounced compared to the unannealed one.

A maximum in the range of 230-380 K in the resistance, is a new feature of these films, which is both concentration and history dependent. Annealing shifts the peak to higher temperatures. The origin of such a peak is not clear yet, but a simple model considering conduction along the perpendicular to the Cr layers accounts for such a peak. Details of such a model will be published elsewhere.

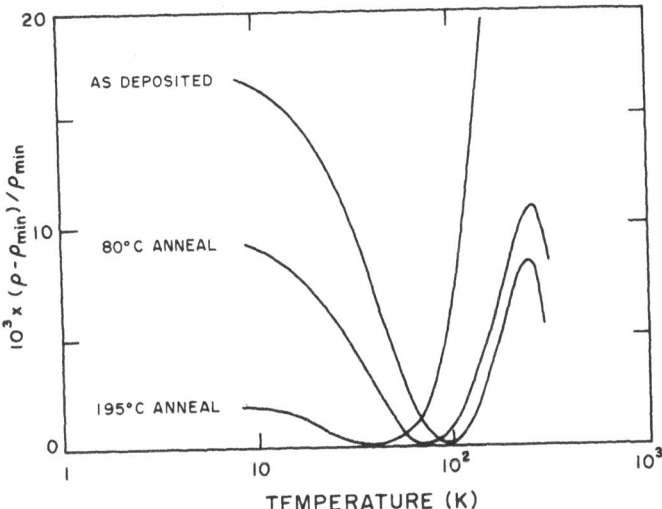

Fig. 5: The fractional change in resistivity for a 60 Å Cr/40 Å SiO$_2$ film at different annealing temperatures.

A zero magnetoresistance coefficient was found for some films in the metallic region and at 4.2K. Due to the fact that Cr is magnetic, probably higher magnetic fields and Hall measurements are needed to be able to distinguish between localization or electron interaction effects at low temperatures (11) in these novel systems.

CONCLUSIONS

Electron transport measurements in the layered Cr/SiO$_x$ films indicated that conduction is an intra and inter-layer mechanism. Analysis of resistivity data in unannealed and annealed samples show a behavior between a 2-D and 3-D system. The fact that annealing changes the proportion of intra and inter-layer conduction to equal weights, it is indicative that annealing for a prolonged time and high temperatures, the system behaves as a 3-D one.

REFERENCES

1. G.J. Dolan and D.D. Osheroff, "Nonmetallic Conduction in Thin Metal Films at Low Temperature", Phys. Rev. Lett. 43 (1979) 721.
2. L. Van den dries, C. Van Haesendonck, Y. Bruynseraede and G. Deutscher, Two-Dimensional Localization in Thin Copper Films, Phys. Rev. Lett. 46 (1981) 565.

506

3. N. Giordano, Experimental study of Localization in thin wires, Phys. Rev. B, 22 (1980) 5635.

4. G. Deutscher, B. Bandyopadhyay, T. Chui, P. Lindenfeld, W.L. McLean and T. Worthington, Transition to Localization in Granular Aluminum Films, Phys. Rev. Lett. 44 (1980) 1150.

5. B.W. Dodson, W.L. McMillan, J.W. Mochel and R.C. Dynes, Metal Insulator Transitions in disordered Germanium−Gold alloys, Phys. Rev. Lett. 46, (1981) 46.

6. E. Abrahams, P.W. Anderson, D.C. Licciardello and T.V. Ramakrishnan, Scaling Theory of Localization: Absence of Quantum Diffusion in Two Dimensions, Phys. Rev. Lett. 42 (1979) 673.

7. B.L. Altshuler, A.G. Aronov and P.A. Lee, Interaction Effects in Disordered Fermi Systems in Two Dimensions, Phys. Rev. Lett. 44, (1980) 1288.

8. G. Deutscher, M. Rappaport and Z. Ovadyahu, Random Percolation in Metal−Ge Mixtures, Sol. Stat. Comm., 28 (1978) 593.

9. N. Savvides, S.P. McAlister, C.M. Hurd and I. Shiozaki, Localization in the Metallic Regime of Granular $Cu-SiO_2$ Films, Sol. Stat. Comm, 42 (1982) 143.

10. B. Abeles, Ping Sheng, M.D. Coutts and Y. Arie, Structural and Electrical Properties of Granular metal films, Adv. Phys. 24 (1975) 407.

11. B.L Altshuler, D. Khmel'nitzkii, A.I. Larkin and P.A. Lee, Magnetoresistance and Hall effect in a disordered two−dimensiona electron gas, Phys. Rev. B, 22 (1980) 5142.

Also Phys. Today p. 19 (May 1981).

CHAPTER 7: SEMICONDUCTOR

SUPERLATTICES

TAILORED SEMICONDUCTORS: COMPOSITIONAL AND DOPING SUPERLATTICES

G.H. Döhler

Max-Planck-Institut für Festkörperforschung, Heisenbergstrasse 1, 7000 Stuttgart 80, F.R.G.

Progress in crystal growth techniques has made feasible the realization of periodic semiconductor structures composed of ultrathin layers. Essentially two types of such "man-made" superlattices may be distinguished: i) Compositional superlattices, consisting of a periodic sequence of two different semiconductor materials with lattice matching (GaAs-Al$_x$Ga$_{1-x}$As, e.g.). A superlattice potential is induced in the conduction and the valence band by the periodic variation of the energy gap in the direction of crystal growth. ii) Doping superlattices, a sequence of n- and p-doped layers, possibly with undoped layers in between ("n-i-p-i-superstructures"), grown in an otherwise homogeneous semiconductor bulk. The periodic potential in n-i-p-i crystals is due to the electrostatic potential of fixed ionized impurities, which may be partly compensated by mobile electrons and holes confined to the n- and p-layers, respectively.

The electronic properties of semiconductor superlattices differ dramatically from those of their components. The energy bands are split into subbands whose separation and band width may be "tailored" by appropriate choice of superlattice period, alloy composition, or doping concentration. Apart from peculiarities of the transport in the direction of periodicity man-made superlattices may also serve as an ideal model system for the study of quasi 2-dimensional systems.

In addition to these features, which are common to both types of superlattices, n-i-p-i crystals show a number of intriguing properties related to the electrostatic origin of the superlattice

potential. The lowest electron and the uppermost hole subband states are spatially separated from each other ("indirect band gap
in real space"). Depending on the "design" parameters of the crystal the electron-hole recombination may be suppressed nearly completely. As a consequence the effective band gap as well as the
electron and hole concentration can be modulated over a wide range
in n-i-p-i crystals either optically or by carrier injection into
the bulk. Some possible device applications related to the tunability the bulk conductivity, of the absorption coefficient, and of
the frequency and the intensity of the photo- and electro-lumines-
cence spectrum will be discussed also.

1. Introduction

The characteristic data on familiar semiconductors cover a
wide variety of values. The energy gap between the top of the valence and the bottom of the conduction band, for instance, may have
practically any value between zero and a few electron volts. The
gap may be a "direct" or an "indirect" one in momentum space. The
carrier concentration may be varied gradually from, say, $10^{20} cm^{-3}$
holes to $10^{20} cm^{-3}$ electrons by appropriate doping of the crystal.
Nevertheless, a solid state physicist playing around with phenomena
principally to be expected from theory but not observed in the real
world or thinking about some new devices may request a bunch of
properties for "new" semiconductors. He might ask for a semiconductor with many more bands, or with a different energy dispersion
$\varepsilon(k)$. A request closely related to the former one would concern a
reduced Brillouin zone and an electronically highly anisotropic,
possibly (dynamically) quasi 2-dimensional, semiconductor. In addition, one might ask for a semiconductor with strongly reduced or
even practically suppressed electron-hole recombination. As a consequence it would be possible to maintain large deviations from the
thermal equilibrium carrier concentration over a long period of time. The electron and the hole concentration in such a semiconductor
would no longer simply be a material parameter, fixed for a given
crystal by the amount of doping, but, would be a tunable quantity
instead. If, finally, also the energy gap itself could be modulated over a wide range for a given semiconductor an extremely flexible solid with intriguing novel properties would be available.

The objective of this paper is to show that all those exotic
properties are, indeed, exhibited by semiconductors with a periodic
superstructure and to explain how they may be "designed" for a given purpose. In section two the basic structure and the growth
technique of two different types of semiconductor superlattices
will be described:
(a) Compositional superlattices, a periodic sequence of different semiconductor layers with lattice matching.
(b) Doping superlattices, a periodic sequence of ultrathin
layers of the same semiconductor, however, with opposite sign of
doping.

These two types fo superlattice differ clearly from each other with respect to their structure. With respect to their electronic properties they have many properties in common, but, nevertheless, there are also drastic differences between them.

In view of the lack of a clear-cut discrimination concerning the electronic behavior it seems appropriate to start with a discussion of the electronic properties of semiconductor superlattices based on the particular case of the conceptionally simpler compositional superlattice (although most of the results apply to the doping superlattices as well) in section three. Section four will then be devoted to the novel properties which are exhibited exclusively by doping superlattices.

2. Basic structure and growth of "man-made superlattices"

2.1 Compositional semiconductor superlattices

This type of superlattice is a periodic sequence of ultrathin layers of different semiconductors, as shown schematically in Fig. 1(a). A necessary requirement for obtaining good crystal quality, of course, is a sufficiently small lattice mismatch between the components (I and II). The most prominent example of compositional superlattices is the combination GaAs-$Al_x Ga_{1-x}$As. From the range of particularly interesting values of the superlattice period d, as given in Fig. 1(a), it is obvious, that the preparation of such sophisticated structures represents serious problems. The method of crystal growth by molecular beam epitaxy (MBE), however, has proved as an extremely successful tool (1-3) for this goal. Actually to a large extent MBE has even been developed to today's high degree of perfection to accomplish the growth of semiconductor superlattices. Figure 2 shows schematically the simple principal arran-

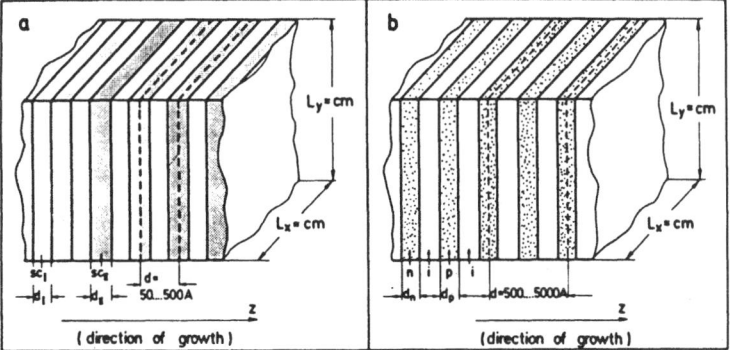

Fig.1. Semiconductor superlattices, schematic. (a) Compositional superlattice consisting of ultrathin layers of semiconductor species SC^I and SC^{II}. (b) Doping superlattice in an otherwise homogeneous semiconductor. n=doping by donors, p,=doping by acceptors, i=intrinsic.

gement of a MBE apparatus. The molecular or atom beams which propagate in an ultra-high vacuum chamber from the Ga-, Al- and As-effusion cells onto the substrate may be manipulated by shutters. In this way it is possible to grow periodically $Al_xGa_{1-x}As$ with the desired Al content or pure GaAs at a rate of typically 1 As^{-1}. Because of this low growth rate and because of an extremely smooth crystal surface a period of a few bulk lattice periods can be obtained. Figure 3 gives an impressing example for the high perfection which has been achieved (2). The dark field transmission electron micrograph shows a GaAs-AlGaAs superlattice with a period of only 28 A! Recently it has been demonstrated that high quality $GaAs-Al_xGa_{1-x}As$ superstructure crystals may also be grown by metal-organic chemical vapor deposition (MO-CVD) (4).

2.2. Doping superlattices

Another kind of periodic structure may be obtained by variation of doping. Whereas it is not surprising that a periodic array of different semiconductors may modify the properties of semiconductors drastically this appears less evident for the case of periodic doping in an otherwise homogeneous semiconductor. Later on we will show, however, that a crystal with a periodic sequence of n- and p-doped layers, possibly with intrinsic layers in between (a so-called "n-i-p-i-crystal" (6,7)). exhibits even more peculiar properties than compositional superlattices. These n-i-p-i crystals (see Fig. 1b) may be grown by MBE as well, if the effusion cell for the second component is replaced by two effusion cells for the two species of dopants.

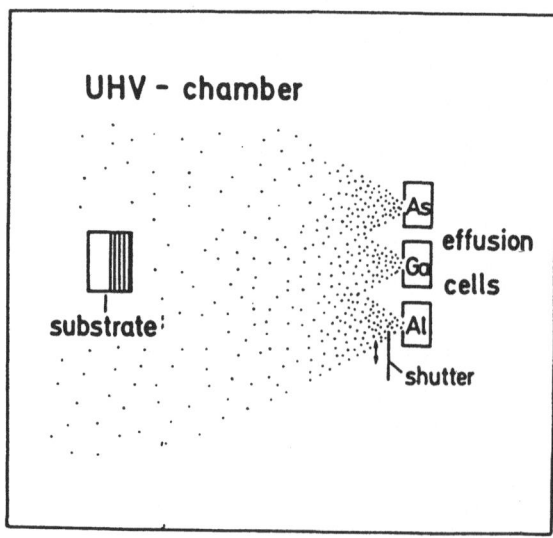

Fig. 2. Principle of a MBE apparatus. For realistic version see (3).

In Fig. 4 a scanning-electron micrograph of a GaAs p-n doping superlattice is shown (8). The dopants used in this structure, Si as donor (8,9) and Be as acceptor (9,10), are particularly favorable since hyperabrupt doping profiles may be obtained under growth conditions which yield high quality GaAs crystals.

There are a few advantages of doping superlattices compared with compositional ones from the crystallographic point of view. The incorporation of dopants, which always concerns only a very small fraction of the atomic sites represents only a minor perturbation of the otherwise homogeneous crystal lattice. Therefore, no problem with interfaces arise and no restrictions in the choice of semiconducting bulk material evolve from problems of lattice matching. The range of superlattice periods of major interest is shifted to higher values, which may permit the use of alternative methods of crystal growth, such as liquid phase epitaxy (Bauser (11), Scheel (12)), or vapor phase exitaxy (13). Another, potential, advantage is the possibility of "writing" a 3-dimensional doping pattern into the crystal during its growth by using a focused ionized dopant beam, for example.

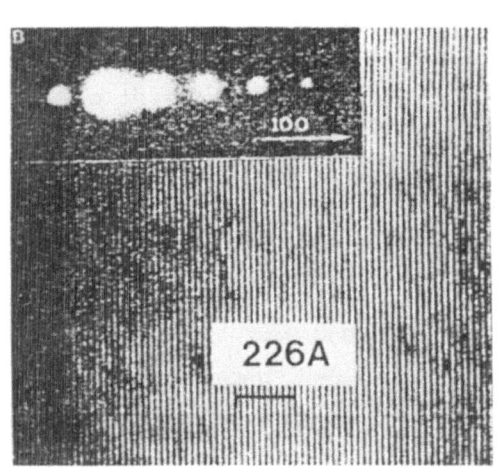

Fig. 3. Transmission electron micrograph of a compositional semiconductor superlattice with a periodic sequence of 4 layers of GaAs and 4 layers of AlAs (from (2)). The inset shows the superlattice satellite diffraction spots.

The superlattices as described above do not occur in nature, in contrast to other superlattices formed by the polytypes of the well known layered compounds such as TaS , TaSe , and NbS or also in ZnS or SiC, where a periodic variation between wurtzite and zincblende structure is found in naturally grown crystals. In order to distinguish the present artificial superstructure crystal from the natural ones they have been called "man-made superlattices" (1). A major advantage of man-made crystals is the chance that one has no longer to rely only on what is provided by nature. Instead of that, the properties of these artificial semiconductors may be "designed" or "tailored" within a very wide range according to the actual requirements, as we shall see in the following sections.

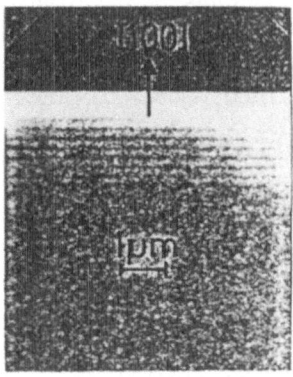

Fig. 4. Scanning electron micrograph of a periodic n-p doping superlattice in GaAs (Si-donor, Be-acceptor, 10 periods of 2000 A), from (8).

3. Electronic structure and properties of semiconductor superlattices

Our discussion of electronic properties in this section will be based on the particular case of a compositional superlattice of the type GaAs-Al$_x$Ga$_{1-x}$As with abrupt changes in the alloy composition for the sake of clarity. We should bear in mind, however, that most of the results apply without qualitative changes to doping superlattices as well. The difference between a superlattice and a familiar semiconductor results from the periodic potential which originates from the superstructure.

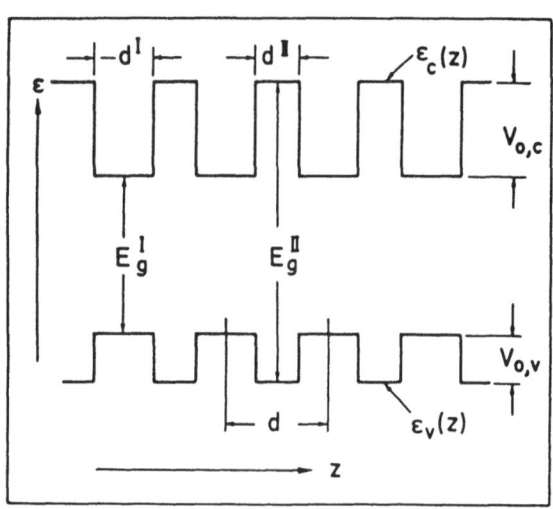

Fig. 5. Band profile of a semiconductor with compositional superlattice showing the periodic potential arising from the band gap difference $E_g^{II} - E_g^I$ between the components CS^I and CS^{II}.

This periodic potential may be understood easily from the following naive picture. Let us assume that the superlattice is characterized by the band gaps E_g^I and E_g^{II} and the layer thickness d^I and d^{II} of the two components of the superlattice (say I = GaAs, II = Al$_x$Ga$_{1-x}$As, for instance).

The gap energy difference

$$V_o = E_g^{II} - E_g^{I} \tag{1}$$

is split into one part $V_{o,c}$ which appears as a discontinuity in the conduction band edge $\varepsilon_c(z)$ and into another part $V_{o,v}$ which modulates the valence band edge $\varepsilon_v(z)$, as shown in Fig. 5. Whereas the motion of charge carriers parallel to the layers will not be strongly affected by the periodic potential, the motion in z-direction (the direction of periodic growth) corresponds to the motion in a system of period d. Consequently, the dispersion curve $\varepsilon(k_z)$ will be folded back into the new reduced "mini Brillouin zone" defined by $|k_z| < \pi/d$. Furthermore, each band $\varepsilon(k_z)$ will be split into mini-bands $\varepsilon_z(k_z)$ separated by mini-gaps at $k_z = 0$ and at the zone boundaries $k_z = \pm \pi/d$, as depicted in Fig. 6.

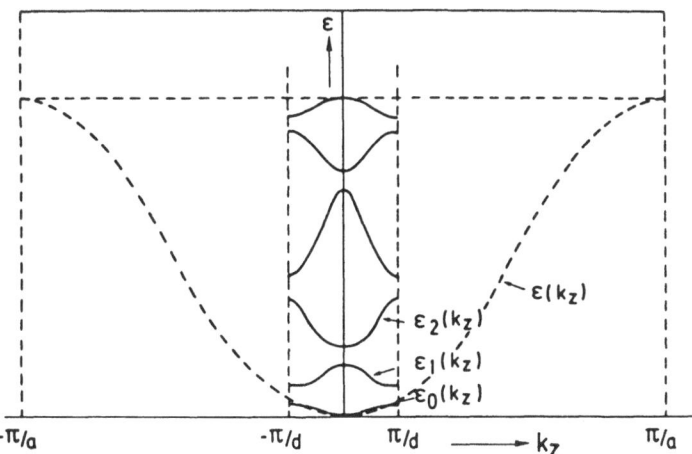

Fig. 6. Splitting of a band $\varepsilon(k_z)$ into subbands, $\varepsilon_v(k_z)$ by a periodic superlattice potential of period d. The number of subbands is d/a.

The gap difference V_o is about 300meV between $Al_{0.3}Ga_{0.7}As$ and GaAs, for example. Thus, the periodic modulation of the band edges $\varepsilon_c(z)$ and $\varepsilon_v(z)$ represents such a strong potential for the motion of electrons and holes, respectively, that a tight binding calculation provides the most appropriate description of the superlattice in most of the cases of interest.

At the same time the tight-binding treatment also exemplifies how one may "tailor" the electronic structure of a superlattice crystal. Let us consider first the square-well potential formed by a layer of semiconductor I buried in the bulk of semiconductor II. The motion of electrons in z-direction will be quantized, yielding a number of discrete energies $E_{c,v}$ (see Fig. 7). The position of these energies depends on the thickness d^I of the layer and on the depth of the well $V_{o,c}$ which increases with increasing Al content x for our specific example.

A crude estimate for $E_{c,}$ gives

$$E_{c,\nu} \approx (\hbar^2\pi^2/2m_c^I(d^I)^2)(\nu + 1)^2; \quad \nu = 0,1,2,\ldots \tag{2}$$

for $E_{c,\nu} V_{o,c}$ which yields $E_{c,0} \simeq 50\text{meV}$ and $E_{c,1} \simeq 200\text{meV}$ for a 100 A thick GaAs layer in $Al_{0.3}Ga_{0.7}As$, for example.

For the free motion parallel to the layer the effective mass approximation (EMA) with the effective mass of the semiconductor I m_c^I represents a reasonable approximation for the low-laying 2-dimensional conduction subbands

$$\varepsilon_{c,v}(k_{\shortparallel}) \simeq E_{c,v} + \hbar^2 k_{\shortparallel}^2/2m_c^I \tag{3}$$

The subband structure remains essentially unchanged (at least for the lowest subbands) in a superlattice with large superlattice period d, i.e., if the type I layer is repeated periodically with a sufficiently thick type II layer in between (Fig. 8). Bands with an appreciable width with respect to the motion in z-direction are only formed if the type-II semiconductor barriers become sufficiently weak to allow for a considerable overlap between the wave function of adjacent wells

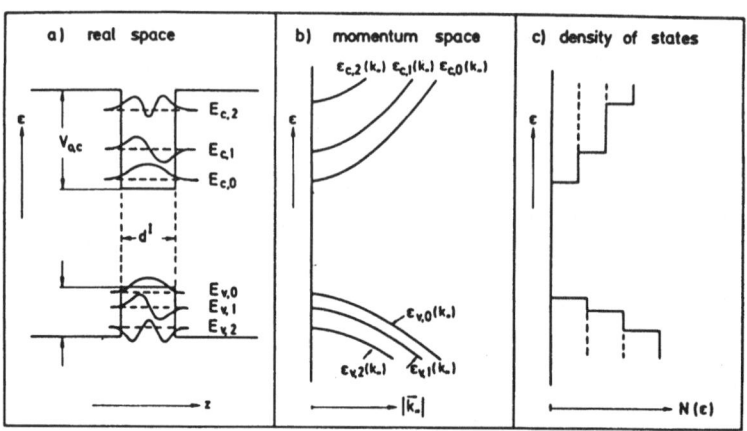

Fig. 7. Electronic structure of a single quantum well structure formed by a thin layer of semiconductor SC^I buried in a SC^{II} crystal, schemtalically (only one valence band is assumed for simplicity).

The superlattice electron subband dispersion in nearest neighbor approximation is then given by

$$\varepsilon_{c,\nu}(\vec{k}) = \varepsilon_{c,\nu}(\vec{k}_{\shortparallel}(+ V_{c,\nu} \cos k_z d; \quad |k_z| < \pi/d \tag{4}$$

with $\varepsilon_{c,v}(k_{\shortparallel})$ from eq. (3). $V_{c,v}$ is the tight binding nearest neighbor matrix element which depends strongly on d^{II}. By appropriate choice of d^{II} the subband width may be designed after one has chosen the subband spacing by choosing a certain value of d^{I}.

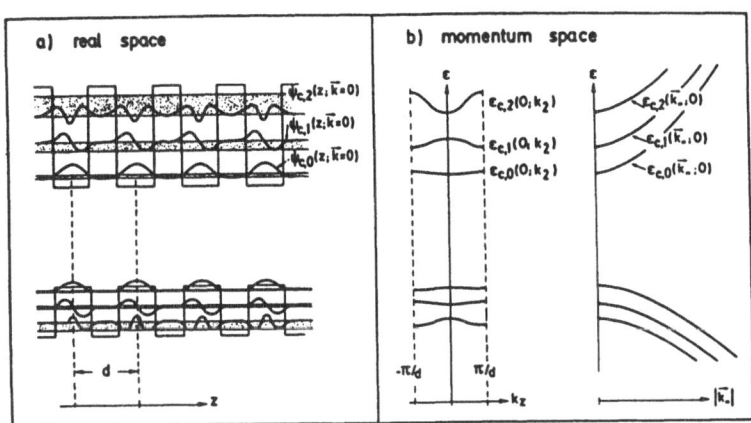

Fig. 8. Electronic structure of the compositional superlattice shown in Fig. 5. The dispersion parallel $(\varepsilon_{v}(k_{\shortparallel}; k_{z} = 0))$ and perpendicular $(\varepsilon_{v}(k_{\shortparallel} = 0; k_{z}))$ is shown schematically.

It turns out that $d^{Al_xGa_{1-x}As}$ has to be quite small, about 50 Å or less, in order to obtain a subband width of about 10meV or more for the lowest conduction subband in the GaAs-Al$_x$Ga$_{1-x}$As superlattice.

The situation, in principle, is similar for the valence subbands, but, unfortunately it is somewhat complicated by the mixing of heavy and light hole states in real semiconductors.

The first experimental answer to the question whether and how good a simple effective mass picture really describes the situation in a superlattice came from optical absorption experiments on an Al$_x$Ga$_{1-x}$As-GaAs superlattice by Dingle and coworkers (15,17). In order to investigate the simplest situation first, they started with a superlattice where the barrier width $d^{Al_xGa_{1-x}As}$ was so large that the k_z dispersion term in eq. (4) was negligibly small for the lowest subbands. Thus, the system represents essentially a periodic repetition of a single quantum well with a 2-dimensional band structure given by eq. (3) and is called a multiple quantum-well (MQW) structure. From the real space picture in Fig. 7(a) it is intelligible that there exists a quasi-selection rule for the optical valence-to-conduction subband transitions. The wave functions shown are actually envelope functions of the full functions which contain the (bulk) lattice-periodic Bloch part $u_c(r)$ and $u_v(r)$, respectively. Thus, only interband dipole matrix elements between electron

518

and hole subbands of the same subband index ν are large, since the small overlap between any other envelope functions makes the total dipole matrix elements very small.

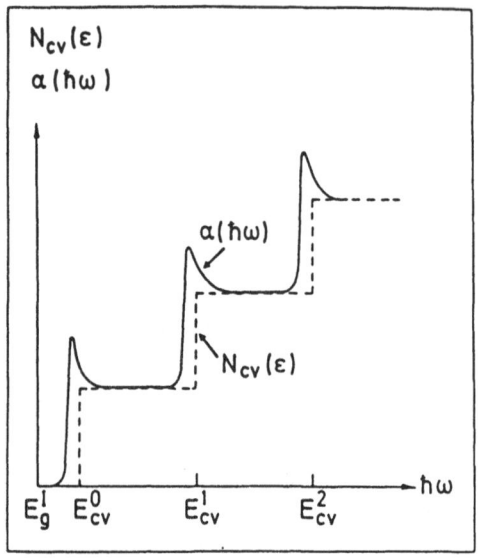

Fig. 9. Energy dependence of the absorption coeffic-ient $\alpha(\omega)$ in a compositio-nal superlattice, schema-tically. Dashed line: one-particle picture, without excitonic effects. Full line: with electron-hole attraction taken into ac-count.

This selection rule allows for the definition of a joint den-sity of states $N_{cv}^{\nu}(\varepsilon)$ for each pair of subbands with the same sub-band index. Because of the two dimensional character of the system $N_{cv}^{\nu}(\varepsilon)$ has a constant value for $\varepsilon > E_{c,\nu} - E_{v,\nu}$ (within the effective mass approximation), which leads to the total joint density of states $N_{cv}(\varepsilon)$ shown in Fig. 9 (dotted line). From a one particle picture one would expect that the optical absorption coefficient should roughly be proportional to $N_{cv}(\varepsilon)$. Due to (2-dimensional) excitonic effects, however, absorption peaks are to be expected slightly below each step of $N_{cv}(\varepsilon)$ as shown schematically in Fig.9 (full line).

The experimental results (see Fig. 10) of Dingle et al. have confirmed the expected behavior quantitatively. The agreement does not only provide a nice experimental demonstration of elementary text-book quantum mechanics (energy levels of a 1-dimensional square well potential) but it also shows that the simple effective mass approximation is quite appropriate to describe the system. More sophisticated microscopic calculations (17), indeed, have con-firmed that the accuracy of the effective mass approximation is satisfactory for this system. Another result of interest is the information that about 85% of the gap difference V_o occurs as a discontinuity of the conduction band edge $V_{o,c}$ whereas $V_{o,v}$ is only 15%.

An interesting device application, which makes use of the pos-sibility of tailoring the properties of a superlattice, is the following: The luminescent recombination between the lowest sub-bands is always shifted to higher energies compared to the gap of the type I semiconductor. The amount of the blue shift depends on the layer thickness d^I. Van der Ziel and coworkers (18) and Rezek and coworkers (19) have taken advantage of that in order to grow multiple quantum-well laser structures with coherent light emission at some energy $\hbar\omega$ between E_g^{GaAs} and $E_g^{Al_xGa_{1-x}As}$ according to their own choice.

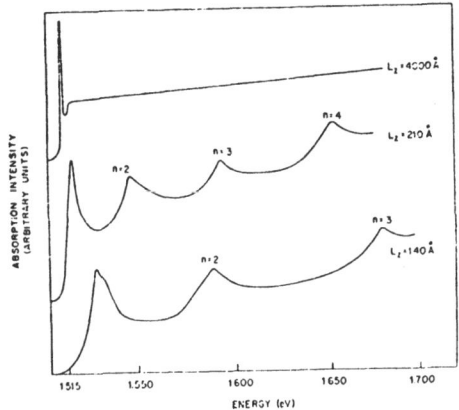

Fig. 10. Absorption spectra (2K) of 4000, 210 and 140 Å thick GaAs layers between $Al_{0.3}Ga_{0.7}As$ barriers. The QSE manifests itself as an increase in the bulk exciton energy and by the appearance of higher energy peaks in the thinner layers (from (15)).

The above given experiments may only exemplify a wide variety of phenomena which can be observed in superlattices. Application of a magnetic field perpendicular to the layers, fully quantizes the system by splitting the subbsnds into Landau levels. Doping of the crystal provides a dynamically 2-dimensional electronic system with finite carrier concentration in the ground state. The 2-dimensional character of the system has been demonstrated by magnetc absorption (20) and cyclotron resonance (21), Shubnikov de Haas (22), and Raman scattering (23,24) experiments, for instance (25).

The following topic demonstrates another aspect of the possibilities to design the properties of a crystal by introducing a superstructure. Störmer, Dingle, Gossard, Wiegmann and Logan (26) applied a simple trick for increasing the mobility of the charge carriers in the layers of a doped superlattice. Instead of doping the crystal uniformly they restricted the incorporation of dopants to the layers of the larger band gap component of the superlattice. By this "modulation doping" the charge carriers and the impurities whose potentials normally act as strong scattering centers are kept apart. As a consequence a large increase of the mobility at low temperatures and a considerable one even at room temperature could be achieved (26). The latter result, of course, has stimulated the interest of researchers working in the field of fast devices.

So far we have discussed examples of multiple quantum-well structures, i.e., systems where the interaction between states in adjacent potential wells plays no essential role because of relatively thick barriers between them. For sufficiently thin barriers, i.e., for values of d^{11} of typically 50 Å or less for the GaAs-$AL_xGa_{1-x}As$ system, the mini-band structure shows a significant, although narrow bandwidth with respect to motion in the direction of periodicity (see eq. (4)). The band structure of such a superlattice provides the ideal conditions for the observation of a number of high-field phenomena to be expected for periodic structures in principle, which, however, most probably are not observable in any "natural" solid because the electron-phonon and/or electron-electron interaction are too strong. All these phenomena are related to the qunatization of the motion in an energy band by a homogeneous electric field ("Wannier-Strak ladder") (27)). In a semiclassical picture the motion of the carriers is described by the well-known Bloch oscillations. The condition for observation of Bloch oscillation related effects is

$$\omega_B \tau > 1 \qquad (5)$$

In superlattices the frequency of the Bloch oscillations

$$\omega_B = eEd/\hbar$$

for a given electric field E is much larger than in familiar semiconductors because it is proportional to the large superlattice constant d rather than the original microscopic lattice parameter a. Moreover, also the (momentum) relaxation time τ at high fields is much longer because of the small kinetic energies acquired by electrons accelerated to the mini Brillouin zone boundary. A band width of less than the LO phonon energy, in particular, prevents for the possibility of any relaxation process by the otherwise very strong interaction of carriers with optical phonons. The transport theory predicts a number of high-field anomalies for conduction in systems with Bloch-oscillations (28-30). Historically, these phenomena, in particular the prediction of negative differential conductivity (NDC) and potential applications for fast devices motivated Esaki and his coworkers to attack the problem of growing semiconductor superlattices.

The experimental work of Esaki and his group provided evidence for NDC in superlattices and for conductivity oscillators related to this NDC (31).

With doday's high degree of perfection of growth technique it would probably be possible to reproduce these genuine superlattice effects more clearly than at the pioneer period when they were first investigated. Unfortunately, the success and the competition in the field of multiple quantum well structures seems to have paralyzed activities in this field although it is of high fundamental interest in several respects. In order to mention just one of them it may be recalled that there has been a long lasting dispute on the question of the conceptional correctness of Wanniers suggestion of the above mentioned "Wannier-Stark ladder" (32).

4. Doping Superlattices

In the preceding section it was shown that a periodic compositional superstructure provides the fascinating possibility of "designing" semiconductors with a new kind of band structure (mini- or subbands) which yield novel optical and transport properties. Going back to the list of properties requested in the introduction we will observe that we still missed to get granted two major wishes: a semiconductor whose energy gap and whose carrier concentration may be modulated.

In this chapter we will show that a nipi superstructure, indeed, represents a system in which these requests can be satisfied even within a very wide range of values for both, energy gap and electron and hole concentration.

4.1 Tunable electronic structure

The basic structure of a doping superlattice was shown in Fig. 1(b). All donors will be positively charged and all acceptors negatively, if the doping concentration is not too high and if the number of donors per n-layer, i.e., the product $n_D d_n$, equals the number of acceptors per p-layer $n_A d_p$. For the moment we neglect the randomness of the distribution of the impurity charges within the respective layers. This periodic space charge causes a periodic space charge potential which modulates the band edges $\varepsilon_c(z)$ and $\varepsilon_v(z)$ of the crystal, as shown in Fig. 11. The amplitude V_o of this potential can be estimated from the field $F_i = 2\pi e\, n_D d_n/\varepsilon$ in the intrinsic layer (ε = static dielectric constant of the unmodulated semiconductor) and the period d of the superstructure, of the doping layers are thin compared with d.

$$V_o \simeq eF_i d/4 \qquad (7)$$

(For a system of homogeneously doped n- and p-layers with $n_D=n_A$ and $d_n=d_p$, without intrinsic layers the value is just 1/2 of the one given by eq. (7), for example).

The space charge potential also quantizes the motion of the carriers in z-direction and leads to a subbandstructure analogue to the case of the compositional superlattice (Fig. 11(b) and (c)). The subband structure differs only qualtitatively from eq. (3) because of the different shape of the superlattice potential. The values of $E_{c,\nu}$ are no longer given by eq. (2) but rather by

$$E_{c,\nu} \simeq \hbar(4\pi n_D\, e^2/m_c\varepsilon)^{1/2}(\tfrac{1}{2} + \nu); \quad \nu = 0,1,2,\ldots \qquad (8)$$

for the empty conduction subband system, if the n-layers are not too thin.

522

The energies $E_{c,\nu}$ are of the order of $(12.5\ldots125)$ $(\nu + \frac{1}{2})$meV for $n_D = 10^{17}$-10^{19}cm^{-3}, whereas the amplitude V_0 is one half of the energy gap of GaAs $E_g^0 \simeq 1.5$eV for $n_D = n_A = 10^{18}$cm^{-3} and $d_n = d_p = d/2 = 650$ Å, for example.

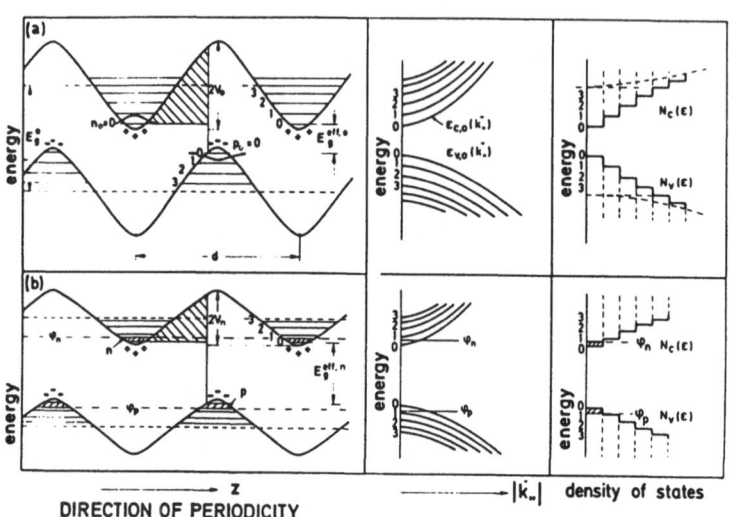

Fig. 11. Electronic structure of a n-i-p-i crystal, schematically. Upper part: ground state; lower part: excited state with increased effective band gap $E_g^{eff,\Delta n}$ and reduced tunneling barrier for electron-hole recombination.

Thus, $E_{c,o}$ (and $E_{v,o}$ as well) is small compared with $2V_o$ and the effective energy gap of the crystal, i.e., the difference between the bottom of the lowest conduction subband and the top of the uppermost valence subband, is apprxomately.

$$E_g^{eff,o} = E_g^o - 2V_o + E_{c,o} + E_{v,o} \simeq E_g^o - 2V_o \qquad (9)$$

The above given example for the value of V_o illustrates that any value of $2V_o$ up to E_g^o, and consequently any value within the range

$$0 < E_g^{eff,o} \leq E_g^o \qquad (10)$$

may be obtained by appropriate choice of the materials parameters n_D, n_A, d_n, d_p, and d_i.

Needless to say that n- or p-type nipi crystals may be obtained by choosing $n_D d_n$, or $n_A d_p$, respectively.

A particular interesting example is the case of high doping, where our estimates for the amplitude would give $2V_o > E_g^o$. In this case, of course, none of both types of doping layers will be completely depleted in the ground state. The electrons and holes in excess of those required to make $E_g^{eff,o} = 0$, according to eqs. (7) and (9) are populating the lowest conduction and the uppermost valence subband(s), respectively, forming dynamically quasi 2-dimensional carrier systems. This system containing electrons and holes in the ground state may be considered as a "n-i-p-i semimetal". With the simple procedure of periodically n- and p-doping which, in principle, is applicable to any semiconductor we have found a tool which permits to modulate the band structure over a wider range than in the case of compositional superlattices. In order to see that there are, in addition to these quantitative differences, also, even more important qualitative differences we will compare Figs. 5 and 11. The band structure in momentum space, indeed, shows only quantitative differences between the compositional and the doping superlattice. The real space picture, however, displays the important qualitative difference. Whereas the periodic modulation of the band edges has opposite sign for the conduction and the valence band edge in the GaAs-Al$_x$Ga$_{1-x}$As system, the space charge potential in the doping superlattice causes a parallel modulation of the conduction and valence band edges. As a consequence the wavefunctions of the electron and of the hole subbands are concentrated in the same region in the former case, but they are spatially shifted against each other by half a superlattice period in the latter one. Therefore, in analogy to the familiar indirect gap in momentum, a n-i-p-i crystal may be called a semiconductor with an "indirect gap in real space".

The effect of an indirect gap on the electron hole recombination, e.g., may be much more drastic in the real space case than in momentum space. The spatial separation of conduction and valence subband states suppresses the recombination nearly completely in the case of a long n-i-p-i superlattice period d. (The hatched area in Fig. 11 indicates the tunneling barrier for c,o → v,o recombination). The momentum selection rule in the case of an indirect gap in momentum space only requires some impurity scattering or the emission of phonons together with the emission of a photon. Therefore, the recombination probably will be reduced by not more than a few orders of magnitude in the latter case. Large deviations from thermal equilibrium are quasistable in a n-i-p-i crystal if the lifetime τ^{nipi} of excess carriers tends to infinity due to a sufficiently large superlattice period. In other words an excited state of the n-i-p-i crystal with different quasi Fermi levels for electrons and holes throughout the crystals, i.e.,

$$\phi_n \neq \phi_p \tag{11}$$

may be maintained over a long period without (or nearly without) any energy dissipation. From the requirement of macroscopic charge neutrality it is clear that the number of excess electrons always equals the number of excess holes

$$\Delta n = \Delta p \tag{12}$$

The excess carriers, nevertheless, influence the microscopic super-lattice potential, as they are not homogeneously distributed. Excess electrons in the n-layers reduce the positive space charge of the ionized donors and the excess holes reduce the negative space charge of the p-layers. As a consequence the superlattice potential amplitude V_n decreases and the effective band gap increases with increasing carrier concentration (see lower part of Fig. 11).

We have just seen that the effective band gap $E_g^{eff,\Delta n}$ in a n-i-p-i crystal can be modulated by a variation $\Delta n \overset{>}{\underset{<}{}} \Delta p$ of the electron and hole concentration. The important question how the carrier concentration can be changed has not yet been answered.

The easiest way to increase the electron and the hole concentration is by optical absorption. The electrons and holes created by the absorption of photons relax towards the edge of the respective subband system within picoseconds. The efficiency for this space-charge separation process is very high, since the electron-hole recombination lifetimes are by order of magnitude larger, even in the unmodulated host material. In this way the carrier concentration and the effective band gap will increase as a function of time of illumination until the electron-hole generation rate is balanced by the recombination rate which becomes large only in the highly excited state of the n-i-p-i crystal.

Another way of tuning the carrier concentration and the effective gap of a n-i-p-i crystal, namely by the application of "selective electrodes", is more convenient for experimentss and more useful for device application. Figure 12 depicts schematically what is meant by the term "selective electrodes". The n^+ zones represent ohmic contacts to any of the n-layers, but they are forming blocking p-n junctions to all the p-layers. Thus, the electron quasi-Fermi level ϕ_n is the same throughout the n-layers and the n^+-contacts in the steady state if there are no appreciable recombination currents. The same applies accordingly to the p^+-zones as hole-contacts and to the hole quasi Fermi level ϕ_p. The difference between the two quasi Fermi levels, of course, corresponds to an external potential

$$eU_{np} = \phi_n - \phi_p \simeq E_g^{eff,\Delta n} \tag{13}$$

Electrons and holes are injected through the respective contacts until the energy gap for the new carrier concentration corresponds to eU_{np}, if the external bias is increased. The "$\simeq$" sign in the expression (13) reflects the fact that the energy difference between the lowest subband edges and the respective quasi Fermi levels is a relatively small quantity. The major difference between a change of carrier concentration by optical excitation or by external bias is the possibility of carrier extraction in the latter case. This is of importance for reversible changes on a short time scale. It is of particular interest for the case of n-i-p-i semimetals. In this case carrier extraction is possible even in the ground state, by applying $eU_{np} < 0$. This, indeed, corresponds to a negative effective band gap, according to eq. (13).

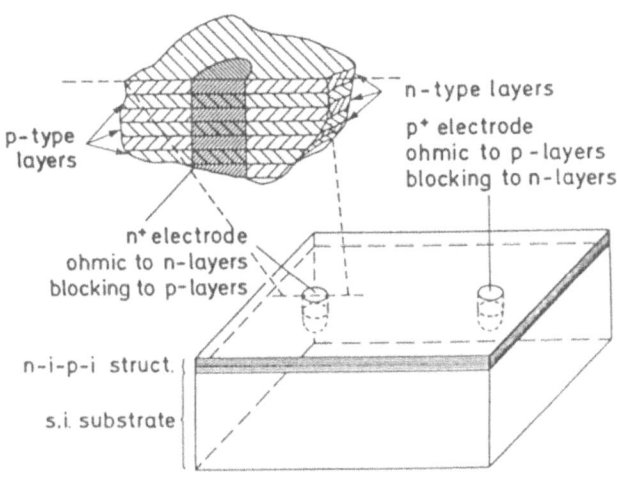

Fig. 12. Concept of selective electrodes in a n-i-p-i crystal. With a voltage U_{np} applied between the n^+ and the p^+-electrode electrons and holes will be injected or extracted from the respective layers until the difference in quasi-Fermi levels $\phi_n - \phi_p$ agrees with the external potential eU_{np}.

The electron and hole concentration as a function of the potential eU_{np} is shown schematically in Fig. 13 for a n-i-p-i semimetal with an excess of free holes and an example. The upper limit for eU_{np}, of course, lies at some value below the gap of the unmodulated crystal because of increasing leakage current. The threshold potential eU_{np}^{th} corresponds to the complete depletion of one carrier type. At eU_{np}^{th} the requirement of macroscopic charge neutrality prohibits any further reduction of carrier concentration of the opposite type as well.

4.2. Properties of n-i-p-i crystals

4.2.1 Tunable conductivity

The modulation of the carrier concentration as a function of the external potential eU_{np} implies a modulation of the conductivity σ_{nn} and σ_{pp} parallel to the layers. These conductivities can be measured by the current I_{nn} (I_{pp}) induced by a small voltage U_{nn} (U_{pp}) between two contacts of the respective type (see Fig. 12). At zero minority-carrier concentration, i.e., at U_{np}^{th}, also the corresponding channel conductivity vanishes.

Thus, the conductivity within the whole n-i-p-i crystal may be varied between zero and values characteristic for conventionally doped crystals by applying a variable voltage U_{np}. One is dealing with a novel bulk field-effect transistor which is free of crystallographic inhomogeneities. In some respect this device can be considered as a 3-dimensional version of a junction field effect transistor (JFET).

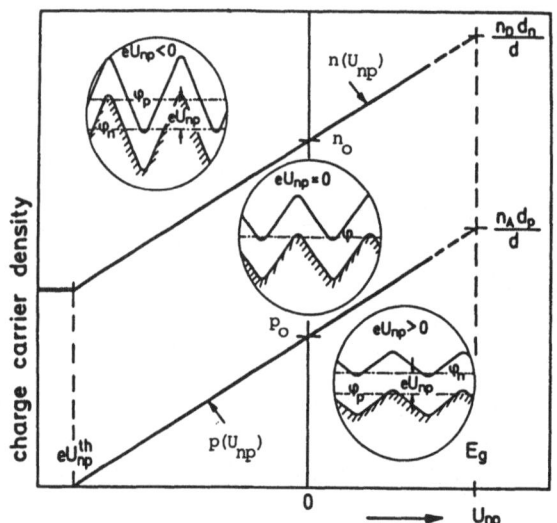

Fig. 13. Electron and hole concentration in a n-i-p-i "semimetal" with $p_o < n_o$ as a function of external voltage $U_{np} = (\phi_n - \phi_p)/e$, schematically. The hole concentration becomes zero at U_{np}^{th}. The insets shows the modulated band edges and the position of the quasi-Fermi levels for electrons and holes for the cases $eU_{np} = 0$.

By an appropriate choice of the doping profile and concentration the threshold voltage U_{np}^{th} and the characteristics such as $\sigma_{nn}(U_{np})$ can be designed according to the special requirements. A particularly interesting extension might be a 3-dimensional doping structure which could be achieved by means of a doping ion-beam lithography, e.g..n-i-p-i crystals of that kind could be designed as 3-dimensional integrated circuits.

The feasibility of the concept of selective electrodes was first tested by Ploog, Künzel, Knecht, Fischer, and Döhler [34]. Small Sn or Sn/Zn balls were alloyed from the surface as n^+- and p^+- electrodes, respectively, on GaAs doping superlattice structures. In Fig. 14 we give an example of the results of the n- and p-layer conductances $G_{nn}(U_{np})$ and $G_{pp}(U_{np})$ as obtained from current measurements in an arrangement with two pairs of n^+- and p^+ electrodes. At the threshold voltage, U_{np}^{th}, the hole layers are depleted completely. Thus $G_{pp}(U_{np} < U_{np}^{th}) = 0$. At U_{np}^{th} there are still electrons in the n-i-p-i structure, since $n^{(2),th} = n_D d_n - n_A d_p > 0$ for this system. The electron concentration can, however, not be reduced below the threshold value $n^{(2)th}$, as explained before. Therefore, $G_{nn}(U_{np})$ becomes constnat for $U_{np} < U_{np}^{th}$.

This example demonstrates the operation of a n-i-p-i crystal as a bipolar multiple-junction field effect transistor. Apart from these device aspects n-i-p-i crystals represent an interesting model system for the study of 2-dimensional transport properties including 2-D subband effects as well as the metal-insulator transition in a disordered (dynamically) two-dimensional system (29).

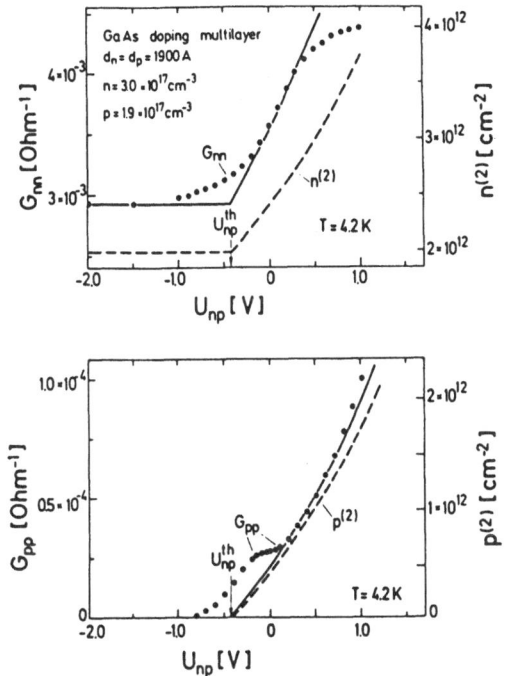

Fig.14. Simultaneous modulation of the conductance G_{nn} and G_{pp} in a GaAs n-i-p-i semimetal, grown by molecular beam epitaxy, as a function of applied bias U_{pp}. Full lines represent the expected conductance behavior as calculated from the sample design parameters. Dashed lines display the variation of free carrier concentration (From Ref.34)

The next example illustrates both, the modulation of carrier concentration and band gap by light and, at the same time, the extremely long recombination lifetimes which can be achieved in n-i-p-i crystals. In Fig. 15 results for the lifetimes as obtained

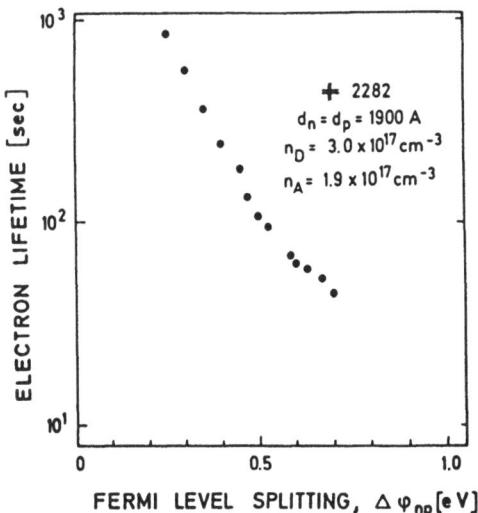

Fig. 15. Lifetimes as a function of Fermi level splitting in a GaAs doping superlattice as determined from the decay of photoconductivity $\sigma_{nn}(\Delta\phi_{np}(t))$ (From Ref. 36).

from measurements of the time resolved decay of photoconductivity (37) are displayed. For $\phi_n - \phi_p < 0.2$eV the lifetimes are in excess of 10^3 sec, or, in other terms, by more than 12 orders of magnitude larger than typical bulk values. These results prove, that the response of a n-i-p-i superlattice photodetector may exceed the response of its unmodulated-bulk counterpart by more than a factor of 10^{12}(!). Note, that the (low temperature) dark conductivity for the carrier type of interest can always be made zero by appropriate design of the n-i-p-i structure.

The large photoconductivity sensitivity provides also an elegant possibility to measure very small values of the absorption coefficient (see next section) and, also, of optical gain (see section 4.2.3).

4.2.2 Tunable optical absorption

Absorption of photons with $\hbar\omega > E_g^{eff}$ is possible in n-i-p-i superlattices. Below E_g^o the absorption coefficient $\alpha^{nipi}(\omega)$ depends strongly on the design parameters of the sample. In samples with large superlattice period and rather low doping concentrations the absorption coefficient $\alpha^{nipi}(\omega)$ at $\hbar\omega > E_g^{eff}$ may be unmeasurably small, because of nearly vanishing overlap between the corresponding initial and final states. In this case an exponentially decreasing absorption tail for $\hbar \quad E_g^o$ extends over a wide range of photon energies.

In Fig. 16 we compare results of $\alpha^{nipi}(\omega)$ which where obtained from measurements of the photoconductive response with our calcula-

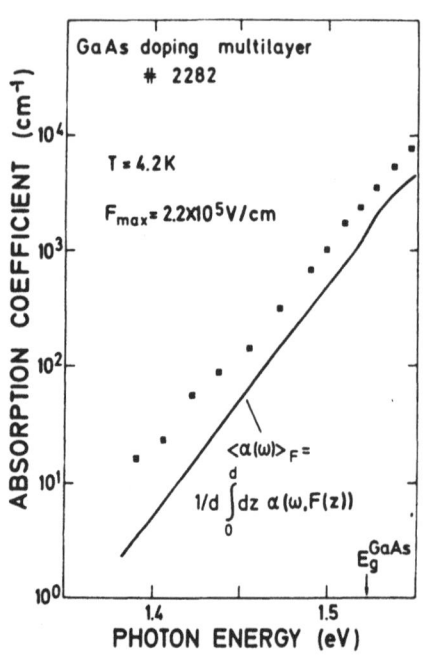

Fig. 16. Comparison between measured and calculated absorption coefficient of a GaAs n-i-p-i crystal with $d_n = d_p = 1900$ Å and $n_D = n_A = 1.9 \times 10^{17}$ cm^3 (from Ref. 37)

lations (37). The agreement between calculated and observed absorption tail is very satisfactory. In Ref. 37 we have also shown theoretically and verified experimentally that $\alpha^{nipi}(\omega)$ is a tunable quantity which varies if the effective gap (or the quasi Fermi level difference $\phi_n - \phi_p$) is changed. This possibility of modulating the transmitted light either electrically or by absorption of light of the same or some other frequency, again offers interesting possible device applications.

In n-i-p-i crystals with small superlattice period and sufficiently strong doping concentrations it should be possible to observe a step-like structure in $\alpha^{nipi}(\omega)$ reflecting the discrete subband structure and its consequences on energetically allowed absorption processes (38,39).

4.2.3 Tunable luminescence

The characteristic luminescence process in n-i-p-i crystals is the recombination across the indirect band gap in real space between electrons in electron subbands with holes in the acceptor impurity band (40,41). The theoretical luminescence spectrum involves the photon energies

$$E_{c,o} - \phi_p < \hbar\omega < \phi_n - \phi_p. \tag{14}$$

Thus, the energetic position of the luminescence signal directly reflects the tunable band gap $E_g^{eff,\Delta n}$ (see Fig. 11). The photoluminescence spectra of Fig. 17 confirm the expected increase of $E_g^{eff,\Delta n}$ with increasing excitation intensity.

We note, that the peak position shifts to higher energies rather exactly with the logarithm of the excitation intensity (42). This is a consequence of an exponential dependence between recombination lifetime and effective band gap $E_g^{eff,\Delta n}$ (33,42).

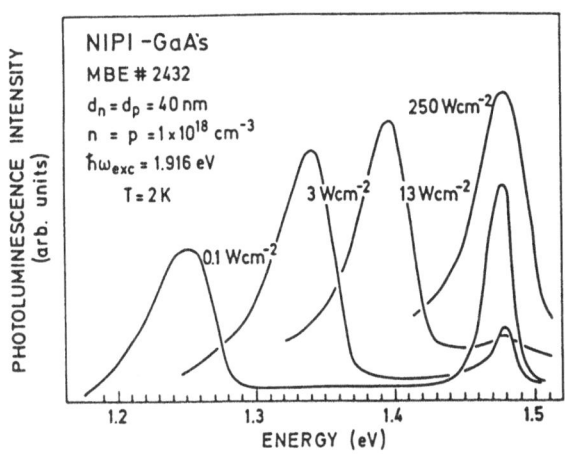

Fig.17. Photoluminescence spectra of a GaAs n-i-p-i superlattice excited with different laser intensities. Note, that the peak position shifts by more than 200meV as a function of excitation intensity.

Recently it was also demonstrated that tunable electroluminescence due to electrons and holes injected via selective electrodes can be observed if the value of $\phi_n-\phi_p$ in the n-i-p-i crystal is changed by variation of the external potential eU_{np} (43).

Finally, we mention another topic of interest, i.e. the stimulated light emission from n-i-p-i structures. A nearly perfect population inversion exists for a rather broad range of photon energies, even at very low optical excitation intensities or injection current densities, because of the long lifetimes and the resulting large steady state carrier concentrations. Tunable gain spectra with peak values of the order of 10^2 cm^{-2} at about 150meV below E_g^o, indeed, recently were detected under low-intensity excitation (44). Thus, a tunable n-i-p-i laser with low threshold intensity appears feasible

4.2.4 Tunable subband structure

There are, in principle, many phenomena in which quantum size effects should appear in n-i-p-i crystals due to the 2-dimensional subband formation in these space charge induced quantum wells. The first observation of subbands and of the tunability of the subband spacing was achieved by resonant spin-flip Raman experiments (40). The spin density intersubband excitation energies which appear as peaks of the Raman spectra differ only slightly from the corresponding subband energy differences $E_{c,\mu'}-E_{c,\mu}$. In Fig. 18 the first and second peak of the observed spin-flip Raman spectra are shown as a function of $n^{(2)}$ and compared with the calculated results for a simple with the design parameters $d_n=d_p=40$nm, $n_D=n_A=10^{18}$ cm^{-3}. The carrier concentration $n^{(2)}$ at a given laser intensity was determined by correlating the position of the corresponding photoluminescence spectra with $n^{(2)}$ using the calculated results (40,45,46). The agree-

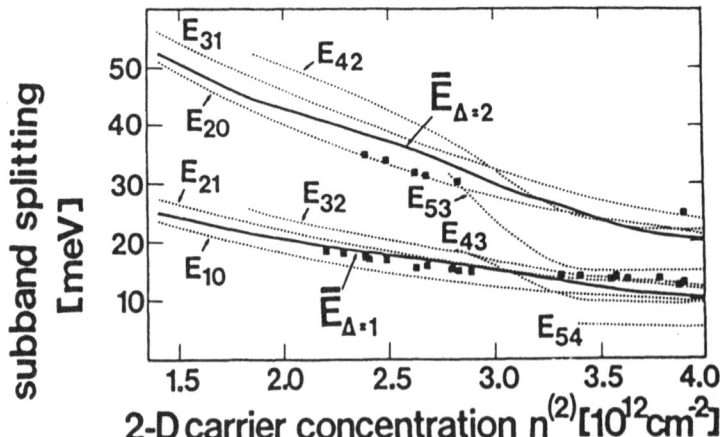

Fig. 18. Comparison between calculated subband spacings $E_{\mu'\mu}=E_{c,\mu'}-E_{c,\mu}$ as a function of 2-dimensional carrier concentration (dotted lines), averaged occupied subband spacing (heavy lines) for $\Delta=\mu'-\mu=1$, and 2 with experimentally determined spin density excitation energies as obtained by spin-flip Raman measurements (From Ref. 40). The design parameters are the same as in Fig. 17.

ment between the calculation, which is based on the design parameters
of the sample only, but does not contain any adjustable parameter,
and the experiment is very good. This example represents an example
that n-i-p-i crystals re-present a very flexible model substance for
the study of quasi-two-dimensional many-body systems.

5. Conclusions

Two different kinds of man-made superlattices have been considered: semiconductors with a periodic variation of composition
(with GaAs-Al$_x$Ga$_{1-x}$As, as the most prominent example) and semiconductors with a periodic variation between n- and p-doping ("n-i-p-i
crystals"). The electronic structure of both types of superlattices
differs drastically from familiar semiconductors:
(a) Each band is split into mini-bands;
(b) The width and the spacing of these minibands may be
designed by the choice of the superlattice parameters;
(c) Semiconductor superlattices are highly anisotropic.

In most cases they are dynamically quasi 2-dimensional. In addition to these novel features the n-i-p-i crystals exhibit peculiarities which result from the fact that n-i-p-i crystals represent
semiconductors with "indirect gap in real space".
(a) The energy gap in a given crystal may be tuned over a
wide range, including a negative gap.
(b) Strong deviations from thermal equilibrium for electron
and hole concentration can be metastable.

We have given a few examples of interesting experiments and
device applications in order to demonstrate that semiconductor superlattices in general, and n-i-p-i crystals in particular, represent both, a versatile model substance for dynamically 2-dimensio-
nal, tunable many-body systems and a versatile basic material for
novel devices.

At present we are still quite at the beginning of an investigation of the fascinating properties of man-made superlattices, in
particular of n-i-p-i structures.

Actually, apart from the two types of superlattices considered
so far, there are other ones which again differ qualitatively from
the former ones. One may think of compositional superlattices, e.g.,
in which the band edge discontinuities at the interface between the
two components have the same sign, in contrast to the GaAs-Al$_x$Ga$_{1-x}$As
system with opposite sign.

An extreme case of this kind of superlattice has first been
proposed (47) and investigated (48) experimentally by Esaki and
coworkers. In their system, an InAs-GaSb superlattice, the discontinuity of the band edges is so large that the conduction band edge

of InAs is lower in energy than the uppermost valence band edge of GaSb. The formation of a superlattice leads to new pecularities of the electronic structure, in particular due to the electron-hole mixing in the energy range where electrons and holes coexist in the respective unmodulated bulk material.

I would like to conclude with the remark that the properties of n-i-p-i crystals and of compositional superlattices also may be combined in a favorable way (49,45). An $Al_x Ga_{1-x}$ As n-i-p-i crystal with undoped GaAs-layers in the middle of each doping layer, e.g., (see Fig. 19) will exhibit the typical features of "familiar" n-i-p-i crystals, like modulation of effective gap and modulation of electron- and hole concentration. But, in addition, it will also show the high mobility resulting from "modulation doping", as mentioned in Section 3 of this paper.

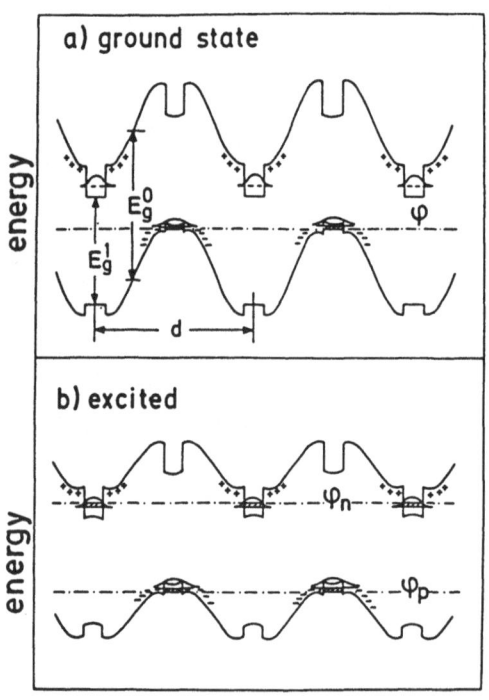

direction of crystal growth

Fig. 19. Doping superlattice, modified by the incorporation of undoped layers of a lower band gap material $(E_g^1 \ll E_g^0)$. The system behaves qualitatively like a normal n-i-p-i crystal. The mobility in the subbands, however, is much higher due to the local separation between free carriers and scattering centers (i.e. the randomly distributed ionized impurities). (a) p-type crystal in the ground state $(p_o; n_o = 0)$. (b) Excited state $(p = p_o^0 + \Delta p; n = \Delta n = \Delta p)$.

These two examples may be sufficient in order to illustrate that this field of physics is one of those rare nearly unexpected domains where some simple ideas may direct your step to fascinating places. Unfortunately, it must be admitted, you never will get to enjoy these marvellous areas if you are not provided with some rather expensive equipment and with a lot of skill in crystal growth.

References

1. Chang, L.L., Esaki, L., Howard, W.E., Ludeke, R. and Schul, G., J. Vac. Sci. Technol. 10, 655 (1973)
2. Petroff, P.M., Gossard, A.C., Wiegmann, W. and Savage, A., J. Crystal Growth 44, 5 (1978)
3. For a recent review see: Ploog, K., in Crystals-Growth, Properties, and Applications (Edited by H.C. Freyhardt), p. 73, Berlin (1980)
4. Holonyak, N., Kolbas, R.M., Dupuid, R.D. and Dapkus, P.D., IEEE J. of Quantum Electronics QE-16, 170 (1980)
5. Kolbas, R.M., Holonyak, N., Jr., Vojak, B.A., Hess, K., Altarelli, M., Dupuis, R.D. and Dapkus, P.D., Solid State Commun., 31, 1033 (1979)
6. Döhler, G.H., Phys. Status Solidi(b) 52, 79 (1972)
7. Döhler, G.H., Phys. Status Solidi(b) 52, 533 (1972)
8. Ploog, K., Döhler, G.H., Fischer, A. and Künzel, H., Inst. Phys. Conf. Ser. 59, 721 (1981)
9. Ploog, K., Fischer, A. and Künzel, H., J. Electrochem. Soc. 128, 400 (1981)
10. Ilegems, M., J. Appl. Phys. 48, 1267 (1977)
11. Bauser, E., Schmidt, L., Lochner, K.S. and Raabe, E., Proc. 8th Conf. (1976) on Solid State Devices, Japan J. Appl. Phys. 16, 457 (1977)
12. Scheel, H.J., J. Crystal Growth 42, 301 (1977)
13. Yoshida, M., Mizutani, T., Watanabe, H. and Seki,Y., Spring Meeting of the Electrochemical Societey Seattle, Washington (1978)
14. Bailay, S.W., et al., Acta Cryst. A33, 681 (1977); Wilson, J.A., DiSalvo, F.J. and Mahajan, S., Advances in Physics 24, 117 (1975)
15. Dingle, R., Wiegmann, W. and Henry, C.H., Phys. Rev. Lett. 33, 827 (1974)
16. Dingle, R., Gossard, A.C. and Wiegmann, W., Phys. Rev. Lett. 34, 1327 (1975)
17. Ivanov, I. and Pollmann, J., Solid State Commun. 32, 869 (1979)
18. van der Ziel, J.P., Dingle, R., Miller, R.C., Wiegmann, W. and Nordland, W.A., Jr., Appl. Phys. Lett. 26, 463 (1975)
19. Rezek, E.A., Holonyak, N.,Jr., Vojak, B.A., Stillman, G.E., Roni, J.A., Keune, D.L. and Fairing, J.D., Appl. Phys. Lett. 31, 288 (1977)
20. Dingle, R., Störmer, H.L., Kopf, L. and Wiegmann, W., Unpublished
21. Chang, L.L., Sakaki, H., Chang, C.A. and Esaki, L., Phys. Rev. Lett. 38, 1489 (1977)
22. Dingle, R., Störmer, H.L., Gossard, A.L. and Wiegmann, W., Surface Science 98, 90 (1980)
23. Manuel, P., Sai-Halasz, G.A., Chang, L.L., Chang,

534

C.A. and Esaki, L., Phys. Rev. Lett. 37, 1701 (1976)
24. Pinczuk, A., Worlock, J.M., Störmer, H.L., Dingle,
    R., Wiegmann, W. and Gossard, A.C., Surface Science
    98, 126 (1980)
25. See also: Surface Science 73(1978), 98 (1980) and 113 (1982)
26. Störmer, H.L., Dingle, R., Gossard, A.C., Wiegmann,
    W. and Logan, R.A., in Physics of Semiconductors
    (Edited by B.L.H. Wilson), Inst. Phys. Conf. Ser.
    43, 557 (1978)
27. Wannier, G.H., Phys. Rev. 117, 432 (1960);
    Kane, O.E., J. Phys. Chem. Solids 12, 181 (1959)
28. Döhler, G.H. and Hacker, K., Phys. Status Solidi
    26, 551 (1968); Hacker, K., Phys. Status Solidi
    33, 607 (1969)
29. Saitoh, M.J., Phys. C5, 914 (1972)
30. Esaki, L. and Tsu, R., IBM J. Res. Develop. 14,
    61 (1970)
31. Esaki, L. and Chang, L.L., Phys. Rev. Lett. 33, 495
    (1974)
32. Zak, J., Phys. Rev. Lett. 20, 1477 (1968); Avron,
    J.E., Phys. Rev. Lett. 37, 1568 (1976)
33. Döhler, G.H., J. Vac. Sci. Technol. B 1, 278 (1983)
34. Ploog, K., Künzel, H., Knecht, J., Fischer, A. and
    Döhler, G.H., Appl. Phys. Lett. 38, 870 (1981)
35. For a review see: Ando, T., Fowler, A.B., Stern, F.,
    Rev. of Modern Physics 43, 437 (1982)
36. Ploog, K. and Künzel, H., Microelectronics Journal
    13, 5 (1982)
37. Döhler, G.H., Künzel, H. and Ploog, K., Phys. Rev.
    B 25, 2616 (1982)
38. Döhler, G.H., Ruden, P., Künzel, H. and Ploog, K.,
    Verhandl. DPG (VI) 16, 161 (1981)
39. Döhler, G.H. and Ruden, P., tc be published
40. Döhler, G.H., Künzel, H., Olego, D., Ploog, K., Ruden, P.,
    Stoltz, H.J. and Abstreiter, G., Phys. Rev. Lett. 47, 864
    (1981)
41. Jung, H., Döhler, G.H., Künzel, H., Ploog, K.,
    Ruden, P. and Stolz, H.J., Solid State Commun. 43,
    291 (1982)
42. Döhler, G.H. in: Festkörperprobleme XXIII, P. Grosse, Ed.
    (Vieweg, 1983), p.
43. Künzel, H., Döhler, G.H., Ruden, P. and Ploog, K.,
    Appl. Phys. Letts. 41, 852 (1982)
44. Jung, H., Döhler, G.H., Göbel, E. and Ploog, K.,
    Appl. Phys. Letts.   (1983)
45. Ruden, P. and Döhler, G.H., Phys. Rev. B 27, 3538 (1983)
46. Ruden, P. and Döhler, G.H., Phys. Rev. B 27, 3547 (1983)
47. Sai-Halasz, G.A., Tsu, R. and Esaki, L., Appl. Phys.
    Lett. 30, 651 (1977)
48. Chang, L.L. and Esaki, L., Surface Science 98, 70
    (1980) and references therein

49. Döhler, G.H. and Ploog, K., Progr. Crystal Growth
     Charact. $\underline{2}$ 145 (1979)

CHAPTER 8: MODULATIONS IN SOLIDS

PREMARTENSITIC BEHAVIOR AND CHARGE DENSITY WAVES IN TiNi ALLOYS

C. M. Hwang, M. Meichle, M. B. Salamon and C. M. Wayman
Materials Research Laboratory
University of Illinois at Urbana-Champaign
Urbana, Illinois 61801 USA

ABSTRACT

Variations of the intermetallic compound TiNi are well-known examples of shape-memory alloys. The addition of Fe, for example, is a method of changing the martensitic start temperature, and thus the shape memory effect, over a wide temperature range. A puzzling aspect of these materials has been the presence of precursive effects. Reports of the appearance of superlattice spots in x-ray studies, of resistivity and susceptibility anomalies, and of changes in crystal structure have appeared without a suitable explanation.

We have recently completed a detailed study of these premartensitic effects in TiNi(Fe) with Fe substituting up to 3.2 at.% for Ni. The premartensitic stage of theses alloys is found to be comprised of two distinct phases, one of which is a rare example of an incommensurate charge density wave (CDW) in a three dimensional metal. Two phase transitions occur, one separating the incommensurate phase from the normal, CsCl high temperature phase, and the other marking the onset of a rhombohedral distortion in which the lattice distorts to accommodate the periodicity of the charge density wave.

At the transition temperature, superlattice spots appear which are approximately 1% removed from the true 1/3 positions. Electron micrographs of a sample using a superlattice spot show antiphase-like domains of diameter ~ 1500 Å which are suggestive of discommensuration domains. However, these domains, once formed do not change size as expected from the theory of discommensurations. Accompanying the appearance of the superlattice spots is a kink in the susceptibility curve, indicating a change in the density of states at the Fermi energy.

Approximately 10 K below the onset of this incommensurate phase, a structural transition occurs to a rhombohedral phase. The nature of the distortion is such that the longer (111) axis now corresponds to exactly one wavelength of the charge density wave. Thus, the lattice appears to distort to accommodate the CDW, rather than the more usual change in CDW periodicity. At this transition, a large peak in the specific heat is observed, which shows some hysteresis. The rhombohedral angle decreases abruptly at this temperature to 89.7°. At lower temerpatures, the rhombohedral distortion increases, with the CDW wavelength remaining locked-in. At this transition, electron micrographs show the appearance of needle-shaped domains, each of which has a single rhombohedral axis and which has a twin relationship with the matrix in which they appear. Examination of the superlattice spots within single needles indicates that a single (111) wavevector occurs along the rhombohedral axis and the (110) superlattice is oriented normal to this rhombohedral axis.

1   INTRODUCTION

The intermetallic compound TiNi is a well-known shape memory alloy which undergoes a martensitic transformation slightly above room temperature. A puzzling aspect of this material is that it exhibits such premonitory or "premartensitic" effects immediately above the martensitic start (M_s) temperature as: an electrical resistivity anomaly (1,2), streaks in electron diffraction patterns (1,3), and "1/3 spots" seen in both x-ray and electron diffraction patterns (1,4).   Other anomalies in physical or mechanical properties have also been reported, such as a decrease in sound velocity (5,6) and elastic modulus (7), internal friction peaks (8) and specific heat peaks (9). The nature of these so-called "premartensitic" phenomena has long been unclarified.

Many of these anomalies are also observed in materials undergoing a structural transition driven by the Fermi surface, to the charge density wave (CDW). CDW phenomena are most commonly observed in quasi-one and two-dimensional materials. Extensive theoretical and experimental studies on the nature of such phase transitions in quasi-one-dimensional organic conductors, such as TTF-TCNQ (10-12) and KCP (13), and quasi-two-dimensional layered compounds, such as transition metal dichalcogenides (14-17), have attributed the observed transitions to CDW formation.

Because of the unlikelihood of favorable Fermi surface nesting, CDW phenomena are rare in three-dimensional materials. Based on a recent study of the "premartensitic" effects in ternary TiNiFe alloys of nominal composition $Ti_{50}Ni_{47}Fe_3$, we conclude that CDW phenomena are involved in the premartensitic behavior of these alloys. The substitution of Fe for Ni in TiNi lowers the temperature at which the martensitic transformation takes place by more than 100°C, and reveals a well-defined " premartensitic" phase. The premartensitic stage of these alloys is found to be comprised of two, rather than one, distinct phases, one of which is a rare example of an incommensurate charge density wave (CDW) in a three dimensional metal. Two phase transitions occur, one separating the incommensurate phase from the normal, CsCl high temperature phase, and the other marking the onset of a rhombohedral distortion in which the lattice distorts to accommodate the periodicty of the change density wave. This paper reports the premartensitic behavior and charge density waves in the TiNiFe alloys with Fe substituting up to 3.2 at.% for Ni.

2   EXPERIMENTAL METHODS

TiNiFe alloys of normal composition $Ti_{50}Ni_{47}Fe_3$ are described in the present report. The alloys were produced by vacuum arc-melting, following which specimens were hot-swaged and then hot-rolled to strip. The strips were heated to 900°C for 10 minutes in vacuum, and then water-quenched. A polycrystalline $Ti_{50}Ni_{47}Fe_3$ alloy was used for most TEM experiments and electrical resistance vs. temperature measurements. Single crystal samples of slightly different composition, $Ti_{50.1}Ni_{46.7}Fe_{3.2}$ were cut from large grains of an ingot. These single crystals were used for TEM observations as well as neutron and x-ray diffraction experiments. These alloys were also used for other physical property measurements, such as specific heat and magnetic susceptibility (18). Strip samples 50 mm x 0.1 mm were used for electrical resistance-temperature measurements using a method described earlier (19). Three millimeter disk TEM specimens were spark-cut from strips and then electropolished. The electrolyte used was 7.5% perchloric acid in acetic anhydride at 5°C. TEM work was carried out using two different microscopes: Hitachi H500 and Philips 400T, both equipped with a cooling-tilting stage which permits specimens to be cooled from room temperature to about -130°C and -165°C, respectively.

3   EXPERIMENTAL RESULTS

3.1 Electrical Resistance As A Function of Temperature

Figure 1 shows a plot of electrical resistance vs. temperature for a full thermal thermal cycle between room temperature and liquid nitrogen temperature (-196°C). Upon cooling below room temperature, the resistance is seen to decrease slightly until $T_p \sim 0°C$, the premartensitic start temperature is reached, after which the resistance starts to rise. At the M_s temperature, the resistance starts to decrease until liquid nitrogen temperature, which is approximately the M_f temperature. On heating from liquid nitrogen temperature, the resistance increases up to A_s where the curve changes its slope. At A_f, the

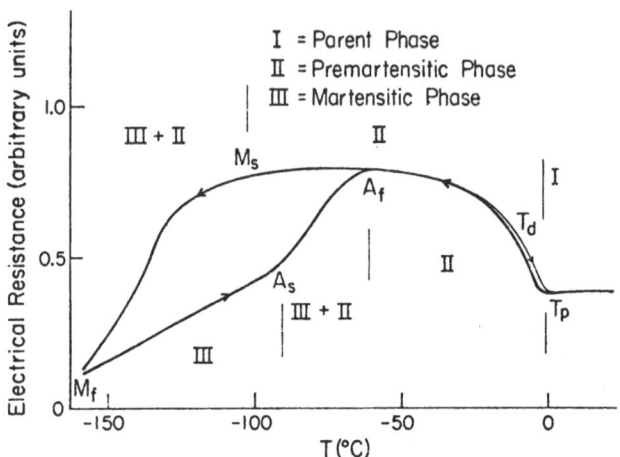

Fig. 1 Electrical resistance as a function of temperature for a full complete thermal cycle: room temperature → liquid nitrogen temperature → room temperature, $Ti_{50}Ni_{47}Fe_3$ alloy.

heating curve crosses the cooling curve and then shows a small hysteresis. The larger thermal hysteresis seen below A_f is typical for a normal martensitic transformation. Preliminary characterization of different phases (I-III) can be inferred from this electrical resistance vs. temperature plot. Above T_p, phase I is the high-temperature parent phase. Between T_p and M_s, a distinct premartensitic phase II is obvious. Between M_s and M_f, the martensitic phase as well as the premartensitic phase coexist. Below M_f, the entire crystal would be in the martensitic phase III. At this stage, one might conclude that there are two phase transformations. As will be shown later, additional features show up in electron microscopy at the temperature T_d, which is about 12°C lower than T_p.

3.2 Electron Diffraction and Microscopy of the Parent Phase and the Premartensitic Phase

The crystal structure of the high temperature parent phase was examined by electron diffraction at room temperature. The selected area diffraction (SAD) patterns show the CsCl (B_2) type structure. The lattice constant of the parent phase was determined to be $a_0 = 3.02$ A. Figure 2 shows a transmission electron micrograph and the corresponding $[001]_{B2}$ zone electron diffraction pattern of the parent phase taken at room temperature.

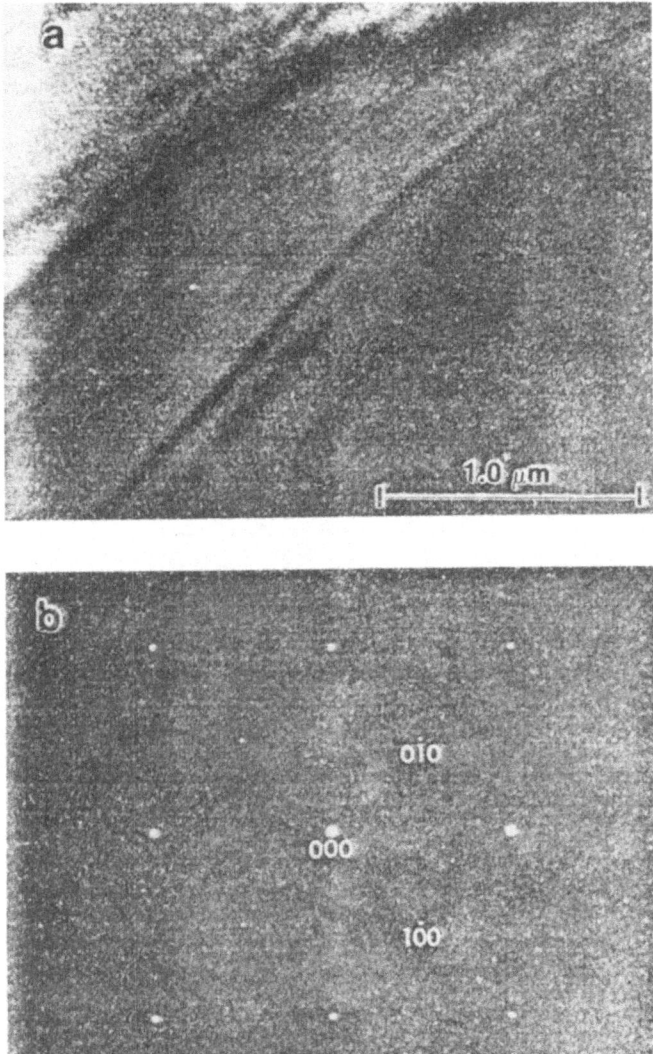

Fig. 2 (a) Transmission electron micrograph and (b) corresponding $[001]_{B2}$ zone electron diffraction pattern of the parent phase of the $Ti_{50}Ni_{47}Fe_3$ alloy taken at room temperature.

As specimens were cooled below room temperature, they developed superlattice reflections in the vicinity of 0°C (T_R). The superlattice reflections were seen to lie along the <110>* reciprocal lattice vectors and to be of the 1/3 (110) type, i.e., they locate at about "1/3" positions of the <110>* reciprocal lattice vectors; 1/3 (111) type superlattice reflections were also seen. Typical electron diffraction patterns of the premartensitic

544

phase showing the 1/3 (110) superlattice reflections in $[001]_{B2}$, $[110]_{B2}$, $[111]_{B2}$ and $[113]_{B2}$ zones are shown in Fig. 3. The "1/3" superlattice reflections were observed to intensify with decreasing temperatures, as shown in Fig. 4. The intensity change is reversible with temperature and is most obvious between 0°C and -10°C and less so at lower temperatures. The positions of the 1/3 (110) superlattice reflections were examined closely by measuring the space between them. The results of such measurements showed that the 1/3 (110) reflections deviated slightly from the exact 1/3 position relative to the parent CsCl structure over a 12°C temperature range. More quantitative measurements on the intensities and positions of the superlattice reflections using neutron or x-ray diffraction gave parallel results, with a shift of approximately 0.5% from commensurate positions.

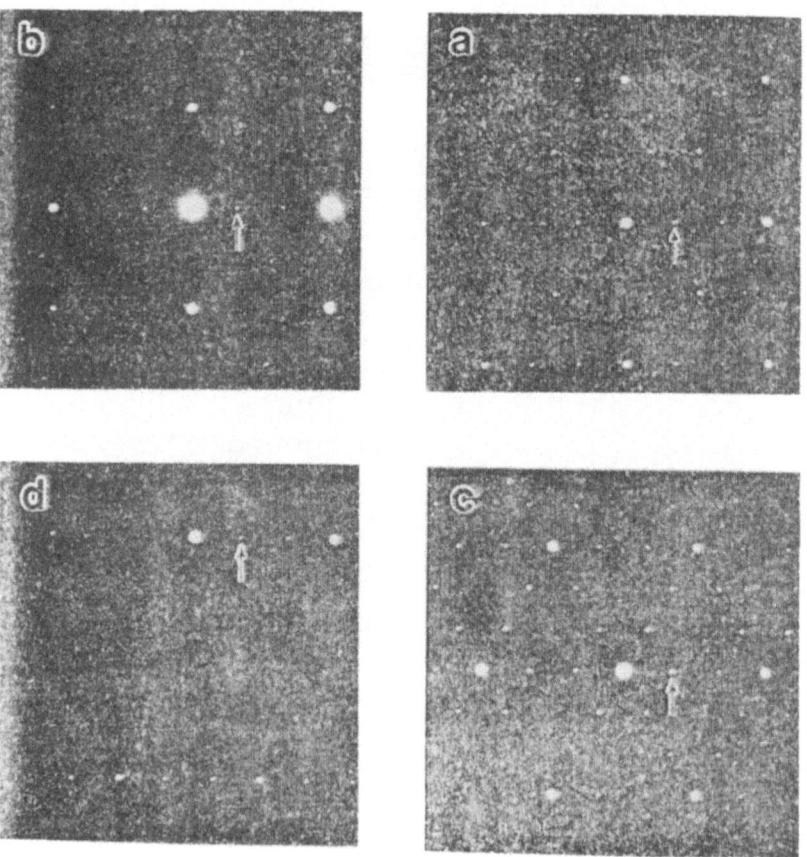

Fig. 3 Typical electron diffraction patterns from the "premartensitic" phase of the $Ti_{50}Ni_{47}Fe_3$ alloy showing 1/3 (110) superlattice reflections. (a) $[001]_{B2}$, (b) $[110]_{B2}$, (c) $[111]_{B2}$ and (d) $[113]_{B2}$ zones.

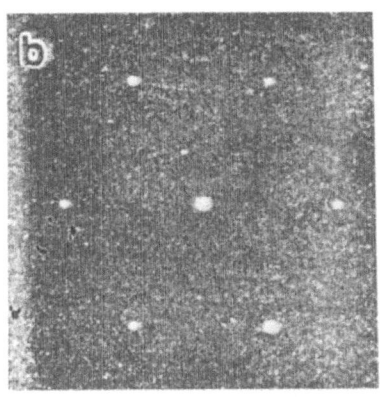
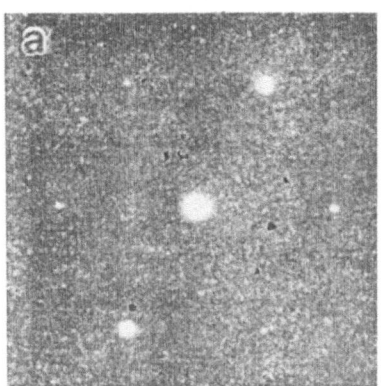
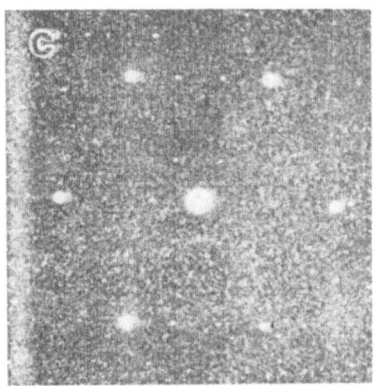

Fig. 4 Electron diffraction patterns showing temperature
dependence of 1/3 (110) superlattice reflections in the $[111]_{B2}$
zone. The superlattice reflections appear around 0°C and
intensified with decreasing temperature. (a) 4°C, (b) 0°C and (c)
-4°C.

Antiphase-like microdomains (APD's) ~ 1500 Å in diameter were
revealed in dark-field images using the "1/3" reflections. A
typical example of these APD's is shown in Fig. 5. These
microdomains are similar to antiphase domains observed in ordered
alloys. However, no APD's are seen in the dark-field images taken
by using $(100)_{B2}$ type superlattice reflections in either the
parent phase or the premartensitic regions. It was noted that
APD's are temperature independent, in the sense that the domain
size and shape do not change with temperature throughout the
premartensitic regime.

At about 12°C below T_p, needle-like microdomains are observed
to form in the $Ti_{50}Ni_{47}Fe_3$ alloy. The temperature corresponding

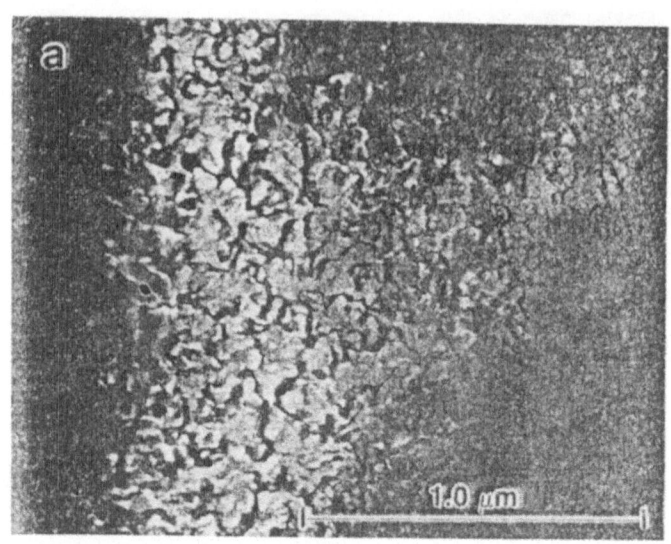

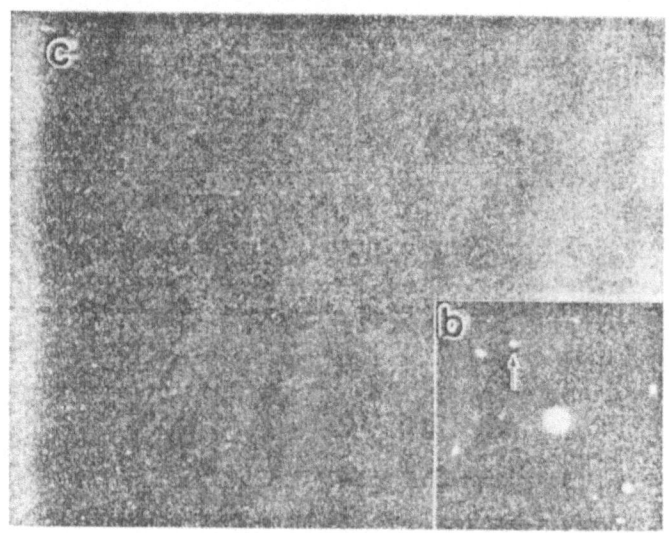

Fig. 5 Typical example showing antiphase-like domains (APD's) in the premartensitic phase of the $Ti_{50}Ni_{47}Fe_2$ alloy. (a) Dark-field electron micrograph showing APD's, (b) corresponding selected area diffraction pattern showing the 1/3 (110) superlattice reflection used for the dark-field imaging, and (c) bright-field image.

to the formation of the "needle" domains is designated T_d. A typical example of the "needle" domains is shown in Fig. 6(a). Figure 6(b) is the corresponding SAD pattern taken from both the needle domains and the matrix in Fig. 6 (a).

Fig. 6 (a) Transmission electron micrograph showing parallel "needle" domains. (b) Corresponding selected area diffraction pattern taken from both the needle domains and the matrix region. This diffraction pattern is interpreted as a superimposed pattern consisting of the $[001]_H$ and $[201]_H$ zones. The inner "split" spots correspond to the $[201]_H$ zone, while the outer "split" spots correspond to the $[001]_H$ zone.

The premartensitic structure of the purely binary TiNi alloy was reported to be rhombohedral (3,20). Electron diffraction patterns of the rhombohedral structure of the martensitic phase in

548

Au-50 at.% Cd are well-established (21). A series of SAD patterns of the $Ti_{50}Ni_{47}Fe_3$ alloy in the premartensitic state between T_d and M_s was taken. A comparison of those results with the SAD patterns from the rhombohedral structure (21), shows that they are identical in symmetry. In addition, neutron and x-ray diffraction (18), as will be discussed later, show that the premartensitic structure of TiNiFe between T_d and M_s is indeed rhombohedral. The SAD patterns are therefore indexed according to the rhombohedral structure in terms of hexagonal indices. Figure 7 shows SAD patterns of the $Ti_{50}Ni_{47}Fe_3$ premartensitic structure below T_d, which are identified as $[\bar{1}00]_H$, $[\bar{2}01]_H$ and $[101]_H$ zones of the rhombohedral structure. The diffraction pattern in Fig. 6(b) is interpreted as a superimposed pattern consisting of two orientations. The inner "split" spots correspond to the $[201]_H$ zone, while the outer "split" spots correspond to the $[001]_H$ zone. It was proven by dark-field imaging that needle domains

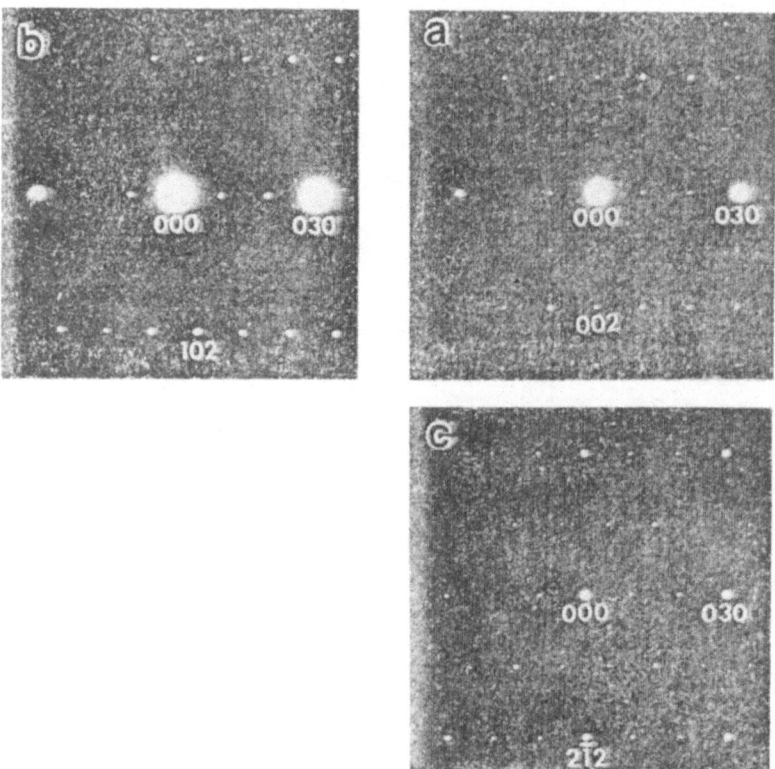

Fig. 7 Selected area diffraction patterns of the $Ti_{50}Ni_{47}Fe_3$ alloy in the premartensitic state between T_d and M_s, identified as corresponding to the rhombohedral structure and indexed in terms of hexagonal indices: (a) $[\bar{1}00]_H$, (b) $[\bar{2}01]_H$ and (c) $[101]_H$.

correspond to $[210]_H$ (inner) spots and the matrix corresponds to $[001]_H$ (outer) spots. Three variants of "needle" domains in addition to the matrix were usually seen to form below T_d. The disappearance of the "needle" domains upon heating shows some thermal hysteresis, i.e., "needle" domains disappear completely only above T_d. A more detailed analysis of the "needle" domains shows that they can be identified as twins with a $\{110\}$ twinning plane.

The positions of 1/3 (110) type superlattice reflections from the premartensitic phase (below T_d) were examined closely. Within the accuracy of the measurements, the spacing between superlattice reflections along <110>* is the same, i.e., the 1/3 (110) reflections are at exact 1/3 positions of the <110>* reciprocal lattice vector. X-ray diffraction (18) also showed that the 1/3 (111) superlattice reflections shift to exact 1/3 positions at T_d and remain commensurate with the Bragg reflections of the rhombohedral structure in the premartensitic state. T_d values are -12°C and -49°C for the $Ti_{50}Ni_{47}Fe_3$ and $Ti_{50.1}Ni_{46.7}Fe_{3.2}$ alloys respectively.

In addition to 1/3 (110)-type superlattice reflections, 1/3 (111)-type superlattice reflections were also observed in different needle domains as shown in Fig. 8(a)-(b). Figure 8(b) is the diffraction pattern taken from the encircled region in Fig. 8 (a). 1/3 (111) and 1/3 (110)-type superlattice reflections in Fig. 8(b) were proved to be from domain B and domain C respectively. A summary of 1/3 (110) and 1/3 (111)-type superlattice reflections within a single needle domain is shown in Fig. 9.

Below T_d, APD's were seen in both the "matrix" and "needle" domains. Figure 10(a) and 10 (b) are the bright field and corresponding dark-field images taken using a 1/3 (110) superlattice reflection in the diffraction pattern from the matrix, as shown in Fig. 10(c). In the bright field image, APD's are seen in both the matrix and needle domains, although they are not in good contrast. In the dark-field image, APD's appear in the matrix only because the superlattice reflection used for dark-field imaging was from the matrix, and not from the needle domains. Thus no APD's are seen in the needle domains, which appear dark.

4   DISCUSSION AND CONCLUSIONS

4.1 Two Premartensitic Transformations

Two phase transformations rather than one are found to be associated with the "premartensitic" behavior of the $Ti_{50}Ni_{47}Fe_3$ alloy. At T_p^-, the formation of "1/3" superlattice reflections is

550

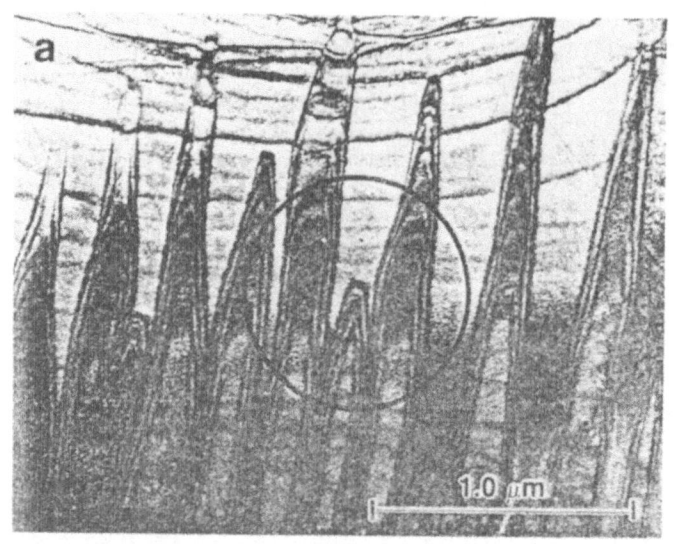

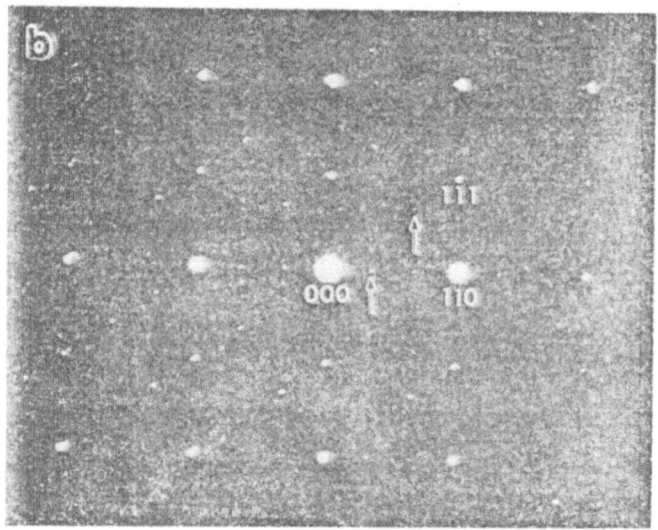

Fig. 8 (a) Electron micrograph showing domains B and C and (b)
corresponding diffraction pattern taken from both domains B and C
1/3 (111) and 1/3 (110)-type superlattice reflections are from
domains B and C respectively.

coincident with the onset of the electrical resistance anomaly
(increase) on cooling. APD's formed simultaneously when the
crystal developed "1/3" superlattice reflections. Accompanying
the appearance of the superlattice reflections at T_p, a kink in
the magnetic susceptibility curve was also observed (18),
indicating a change in the density of states at the Fermi

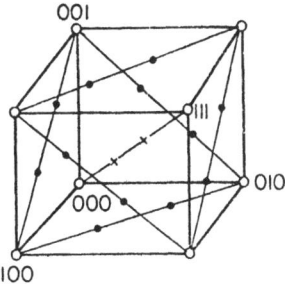

○ Fundamental

● $\frac{1}{3}$ (110)

× $\frac{1}{3}$ (111)

Fig. 9 Summary of 1/3 (110) and 1/3 (111)-type superlattice reflections within a single needle domain. The 1/3 (111) superlattice reflections lie along only one [111]* direction, and the 1/3 (110) superlattice reflections lie only along those <110>* directions which are normal to [111]*.

energy. At a still lower temperature, T_d, needle-like domains were observed to form. A specific heat peak which shows some hysteresis around T_d has also been observed (18). The microstructure change as well as physical property anomalies at T_d indicate that a second possible "premartensitic" transformation occurs at a lower temperature than that of the first. The transformation temperatures are $T_p \sim 0°C$, $T_d \sim -12°C$ and $T_p \sim -41°C$, $T_d \sim -49°C$ for the $Ti_{50}Ni_{47}Fe_3$ and $Ti_{50.1}Ni_{46.7}Fe_{3.2}$ alloys respectively.

4.2 The Normal-to-Incommensurate Transition

The positions of the 1/3 (110) superlattice reflections were found not to be at exact 1/3 positions between two fundamental reflections. Neutron diffraction experiments (22) showed quantitatively that the 1/3 (110) reflections are displaced slightly toward the nearest fundamental reflections by about 0.45% at the onset. The significance of this finding is that it proves the formation of a 1/3 ($1-d_1$) (110) incommensurate superlattice with $d_1 \sim 0.0045$. The appearance of an incommensurate superlattice is strong evidence that a Fermi-surface driven phenomena, such as the formation of a charge density wave, occurs in the premartensitic state of the TiNiFe alloys. The formation of the superlattice reflections is due to the CDW superstructure forming "on" the parent cubic structure. The fact that the

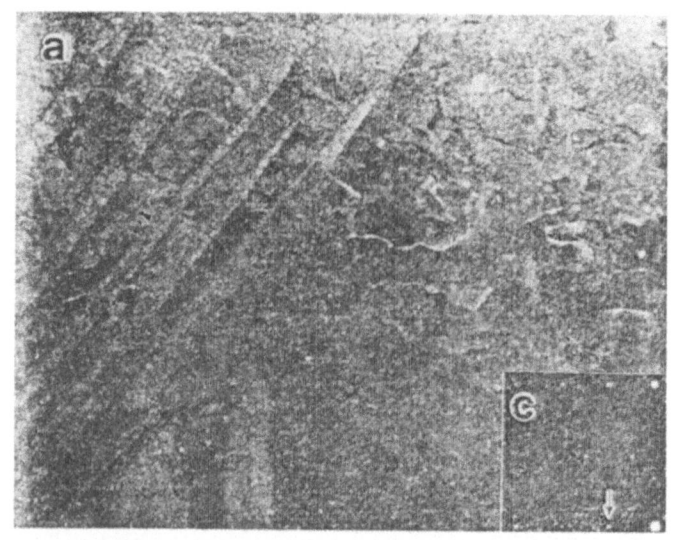

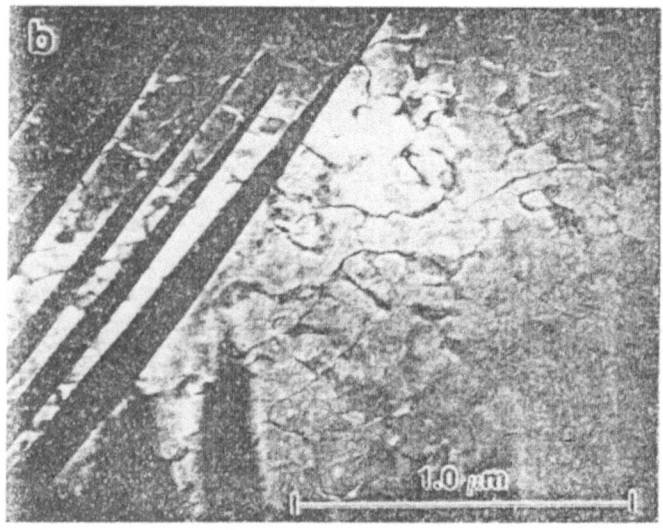

Fig. 10 (a) Bright-field image showing APD's in needle-domains as well as in the matrix region. (b) Corresponding dark-field image taken by using one of the 1/3 (110) superlattice reflections shown in (c).

intensity of the $1/3$ $(1-d_1)$ (110) superlattice reflections increases continuously from nil on cooling implies a continuous change of order parameter, which is an indication of a second order normal-to-incommensurate phase transition. $1/3$ (111) type superlattice reflections were also found in the $[110]_{B2}$ zone electron diffraction patterns. X-ray diffraction (18) showed quantitatively the incommensurate nature of these $1/3$ $(1-d_2)$ (111) superlattice reflections with $d_2 \sim 0.012$ at the onset. Neutron scattering studies (22,24) show that the distortion giving rise to the $1/3$ (110) and $1/3$ (111) superlattice are associated with soft TA (110) and LA (111) phonons respectively. The nature of the first premartensitic phase transformation is therefore a second order normal-to-incommensurate transition involving the formation of two possible CDW's. The APD's appearing in the incommensurate state of the TiNiFe alloy, are suggestive of discommensuration-type domains, although, anomalously, the APD size is temperature independent, being inconsistent with the defect melting hypothesis (17).

4.3 The Incommensurate-to-Commensurate Transition

A system in an incomensurate state can lower its total energy by further transforming to a commensurate state, the so-called "locked in" state. For the $Ti_{50}Ni_{47}Fe_3$ alloy, the $1/3$ (110) type superlattice reflections in the electron diffraction patterns were found to become situated at $1/3$ positions along $\langle 110 \rangle^*$ at lower temperatures in the premartensitic region. After the rhombohedral distortion, $1/3$ (111) type superlattice reflections were found by x-ray diffraction (18) to become commensurate with the Bragg reflections at T_d. The $1/3$ (111) reflections remain commensurate with the new Bragg peaks even though the rhombohedral distortion gradually continues on further cooling. The high temperature parent phase has a cubic (CsCl) structure which becomes an incommensurate phase between T_p and T_d. Thus, the second "premartensitic" transformation involves a "distorted" cubic to a rhombohedral product. The rhombohedral distortion is accomplished by a homogeneous expansion along $\langle 111 \rangle$ direction, allowing the lattice and CDW's to lock-in and become commensurate. This behavior is different from the lock-in mechanism in layered compounds (14). The hysteresis effect in the appearance and disappearance of "needle" domains indicates that the lock-in transition is first order. Therefore, the nature of the second "premartensitic" transformation in the $Ti_{50}Ti_{47}Fe_3$ alloy is established as a first order incomensurate-to-commensurate CDW transition and is associated with a "distorted" cubic to rhombohedral structural change.

Similar CDW phenomena and associated phase transitions were also observed in TiNiAl and TiNiFe alloys of various compositions (23). The CDW transitions can take place with or without a subsequent martensitic transformation in both TiNiAl and TiNiFe alloys. It is concluded that the so-called "premartensitic" effects found in the TiNiFe and TiNiAl alloys are actually independent, electronically driven structure changes and not precursory effects insofar as the martensitic transformation per se is concerned. The observations reported here represent a well

554

documented case of a rare but nevertheless simple example of the occurence of charge density waves in a three-dimensional solid system.

ACKNOWLEDGEMENTS

The work described herein has been supported by the National Science Foundation through the Materials Research Laboratory at the University of Illinois. Their support is gratefully acknowledged.

REFERENCES

1. Sandrock, G.D., Perkins, A.J. and R.F. Hehemann. Met. Trans. 2, 2769 (1971).
2. Wayman, C.M., Cornelis, I. and K. Shimizu. Scripta Met. 6, 115 (1972).
3. Chandra, K. and G.R. Purdy. J. Appl. Phys. 39, 2176 (1968).
4. Otsuka, K., Sawamura, T. and K. Shimizu. Phys. State. Sol. 5, 457 (1971).
5. Wang, F.E., DeSavage, B.F., Buehler, W.J. and W.R. Hosler. J. Appl. Phys. 39, 2166 (1968).
6. Bradley, D. J. Acoust. Soc. Am. 37, 700 (1965).
7. Hasiguti, R.R and K. Iwusaki. J. Appl. Phys. 39, 2182 (1968).
8. Wasilewski, R.J. Trans. Met. Soc. AIME 233, 1691 (1965).
9. Dautovich, D.P., Melkvi, Z., Purdy, G.R. and C.V. Stager. J. Appl. Phys 37, 2513 (1966).
10. Denoyer, F., Comes, R., Garito, A.F. and A.J. Heeger. Phys. Rev. Lett. 35, 445 (1975).
11. Shirane, G., Shapiro, S.M., Comes, R., Garito, A.F. and A.J. Heeger. Phys. Rev. B 24, 2325 (1976).
12. Bak, P. and V.J. Emery. Phys. Rev. Lett. 36, 978 (1976).
13. Eagen, C., Werner, S. and R. Saillant, Phys. Rev. B 12, 2036 (1975).
14. Wilson, J.A., DiSalvo, F.J. and S. Mahajan. Adv. Phys. 24, 117 (975).
15. Moncton, D.E., Axe, J.D. and F.J. DiSalvo. Phys. Rev. Lett. 34, 734 (1975).
16. Craven, R.A., DiSalvo, F.J. and F.S.L. Hsu. Solid St. Commun., 25, 39 (1978).
17. McMillan, W. Phys. Rev. B 12, 19975 (1975); Ibid., 14, 1496 (1976); Ibid., 16, 643, 4655 (1977).
18. Meichle, M. Ph.D. Thesis, 1981, Physics Dept., Univ. of Illinois at Urbana-Champaign.
19. Thornburg, K.A., Dunne, D.P. and C.M. Wayman. Met. Trans. 2, 2302 (1971).
20. Dautovich, D.P. and G.R. Purdy. Canadian Met. Quarterly 4, 129 (1965).

21. Tadaki, T., Katano, Y. and K. Shimizu. Acta. Met. 26, 883 (1978).
22. Salamon, M., Meichle, M., Wayman, C.M., Hwang, C.M. and M. Shapiro. AIP Conf. Proc. 53, 223 (1979).
23. Hwang, C.M. and C.M. Wayman. Scripta Met. 17, 381, 385 (1983).
24. Satija, S.K., Shapiro, S.M. and M.B. Salamon. Bull. Am. Phys. Soc. 26, 381 (1981).

ELECTRONIC CONTRIBUTIONS TO MIXING- AND GRADIENT-ENERGY
OF COMPOSITION-MODULATED ALLOY SYSTEM

H. Yamauchi

Department of Engineering Materials
University of Windsor
Windsor, Ontario
Canada   N9B 3P4

Electronic contributions to mixing- and gradient-energy
of a composition-modulated binary alloy system are obtained.
Hohenberg and Kohn's formula for the ground-state energy of an
inhomogeneous electron gas is utilized in a sinusoidally
modulated positive-background model.   Origins of electronic
mixing- and gradient-energy are explored. If electronic contributions are dominant in the energetics of a low electron-density
system such as Cs-Rb, the system may well be of clustering type.
Model dependency of calculated electronic mixing- and gradient-
energy is discussed.

1. INTRODUCTION

Two major contributions to the free energy of a non-uniform
mixture, according to Cahn and Hilliard (1), are: (a) free energy
density, $f(c)$, as a function of local composition $c(\vec{r})$ at position
$\vec{r}$, and (b) gradient energy density, $\kappa(\nabla c)^2$, being proportional to
the square of composition gradient ∇c. [Constant κ was termed the
gradient-energy coefficient (1)]. Thus, the free energy of a nonuniform mixture can be obtained as a functional of the local
composition c. Cahn and Hilliard (2) and Cahn (3), respectively,
developed theories of nucleation and of spinodal decomposition in
nonuniform mixtures based on their free energy functional.

The purpose of the present work is to obtain electronic
contributions to the mixing energy (Sec. 3.1) and to the gradient
energy (Sec. 3.2) by means of Hohenberg and Kohn's formula (4),
for the ground-state energy functional of a nonuniform electron
gas system, applied to a composition-modulated positive-background

model (5,6),

2. MODEL

A binery metallic mixture of A and B atoms is considered. It is assumed that the local composition, c, of component B is modulated along x-axis with a sinusoidal plane wave:

$$c(x) = c_0 + A \cos(kx), \qquad (1)$$

where c_0 is the average composition of B atoms, A is the amplitude and k is the wave number, being equal to $2\pi/\lambda$, when the wavelength is λ. When both the atomic volume difference defined by $\epsilon = 2(v_A - v_B)/(v_A + v_B)$ [v_A and v_B are atomic volumes of an A and a B atom] and the modulation amplitude A are small compared with unity, the positive charge distribution function may be given by

$$n_+(x) = n_0[1 + \alpha\cos(kx)], \qquad (2)$$

where

$$n_0 \equiv n_A(1-c_0) + n_B c_0 + (n_A - n_B)c_0(1-c_0)\epsilon ,$$

$$\alpha \equiv A(n_B - n_A)[1 - \epsilon(1-2c_0)]/n_0,$$

in which n_I (I=A,B) is the average density of free electrons in pure metal of I kind. Thus, we may term our model, in which the positive change is distributed according to Eq. (2), modulated positive-background model. In this model described by Eq. (2), the free electron distribution n(x) may be given by:

$$n(x) = n_0[1 + \beta\cos(kx)], \qquad (3)$$

where the normalized amplitude β is an unknown parameter to be determined by the minimum energy principle. A usable formula for the ground state energy U_e of the nonuniform electron-gas system described by Eq. (3) was given by Hohenberg and Kohn (4):

$$U_e[n] = \frac{1}{2}\iint [n(\vec{r}) - n_+(\vec{r})][n(\vec{r}') - n_+(\vec{r}')]d\vec{r}d\vec{r}'/|\vec{r}-\vec{r}\,| $$
$$+ G[n]. \qquad (4)$$

Here atomic units (a.u.) have been used and will be employed throughout. In G[n], we use Thomas-Fermi-von Weiszäcker formula for the kinetic energy and Gaspar's expression for the exchange energy:

$$G[n] = \frac{3}{10}(3\pi^2)^{2/3}\int n^{5/3} \, d\vec{r} + w\int \frac{[\nabla n(r)]^2}{n(r)} d\vec{r} -$$

$$-\frac{3}{4}(\frac{3}{\pi})^{1/3} \int n^{4/3} \ d\vec{r} + \int n\epsilon_c(n) \ d\vec{r}, \tag{5}$$

where $\epsilon_c(n)$ is the correlation energy density. Among various formulae proposed for $\epsilon_c(n)$, we employ the following three expressions for the three different cases. For a low electron density system where $r_0 \gtrsim 20$ a.u. [$r_0 \equiv (3/4\pi n_0)^{1/3}$: average Wigner-Seitz radius], the interpolation formula of Wigner (7) may be employed:

$$\epsilon_c^W(n) = -0.44/[7.8+(3/4\pi n)^{1/3}] \ . \tag{6}$$

For the intermediate range of r_0 (=1.8∿6.5 a.u.) for actual metals, the Nozières-Pines formula (8) may be used:

$$\epsilon_c^{NP}(n) = -[0.0575 + (0.0155/3)\ell n(4\pi n/3)]. \tag{7}$$

For a very high electron density where $r_0 \lesssim 1$ a.u., Gell-Mann and Brueckner's formula (9) may be utilized:

$$\epsilon_c^{GB}(n) = -[0.048 + (0.031/3)\ell n(4\pi n/3)]. \tag{8}$$

It should be noted that w in Eq. (5) represents the von Weiszäcker coefficient and may be located in the range of 1/72∿1/8 a.u. (10).

3. RESULTS AND DISCUSSION

The difference in electronic energy per unit volume, $\Delta U_e/V$ between the homogeneous solution in which $c(\vec{r}) = c_0$ and one with a modulated composition having the positive charge distribution as given by Eq. (2) can be obtained using Eqs. (2) and (3) in Eq. (4). [See Refs. (5, 6)]. Restricting ourselves to the case where the relative amplitude of electron distribution, β [defined by Eq. (3)], is small, we may approximate $\Delta U_e/V$ by

$$\frac{\Delta U_e(\beta)}{V} \cong \frac{\omega_{es}(\alpha-\beta)^2}{k^2} + (\omega_0 + \omega_1 k^2)\beta^2 \ . \tag{9}$$

Energy parameters, ω_{es}, ω_0 and ω_1, are defined as follows:

$$\omega_{es} \equiv \pi n_0^2 \tag{10}$$

$$\omega_0 \equiv (1/12)[(3\pi^2)^{2/3}n_0^{5/3} - (3/\pi)^{1/3}n_0^{4/3}] + \omega_{0c}, \tag{11}$$

in which ω_{0c} depends on the correlation energy density ϵ_c:

$$\omega_{oc}^{W} = -(0.056/18)n_0 \, \xi(1-\xi)(2-\xi),$$

where $\xi \equiv n_0^{1/3}(0.079+n_0^{1/3})$,

$$\omega_{oc}^{NP} = -(0.0155/12)n_0,$$

and

$$\omega_{oc}^{GB} = -(0.031/12)n_0,$$

corresponding to ϵ_c^{W}, ϵ_c^{NP} and ϵ_c^{GB} given by Eqs. (6), (7) and (8), respectively, and

$$\omega_1 \equiv (W/2)n_0.$$

Minimizing $\Delta U_e(\beta)/V$ with respect to β, we obtain the most probable value for β:

$$\beta/\alpha \cong 1/[1+(\omega_0/\omega_{es})k^2 + (\omega_1/\omega_{es})k^4]. \tag{13}$$

The corresponding minimum value for $\Delta U_e/V$ is given by

$$\Delta U_e/V \cong (\omega_0+\omega_1 k^2)\alpha^2/[1+(\omega_0/\omega_{es})k^2+(\omega_1/\omega_{es})k^4]. \tag{14}$$

It should be noted (6) that, in most cases, the maximum error in $\Delta U_e/V$ given by Eq. (14) due to the approximation employed in Eq. (9) is within ±4%. Nonetheless, it may be safe for us to restrict ourselves only to the case where the modulation wavelength is long, or $|k|$ is small, since both (a) electron density-gradient terms higher than the lowest order one and (b) discrete positive ion cores are not taken into account in our energy functional given by Eqs. (4) and (5). Thus:

$$\Delta U_e/V \cong [\omega_0 + (\omega_1 - \omega_0^2/\omega_{es})k^2]\alpha^2. \tag{15}$$

As stated in Section 1, Cahn and Hilliard (1) obtained a free energy functional for a nonuniform mixture. Excess free energy, $\Delta F/V$, in the system with composition modulation given by Eq. (1) is written as (3):

$$\Delta F/V = (1/4)(f_0'' + 2\kappa k^2)A^2. \tag{16}$$

Here, f_0'' is the second derivative of the homogeneous free energy density, $f(c)$, with respect to the local composition c:

$f_0'' \equiv [d^2 f(c)/dc^2]_{c=c_0}$ and κ is the gradient energy coefficient (1). In the ground state at zero temperature, no contributions from entropy should be included in the excess free energy which may now be termed excess energy, ΔU:

$$\Delta U/V = (1/4)\,(u_o{}'' + 2\kappa k^2)A^2, \tag{17}$$

where $u_o{}''$ is the second derivative of the mixing energy with respect to c and evaluated at $c = c_o$ and is termed the mixing-energy coefficient. It should be noted that no entropy contributions are included in the gradient-energy coefficient, κ.[1] Comparing Eq. (15) with Eq. (17), we may obtain electronic contributions to the mixing- and the gradient-energy coefficients:

$$u_e{}'' \equiv 4\theta\omega_o\,, \tag{18}$$

and

$$\kappa_e \equiv 2\theta\,(\omega_1 - \omega_o{}^2/\omega_{es}), \tag{19}$$

where

$$\theta \equiv [(n_A - n_B)(1 - \epsilon\,(1 - 2c_o))/n_o]^2.$$

3.1. Mixing-Energy

The electronic mixing-energy coefficient obtained in Eq. (18) may be rewritten as

$$u_e{}'' = \theta n_o{}^2\,[d^2 u_e(n)/dn^2]_{n=n_o}, \tag{20}$$

if the homogeneous mixing energy is defined by

$$u_e(n) = (3/10)(3\pi^2)^{2/3}n^{5/3} - (3/4)(3/\pi)^{1/3}n^{4/3} - n\epsilon_c(n). \tag{21}$$

Thus, as shown Eq. (20), quantity $u_e{}''$ which was defined by Eq. (18) via comparison of Eq. (15) with Eq. (17) is proportional to the second derivative of $u_e(n)$ with respect to (n/n_o). The mixing energy, u_e, defined by Eq. (21) receives contributions from kinetic, exchange and correlation energies of a homogeneous electron gas with density of n. It should be noted that u_e receives no contributions from electrostatic interactions (ω_{es}) and von Weiszäcker term (ω_1).

It is worthwhile to see if electronic contributions work for ordering or clustering in an alloy system. Since $u_e{}''$ is proportional to ω_o, electronic contributions promotes ordering when $\omega_o > 0$. Figure 1 shows ω_o plotted with respect to the average Wigner-Seitz radius, $r_o [= (3/4\pi n_o)^{1/3}]$. For Curves 1, 2 and 3, different correlation energy expressions, $\epsilon_c{}^W$, $\epsilon_c{}^{NP}$ and $\epsilon_c{}^{GB}$, as given in Eqs. (6), (7) and (8), are respectively used. It is seen in

Fig. 1 that each curve crosses the zero level at $r_0 = r_0^*$ so that ω_0 may be negative for r_0 larger than r_0^*. That is, in alloy systems whose r_0's are smaller than r_0^*, i.e. high electron-density systems, electronic contributions work for promoting ordering. On the other hand, in alloy systems whose r_0's are larger than r_0^*, i.e. low electron-density systems, electronic contributions promote clustering. The value of r_0^* depends on the correlation energy expression: 5.44, 5.35 and 4.88 a.u. for Curves 1, 2 and 3, respectively.

3.2 Gradient Energy

The electronic contribution to the gradient energy coefficient, κ_e, is obtained in Eq. (19). Thus the electronic gradient energy for a composition wave with wave number equal to k and amplitude equal to A is given by $(1/2)\kappa_e k^2 A^2 = \theta(\omega_1 - \omega_0^2/\omega_{es})k^2 A^2$. It is immediately seen that κ_e depends not only on the von Weiszäcker term (ω_1) but also on ω_0 (which consists of kinetic, exchange and correlation energy of electrons) and electrostatic term (ω_{es}).

In Fig. 2, Quantity $2(\omega_1 - \omega_0^2/\omega_{es})$, which is proportional to κ_e, is plotted with respect to r_0. Curve 1 is obtained for the von Weiszäcker coefficient $w = 0.45/8$ a.u. as suggested by Robinson and de Chatel (10) and Curve 2 is for $w = 1/72$ a.u. as utilized by Hohonberg and Kohn (4). Both curves were obtained using ϵ_c^{NP} given by Eq. (7) for the correlation energy. It is observed in Fig. 2 that $\kappa_e > 0$ for $r_0 > r_0^G$, i.e. in low electron-density systems, and $\kappa_e < 0$ for $r_0 < r_0^G$, i.e. in high electron-density systems. The value of r_0^G depends on the von Weiszäcker coefficient: 1.9 a.u. for $w = 0.45/8$ a.u. (Curve 1) and 3.3 a.u. for $w = 1/72$ a.u. (Curve 2). If $w = 0.45/8$ a.u. is employed, κ_e is positive for almost all actual metals since r_0 of the actual metal is in the range of 1.8~6.5 a.u.

In Cahn and Hilliard's nonuniform regular solution model (1), the sign of the gradient energy coefficient κ is always opposite to that of the mixing energy coefficient u''. It is apparent from Figs. 1 and 2 that such a relation between signs of κ_e and u_e'' is not always held, that is, the two characteristic Wigner-Seitz radii r_0^* and r_0^G defined above are not coincident. Nontheless, signs of κ_e and u_e'' are opposite for $r_0 > 5.0$ a.u. This may indicate that binary alloy systems such as Cs-Rb, Cs-K and Rb-K which have

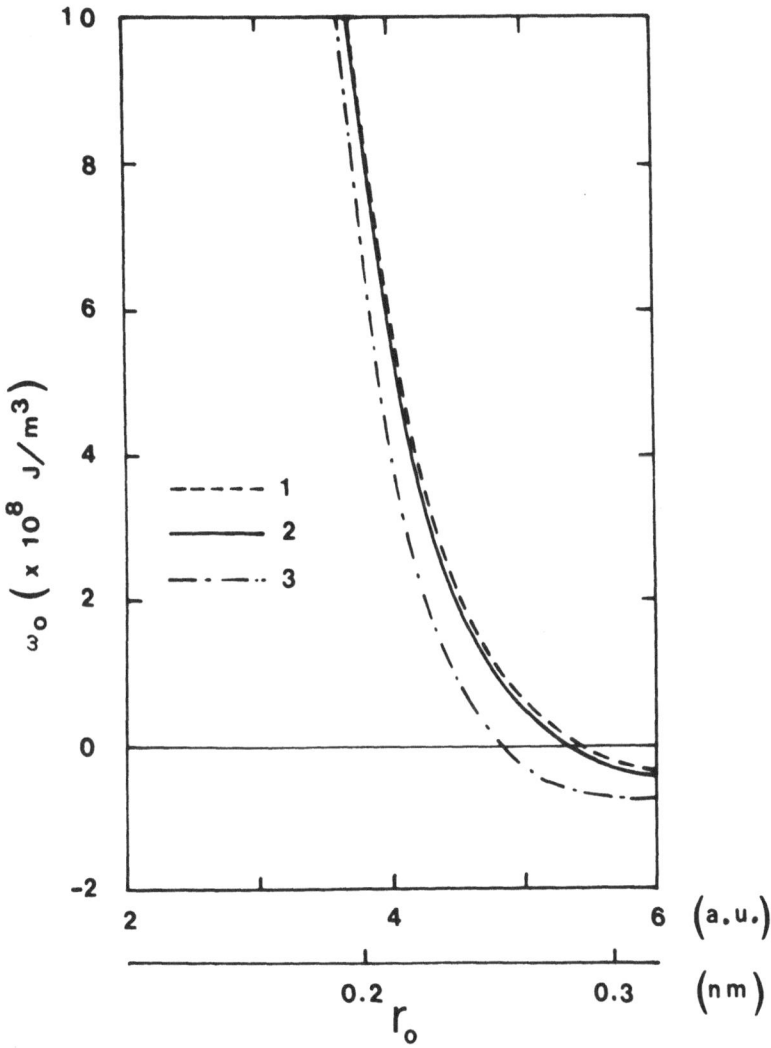

Figure 1.   Quantity ω_0 ,which is defined by Eq. (11) and is proportional to the electronic mixing-energy coefficient as shown in
Eq. (18),depends on the average Wigner-Seitz radius, r_0.   Curves
1, 2 and 3 are obtained for three different expressions for the
electron correlation energy given in Eqs. (6), (7) and (8),
respectively.

the average Wigner-Seitz radius r_0 larger than 5.0 a.u. could be
of clustering type, since the electron density functional approach
(11) parallel to the present method may well be applicable to low
electron-density systems.   It should be noted that, if Curve 3 in
Fig. 1 is employed for the calculation of u_e'' and the relation that

564

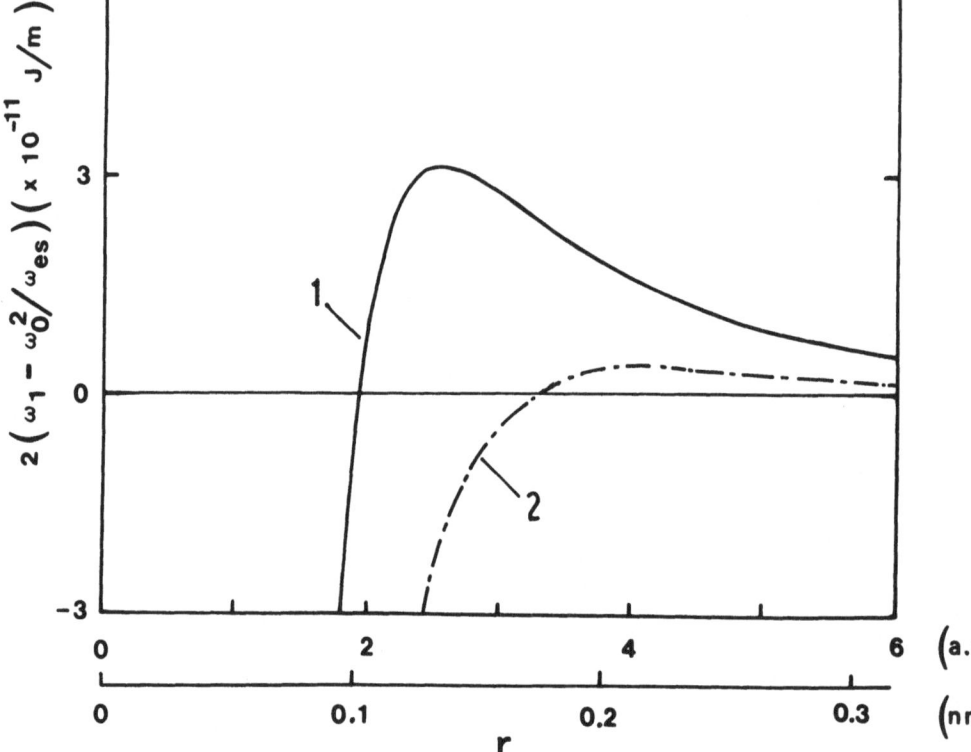

Figure 2. Quantity, $2(\omega_1-\omega_0^2/\omega_{es})$, which is proportional to the electronic gradient energy coefficient as shown in Fig. (19), is plotted vs. the average Winger-Seitz radius, r_0. Quantities, ω_{es}, ω_0 and ω_1 are defined in Figs. (10)-(12). Curves 1 and 2 are obtained with von Weiszäcker coefficient, $w = 0.45/8$ a.u. and $1/72$ a.u., respectively.

$u''_0 \tilde{=} u''_e$ is assumed for the Cs-Rb system, the critical temperature for the expected miscibility gap is 27K in a regular solution model. This indicates that it is practically impossible to observe such miscibility gaps in systems with r_0 larger than 5.0 a.u.

4. CONCLUSION

A simple model has been discussed for obtaining electronic contributions to mixing- and gradient-energy in a composition-modulated alloy system. Electronic contribution to the mixing energy decreased as the average Wigner-Seitz radius, r_o, increased. Sign of the electronic mixing energy changed from positive to negative around r_o = 5.0 a.u. Electronic contribution to the gradient energy was highly model dependent. Sign of the contribution might well be positive for systems with r_o >3.3 a.u. Low

electron-density systems such as Cs-Rb, Cs-K and Rb-K with r_o > 5.0 a.u. were predicted to be clustering systems with very low critical temperatures for the miscibility gap.

ACKNOWLEDGMENT

The author would like to thank Professor H. Ogata of the University of Windsor and Dr. H. E. Cook of Ford Scientific Research Laboratory for their valuable discussions. The present work was supported by the Natural Science and Engineering Research Council Canada under Grant No. A1290.

REFERENCES

1. Cahn, J. W. and J. E. Hilliard, Free Energy of a Nonuniform System, I. Interfacial Free Energy, J. Chem. Phys. 28 (1958) 258-267.
2. Cahn, J. W. and J. E. Hilliard, Free Energy of a Nonuniform System, III. Nucleation in a Two-Component Incompressible Fluid, J. Chem. Phys. 31 (1959) 688-699.
3. Cahn J. W., On Spinodal Decomposition, Acta Met. 9 (1961) 795-801.
4. Hohenberg, P. and W. Kohn, Inhomogeneous Electron Gas, Phys. Rev. 136 (1964) B864-B871.
5. Yamauchi, H., An Electronic Contribution to Gradient Energy of a Nonuniform Metallic Mixture, Scripta Met. 17 (1983) 83-86.
6. Yamauchi, H., Wavelength Dependence of Electronic Excess Energy of a Composition-Modulated Metallic Mixture, Phys. Rev. B 27 (1983) 905-909.
7. Pines D., Elementary Excitations in Solids (W. A. Benjamin, New York, 1964), p. 94.
8. Nozières and D. Pines, Correlation Energy of a Free Electron Gas, Phys. Rev. 111 (1958) 442-454.
9. Gell-Mann, M. and K. A. Brueckner, Correlation Energy of an Electron Gas at High Density, Phys. Rev. 106 (1957) 364-368.
10. Robinson, G. G. and P. F. de Chatel, Charge Distribution in Heterovalent Alloys, J. Phys. F: Metal Phys. 5 (1975) 1502-1511.

11. Lang, N. D., The Density-Functional Formalism and the Electronic Structure of Metal Surfaces, in Solid State Physics, Vol. 28, Ed. H. Ehrenreich, F. Seitz and D. Turnbull (Academic Press, New York, 1973) pp 225-300.

MECHANICAL BEHAVIOR OF SOLID FILM ADHESIVES
WITH SCRIM CARRIER CLOTHS

Erol Sancaktar

Mechanical and Industrial Engineering Department
Clarkson College of Technology
Potsdam, New York 13676, U.S.A.

Recent advances in composite and lightweight material technolo-
gies have led to the increased use of adhesives in aerospace, auto-
motive, and naval industries. Adhesive bonding processes are being
adapted in assembly-line manufacturing procedures. Even where rivet-
ing is possible, adhesives are sometimes used in conjunction with it
as a secondary means of load support. Structural adhesives are often
provided in the solid film form with scrim carrier cloths and latent
hardeners (particularly for the aerospace industry) to facilitate the
bonding procedure. The scrim cloth is usually ~0.01 cm thick and is
used as a carrier for the adhesive as well as for bondline thickness
control. The polymeric adhesive is applied onto this scrim cloth
building its thickness up to 0.020-0.025 cm. A variety of materials
such as polyester and glass can be used as carrier cloth. The adhesive material is either a thermosetting or thermoplastic polymer.
The bonding procedure usually requires the use of chemical primers on
adherend and carrier cloth surfaces to avoid interfacial failures.

This paper presents a general overview on the mechanical behavior of solid film adhesives with carrier cloths by focusing on
four aspects of their mechanical behavior:

1 - Constitutive equations for solid film adhesives with carrier
      cloths applied on single lap specimens with 0.127-0.145 cm thick,
      2.54 cm wide metal (aluminum or titanium alloys) adherends with
      0.76-1.27 cm overlap length;
2 - The effects of cure conditions on the mechanical behavior of
      solid film adhesives;
3 - Adhesive fracture considerations with attention to the effects
      of cure conditions on adhesive fracture;
4 - The effects of carrier cloth and interfacial weaknesses on the
      failure of adhesively bonded joints.

Experimental data which will be presented or referred to in this paper were obtained with the use of following model adhesives: Metlbond, FM 73, FM 300, LARC 3, LARC 13, Thermoplastic Polyimide-sulfone.

1 - Metlbond is a solid film, modified nitrile epoxy resin, supplied in two forms: Metlbond 1113 and 1113-2, the former of which is supported with synthetic fabric carrier cloth. It has a service temperature range of 93°C to 143°C. Metlbond is manufactured by Whittaker, Narmco Materials Division (1).

2 - FM 73 is a modified epoxy adhesive film with polyester knit fabric carrier cloth. It is manufactured by the American Cyanamid Company. Its product information brochure reports a service temperature range of -55°C to 120°C with high moisture resistance. The same adhesive is also available with random polyester mat carrier under the FM 73M brand name (to offer better handling characteristics) (2).

3 - FM 300 is a modified epoxy adhesive manufactured by the American Cyanamid Company. It is supported with a "tricot-knit" polyester carrier. Manufacturer's product information brochure (2) reports superior metal-to-metal peel strength and service temperatures to 150°C along with moisture and corrosion resistance. It is opaque to x-ray. The same adhesive is also available with "random mat" polyester carrier under the brand name FM 300M and with "wide-open knit" polyester carrier under the brand name FM 300K. The use of "random mat" carrier reduces tendency to trap air during cure and provides good bondline and flow control. The manufacturer's brochure (2) also reports that the highest overall performance is obtained with the use of "wide-open knit" carrier.

4 - LARC 3 is a linear high molecular weight polyimide adhesive developed at NASA Langley Research Center for high temperature applications. It is prepared in tape form with style 112 E glass cloth (A-1100 finish) and aluminum powder filler. It can be used in thermal environments with temperatures up to 288°C (3).

5 - LARC 13 is a thermosetting addition-polyimide adhesive with service temperatures in excess of 260°C. It is prepared in tape form on type 112 E-glass carrier cloth with an A-1100 finish. The resin contains 30% w/w (325 mesh) aluminum powder filler (4).

6 - Thermoplastic Polyimidesulfone is a novel adhesive developed at NASA Langley Research Center. It has thermoplastic properties and solvent resistance. It is prepared in tape form on 112 E-glass which has an amino-silane surface treatment. It can be used in thermal environments with temperatures up to 232°C (5).

The experimental data which will be presented in this paper were obtained during previous investigations. Information on the experimental procedures involved should be obtained from the references.

CONSTITUTIVE EQUATIONS

Adhesives are usually molecular high polymers. The mechanical stress-strain properties of most polymers are strongly influenced by time and temperature. This implies that either bulk or bonded mechanical stress-strain properties of adhesives would be expected to be similarly viscoelastic. Previous investigations by Sancaktar et al. (6,7,8,9) revealed that the stress-strain behavior of thermosetting adhesives with carrier cloths (such as Metlbond, FM 73 and 300, LARC 3 and 13 adhesives) could all be characterized with the use of viscoelastic models. The effect of carrier cloth on the mechanical behavior of Metlbond thermosetting adhesive was first investigated by Brinson et al. (10) with the use of bulk tensile coupons. Their results revealed that addition of the carrier cloth increased the bulk adhesive strength. An average of ~17% increase was observed in the maximum strength of the adhesive with the addition of carrier cloth (Metlbond 1113). The maximum strain values, however, were reduced by ~20% when the carrier cloth was used. Brinson et al. also concluded that carrier cloth tended to stabilize the mechanical behavior of the neat resin. This conclusion was based on the observations of consistent failure strains and constant values of the Poisson's ratios above the stress-whitening level. Typical viscoelastic phenomena such as relaxation, creep, and creep to failure (delayed failure) were observed for the adhesive with the carrier cloth. Creep failure strain and creep rate values, however, were both reduced with the addition of carrier cloth.

The effect of carrier cloth on toughness of Metlbond adhesive is illustrated in Figure 1, where the area under the tensile stress-strain curve is plotted as a function of the elastic strain rate. Maximum in toughness appear for both adhesive forms (with and without carrier cloth), but shifts from ~10^{-3}% sec^{-1} for epoxy alone to ~10^{-1}% sec^{-1} for the epoxy on carrier cloth. The high fiber stiffness apparently contributes resistance to rate effects. On the other hand, interfacial sites for crack initiation lower toughness by ~20% on the average. The maxima indicate ductile-to-brittle transition phenomena, confirmed by our previous fractographic investigation (7).

Previous investigations by Sancaktar et al. on five different thermosetting adhesives with carrier cloths revealed that viscoelastic modified Bingham and Chase-Goldsmith models could be used successfully to characterize the constant strain rate stress-strain behavior of the adhesives in the bonded form. The modified Bingham model was first used by Brinson et al. for characterizing the bulk tensile behavior of polycarbonate and Metlbond adhesives (10). The same model was later used by Sancaktar for characterizing the bulk and bonded shear behavior of Metlbond adhesives (6,7) and the bonded shear behavior of FM 73 adhesive (9). In order to be valid for the pure shear mode, the one-dimensional constitutive equation for the modified Bingham model was applied by Sancaktar in the form:

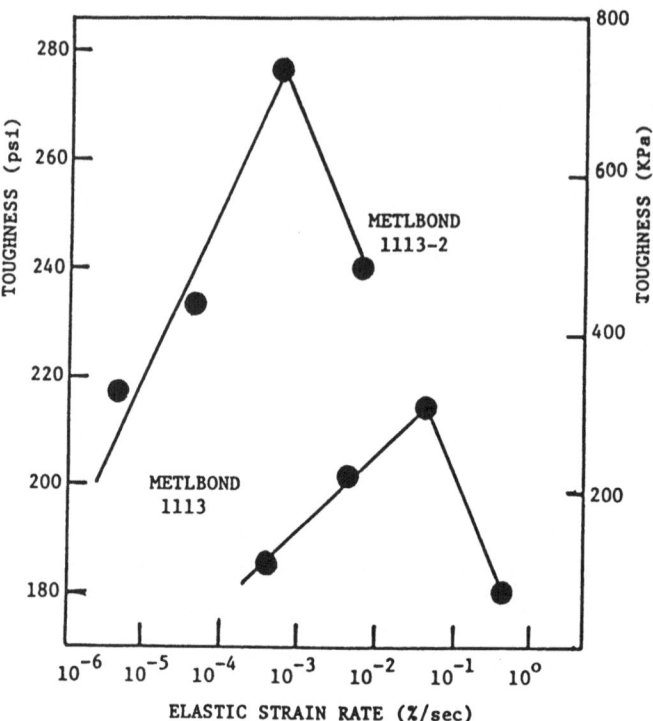

Fig. 1: Area under the stress-strain curve as a function of test rate (7).

$$\overset{\circ}{\gamma} = \overset{\circ}{\tau}/G \qquad\qquad\qquad \tau \leq \theta$$
$$\overset{\circ}{\gamma} = (\overset{\circ}{\tau}/G)+(\tau-\theta)/\mu \qquad\qquad \theta < \tau \leq \tau_{max} \qquad (1)$$

where τ and γ are shear stress and strain respectively, and θ is the elastic limit shear stress. A sketch of the mechanical model is shown in Figure 2A. The constant strain rate stress-strain relation based on the modified Bingham model is obtained by solving the constitutive equations 1 according to the condition:

$$\overset{\circ}{\gamma} = R = constant, \qquad\qquad\qquad (2)$$

to result in:

$$\tau = G\gamma \qquad\qquad\qquad \tau \leq \theta$$
$$\tau = \theta + \mu \overset{\circ}{\gamma}\{1-\exp[-(\gamma-\phi)G/\mu\overset{\circ}{\gamma}]\} \qquad \theta < \tau \leq \tau_{max} \qquad (3)$$

where ϕ is the elastic limit strain. The creep relation is obtained by using a step loading

$$\tau(t) = \tau_o H(t) \tag{4}$$

where $H(t)$ represents the unit step function and τ_o is the level of constant stress, with the initial condition

$$\gamma(t=o) = \tau_o/G_o \quad , \tag{5}$$

to obtain:

$$\gamma(t) = (\tau_o/G) + (\tau_o-\theta)t/\mu \qquad \tau > \theta \ . \tag{6}$$

Figure 2A shows the application of modified Bingham model (equations 3) to describe the constant strain rate stress-strain behavior of FM 73 adhesive in the bonded form. Examination of Figure 2A reveals the presence of a perfectly plastic flow region depicted by the horizontal portion of the stress-strain curve. Some thermosetting adhesives do not exhibit such perfectly plastic flow behavior even though they are viscoelastic in nature. Their stress-strain curves usually contain a monotonically increasing viscoelastic region (Figure 2B). Previous investigations by Sancaktar et al. on two different types of thermosetting adhesives with carrier cloths (LARC 3 and FM 300) revealed that Chase-Goldsmith model could be used to characterize the (room temperature) mechanical behavior of such adhesives (8).

K. W. Chase and W. Goldsmith modified the three parameter solid model in 1974 to describe the constant strain rate stress-strain behavior of a polyester-styrene copolymer in the bulk tensile mode (11). The same model was later used by Sancaktar et al. to describe the bonded shear behaivor of LARC 3 and FM 300 adhesives (8). The Chase-Goldsmith model (Figure 2B) constitutive equations to apply to shear are:

$$
\begin{aligned}
\overset{o}{\gamma} &= \overset{o}{\tau}/G_o & \tau &\leq \tau_s \\
\overset{o}{\gamma} &= (\overset{o}{\tau}/G_o) + (\tau-\tau_s)/\bar{\eta} & \tau &> \tau_s
\end{aligned}
\tag{7}
$$

where $\tau_s = G_o(\tau_y+G_1\gamma)/(G_o+G_1)$ and $\bar{\eta} = G_o\eta/(G_o+G_1)$. To obtain the constant strain rate relation, the above constitutive equations are solved according to the condition given by equation 2 to result in:

$$\tau = G_o\gamma \qquad\qquad\qquad \tau \leq \tau_s$$

$$\tau = G_o\{(\tau_y+\bar{\eta}R+G_1\gamma)(1-\exp[-\alpha(\gamma-\gamma_y)])\}/(G_o+G_1) \tag{8}$$

$$+ [(\tau_y)+G_oG_1(\gamma-\gamma_y)/(G_o+G_1)]\exp[-\alpha(\gamma-\gamma_y)], \qquad \tau > \tau_s$$

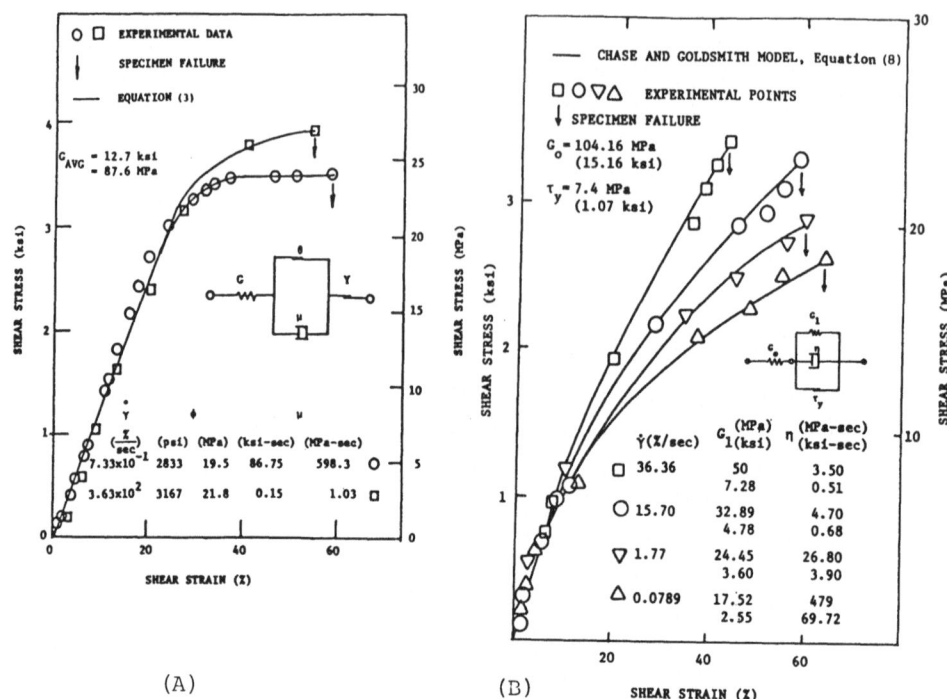

Fig. 2: Constant strain rate stress-strain behavior of (A) FM 73 (9), and (B) LARC 3(8) adhesives.

where $\gamma_y = \tau_y/G_o$ and $\alpha = G_o/\bar{\eta}R$. The creep relation is obtained by applying the conditions described in equations 4 and 5 to result in:

$$\gamma(t) = \{\tau_y[\exp(-G_1 t/\eta)-1]/G_1\} + \{\tau_o[1-\exp(-G_1 t/\eta)]/G_\infty\}$$
$$+ \{\tau_o[\exp(-G_1 t/\eta)]/G_o\} \qquad \tau > \tau_s \tag{9}$$

Figure 2B shows the application of Chase-Goldsmith model (equations 8) to describe the constant strain rate stress-strain behavior of LARC 3 adhesive in the bonded form. It should be noted that the use of rate dependent viscosity coefficient (μ and η) as shown in Figures 2A and 2B renders both modified Bingham and Chase-Goldsmith models nonlinear.

The rate dependence of the shear rupture stresses in constant strain rate tests can be described with a semi-empirical equation proposed by Ludwik (10) in the form

$$\tau_{max} = \tau' + \tau'' \text{ Log } (\mathring{\gamma}/\mathring{\gamma}') \tag{10}$$

where τ_{max} is the maximum shear stress, $\overset{o}{\gamma}$ is the initial elastic shear strain rate and τ', τ'' and $\overset{o}{\gamma}'$ are material constants. This equation was used successfully for metals and polymers in many previous investigations, some of which included structural adhesives in the bulk tensile, shear and bonded lap shear modes (6,7,8,9,10). It should be noted that equation 10 can also be used for describing the rate dependence of the elastic limit stresses (6,10).

Viscoelastic adhesives are likely to fail as a result of creep rupture when subjected to a high level of constant load. A rational mathematical characterization of creep rupture phenomenon has been proposed by Crochet (12). He used a decaying exponential relation of the form

$$Y(t) = A + B \exp(-CX) \tag{11}$$

to describe the delayed failure phenomenon. In equation 11, $Y(t)$ is the time dependent maximum stress, A, B, and C are material constants and X is a time dependent material property given by

$$X = [(\varepsilon_{ij}^{V} - \varepsilon_{ij}^{E})(\varepsilon_{ij}^{V} - \varepsilon_{ij}^{E})]^{1/2} \tag{12}$$

where ε_{ij}^{V} and ε_{ij}^{E} refer to viscoelastic and elastic strains respectively. Equation 11 can be interpreted for the pure shear condition. Use of a viscoelastic model is necessary, however, in order to obtain the material property X. Since the delayed failure occurs in those elements loaded up into the viscoelastic region, one needs to subtract the elastic strain (defined by equation 5) from the model's creep equation (equation 6 or 9), which describes viscoelastic strains, to subsequently obtain X. Figure 3 shows the application of equation 11 based on modified Bingham (or Maxwell) model to describe the delayed failure behavior of FM 73 adhesive in the bonded form and at various environmental temperatures. The asymptotic value A of equation 11 (Figure 3) is an important design parameter, as it represents the maximum safe stress value below which creep failures are not expected to occur. Crochet's equation does not contain any terms to account for thermal effects. One can, however, express the asymptotic value (A) as a function of the applied temperature to represent experimental data in an empirical fashion.

If the viscoelastic effects on adhesive behavior is determined to be weak, then a nonlinear elastic relation can be used to fit the stress-strain data. The power function relation expressed for the state of pure shear in the form:

$$\tau = K \gamma^{n} \tag{13}$$

where K and n are material constants, is proposed by the author for this purpose. The Thermoplastic Polyimidesulfone adhesive provides a good example for this type of behavior. A previous investigation

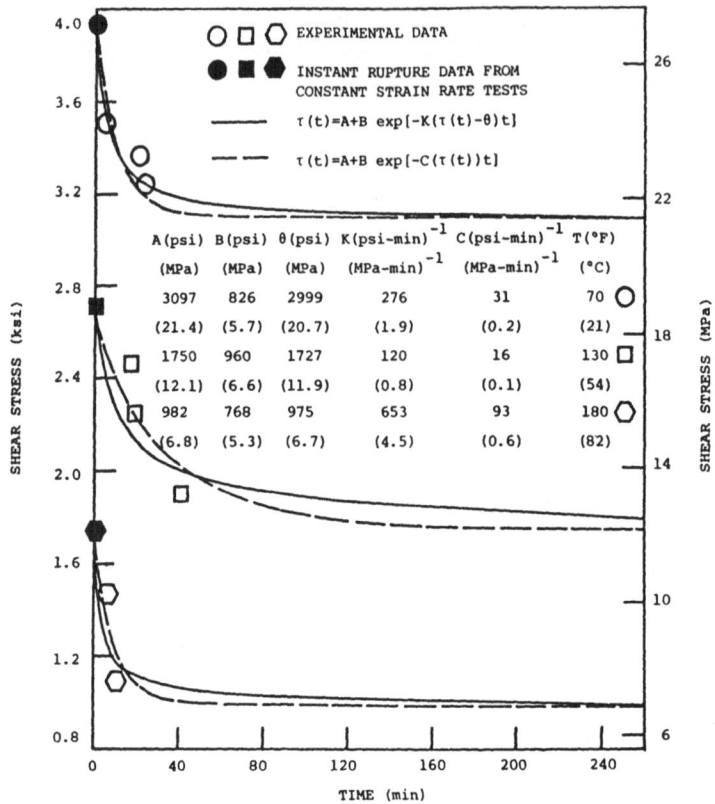

Fig. 3: Creep rupture data and comparison with Crochet's equation for FM 73 adhesive (9).

by Sancaktar et al. revealed that Thermoplastic Polyimidesulfone exhibits weak rate and time dependence (9). Because of this reason, the observed mechanical behavior of Polyimidesulfone can be considered to be closely (nonlinear) elastic.

EFFECTS OF CURE CONDITIONS

The tensile strength and rigidity of adhesives are affected by cure temperature, time and cool-down conditions. The effect of cure temperature and time on the bulk tensile strength of Metlbond 1113 is shown in Figure 4. A cure temperature vs. tensile strength curve is shown for each cure time. The typical "bell-shaped" temperature-strength curves exhibit the increasing-decreasing behavior of the tensile strength with respect to the cure temperature and reveal the existance of a maximum in such behavior. Figure 4 also shows the "optimization" (envelope) curve for the adhesive strength, with respect to the cure temperature and time. The optimization curve is obtained by connecting the maximum points on each bell-shaped curve. It exhibits a decrease in tensile strength with increasing cure temperatures and decreasing cure times. The author attributes this behavior to the increase in randomness of the adhesive crosslink network resulting in lower values of tensile strength. Apparently, high temperature short time conditions which were applied are far from approximating steady conditions which are better represented by the low temperature long time conditions. Degradation of the latent hardener and the polymer structure itself at high temperatures is also believed to result in lower values of tensile strength. The author also believes that the magnitudes of the thermal residual stresses which are intensified due to the presence of the carrier cloth are increased at high temperature short time cure conditions, resulting in lower values of tensile strength. The effects of cool-down conditions on the bulk tensile strength-cure optimization behavior of Metlbond 1113 and 1113-2 are shown in Figure 5. Evidently, the slow cool-down condition yields a greater tensile strength in comparison to the fast cool-down condition. The thermal residual stresses that develop during the initial cure process are relieved during the slow cool-down post-cure period, resulting in a more stable and favorable

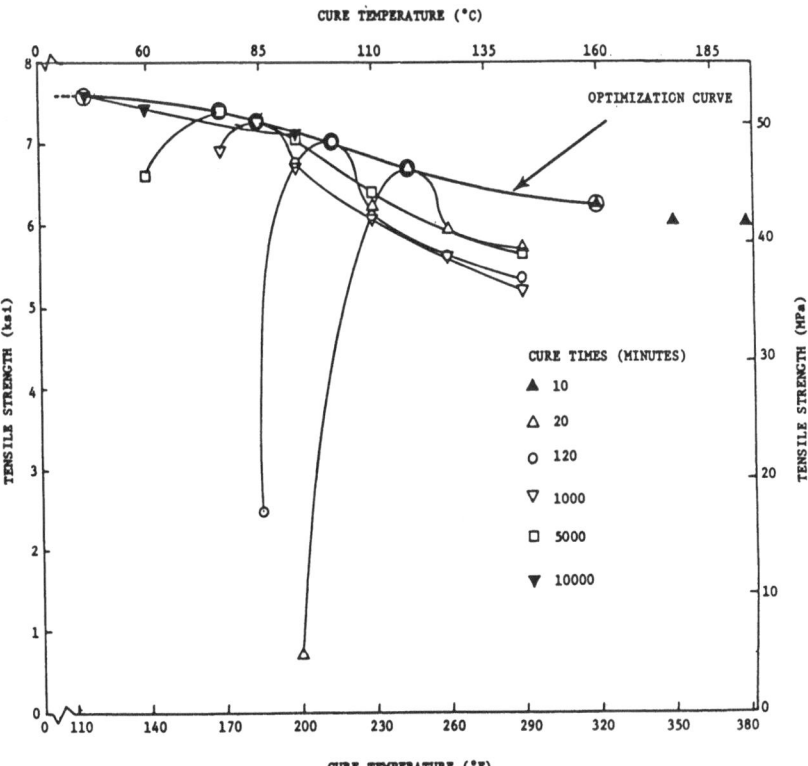

Fig. 4: Metlbond 1113 strength-cure optimization curve for the fast cool-down condition (13).

576

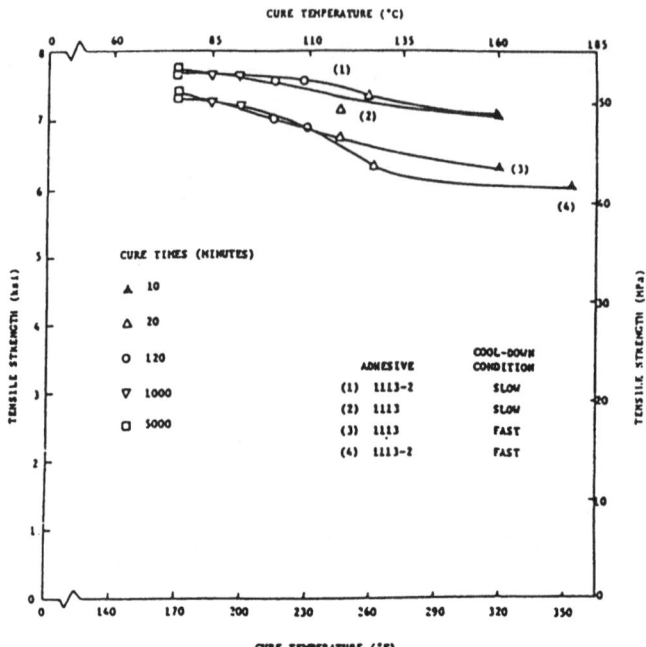

Fig. 5: Effect of carrier cloth on the strength-cure optimization
behavior of Metlbond adhesive (13).

molecular network structure. Figure 5 also shows that the difference
in the tensile strengths obtained with the fast and slow cool-down
conditions is greater at high cure temperatures and short cure times
than at low cure temperatures and long cure times. Apparently, the
thermal residual stresses that develop during the initial cure process are lower to start with at low temperature and long time cure
conditions. In other words, the effects of low temperature-long time
cure and slow cool-down are similar.

Figure 5 also reveals the effects of carrier cloth on adhesive
tensile strength with the fast and slow cool-down conditions. For
the slow cool-down condition the carrier cloth does not show any
significant effect on tensile strength. This may be attributed to
the effect of the slow cool-down condition in the elimination of
thermal residual stresses. For the fast cool-down condition the
carrier cloth again does not have any effect on the tensile strength
up to (about) 110°C cure temperature. This behavior is similar to the
above mentioned behavior with slow cool-down as low temperature-long
time conditions are similar to slow cool-down conditions. Past 110°C
cure temperature, (with fast cool-down) the carrier cloth yields
higher tensile strength as its stiffening effect becomes dominant.
This behavior is in agreement with the results of Brinson (10).

FRACTURE CONSIDERATIONS

Previous work by Sancaktar (6,7,8) showed that the failure of adhesively bonded joints is generally caused by catastrophic crack propagations originating from inherent flaws and impurities. Consequently, it becomes necessary to characterize adhesive failures with the use of fracture stresses rather than the yield stresses. A measure of the fracture stress that occurs in the vicinity of a crack tip is the strain energy release rate. The most common mode of failure for bonded joints (particularly for the single lap joint geometry) is the opening (cleavage) mode (Mode I) (14). The load level which causes a spontaneous crack propagation in cleavage mode is called the Mode I critical load, and the corresponding strain energy release rate is called the Mode I critical strain energy release rate (G_{IC}). Inherent flaws usually cause the bondline to fail in a brittle manner (Figure 6). For such failures, the use of a linear elastic fracture mechanics (LEFM) approach can be considered appropriate for characterizing the fracture stresses. An elastic-plastic material behavior assumption has been used widely for structural adhesives by investigators such as Bascom (14). Bascom used a crack tip plane strain assumption to relate the adhesive yield stress (σ_y), tensile modulus (E), Poisson's ration (ν), and the radius of the crack tip deformation zone (r_y) to the Mode I critical strain energy release rate with the following equation:

$$G_{IC} = 6\pi(1-\nu^2)r_y\sigma_y^2/E \tag{14}$$

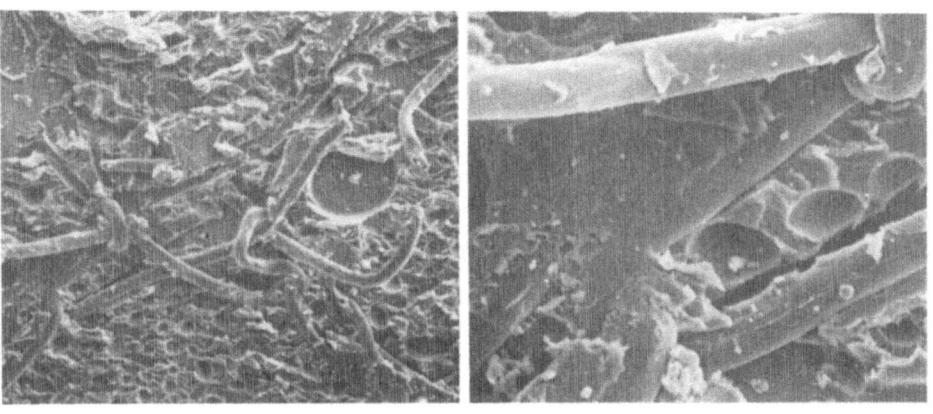

4.1x10^{-1}mm^2 4.1x10^{-3}mm^2

Fig. 6: Scanning electron microscope photomicrographs showing brittle post-fracture surfaces of Metlbond 1113 adhesive layer in the bonded form (7).

One can calculate the G_{IC} values directly from the tensile properties, based on equation 14, if information on the radius of the crack tip deformation zone (r_y) is available. Bascom reports the magnitude of r_y as 1.78×10^{-1} mm for a carboxy-terminated butadinene-acryloni-trile epoxy (14).

The involvement of bulk tensile properties σ_y and E in equation 14 indicates that the adhesive fracture stress is affected by the cure conditions. The presence of carrier cloth also affects the fracture process. A previous investigation on the Metlbond adhesive (15) reported that slow crack propagation (a result of ductile fracture) occured along a plane of fiber while fast crack propagation (a result of brittle fracture) was confined to planes without fiber. Table 1 shows the critical strain energy release rate values for the extreme cure conditions of Figure 4, calculated with the use of equation 14 and by assuming $r_y \approx 1.78 \times 10^{-1}$ mm (14). Apparently, the optimum G_{IC} value (i.e. the highest G_{IC} value representing the least favorable condition for a crack to propagate) is obtained with the high temperature-short time cure condition and in the absence of carrier cloth. This means that the presence of carrier cloth increases the likelihood of catastrophic crack propagation. Examination of the G_{IC} values obtained with 160°C, 10 minutes cure and slow cool-down conditions indicates a 15% increase in G_{IC} for the adhesive without carrier cloth. Also note that the slow cool-down condition does not improve G_{IC} values in the presence of carrier cloth as much as it does in the absence of it.

A BRIEF DISCUSSION ON INTERFACIAL FAILURES

In adhesively bonded joints interfacial failures may occur in the form of fiber/matrix or adhesive/adherend failures. As mentioned in the previous section, the presence of carrier cloth creates a more favorable condition for catastrophic crack propagation. Such crack

1113 Cured at 46°C for 5000 Min. Fast Cool: $G_{IC} = 5023$ J/m^2 Slow Cool: $G_{IC} = 5382$ J/m^2	1113 Cured at 160°C for 10 Min. Fast Cool: $G_{IC} = 5280$ J/m^2 Slow Cool: $G_{IC} = 5839$ J/m^2
1113-2 Cured at 77°C for 5000 Min. Fast Cool: $G_{IC} = 5237$ J/m^2 Slow Cool: $G_{IC} = 6018$ J/m^2	1113-2 Cured at 160°C for 10 Min. Fast Cool: $G_{IC} = 5111$ J/m^2 Slow Cool: $G_{IC} = 6699$ J/m^2
Calculated G_{IC} Values ($r_y \approx 1.78 \times 10^{-1}$ mm Assumed)	

Table 1

propagations are likely to occur at fiber/matrix interfaces. Figure 7 illustrates physical consequences of this phenomenon. The large strain discontinuities observed are likely the result of crack propagation and arrest mechanisms as rupture did not follow.

The effect of chemical surface treatment on the carrier cloth and/or the adherends may deteriorate at elevated temperatures. Figure 8 illustrates the consequence of such deterioration. The extent of interfacial separation increases with an increase in the environmental temperature. The gradual darkening of the post-failure overlap areas with increasing temperatures is an indication of adhesive matrix layers which have separated from the adherend and/or the carrier cloth.

One last effect worth mentioning here is the effect of interfacial impurities. They may be induced prior or subsequent to the cure process. Those that appear after the cure processes are the results of chemical environments causing catalysis at the interface. The most common impurity induced in that manner is corrosion. Impurities may also be present on the adherend or adhesive surface (this is more likely, as the adherend surfaces are almost always cleaned) prior to bonding. During a previous investigation by

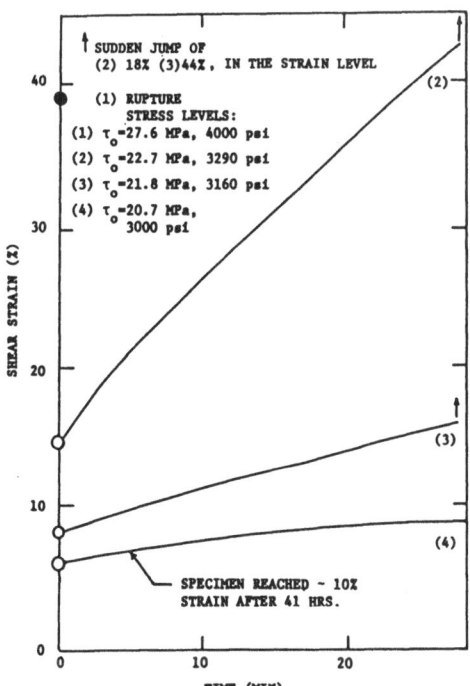

Fig. 7: Symmetric single lap creep response of Metlbond-1113 with inherent voids (6).

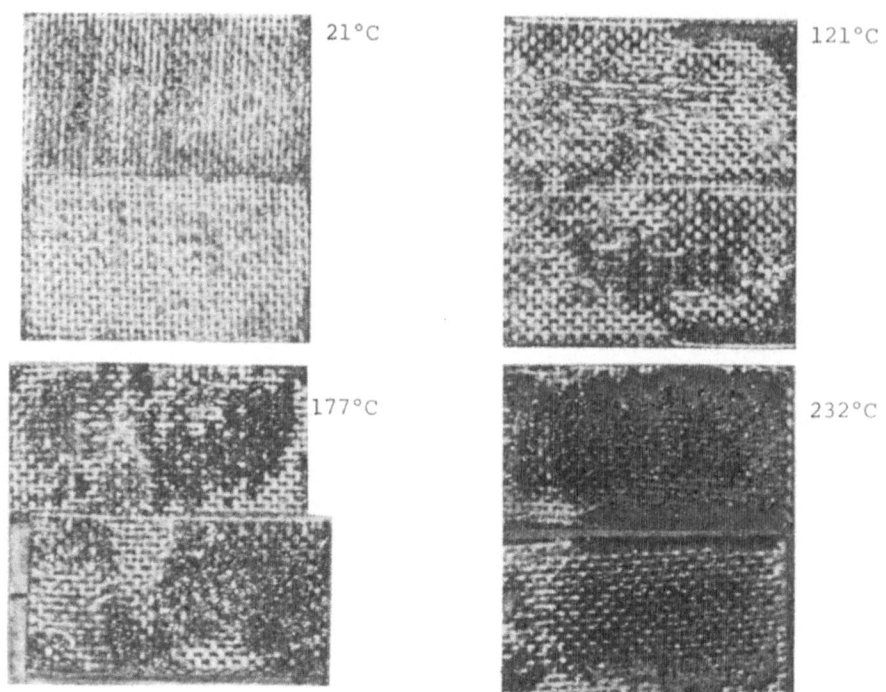

Fig. 8: Scanning electron microscope photomicrographs showing
interfacial failure of Metlbond 1113 adhesive (9).

Sancaktar EDAX (energy dispersive analysis of x-ray fluorescence)
examination of the fracture surfaces of aluminum-Metlbond 1113 specimens indicated the presence of silicon on some of them (7). Silicon
caused weak-interfacial failures in these specimens.

An engineering mechanics approach to the analysis of adhesive
joint behavior was presented in this paper. The challenging problem
of relating such methods and findings to the adhesive's chemical
structure and the intrinsic adhesion forces acting across interfaces
still awaits the chemists. The establishment of such correlations
will enable the development of better adhesives and improved surface
pretreatments.

REFERENCES

1. Metlbond Adhesive Product Information Brochure, Narmco Materials Division, Whittaker Corp., Costa Mesa, California,
   U.S.A.

2. FM 73 and FM 300 Adhesive Film Information Brochures, American Cyanamid Company. Bloomingdale Products Havre de Grace, Maryland, U.S.A.

3. Progar, D.J., and St. Clair, T., Preliminary Evaluation of a Novel Polyimide Adhesive for Bonding Titanium and Reinforced Composites," NASA Langley Research Center internal report.

4. St. Clair, T.L., and Projar, D.J., "LARC 13 Polyimide Adhesive Bonding," 24th National SAMPE Symposium, San Francisco, California, May 1979.

5. St. Clair, T.L., and Tomahi, D.A., "A Thermoplastic Polyimidesulfone," NASA Technical Memorandum 84574, Nov. 1982.

6. Sancaktar, E., and Brinson, H.F., "The Viscoelastic Shear Behavior of a Structural Adhesive," Polymer Science and Technology Series, Vol. 12-A, pp. 279-300, Plenum Press, 1980.

7. Dwight, D.W., Sancaktar, E., and Brinson, H.F., "Failure Characterization of a Structural Adhesive," Polymer Science and Technology Series, Vol. 12-A, pp. 141-164, Plenum Press, 1980.

8. Sancaktar, E., and Padgilwar, S., "The Effects of Inherent Flaws on The Time and Rate Dependent Failure of Adhesively Bonded Joints," Transactions of the ASME, Journal of Mechanical Design, Vol. 104, No. 3, pp. 643-650, 1982.

9. Sancaktar, E. and Schenck, S.C., "Material Characterization of Structural Adhesives in the Lap Shear Mode," Clarkson College of Technology, Report No. MIE-89, Jan. 1983.

10. Brinson, H.F., Renieri, M.P., and Herakovich, C.T., "Rate and Time Dependent Failure of Structural Adhesives," Fracture Mechanics of Composites, ASTM STP 593, pp. 177-199, 1975.

11. Chase, K.W., and Goldsmith, W., "Mechanical and Optical Characterization of an Anelastic Polymer at Large Strain Rates and Large Strains," Experimental Mechanics, p. 10, Jan. 1974.

12. Crochet, M.J., "Symmetric Deformation of Viscoelastic-Plastic Cylinders," Journal of Applied Mechanics, No. 33, p. 326, 1966.

13. Sancaktar, E., Jozavi, H., and Klein, R.M., "The Effects of Cure Temperature and Time on the Bulk Tensile Properties of a Structural Adhesive," Journal of Adhesion, Vol. 15, No. 3, 1983.

14. Bascom, W.D., Jones, R.L., and Timmons, C.O., "Mixed-Mode Fracture of Structural Adhesives," Adhesion Science and Technology, Vol. 9B, L.H. Lee, Editor, Plenum Press, 1975.

15. O'Connor, D.G., "Factors Affecting the Fracture Energy for a Structural Adhesive," Master's Thesis, Virginia Polytechnic Institute and State University, Blacksburg, VA, Aug. 1979.

ORIENTATIONAL PHASE TRANSITIONS IN A QUASI-ONE-DIMENSIONAL
CONDUCTOR

Constantine Mavroyannis

Division of Chemistry, National Research Council of
Canada, Ottawa, Ontario, Canada K1A 0R6

ABSTRACT. A method is presented by calculating orientational phase
transitions in quadrupolar strands with itinerant-electrons, which
arise from the coupling of the conduction electrons with librons
(the small tortional oscillations about the equilibrium orientations of the molecules). The Peierls instability has been studied
for a model of one-dimensional conductor in which the rotational
motion is described in terms of a pseudo spin S=1 Hamiltonian,
where the molecules in the solid interact through quadrupole-
quadrupole and dipole-dipole interactions. The electron and libron
fields are expressed in the second quantization representation and
the Green's function formalism is used to calculate the excitation
spectrum of the coupled system. The derived expressions for both
the electron and libron Green's functions for the coupled system
give a simple recursion formula for the Peierls transition
temperature T_c, which is driven by the loss of librons from the
Peierls condensate through thermal excitation. The recursion
formula is solved numerically for a half- filled tight-binding
conduction band and the dependence of T_c on the values of the
different parameters in the problem is studied; the numerical
results are graphically pre-sented. An expression for the
electrical conductivity is derived, which describes the collective
contribution arising from the phase oscillations of the lattice
distortion.

1. INTRODUCTION

A great deal of interest in recent years has focused on the problem of high anisotropic conductivity of certain quasi-one-dimensional charge-transfer organic salts such as TTF-TCNQ (tetra-thiafulvalene tetracyanoquinodimethane) [1-8]. Since these organic charge-transfer salts consist of molecules, which possess internal structure, the Peierls transition in these quasi-one-dimensional systems is more complicated than the simple longitudinal distortion originally considered by Peierls [9] and Frohlich [10]. The motion of the molecules in these crystals may be classified into translations, which can be either longitudinal or transverse, intermolecular distortions and librations. Unlike the first two types of motion, the Peierls transition caused by the molecular librational modes was first proposed by Morawitz [11]. Merrifield and Suna [12] pointed out that there would not be a Peierls transition if the coupling is quadratic in the libron amplitude; however, this difficulty was avoided when electronic degeneracies were taken into consideration [11].

Since Morawitz [11] suggestion, the possibility of an orienta-tional Peierls distortion due to the coupling of the conduction electrons with librons has received considerable attention [13-19]. In this model, it is assumed that at low temperatures, a one-dimensional electron system in a deformable lattice is unstable against a modulation of the lattice of wavenumber $2k_F$ (k_F is the Fermi wavenumber) which creates a gap in the excitation spectrum at the Fermi energy. The presence of the lattice modulation causes the electrons to scatter from the states k to $k \pm 2k_F$ and the libron mode of wavenumber $q=2k_F$ goes soft at the metal-insulator transition temperature.

A method is presented here by considering orientational phase transitions in one-dimensional conductors which arise from the coupling of the conduction electrons with librons [20-22]. In section 2, the rotational motion of the molecules is described in terms of a pseudo-spin S=1 Hamiltonian where the molecules in the solid interact through quadrupole-quadrupole and dipole-dipole interactions. The Hamiltonian also takes into account of the conduction-electron-band motion as well as the coupling of the conduction band to the rotation of the molecules. The electron and libron fields are then expressed in the second quantization presentation. The equations of motion for the electron and libron Green's functions are considered in sections 3 and 4, respectively. The numerical results in the mean field approximation are derived in section 5 and they are graphically presented as well. An expression for the electrical conductivity for a strand is derived in section 6 while a summary of the derived results is given in section 7.

2. THE MODEL HAMILTONIAN

The crystal is assumed to consist of neutral molecules, which are located at the lattice sites with nonoverlapping charge distributions between neighbouring lattice sites. The molecules are free to rotate about the centers of electrical charge distributions with angular momentum S. In such a system, the interaction energy between the two molecules at the neighbouring lattice sites i and j can be expressed as a sum of terms describing multipole-multipole interactions at these two lattice sites [23]. The terms of the expansion are a product of spherical harmonics whose arguments depend on the orientation of the molecules while the coefficients depend upon the details of the charge distributions. Then the Hamiltonian describing the spin-spin interaction for a d-dimensional lattice takes the form

$$H_r = -\tfrac{1}{2} \sum_{ij} \sum_{\ell\ell'>0} \sum_{mm'} V_{\ell\ell'}^{mm'}(i,j) L_\ell^m(i) L_{\ell'}^{m'}(j)$$

$$+ \sum_{ij} \sum_{\ell} V_\ell(i,j) L_\ell^0(i) \rho(j) \qquad (1)$$

where $V_{\ell\ell'}^{mm'}(i,j)$ is a generalized interaction between spins at lattice sites R_i^0 and R_j^0 and is a function only of the intermolecular distance $R_{ij} = R_i^0 - R_j^0$; it is zero if the magnitude of m and m' is bigger than the smaller of ℓ and ℓ'. $V_\ell(i,j)$ is the interaction between the 2^ℓ pole at the lattice site i and the charge density $\rho(j)$. The spherical tensor operator [24] $L_\ell^m(i)$ of rank ℓ ($\ell = 0,1,\ldots,2S$) associated with the site i transforms under rotations like a spherical harmonic and, for $\ell = 1,2$ is given by [25-27]

$$L_1^0(i) = S_i^z/\sqrt{f_1}, \quad L_1^1(i) = S_i^+/\sqrt{2f_1}, \quad L_2^0(i) = \left[(S_i^z)^2 - \tfrac{1}{3}S(S+1)\right]/\sqrt{f_2},$$

$$L_2^1(i) = (S_i^z S_i^+ + S_i^+ S_i^z)/\sqrt{6f_2}, \quad L_2^2(i) = (S_i^+)^2/\sqrt{6f_2}, \quad (L_\ell^m)^+ = L_\ell^{-m},$$

$$f_1 = S(S+1), \quad f_2 = S(S+1)(2S+3)(2S-1)/15, \quad S_i^\pm = S_i^x \pm i S_i^y, \qquad (2)$$

where S_i^x; S_i^y and S_i^z are the x,y and z components of the spin operators at the site i, respectively.

The interaction $V_{\ell\ell'}^{mm'}(i,j)$ is anisotropic when the dependence on m and m' is taken into account as it is the case of system consisting of molecular hydrogens [28]; however, in our calculation, for the sake of simplicity, only isotropic quadrupole-quadrupole as well as dipole-dipole interactions will be considered. Then eq. (1) may be written as

$$H_r = H_{EQQ} + H_{EDD} + H_{CQ} + H_{CD} \qquad (3)$$

where H_{EQQ} describes the quadrupole-quadrupole interactions

$$H_{EQQ} = -\tfrac{1}{2} \sum_{<ij>} \sum_{m=-2}^{2} J(i,j) L_2^m(i) L_2^{-m}(j) = \frac{1}{6f_2} \sum_{<ij>} J(ij) \left[\vec{S}_i \cdot \vec{S}_j \right.$$

$$\left. +2(\vec{S}_i \cdot \vec{S}_j)^2 \right] + \frac{Nf_1^2}{9f_2} J_0 . \qquad (4)$$

$\vec{S}_i$ and $\vec{S}_j$ are the vector spin operators at the lattice sites i and
j, respectively, J(ij) is the isotropic coupling between nearest
neighbours <ij>, N is the number of lattice sites and J_0 is the
zero wave vector Fourier transform of J(i,j). The interaction
Hamiltonian (4) for H_{EQQ} is rotationally invariant in spin space
and, hence for d=1 and d=2, H_{EQQ} cannot have rotational ordering
[29-31]. The isotropic dipole-dipole part of the Hamiltonian

$$H_{EDD} = -\tfrac{1}{2} \sum_{<ij>} \sum_{m=-1}^{1} \Gamma(i,j) L_1^m(i) L_1^{-m}(j) = -\frac{1}{2f_1} \sum_{<ij>} \Gamma(i,j) \vec{S}_i \cdot \vec{S}_i , \qquad (5)$$

where $\Gamma(i,j)$ is the nearest neighbour interaction energy for two
dipoles at i and j. The isotropic dipole-dipole interaction H_{HDD}
is characterized by the absence of long-range order [29-32]. The
charge-quadrupole (CQ) interaction

$$H_{CQ} = (\tfrac{2}{3})^{\frac{1}{2}} \sum_{ij} g(i,j) L_2^0(i) \rho(j) = (\frac{2}{3f_2})^{\frac{1}{2}} \sum_{ij} g(i,j) \left[(s_i^z)^2 \right.$$

$$\left. - \tfrac{1}{3} f_1 \right] \rho(j) , \qquad (6)$$

and the charge-dipole (CD) interaction

$$H_{CD} = \sqrt{2} \sum_{ij} \lambda(i,j) L_1^0(i) \rho(j) = (2/f_1)^{\frac{1}{2}} \sum_{ij} \lambda(i,j) s_i^z \rho(j) , \qquad (7)$$

break the rotational symmetry of our model and play important roles
in our problem. The functions g(i,j) and λ(i,j) describe the CQ
and CD interaction energies, respectively.

In addition to the electrostatic interactions between the
molecules, there is the electron Hamiltonian

$$H_e = \sum_{<ij>\sigma} t(i,j) \alpha_{i\sigma}^+ \alpha_{j\sigma} , \qquad (8)$$

which describes the band motion of the electrons in the lattice;
t(i,j) is the intermolecular electron-transfer integral which takes
an electron between a pair of nearest-neighbour sites <ij> while
the Fermi-operators $\alpha_{i\sigma}^+$ and $\alpha_{i\sigma}$ create and annihilate an electron

of spin σ on the lattice i. Hence, our total anisotropic Hamiltonian is given by

$$H = H_e + H_r = H_e + H_{EQQ} + H_{EDD} + H_{CQ} + H_{CD} , \tag{9}$$

where H_{CQ} and H_{CD} may be considered as describing the anisotropy of our system.

For molecules of spin S=1, we may express the spherical tensor operators $L_1^m(i)$ and $L_2^m(i)$ in terms of the pseudo-spin operators a_i and b_i, which cause transitions from the excited states to the ground state, defined as (33,34)

$$a_i = [L_1^{-1}(i) + L_2^{-1}(i)]/\sqrt{2} , \quad b_i = [L_1^1(i) - L_2^1(i)]/\sqrt{2} , \tag{10}$$

while the creation operators a_i^+, b_i^+ are determined from the Hermitian conjugates of a_i and b_i, respectively, given by (10). In terms of these operators, we have

$$L_1^1(i) = (a_i^+ + b_i)/\sqrt{2} , \quad L_1^0(i) = (n_i - m_i)/\sqrt{2} = S_i^z/\sqrt{2} ,$$

$$L_2^0(i) = (n_i + m_i - 2/3)\sqrt{3/2} , \quad L_2^2(i) = a_i^+ b_i , \quad n_i = a_i^+ a_i , \quad m_i = b_i^+ b_i \tag{11}$$

The assumption that the charge is distributed with cylindrical symmetry about the z axis leads to the result that the average values of the number operators n_i and m_i are equal, namely, $\langle n_i \rangle = \langle m_i \rangle$. Then the Hamiltonian (9) in terms of the pseudo spin operators (10) and (11) takes the form [31]

$$H_{EQQ} = \sum_{\langle ij \rangle} J(ij)[n_i n_j/3 + (a_i^+ - b_i)(a_j - b_j^+)/2 + a_i^+ b_i b_j^+ a_j] , \tag{12}$$

$$H_{EDD} = -\tfrac{1}{2} \sum_{\langle ij \rangle} \Gamma(ij)[(a_i^+ + b_i)(a_j + b_j^+) + \tfrac{1}{2}(n_i - m_i)(n_j - m_j)] , \tag{13}$$

$$H_{CQ} = \sum_{ij} g(i,j)(n_i + m_j - 2/3)\rho(j) = -\frac{2}{3} \sum_{ij} g(i,j)n_i \rho(j) , \tag{14}$$

$$H_{CD} = \sum_{ij} \lambda(i,j)(n_i - m_i)\rho(j) , \tag{15}$$

where

$$n_i = -\left(\frac{3}{2}\right)^{\frac{1}{2}} L_2^0(i) = 1 - 3(n_i + m_i)/2 . \tag{16}$$

We shall use (8),(9) and (12)-(15) to calculate the equation of motion for the Green functions of the system.

3. EQUATIONS OF MOTION FOR THE ELECTRON GREEN FUNCTION

To evaluate the equation of motion for the single-particle electron Green's function we have to consider equations of motion of the form $id\alpha_{i\sigma}/dt = [\alpha_{i\sigma}, H]$, where H is the total Hamiltonian (9) whose terms are given by (8) and (12)-(15); units in which $\hbar=1$ are used throughout. Thus in Fourier space, we have

$$id\alpha_{k\sigma}/dt = \varepsilon_k \alpha_{k\sigma} + \sum_p g_p \alpha_{k-p\sigma}\left(n_p + m_p + \frac{2}{3}\delta_{p,o}\right) + \sum_p \lambda_p \alpha_{k-p\sigma}\left(n_p - m_p\right) ,$$

$$(17)$$

where δ is the Kronecker delta, k and p are vectors in the first Brillouin zone, and

$$\alpha_{k\sigma} = (1/\sqrt{N})\sum_i e^{-ikR_i^0}\alpha_{i\sigma}, \quad \varepsilon_k = \sum_j e^{-ikR_{ij}} t(i,j), \quad n_k = (1/N)\sum_i e^{-ikR_i^0} n_i ,$$

$$(18)$$

and the definition of the Fourier transform of $g(i,j)$ and $\lambda(i,j)$ is similar to that of $t(i,j)$ in (18). We introduce the retarded double-time electron Green's function as (35,36)

$$\langle\langle \alpha_{k\sigma}(t); \alpha_{k'\sigma'}^+(t')\rangle\rangle = -i\theta(t-t')\langle[\alpha_{k\sigma}(t), \alpha_{k'\sigma'}^+(t')]\rangle , \qquad (19)$$

where the angular brackets denote the average over the canonical ensemble appropriate to the total Hamiltonian H, $\theta(t)$ is the usual step function and the operator $\alpha_{k\sigma}(t)$ and $\alpha_{k'\sigma'}^+(t')$ are in the Heisenberg representation. From (17), we find that the Fourier transform of eq. (19), $\langle\langle \alpha_{k\sigma}; \alpha_{k'\sigma'}^+\rangle\rangle \equiv \langle\langle \alpha_{k\sigma}; \alpha_{k'\sigma'}^+\rangle\rangle(\omega)$, satisfies the equation

$$(\omega - \varepsilon_k)\langle\langle \alpha_{k\sigma}; \alpha_{k'\sigma'}^+\rangle\rangle = \delta_{kk'}\delta_{\sigma\sigma'} + \sum_p g_p \langle\langle \alpha_{k-p\sigma}\left(n_p + m_p - \frac{2}{3}\delta_{p,o}\right); \alpha_{k'\sigma'}^+\rangle\rangle$$

$$+ \sum_p \lambda_p \langle\langle \alpha_{k-p\sigma}\left(n_p - m_p\right); \alpha_{k'\sigma'}^+\rangle\rangle , \qquad (20)$$

where the last two terms describe higher-order Green's functions than the single-particle one. These higher-order Green's functions may be decoupled as follows:

$$\langle\langle \alpha_{k-p\sigma}\left(n_p + m_p - \frac{2}{3}\delta_{p,o}\right); \alpha_{k'\sigma'}^+\rangle\rangle \approx \langle n_p + m_p - \frac{2}{3}\delta_{p,o}\rangle \langle\langle \alpha_{k-p\sigma}; \alpha_{k'\sigma'}^+\rangle\rangle,$$

$$(21)$$

$$\langle\langle \alpha_{k-p\sigma}\left(n_p - m_p\right); \alpha_{k'\sigma'}^+\rangle\rangle \approx \langle n_p - m_p\rangle \langle\langle \alpha_{k\sigma}; \alpha_{k'\sigma'}^+\rangle\rangle = 0 , \qquad (22)$$

where use has been made that because of the cylindrical symmetry of the molecules $\langle n_p\rangle = \langle m_p\rangle$ and, hence, the average field arising from the dipole-dipole interaction vanishes. The decoupling approximations (21) and (22) imply that all dynamic effects arising from the interactions of the electrons with the quadrupoles are discarded and that an electron sees only the average (static) field produced by this interaction. Using (20) and (21), we rewrite (20) as

$$\left(\omega-\varepsilon_k\right)<<\alpha_{k\sigma};\alpha^+_{k'\sigma'}>>=\delta_{kk'}\delta_{\sigma\sigma'}+\sum_p g_p<n_p+m_p - \frac{2}{3}\delta_{p,0}><<\alpha_{k-p\sigma};\alpha^+_{k'\sigma'}>>.$$

$$(23)$$

The expression (23) implies that the Green functions $<<\alpha_{k\sigma};\alpha^+_{k'\sigma'}>>$ and $<<\alpha_{k-p\sigma};\alpha^+_{k'\sigma'}>>$ are coupled, hence, we shall need to consider the electron Green function in a 2×2 matrix form as

$$G(k,\omega)=<<\begin{pmatrix}\alpha_{k\sigma}\\\alpha_{k+q\sigma}\end{pmatrix};\begin{pmatrix}\alpha^+_{k\sigma} & \alpha^+_{k+q\sigma}\end{pmatrix}>> = \begin{pmatrix}<<\alpha_{k\sigma};\alpha^+_{k\sigma}>> & <<\alpha_{k\sigma};\alpha^+_{k+q\sigma}>>\\<<\alpha_{k+q\sigma};\alpha^+_{k\sigma}>> & <<\alpha_{k+q\sigma};\alpha^+_{k+q\sigma}>>\end{pmatrix}.$$

$$(24)$$

We now assume that only the libron modes with wavenumber $k=0,\pm q$ are coupled to the conduction electrons in a half-filled band. $2q$ is a reciprocal-lattice vector so that q and $-q$ are equivalent. Therefore, in this case we need only to consider correlation between the electron states of wavenumber k and $k\pm q$, which are described by the Green functions given in (24). Using (23), we calculate the complete expression for the Green function $G(k,\omega)$ whose inverse is equal to

$$G^{-1}(k,\omega) = \begin{pmatrix}\omega-\varepsilon_k & -\Delta\\-\Delta & \omega-\varepsilon_{k+q}\end{pmatrix},$$

$$(25)$$

and the poles of $G(k,\omega)$, which determine the energies of excitation $E_k^{\pm}$ are given by

$$2E_k^{\pm} = \varepsilon_k+\varepsilon_{k+q}\pm\left[\left(\varepsilon_k-\varepsilon_{k+q}\right)^2+4\Delta^2\right]^{\frac{1}{2}}$$

$$(26)$$

Thus the energy levels of the electrons are split for wavenumber $k=\pm q/2$ and the lattice potential has created a gap of size 2Δ in the conduction band. For the Peierls distortion, the potential acting on the electrons has a period π/k_F and is due to librons.

The energy gap occurs at the Fermi energy and is given by

$$\Delta = 2g_q<n_q> = -2q_q\int_{-\infty}^{+\infty}d\omega\,Im\sum_k <<a_{k+q};a^+_k>>\left(e^{\omega/2T}-1\right)^{-1},$$

$$(27)$$

where Im stands for the imaginary part of the expression in question. Equation (27) implies that in order to determine the gap parameter Δ and its temperature dependence, we have to calculate the libron Green function $<<a_{k+q};a_k^+>>$ by means of the total Hamiltonian (9).

4. EQUATIONS OF MOTION FOR THE LIBRON GREEN FUNCTION

The equations of motion for the libron Green function are determined by commuting the operators a_i and b_i with the Hamiltonian (9) with the result

$$ida_k/dt = \nu_0(k)a_k + \omega_0(k)b_{-k}^+ + A_k \ , \tag{28}$$

$$-idb_{-k}^+/dt = \omega_0(k)a_k + \nu_0(k)b_{-k}^+ + B_k \ , \tag{29}$$

where $\nu_0(k)=J_0-J_k/2$, $\omega_0(k)=J_k/2$ and J_k is the Fourier transform of the EQQ interaction $J(i,j)$. The terms A_k and B_k are nonlinear functions of the pseudo spin operators, their explicit expressions are lengthy and given by eqs. (4.5)-(4.10) of (21).

The libron Green functions arising in this problem consists of a 4×4 matrix which is conveniently defined as $K(\omega)=<<y_k;y_k^+>>$, where y_k^+ is a row vector given by $y_k^+=(a_k^+b_{-k}a_{k+q}^+b_{q-k})$. Then using (28) and (29), we derive the Dyson equation as

$$K(k,\omega)=K_0(k,\omega)+K_0(k,\omega)\Pi(k,\omega)K_0(k,\omega)=K_0(k,\omega)+K_0(k,\omega)\Sigma(k,\omega)K(k,\omega) \ ,$$

$$\tag{30}$$

with the proper self-energy $\Sigma(k,\omega)$ is determined by

$$\Sigma(k,\omega) = \Pi(k,\omega)\left(I_4+K_0(k,\omega)\Pi(k,\omega)\right)^{-1} \ , \tag{31}$$

where I_4 is the unit 4×4 matrix and $K_0(k,\omega)$ is the unperturbed Green function defined as

$$K_0^{-1}(k,\omega) = \begin{pmatrix} M(k,\omega) & 0 \\ 0 & M(k,\omega) \end{pmatrix}, \quad M(k,\omega) = \frac{1}{\bar{\eta}}\begin{pmatrix} \omega-\nu_0(k) & -\omega_0(k) \\ -\omega_0(k) & -\omega-\nu_0(k) \end{pmatrix}. \tag{32}$$

In (32), 0 is the null 2×2 matrix and $\bar{\eta}$ is equal to

$$\bar{\eta} = 1-3<n_0> = (1/N)\sum_i \bar{\eta}_i \ , \quad \bar{\eta}_i = -(3/2)^{\frac{1}{2}}<L_2^0(i)> \ ,$$

$$<n_0> = (1/N)\sum_k <a_k^+a_k> = (1/N)\sum_k <b_k^+b_k> \ , \tag{33}$$

$<n_0>$ being the average number of librons in the $k=0$ wavenumber state. $K_0(k,\omega)$ given by (32) may be considered as the linear approximation $K(k,\omega)$, while $\bar{\eta}$ is a measure of the rotational ordering of the molecules and it is assumed to be different than zero. $\Pi(k,\omega)$ is a complicated expression and is given by eq.

(4.22) of [21]; it is made up of two parts: one of them consists of frequency independent terms which describe static effects, while the other part is frequency dependent involving higher order libron Green's functions describing libron-libron scattering processes [21].

It is easily seen from (32) that if $K(k,\omega)$ is approximated by $K_0(k,\omega)$ then the band gap Δ defined by (27) vanishes since the off-diagonal matrices of $K_0(k,\omega)$ are equal to the null matrix. Therefore, in order to obtain a Peierls transition, fluctuations in the libron motion have to be included. However, to complete calculation of $\Pi(k,\omega)$ is very complicated and it will not be given here. Instead, only the static part of $\Pi(k,\omega) \approx \Pi(k)$ will be taken into account which is equivalent of calculating $\Pi(k,\omega)$ or $\Sigma(k,\omega)$ in the mean-field approximation; this approach will greatly facilitate the numerical calculation.

5. NUMERICAL RESULTS IN THE MEAN-FIELD APPROXIMATION

If we neglect correlations between the spins so that each spin precesses in the mean-field produced by its neighbours, the numerical calculation of Δ and $\bar{n}$ is considerably simplified. Thus in the mean-field approximation, the libron Green function is given by [20,21]

$$K^{-1}(k,\omega) = \frac{1}{\bar{n}} \begin{pmatrix} \Lambda_1(k,\omega) & -\Lambda_2(k) \\ -\Lambda_2(k+q) & \Lambda_1(k+q,\omega) \end{pmatrix} , \qquad (34)$$

where

$$\Lambda_1(k,\omega) = \begin{pmatrix} \omega-\nu_k & -\omega_k \\ -\omega_k & -\omega-\nu_k \end{pmatrix} , \quad \Lambda_2(k) = \begin{pmatrix} \tilde{\Omega}_k & \hat{\Omega}_k \\ \hat{\Omega}_k & \tilde{\Omega}_k \end{pmatrix} . \qquad (35)$$

The frequencies $\nu_k, \omega_k, \tilde{\Omega}_k$ and Ω_k are defined as

$$\nu_k = \bar{n}\nu_0(k)+(g_0+\lambda_0)<\rho_0> + \frac{3}{2}\gamma_k<n_0>, \quad \omega_k = \bar{n}\omega_0(k) + \frac{3}{2}\gamma_k<n_0>,$$

$$\tilde{\Omega}_k = 3\left[\omega_0(k+q)-2\omega_0(q)+\tfrac{1}{2}\gamma_{k+q}\right]<n_q>+(g_q+\lambda_q)<\rho_q>,$$

$$\hat{\Omega}_k = -3\left[\omega_0(k+q)-\tfrac{1}{2}\gamma_{k+q}\right]<n_q>.$$

Equations (27) and (34) determine the gap function of the conduction band. For small gap, we may expand the determinants in (34) to the order of Δ with result

$$<<a_{k+q};a_k^+>> \approx \frac{\bar{n}}{\left(\omega^2-\Omega_k^2\right)\left(\omega^2-\Omega_{k+q}^2\right)} \{(\omega+\nu_k)(\omega+\nu_{k+q})\tilde{\Omega}_{k+q} -\omega_{k+q}\left[(\omega+\nu_k)\hat{\Omega}_{k+q}\right.$$

$$\left. -\omega_k\tilde{\Omega}_{k+q}\right]-\omega_k\hat{\Omega}_{k+q}(\omega+\nu_{k+q})\} , \qquad (36)$$

where $\Omega_k = (v_k^2 - \omega_k^2)^{\frac{1}{2}}$ is the eigenfrequency of the libron mode in the absence of a conduction–electron band gap $(20, 21, 26)$. Then from (27) and (36) after integrating over frequencies and wavevector, we derive the expression

$$\frac{\bar{n}}{N} \sum_k (\Omega_k^2 - \Omega_{k+q}^2)^{-1} \left[F(k, \Omega_k) \coth\left(\frac{\Omega_k}{2T_c}\right) - F(k, \Omega_{k+q}) \coth\left(\frac{\Omega_{k+q}}{2T_c}\right) \right] = 1, \quad (37)$$

which determines the mean–field Peierls transition temperature T_c. The function $F(k,z)$ is defined as

$$F(k,z) = z^{-1} \left[(z^2 + v_k v_{k+q} + \omega_k \omega_{k+q}) \tilde{\theta}_{k+q} - (v_k \omega_{k+q} + \omega_k v_{k+q}) \hat{\theta}_{k+q} \right], \quad (38)$$

where

$$\tilde{\theta}_k = \frac{3}{2} \left[\omega_0(k+q) - 2\omega_0(q) + \frac{1}{2}\gamma_{k+q} \right] - g_q (g_q + \lambda_q) \sum_{k'} \varepsilon_{k'}^{-1} \tanh \frac{\varepsilon_{k'}}{2T_c}, \quad (39)$$

and $\hat{\theta}_k = (-3/2) \left[\omega_0(k+q) - \frac{1}{2}\gamma_{k+q} \right]$. At T_c, $\Delta = 0$ then from (34) we obtain

$$\langle\langle a_k; a_k^+ \rangle\rangle = \bar{n}(\omega + v_k) / (\omega^2 - \Omega_k^2), \quad (40)$$

which will be used to derive the expression for $\langle n_0 \rangle$ or that for $\bar{n} = 1 - 3 \langle n_0 \rangle$ defined by (33). After integrating over frequencies and wavevector k, we obtain from (40) the result

$$\frac{3\bar{n}}{2N} \sum_k \left[\left(\frac{\omega_k}{\Omega_k} \right) \coth\left(\frac{\Omega_k}{2T_c}\right) - 1 \right] = 1 - \bar{n}, \quad (41)$$

Thus $\bar{n}$ must satisfy both (37) and (41), which have to be solved to determine T_c and $\bar{n}$. The essential factors which determine T_c are contained in the expressions (37) and (41). These are the effective libron–libron interaction, the electron–libron coupling parameter, the libron energy Ω_k, the band structure of the electrons which defines ε_k and the occupation number of the $k=0$ libron modes. In (37) and (41), the hyperbolic cotangent is due to the bosonlike nature of the libron modes. The contangent factors arise from the loss of librons from the Peierls condensate through thermal excitation.

Equations (37) and (41) have been solved numerically for T_c and $\bar{n}$. We have assumed nearest–neighbour EQQ and EDD interactions on a strand with $J_k = J \cos(kc)$ and $\gamma_k = \gamma \cos(kc)$ where c is the lattice spacing. The nearest–neighbour hopping of the electrons between the molecules is taken as $\varepsilon_k = -(W/2)\cos(kc)$ where W is the bandwidth. In Figs. 1a–d, we have plotted T_c as a function of g,W, g_q and g_0. In these plots, the dipole-dipole interaction parameter γ as well as the electron-dipole interaction λ have been taken equal to zero so as to isolate contributions arising from the quadrupoles alone.

The physical picture is as follows: The Peierls condensate is formed from the macroscopic occupancy of the states of vectors k

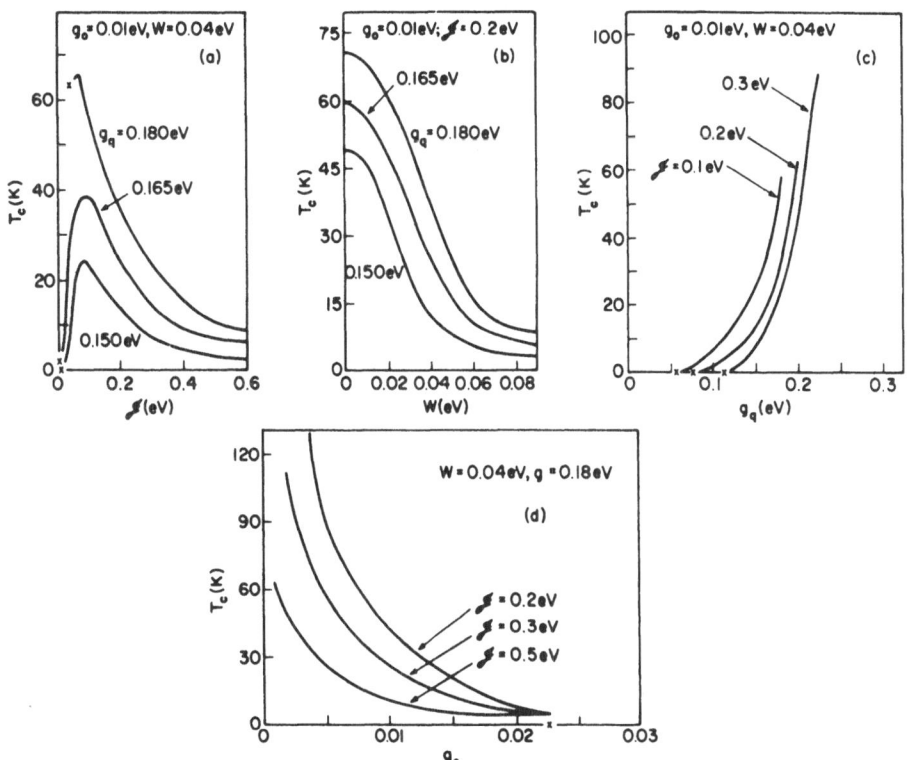

Figure 1. Mean-field Peierls transition temperature T_c is plotted versus: (a) the isotropic quadrupole-quadrupole coupling J, (b) the conduction electron bandwidth W and the Fourier components, (c) g_q and (d) g_0 of the coupling of the conduction band to the quadrupole moments. The dipole-dipole and the electron-dipole interaction parameters are equal to zero.

and k±q. The gap parameter comes from the interaction of the electrons with the librons and it is this interaction which distorts the lattice. At low temperatures, most of the librons are in the condensate. As the temperature increases, some of them are thermally excited from the condensate. The dependence of T_c on g_q

594

is like the exponential variation of the Peierls transition temperature with the electron-phonon coupling in Fröhlich's Hamiltonian with only two (zone-boundary) phonon modes [10]. Figure 1c indicates that there is an upper and lower cutoff limit in the range of values of g_q for which there is a T_c. Comparison between Figs. 1c and 1d implies that T_c depends quite differently on the Fourier components g_0 and g_q.

Calculations have been done with finite values of γ and λ and it is found that their effect is to reduce the Peierls critical temperature. In Figs. 2a and b, $\bar{\eta}$ and T_c have been plotted as a

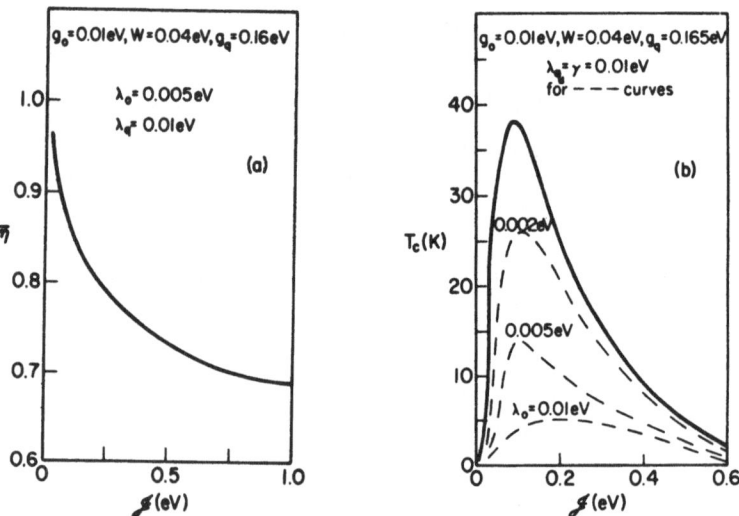

Figure 2. (a) The alignment parameter $\bar{\eta}$ and (b) Peierls transition temperature T_c as a function of the isotropic quadrupole-quadrupole interaction J within the mean-field approximation. The dashed curves in (b) show the effect of the electron-dipole and dipole-dipole interactions on the solid curve for which γ and λ are equal to zero.

function of J, respectively. The effect of increasing values of λ_0 is illustrated in Fig. 2b. The calculation indicates that T_c is more sensitive to changes in λ_0 than it is to λ_q and γ.

6. ELECTRICAL CONDUCTIVITY FOR A STRAND

The derived expressions for the Green functions will be used here to calculate the electrical conductivity $\sigma(k,\omega)$ of one-dimensional conductor, which is described by the Hamiltonian (9). In the long-wavelength limit $\sigma(\omega)=\sigma(\omega,k=0)$ and for a linear conductor with half-filled tight-binding band (37)

$$\sigma(\omega) = \frac{e^2 v_F^2}{i\omega} \left[Q^R(\omega) - Q(0) \right] , \quad Q^R(\omega) = \sum_{pp'\sigma\sigma'} Q^R_{pp'}(0,\omega) , \qquad (42)$$

where $Q(\omega)$ is obtained from $Q_{pp'}(0,0)$ by summing over p,p' and σ,σ', while $Q^R_{pp'}(k,\omega)$ is the Fourier transform of the retarded Green function, whose definition is similar to that given by (19), namely

$$Q^R_{pp'}(k,t-t') = -i\theta(t-t')<\left[\rho_{p'\sigma'}(k,t),\rho^+_{p\sigma}(k,t')\right]> , \qquad (43)$$

with $\rho_{p\sigma}(k,t)=\exp(iHt)\alpha^+_{p-k\sigma/2}\alpha_{p+k\sigma/2}\exp(-iHt)$. The unperturbed electron energy near the Fermi level is approximated by $\varepsilon_k = v_F(|k|-k_F)$, where v_F and k_F are the Fermi velocity and wavevector, respectively. The single particle $Q^{0,R}(\omega)$ and the collective $Q^{c,R}(\omega)$ contributions to $Q^R(\omega)$, which are described by the diagrams in Figs. 3a and b, respectively are found to be

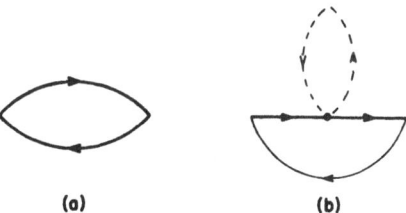

(a)   (b)

Figure 3. Diagrams giving (a) the single-particle and (b) the collective parts of the conductivity. An electron Green's function is denoted by a continuous line and a libron Green's function by a broken line.

$$Q^{0,R}(\omega) = (1/N) \sum_k \tanh\left(\frac{\zeta_k}{2T}\right)\left(\frac{\Delta^2/\zeta_k}{\omega^2-4\zeta_k^2}\right), \qquad (44)$$

$$Q^{c,R}(\omega) = (1/T) \sum_k \frac{f(\zeta_k)f(-\zeta_k)}{\omega+2\zeta_k}\left(\bar{\eta}g_0 + \frac{\Delta^2}{\zeta_k}\right), \qquad (45)$$

where $\zeta_k=\mathrm{sgn}(\varepsilon_k)(\varepsilon_k^2+\Delta^2)^{\frac{1}{2}}$ and $f(\zeta_k)$ is the Fermi-Dirac function. As expected, $Q^{0,R}(\omega)$ has the same form at T=0K as that obtained by Lee, Rice and Anderson [38] using Fröhlich's Hamiltonian. The collective contributions to the conductivity are associated with oscillations in the charge and lattice distortions about their equilibrium values. They are optically active along the chain direction since they involve a displacement of condensed charge.

It is straightforward to show that $Q^{c,R}(\omega)$ vanishes at zero temperature. But for the Fröhlich Hamiltonian with electron-phonon interactions the collective contribution to $\sigma(\omega)$ is not frozen out at T=0K (38,39). This result illustrates the quantitative difference between an orientational and translational Peierls instability for a one-dimensional stack of molecules. Substituting (44) and (45) into (42), we obtain the total conductivity $\sigma(\omega)= \sigma_0(\omega)+\sigma_c(\omega)$ as

$$\sigma(\omega) = e^2 v_F^2 [C(\omega)-C(0)]/i\omega , \qquad (46)$$

where

$$C(\omega) = (1/N) \sum_k \left(\omega+2\zeta_k\right)^{-1}[\frac{\Delta^2}{\zeta_k} \tanh(\zeta_k/2T) + \frac{1}{T} (\frac{\Delta^2}{\zeta_k} + g_0\bar{n})f(\zeta_k)(-\zeta_k)]. \qquad (47)$$

7. CONCLUDING REMARKS

We have made use of a model for one-dimensional conductors, which is based on the multipole expansion of the electrostatic interaction between a pair of charge distributions. While the isotropic EQQ and EDD interactions do not give rise to long range order in one or two-dimensions (30-32), the coupling of the conduction band of the electrons to the librational motion of the molecules gives rise to anisotropic effects which produce an orientational Peierls distortion. Figures 1a-d and 2a;b indicate the dependence of the Peierls transition temperature upon the various parameters of the system. The numerical results suggest that for a one-dimensional conductor to have an orientational Peierls transition at finite temperature it is important that certain molecular features such as the strength of the electron coupling to the rotations of the molecules have to be carefully chosen.

In conclusion, we would like to point out the following: Librons are quantum-mechanical elementary excitations which should not be confused with the classical rigid rotations of the molecules about their centers of mass. Each eigenstate of the molecules is determined by the two-body forces arising from the molecule-molecule and electron-molecule interactions. Basic to our calculations, like several others in the literature (24-27,33), is that for weakly interacting systems S=1 can be taken as nearly a good quantum number. The theory should apply to materials, like liquid crystals or some other solids, where free rotations are possible.

REFERENCES

1. Keller, H.J.: 1975, Low-Dimensional Cooperative Phenomena, Plenum, New York: 1976, Chemistry and Physics of One-Dimensional Metals, Plenum, New York.

2. Berlinsky, A.J.: 1976, Contemp. Phys. 17, p. 331.

3. Pal, L., Grüner, G., Janossy, A. and Solyom, J.: 1977, Lecture Notes in Physics 65, Organic Conductors and Semi-conductors, Springer, Berlin.

4. Sy, H.K. and Mavroyannis, C.: 1977, Solid State Commun. 23, p. 79.

5. Tombs, G.A.: 1978, Phys. Pep. 40C, p. 181.

6. Gumbs, G. and Mavroyannis, C.: 1979, J. Low Temp. Phys. 35, p. 593.

7. Solyom, J.: 1979, Adv. Phys. 28, p. 201.

8. Jerome, D. and Schulz, H.J.: 1982, Adv. Phys. 31, p. 299.

9. Peierls, R.E.: 1955, Quantum Theory of Solids, Clarendon Press, Oxford.

10. Frohlich, H.P.: 1954, Proc. R. Soc. A223, p. 296.

11. Morawitz, H.: 1975, Phys. Rev. Lett., 34, p. 1096: 1977, ref. 3, p. 303.

12. Merrifield, R.E. and Suna, A.: 1976, Phys. Rev. Lett. 36, p. 826.

13. Weyl, C., Engler, E.H., Bechgard, K., Jehanno, G. and Etemad, E.: 1976, Solid State Commun. 19, p. 925.

14. Johnson, C.R. and Watson, C.: 1976, J. Chem. Phys. 64, p. 2271.

15. Weger, M. and Friedel, J.: 1977, J. Phys. (Paris) 38, p. 241, p. 881.

16. Gutfreund, H. and Weger, M.: 1977, Phys. Rev. B16, p. 1753.

17. Gutfreund, H., Weger, M. and Koveh, M.: 1978, Solid State Commun. 27, p. 53.

18. Weger, M. and Gutfreund, H.: 1978, Comments Solid State Phys. 8, p. 135.

19. Zallen, R. and Conwell, E.M.: 1979, Solid State Commun. 31, p. 557.

20. Gumbs, G. and Mavroyannis, C.: 1979, Solid State Commun. 32, p. 569.

21. Gumbs, G. and Mavroyannis, C.: 1980, Phys. Rev. B21, p. 5455.

22. Gumbs, G. and Mavroyannis, C.: 1982, Phys. Rev. B25, p. 7467.

23. Rowlinson, J.S.: 1969, Liquids and Liquid Mixtures, 2nd ed., Butterworths, London, p. 234.

24. Nakamura, T.: 1955, Prog. Theor. Phys. 14, p. 135.

25. Barma, M.: 1974, Phys. Rev. B10, p. 4650.

26. Ritchie, D.S. and Mavroyannis, C.: 1978, Phys. Rev. B17, p. 1679.

27. Fedders, P.A. and Myles, C.W.: 1979, Phys. Rev. B19, p. 1331.

28. Hardy, W.N., Berlinsky, A.J. and Harris, A.B.: 1977, Can. J. Phys. 55, p. 1150, p. 1180.

29. Bogliubov, N.N.: 1962, Physik. Abhandl. Sowjetunion 6, p. 113.

30. Thorpe, M.F.: 1971, J. Appl. Phys. 42, p. 1410.

31. Ritchie, D.S. and Mavroyannis, C.: 1978, J. Low Temp Phys. 32, p. 813.

32. Mermin, N.D. and Wagner, H.: 1966, Phys. Rev. Lett. 17, p. 1133.

33. Raich, J.C. and Etters, R.D.: 1968, Phys. Rev. 168, p. 425.

34. Fittipaldi, I.P. and Tahir-Kheli, R.A.: 1975, Phys. Rev. B12, p. 1839.

35. Zubarev, D.N.: 1974, Nonequilibrium Statistical Thermo-dynamics, Plenum, New York, Ch. 16.

36. Mavroyannis, C.: 1975, Physical Chemistry, Academic, San Francisco, 11A, p. 487.

37. Perel, V.I. and Eliashberg, G.M.: 1962, Sov. Phys.-JETP 14, p. 633.

38. Lee, P.A., Rice, T.M. and Anderson, P.W.: 1974, Solid State Commun. 14, p. 703.

39. Rice, M.J.: 1976, Phys. Rev. Lett. 37, p. 36.

ELECTRON-LIBRON PAIRING IN QUASI-ONE-DIMENSIONAL CONDUCTORS

Constantine Mavroyannis

Division of Chemistry, National Research Council of
Canada, Ottawa, Ontario, Canada K1A 0R6

ABSTRACT. A theory is presented concerning the formation of
electron-libron bound states in quasi-one-dimensional conductors.
The solid is assumed to consist of equally spaced atoms, which are
free to rotate about their center of mass and behave like electrically neutral dipoles of spin 1/2. A conduction electron and a
libron (the quantum of libration), which are located at neighbouring lattice sites, interact through the Coulomb interaction to form
a bound state. The spectrum of the quasi particle consisting of an
electron and a libron has two branches $\omega_k^{(\pm)}$, which depend on the
values of the parameters Δ_k and η. A set of coupled equations are
derived to determine the expressions for the gap function Δ_k and
the parameter η, which describe the strength of the electron-libron
pairing and the orientational order of the atomic system, respectively. Numerical calculations have been made to establish conditions under which the formation of electron-libron modes becomes
possible. The derived results are graphically presented and
discussed.

1. INTRODUCTION

Considerable interest in recent years has been given to the
class of solids known as organic charge transfer salts. This
interest arises from the novelty of the systems as well as from the
fact that the flat planar molecules involved lead to anisotropic
structures and hence to pseudo-one-dimensional electronic
properties (1,2). In the quasi-one-dimensional charge transfer

salts such as TTF-TCNQ (tetrathiafulvalene tetracyanoquinodimethane), the molecular degrees of freedom can be classified in
translations, librations, and intramolecular distortions (3). All
three types of motion have received considerable attention, with
the librational modes being discussed after a suggestion by
Morawitz (4) that they may play a role in the Peierls transition.
Since then, the possibility of an orientational Peierls distortion
due to the coupling of the conduction electrons with librons has
received much attention (5-10).

We shall describe here the excitation spectrum (11) arising
from electron-libron interactions in a one-dimensional model and we
will show that such excitations are well defined within the framework of our recently developed formalism (12-14). In our recent
study of libron modes in crystals (11,12), the molecules are free
to rotate with angular momentum S=1, and thus the formalism is
applicable to crystals with large amplitude librations. The
angular momentum is thus a good quantum number for this model and
we proceed by working with these eigenstates. In a solid material
such as TTF-TCNQ, the amplitude of libration is small, between 1°
and 4° at low temperature, and hence for such materials it would
not be accurate to treat the angular momentum as a good quantum
number in our model. However, for small amplitude librations there
is still a coupling arising from the electron-quadrupole or the
electron-dipole interactions (13) when the molecules are tilted
with respect to the chain axis as they are in TTF-TCNQ. The
crystallographic studies by Johnson and Watson (15) have shown that
for compounds such as $(TTF)_7 I_5$ the rotations are indeed large; the
molecules are perpendicular to the chain axis and there is no
tilt.

The molecules of mesogenic compounds are elongated and the
liquid-crystalline phase of the substances is characterized by
long-range order in the orientation but not in the position of the
molecules. The orientational order parameter deviates from unity
(perfect order) owing to the thermal motions of the molecules. In
the case of complete disorder, the order parameter is zero. A
distinction is made between nematic, cholesteric and smectic
crystalline phases (16,17). These can be best characterized by
considering a liquid single crystal, that is, a region over which
the long-range order is ideal, apart from the thermal fluctuations. In particular, in the smectic phase one finds that in
addition to orientational order there is a partial positional
order, with the molecular centers of mass arranged in equidistant
planes. There are two different possibilities for smectic phases.
In one case the molecular centers do not have any long-range order
within the layers, and thus correspond to a two-dimensional liquid.
In the second case the layers are built up regularly so that the
positions of the molecular centers lie on a two-dimensional
lattice. "If a conducting liquid crystal could be made which has a
smectic-like phase with the molecules being regularly spaced along
a linear chain, then we feel that this would be a system in which

to look for electron-libron excitations of the type discussed here. In section 2 the model Hamiltonian describing our system is discussed. The excitation spectrum and numerical results for the dispersion relations of the electron-libron modes are presented in sections 3 and 4, respectively while the concluding remarks are given in section 5.

2. THE MODEL HAMILTONIAN

The Hamiltonian describing our system is taken as (11)

$$H = H_e + H_{DD} + H_{eD} , \tag{1}$$

where H_e is the electron Hamiltonian representing the motion of the electrons in the lattice and is given by

$$H_e = \sum_{\ell\ell'} t(\ell,\ell') \, \alpha_\ell^+ \alpha_{\ell'} . \tag{2}$$

The function $t(\ell,\ell')$ is the hopping energy of an electron from the lattice site $R_{\ell'}^0$, to the lattice site R_ℓ^0 and $\alpha_\ell^+, \alpha_\ell$ are the Fermi creation and annihilation operators describing an electron at lattice site R_ℓ^0. For the sake of simplicity, all considerations due to electron spins will be neglected. The rotating molecules interact through the dipole-dipole interaction Hamiltonian which may be taken as

$$H_{DD} = - \sum_{\ell\ell'} \sum_{mn} \Gamma_{nm}(\ell,\ell') L_1^m(\ell) L_1^n(\ell') , \tag{3}$$

where $\Gamma_{nm}(\ell,\ell')$ is the dipole-dipole interaction energy which depends on the dipole moment μ between the molecules, the distance $R_{\ell\ell'}$ between the lattice sites ℓ and ℓ', and on the orientation of the molecules at the lattice site ℓ relative to the crystal axis as well as on the angle between the molecules at ℓ and ℓ'; its explicit form is given by eq. (2.6) of (11). The interaction between an electron and the rotating molecule may be also expressed in terms of spherical-tensor operators as

$$H_{eD} = \sum_{\ell\ell'} \lambda(\ell,\ell') L_1^0(\ell) \rho(\ell') , \tag{4}$$

where $\rho(\ell)$ is the electron-density operator in the Wannier representation and $\lambda(\ell,\ell')$ is the coupling between a molecule and the conduction electrons.

In (3) and (4), $L_1^m(\ell)$ is a spherical-tensor operator, which for a molecule of spin S may be expressed in terms of the spin components as

$$L_1^{\pm 1}(\ell) = S_\ell^\pm / \sqrt{2f_1} , \quad L_1^z(\ell) = S_\ell^z \sqrt{f_1} , \quad f_1 = S(S+1) . \tag{5}$$

If we set $a_\ell = S_\ell^+$ and $a_\ell^+ = S_\ell^-$, then the following relations

$$S_\ell^z = \tfrac{1}{2} - n_\ell , \quad [a_\ell, a_{\ell'}^+]_- = (1-2n_\ell)\delta_{\ell\ell'}, \quad [a_\ell, a_{\ell'}]_- = [a_\ell^+, a_{\ell'}^+]_- = 0$$

are satisfied, where $n_\ell = a_\ell^+ a_\ell$. The electron operators (α, α^+) and the libron operators (a, a^+) anticommute with each other. In terms of the libron operators the total Hamiltonian may be written as

$$H = \sum_{\ell\ell'} t(\ell,\ell')\alpha_\ell^+\alpha_{\ell'} , -\frac{1}{3}\sum_{\ell\ell'} \{\Gamma_{00}(\ell,\ell')(1-2n_\ell)(1-2n_{\ell'})$$

$$+[2\Gamma_{+1-1}(\ell,\ell')a_\ell^+ a_\ell + H.C.] + [2\sqrt{2}\ \Gamma_{+10}(\ell,\ell')a_\ell(1-2n_{\ell'}) - H.C.]\}$$

$$+\frac{1}{\sqrt{3}}\sum_{\ell\ell'} \lambda(\ell,\ell')(1-2n_\ell)\rho(\ell') , \tag{6}$$

where the explicit form of the coupling functions can be found in (11). In deriving (6), we have retained only terms, which contribute for the electron-libron pairing and all effects due to electronic spin are ignored. The Hamiltonian (6) will be used in the next section to calculate the spectra due to the electron-libron bound states.

3. EXCITATION SPECTRUM

To study the excitation spectrum arising from the electron-libron bound states, we shall make use of a Green's function method similar to that used in the theory of superconductivity (18) as well as in the studies of electron-exciton (14) and exciton-exciton (19-21) bound systems. The retarded double-time single electron Green's function is defined as

$$\langle\langle\alpha_\ell(t);\alpha_{\ell'}^+(t')\rangle\rangle = -i\theta(t-t')\langle[\alpha_\ell(t),\alpha_{\ell'}^+(t')]\rangle , \tag{7}$$

where $\theta(t)$ is the usual step function, the operators are in the Heisenberg representation and the angular brackets denote the average over the canonical ensemble appropriate to the total Hamiltonian H.

We consider the equations of motion for the electron libron operators

$$[\alpha_\ell^+, H]_- = -\sum_{\ell'} t(\ell,\ell')\alpha_{\ell'}^+ - (n/\sqrt{3})\sum_{\ell'} \lambda(\ell,\ell')\alpha_\ell^+$$

$$+(2/\sqrt{3})\sum_{\ell'} \lambda(\ell,\ell')\langle\alpha_\ell^+, a_\ell^+\rangle a_{\ell'} , \tag{8}$$

$$[a_\ell, H]_- = (4\eta/3) \sum_{\ell'} \{\Gamma_{00}(\ell,\ell')a_\ell - \tfrac{1}{2}[\Gamma_{+1-1}(\ell',\ell)+\Gamma^*_{+1-1}(\ell,\ell')]a_{\ell'}$$

$$-(\sqrt{3/2})\lambda(\ell,\ell')<\alpha_{\ell'}a_\ell>\alpha^+_{\ell'}\} \ . \tag{9}$$

In deriving (8) and (9), we have applied the following decoupling approximations

$$\eta = 1-2<n_\ell> \ , \tag{10}$$

$$a^+_\ell, a^+_\ell, \alpha^+_\ell \approx <a^+_\ell, \alpha^+_\ell>a_{\ell'}, \ , \quad \alpha_\ell, \alpha^+_\ell, a_\ell \approx <\alpha_\ell, a_\ell>\alpha^+_{\ell'}, \ , \tag{11}$$

where η is the thermal average value of the dipolar operator which is independent of the lattice sites since all sites are equivalent. The decoupling approximation (11) implies that only coherent pairing between an electron and a libron is taken into account while all higher order terms describing libron-libron and electron-libron scattering effects are ignored. This approximation is expected to be valid in the coherence approximation where scattering effects are supposed to be negligible. The last two terms in (8) and (9) describe the pairing between an electron and a libron located at different lattice sites of the crystal.

Going into the momentum space for the coupled equations (8) and (9) and then making use of the Fourier transform of the Green function (7) for the electron and libron operators, we find

$$(-\omega-E_k)<<\alpha^+_{-k};\alpha_{-k}>>+\Delta^*_k<<a_k;\alpha_{-k}>> = 1 \ , \tag{12}$$

$$(\omega-\xi_k)<<a_k;\alpha_{-k}>>+\eta\Delta_k<<\alpha^+_{-k};\alpha_{-k}>> = 0 \ , \tag{13}$$

where

$$\Delta_k = \frac{2}{N\sqrt{3}} \sum_{k'} \lambda(k')<a_{k'-k}\alpha_{k-k'}> \ , \tag{14}$$

is the energy-gap function for electron-libron bound pairs. N is the total number of atoms on the chain, with lattice spacing a_0 and the wavenumbers are in the first Brillouin zone, $-\pi/a_0 \leqslant k \leqslant \pi/a_0$ and units with $\hbar=1$ is used throughout. In (12) and (13), we have introduced the notation

$$\xi_k = \frac{2}{3} \eta[2\gamma_{00}(0)-\gamma_{+1-1}(k)-\gamma^*_{+1-1}(-k)] \ , \tag{15}$$

$$E_k = \varepsilon_k+\lambda(0)/\sqrt{3} \ , \tag{16}$$

where $\gamma_{mn}(k)$, $\lambda(k)$ and ε_k are the Fourier transforms of $\Gamma_{mn}(\ell,\ell')$, $\lambda(\ell,\ell')$ and $t(\ell,\ell')$, respectively. ξ_k and ε_k represent the rotational energy of a molecule and the band energy for a free electron, respectively. In deriving (12) and (13), we have discarded the possibility of a libron and an electron scattering off each other into different momentum states.

Solving the two coupled equations (12) and (13), we have

$$<<\alpha^{+}_{-k};\alpha_{-k}>> = \frac{(\omega-\xi_k)}{(\omega-\xi_k)(-\omega-E_k)-\eta|\Delta_k|^2}, \tag{17}$$

$$<<a_k;\alpha_{-k}>> = \frac{-\eta\Delta_k}{(\omega-\xi_k)(-\omega-E_k)-\eta|\Delta_k|^2}. \tag{18}$$

The thermal average $<a_{-k}\alpha_{-k}>$ in the expression (14) for the gap function Δ_k may be obtained by taking the imaginary part of the Green function $<<a_k;\alpha_{-k}>>$ across the branch cut along the real axis and integrating over frequency, namely,

$$<\alpha_{-k}a_k> = -2 \int_{-\infty}^{\infty} \left(\frac{d\omega}{2\pi}\right) f_0(\omega) \text{Im}<<a_k;\alpha_{-k}>> , \quad f_0(\omega)=\left(1+e^{\frac{\omega-E_F}{}}\right)^{-1}, \tag{19}$$

where $f_0(\omega)$ is the Fermi distribution function, E_F is the Fermi energy and T denotes temperature and units with $k_B=1$ are used. Substituting (18) into (19) and after integrating over ω, we obtain the following result for Δ_k as

$$\Delta_k = \frac{4\eta}{N\sqrt{3}} \sum_{k'} \lambda_{k'}\Delta_{k-k'} \frac{f_0\left(\omega^{(+)}_{k-k'}\right)-f_0\left(\omega^{(-)}_{k-k'}\right)}{\omega^{(+)}_{k-k'}-\omega^{(-)}_{k-k'}} \tag{20}$$

where the energies of excitation for the electron-libron bound states are equal to

$$\omega^{(\pm)}_k = \frac{1}{2}\left\{(\xi_k-E_k)\pm\left[(\xi_k+E_k)^2-4\eta|\Delta_k|^2\right]^{\frac{1}{2}}\right\} . \tag{21}$$

The equation of motion method can be also used to calculate the libron-libron Green funtion $<<a_k;a^{+}_k>>$. Using (9), one easily finds that the equation for $<<a_k;a^{+}_k>>$ is coupled to the Green function $<<\alpha^{+}_{-k};a^{+}_k>>$. Therefore, using (8) and (9), we derive a set of coupled equations for $<<a_k;a^{+}_k>>$ and $<<\alpha^{+}_{-k};a^{+}_k>>$, the solution of which yields

$$<<a_k;a^{+}_k>> = \frac{-\eta(\omega+E_k)}{(\omega-\xi_k)(-\omega-E_k)-\eta|\Delta_k|^2}. \tag{22}$$

The average value of the libron occupation number $<a^{+}_k a_k>$ may be evaluated from the expression

$$\langle a_k^+ a_k \rangle = -2 \int_{-\infty}^{\infty} \left(\frac{d\omega}{2\pi}\right)(e^{\omega}-1)^{-1} \text{Im} \ll a_k ; a_k^+ \gg . \tag{23}$$

Then using the expression (10) for η and (22) and (23), we derive the following expresion for η,

$$\frac{\eta}{N} \sum_k \left[\frac{E_k + \omega_k^{(+)}}{\omega_k^{(+)} - \omega_k^{(-)}} \coth \frac{\omega_k^{(+)}}{2} - \frac{E_k + \omega_k^{(-)}}{\omega_k^{(+)} - \omega_k^{(-)}} \coth \frac{\omega_k^{(-)}}{2} - 1 \right] = 1 - \eta . \tag{24}$$

Simultaneous solution (20) and (24) determines self-consistently the values for the energy gap function Δ_k and the average value of the dipolar operator η.

4. DISPERSION RELATION FOR THE ELECTRON-LIBRON MODES

For an undistorted linear chain we have $\Delta_k = \Delta_0 \cos k a_0$. In our calcultion we use $\Delta_0 = 0.02$ eV and take $\lambda(k) = \lambda \cos k a_0$, $\varepsilon_k = \varepsilon_B \cos k a_0$, which correspond to nearest-neighbour electron-libron coupling and electron hopping between nearest-neighbour lattice sites, respectively. For the sake of convenience, the k-dependence of ξ_k defined by (15) will be ignored and ξ_k will be taken as a constant. For the numerical calculations all energies as well as temperature are measured in terms of Δ_0. In Figure 1, the electron band energy E_k in units of Δ_0 is plotted as a function of the wavevector.

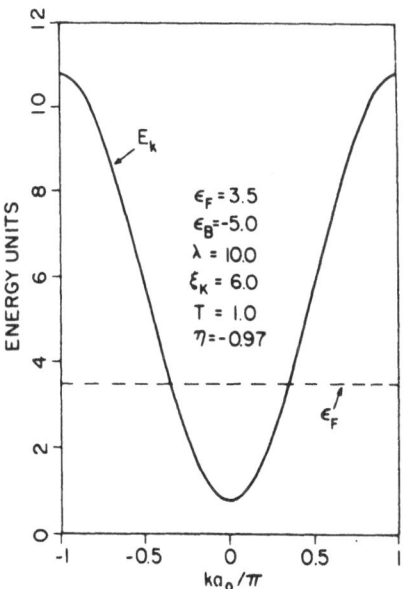

Figure 1. Electron-band energy as a funtion of the wavevector. Δ_0 is the unit of energy.

Then using the same values for the parameters ϵ_F, ϵ_B, ξ_k, T and η in units of Λ_0, the excitation energies $\omega_k^{(\pm)}$ determined by (21) are plotted against the wavevector in Figure 2. The energy of the lower $\omega_k^{(-)}$ branch in Figure 2 shows more variation with k than that

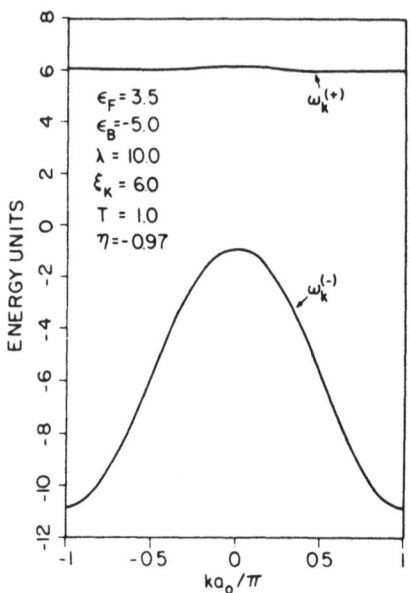

Figure 2. Plot of the two branches $\omega_k^{(\pm)}$ for the electron-libron excitation energies for the same values of the parameters used in Fig. 1. Λ_0 is the unit of energy.

of the upper $\omega_k^{(+)}$ branch of the excitation spectrum. The value of the ordering parameter η must lie between 0 and -1. Keeping all other parameters fixed, a state of total disorder ($\eta=0$) is achieved by increasing the coupling parameter λ. For the electron and libron energies which we have chosen, numerical calculation indicates that the excitation spectrum at T=0 differs from its value at T=0.1 (in units of Λ_0) by only a few percent because of the presence of the exponential in the Fermi distribution function. Comparing the value of η in Figures 2 and 3a, we see that this parameter is not very sensitive to temperature as well. From our results in Figures 3 and 4, we have that for $-\eta$ to be reduced from 0.9 to 0.6, λ has to be increased by a factor of 5. Therefore, since our results are not very sensitive to changes in either temperature or electron-libron coupling, we believe that the experimental verification is feasible.

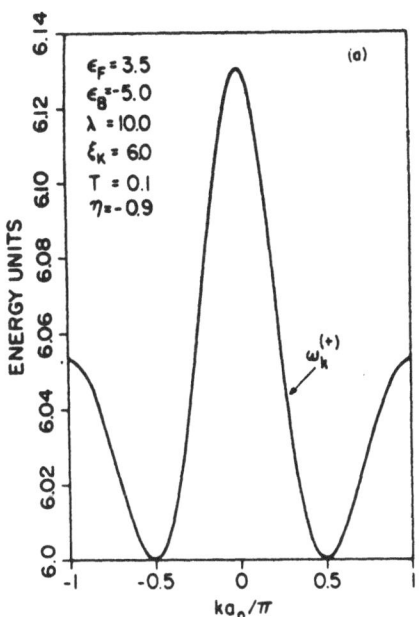
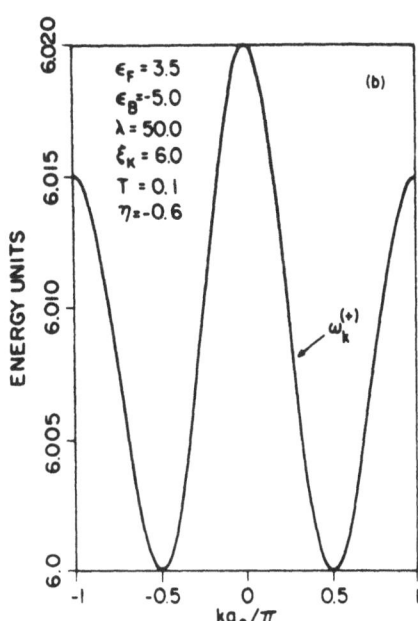

Figure 3. Upper branch $\omega_k^{(+)}$ of the electron-libron excitation spectrum is plotted for two different values of the coupling parameter λ. The Fermi energy (ε_F), the electron bandwidth (ε_B), the libron energy (ξ_k), and the temperature (T) are the same in (a) and (b). Δ_0 is the unit of energy.

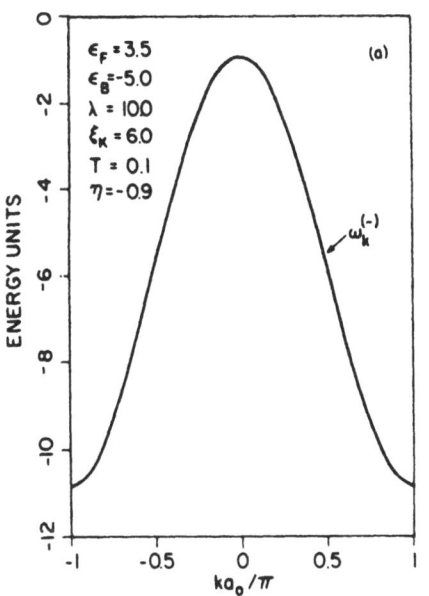
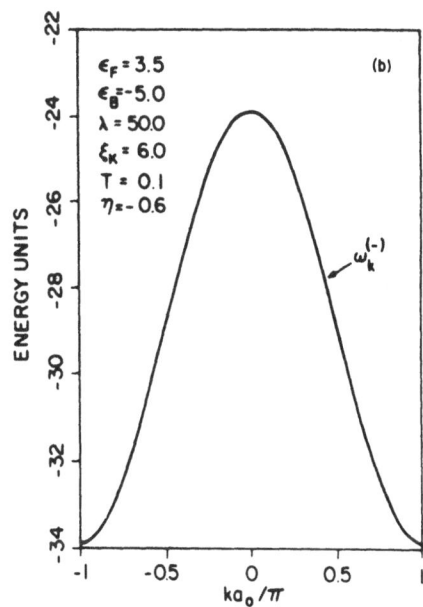

Figure 4. Same as Figure 3 but with the lower branch $\omega_k^{(-)}$ being plotted.

608

5. CONCLUDING REMARKS

The model calculation which we have presented here should be understood as the first steps towards a general model theory for librons with large amplitudes in crystals which are electrically conducting. This microscopic treatment in terms of a model whose molecules have rigidly fixed centers of mass illuminates some essential features of a more general approach. Work is in progress to incorporate phonons into the present model. As mentioned before, the present theory concerning electron-libron pairing is anticipated to be applicable to a class of conducting liquid crystals which have the smectic-like phase with the molecules being regularly spaced along a linear chain.

REFERENCES

1. Bardeen, J.: 1979, in Highly Conducting One-Dimensional Solids, eds. DeVreese, J.T., Evrard, R.P. and van Doren, V.E.: Plenum, New York, Chap. 8.

2. Jérome, D. and Schulz, H.J.: 1982, Adv. in Phys. 31, p. 299.

3. Torrance, J.B.: 1978, Phys. Rev. B17, p. 3099.

4. Morawitz, H.: 1975, Phys. Rev. Lett. 34, p. 1096.

5. Merrifield, R.E. and Suna, A.: 1976, Phys. Rev. Lett. 36, p. 826.

6. Weger, M. and Friedel, J.: 1977, J. Phys. (Paris), 38, p. 241, 881.

7. Gutfreund, H. and Weger, M.: 1977, Phys. Rev. B16, p. 1753: 1978, comments Solid State Phys. 8, p. 135.

8. Weger, M.: 1979, in The Physics and Chemistry of Low-Dimensional Solids, ed. Alc'acer, L., Reidel, Dordrecht, p. 77.

9. Weger, M., Gutfreund, H., Hartzstein, C. and Kavek, M.: 1981, Chem. Soc. 17, p. 51.

10. Conwell, E.M.: 1980, Phys. Rev. B22, p. 1761: 1981, Chem. Soc. 17, p. 69.

11. Gumbs, G. and Mavroyannis, C.: 1982, Phys. Rev. B25, p. 7467.

12. Gumbs, G. and Mavroyannis, C.: 1979, Solid State Commun. 32, p. 569.

13. Gumbs, G. and Mavroyannis, C.: 1980, Phys. Rev. B21, p. 5455.

14. Gumbs, G. and Mavroyannis, C.: 1982 Solid State Commun. 41, p. 237.

15. Johnson, C.K. and Watson, C.R.: 1976, J. Chem. Phys. 64, p. 2271.

16. de Gennes, P.G.: 1976, The Physics of Liquid Crystals, Oxford Univ. Press, London, Chap. 1.

17. Saupe, A.: in Liquid Crystals and Plastic Crystals, eds. Gray, G.W. and Winsor, P.A., Wiley, New York, Vol. 1, p. 18.

18. Schrieffer, J.R.: 1964, Theory of Superconductivity, Benjamin, Reading, Mass., Chap. 7.

19. Gumbs, G. and Mavroyannis, C.: 1981, Phys. Rev. B24, p. 7258.

20. Gumbs, G. and Mavroyannis, C.: 1981, J. Phys. C: Solid State 14, p. 2199.

21. Gumbs, G. and Mavroyannis, C.: 1982, J. Phys. C: Solid State 15, L465.

AUTHOR INDEX

Aifantis, E.C................357
Amelinckx, S.........173,183,223
Raral, D.....................485
Brenner, S.S................309
Clapp, P.C..................475
Cohen, J.B..................265
Currat, R...................285
deFontaine, D................43
deWolff, P.M................133
Dohler, G.H.................529
Dugan, M.P..................425
Flevaris, N.K...............107
Georgopoulos, P.............265
Gonis, A....................107
Greer, A.L..................511
Gregoriades, P..............223
Henein, G.E.................495
Heine, V.....................95
Hilliard, J.E...........485,495
Hwang, C.M..................559
Jankowski, A................387
Kalos, M.H..................125
Ketterson, J.B..............485
Khatchaturyan, A.G..........327
Madariaga, G................151
Makovicky, E................159
Manikopoulos, C.N...........459
Marro, J....................125
Mavroyannis, C.........603,619

Meichle, M..................559
Miller, M.K.................309
Monfroy, G..................521
Morris, J.W., Jr..........327
Moss, S.C....................11
Niarchos, D.................521
Panatheofanis, B.J.........521
Pérez-Mato, J.M.............151
Salamon, M.B................559
Sancaktar, E................587
Schwartz, L.H...............411
Selke, S.....................23
Smith, J.....................95
Soffa, W.A..................309
Soukoulis, C.M...............81
Tanielian, M................521
Tello, M.J..................151
Terasaki, O.................247
Tsakalakos, T....387,425,459
Van Landuyt, J..............183
Van Tendeloo, G......183,223
Watanabe, D.................247
Wayman, C.M.................559
Wen, S.H....................327
Willhite, J.................521
Yamauchi, H.................577
Ye, H.Q.....................173
Yeomans, J...................95